Methods in Pharmacology and Toxicology

Series Editor
Y. James Kang
University of Louisville
School of Medicine
Prospect, Kentucky, USA

For further volumes:
http://www.springer.com/series/7653

Computational Systems Toxicology

Edited by

Julia Hoeng and Manuel C. Peitsch

Philip Morris International R&D, Neuchâtel, Switzerland

Humana Press

Editors
Julia Hoeng
Philip Morris International R&D
Neuchâtel, Switzerland

Manuel C. Peitsch
Philip Morris International R&D
Neuchâtel, Switzerland

ISSN 1557-2153 ISSN 1940-6053 (electronic)
Methods in Pharmacology and Toxicology
ISBN 978-1-4939-2777-7 ISBN 978-1-4939-2778-4 (eBook)
DOI 10.1007/978-1-4939-2778-4

Library of Congress Control Number: 2015943591

Springer New York Heidelberg Dordrecht London

Cover illustration: Art by Dr. Stephanie Boue

Printed on acid-free paper

Humana Press is a brand of Springer
Springer Science+Business Media LLC New York is part of Springer Science+Business Media (www.springer.com)

Preface

Humans are exposed to a multitude of substances such as manufactured products, therapeutic agents, or environmental toxicants that perturb essential biological mechanisms and can thereby cause adverse effects or disease. Assessing the potential risks to human health posed by exposure to the multitude of synthetic as well as naturally occurring chemicals of industrial interest is a daunting task which is often hampered by the limitations of the classical toxicology paradigm. Indeed, classical toxicity testing primarily relies on a collection of studies that evaluate observable apical endpoints in whole animals and a limited number of easy to handle cellular systems. The approach is based primarily on a coarse quantification of clinical signs or pathologic changes that can be indicative of a disease state. In addition, this approach is resource-intensive and time-consuming and has difficulty meeting today's challenges such as increased biological relevance and throughput [1].

Society demands increasingly close scrutiny of the potential health risks associated with exposure to chemicals present in our everyday life, leading to an increasing need for more predictive and accurate risk-assessment approaches. Developing such approaches requires a detailed mechanistic understanding of the ways in which xenobiotic substances perturb biological systems and lead to adverse outcomes. The shortcomings described above and the increased need for toxicity testing that is more accurate have triggered fresh thinking which is leading to the emergence of a new paradigm in toxicity testing called Systems Toxicology. Systems Toxicology [2] resides at the intersection of Systems Biology [3] with toxicology and chemistry and integrates classic toxicology approaches with network models and quantitative measurements of molecular and functional changes occurring across multiple levels of biological organization. Systems Toxicology is highly interdisciplinary and leverages the most recent developments in molecular measurement and high-throughput screening methods, computational toxicology, and bioinformatics [4, 5].

Systems Toxicology Research is primarily aimed at decoding the mechanisms by which substances cause toxic effects. This consists in creating the knowledge about the causal chain of key molecular, cellular, and functional events, which link exposures to their adverse effects and diseases consequences [2]. Second, this knowledge needs to be captured as biological network models. Systems Toxicology-based assessment then leverages this detailed mechanistic knowledge to perform a mechanism-by-mechanism assessment of the biological impact of an active substance [2, 6]. This new mechanism-based paradigm for toxicity testing has the potential to help address important challenges such as a better extrapolation between (1) short-term studies and long-term effects, (2) in vitro and in vivo systems, and (3) across species. Furthermore, Systems Toxicology enables the identification and application of novel biomarkers for improved safety assessments and monitoring. Systems Toxicology is based on a strong scientific foundation that offers the prospect of improved risk-based regulatory decisions and possibly greater public confidence in these decisions [7].

The recent technological developments in experimental biology and multi-omics measurements that enable Systems Biology and Systems Toxicology can only be fully leveraged by the application of a broad range of computational approaches ranging from data management to mathematical modelling. The aim of this book is to provide descriptions for some of the key state-of-the-art computational applications that are crucial in Systems Toxicology. These include data management and processing, data analysis, biological network building and analysis, and the application of computational methods to toxicological assessment.

Neuchâtel, Switzerland

Julia Hoeng
Manuel C. Peitsch

References

1. Krewski D, Westphal M, Al-Zoughool M, Croteau MC, Andersen ME (2011) New directions in toxicity testing. Annu Rev Public Health. 32: 161–178
2. Sturla SJ, Boobis AR, FitzGerald RE, Hoeng J, Kavlock RJ, Schirmer K, et al. (2014) Systems toxicology: from basic research to risk assessment. Chem Res Toxicol 27(3):314–329
3. Peitsch MC, de Graaf D (2014) A decade of systems biology: where are we and where are we going to? Drug Discov Today 19(2):105–107
4. Andersen ME, Krewski D (2009) Toxicity testing in the 21st century: bringing the vision to life. Toxicol Sci 107(2):324–330
5. EPA US (2013) Next generation risk assessment: incorporation of recent advances in molecular, computational, and systems biology (external review draft). In: Agency U (ed). Washington.
6. Hoeng J, Deehan R, Pratt D, Martin F, Sewer A, Thomson TM, et al. (2012) A network-based approach to quantifying the impact of biologically active substances. Drug Discov Today 17(9–10):413–418
7. Hartung T, van Vliet E, Jaworska J, Bonilla L, Skinner N, Thomas R (2012) Systems toxicology. Altex 29(2):119–128

Contents

Contributors

GERALD ANKLEY • *Mid-Continent Ecology Division, US Environmental Protection Agency, Duluth, MN, USA*
GORDANA APIC • *Cambridge Cell Networks Ltd., Cambridge, UK*
NIKO BEERENWINKEL • *Department of Biosystems Science and Engineering, ETH Zurich, Basel, Switzerland; SIB Swiss Institute of Bioinformatics, Basel, Switzerland*
VINCENZO BELCASTRO • *Telethon Institute of Genetics and Medicine, Pozzuoli (Napoli), Italy*
STEPHANIE BOUE • *Philip Morris International R&D, Philip Morris Products S.A., Neuchâtel, Switzerland*
NATALIA BOUKHAROV • *Selventa, Cambridge, MA, USA*
SIMONA CONSTANTINESCU • *Department of Biosystems Science and Engineering, ETH Zurich, Basel, Switzerland; SIB Swiss Institute of Bioinformatics, Basel, Switzerland*
SOPHIE DIJON • *Philip Morris International R&D, Philip Morris Products S.A., Neuchâtel, Switzerland*
STEPHEN EDWARDS • *National Health and Environmental Effects Research Laboratory, US Environmental Protection Agency, Research Triangle Park, NC, USA*
ASHRAF ELAMIN • *Philip Morris International R&D, Philip Morris Products S.A., Neuchâtel, Switzerland*
JULIANE FLUCK • *Fraunhofer Institute SCAI, Sankt Augustin, Germany*
KERRY FROST • *Cyprotex Discovery Ltd., Macclesfield, Cheshire, UK*
KAZUHIRO FUJITA • *The Systems Biology Institute, Tokyo, Japan*
NATÀLIA GARCIA-REYERO • *Institute for Genomics, Biocomputing, and Biotechnology, Mississippi State University, Starkville, MS, USA*
SAMIK GHOSH • *The Systems Biology Institute, Tokyo, Japan*
FRANCESCO GREGORETTI • *Institute of High Performance Computing and Networking, Naples, Italy*
JULIA HOENG • *Philip Morris International R&D, Philip Morris Products S.A., Neuchâtel, Switzerland*
SIMONA IACOBACCI • *Telethon Institute of Genetics and Medicine, Pozzuoli (Napoli), Italy*
ANITA R. ISKANDAR • *Biological Systems Research, Philip Morris International R&D, Philip Morris Products S.A., Neuchâtel, Switzerland*
NIKOLAI V. IVANOV • *Philip Morris International R&D, Philip Morris Products S.A., Neuchâtel, Switzerland*
SARAH KELLY • *Cyprotex Discovery Ltd., Macclesfield, Cheshire, UK*
HIROAKI KITANO • *The Systems Biology Institute, Tokyo, Japan*
ARKADIUSZ K. KUCZAJ • *Philip Morris International R&D, Philip Morris Products S.A., Neuchâtel, Switzerland*
HÉCTOR DE LEÓN • *Philip Morris International R&D, Philip Morris Products S.A., Neuchâtel, Switzerland*
DAVID LYONS • *National Health and Environmental Effects Research Laboratory, US Environmental Protection Agency, Research Triangle Park, NC, USA*

GIANCARLO MARRA • *Institute of Molecular Cancer Research, University of Zurich, Zurich, Switzerland*
FLORIAN MARTIN • *Biological Systems Research, Philip Morris International R&D, Neuchâtel, Switzerland*
YUKIKO MATSUOKA • *The Systems Biology Institute, Tokyo, Japan*
MARKUS NORDLUND • *Philip Morris International R&D, Philip Morris Products S.A., Neuchâtel, Switzerland*
GENNARO OLIVA • *Institute of High Performance Computing and Networking, Naples, Italy*
JENNIFER S. PARK • *Selventa, Cambridge, MA, USA*
MANUEL C. PEITSCH • *Philip Morris International R&D, Philip Morris Products S.A., Neuchâtel, Switzerland*
EDWARD PERKINS • *Environmental Laboratory, US Army Engineer Research and Development Center, Vicksburg, MS, USA*
ROMAN POLISHCHUK • *Telethon Institute of Genetics and Medicine, Pozzuoli (Napoli), Italy*
DRAGANA MITIC POTKRAJAC • *Cambridge Cell Networks Ltd., Cambridge, UK*
BRANKA RAKIC • *Cambridge Cell Networks Ltd., Cambridge, UK*
ROBERT B. RUSSELL • *Cambridge Cell Networks Ltd., Cambridge, UK*
WALTER K. SCHLAGE • *Philip Morris International R&D, Philip Morris Products S.A., Neuchâtel, Switzerland*
THOMAS SCHNEIDER • *Philip Morris International R&D, Philip Morris Products S.A., Neuchâtel, Switzerland*
ALAIN SEWER • *Biological Systems Research, Philip Morris International R&D, Neuchâtel, Switzerland*
TIM SMITH • *Cyprotex Discovery Ltd., Macclesfield, Cheshire, UK*
MIRANDA D. STOBBE • *Bioinformatics Laboratory, Academic Medical Center, University of Amsterdam, Amsterdam, The Netherlands; Netherlands Bioinformatics Centre, Nijmegen, The Netherlands*
SHANA J. STURLA • *Department of Health Sciences and Technology, ETH Zurich, Zurich, Switzerland*
IGNACIO GONZALEZ SUAREZ • *Philip Morris International R&D, Philip Morris Products S.A., Neuchâtel, Switzerland*
MARJA TALIKKA • *Philip Morris International R&D, Philip Morris Products S.A., Neuchâtel, Switzerland*
ADI LAURENTIU TARCA • *Department of Obstetrics and Gynecology, Wayne State University, Detroit, MI, USA; Bioinformatics and Computational Biology Unit, Perinatology Research Branch, NICHD/NIH/DHHS, Wayne State University, Bethesda, MD, USA*
BJOERN TITZ • *Philip Morris International R&D, Philip Morris Products S.A., Neuchâtel, Switzerland*
DANIEL VILLENEUVE • *Mid-Continent Ecology Division, US Environmental Protection Agency, Duluth, MN, USA*
PAUL WALKER • *Cyprotex Discovery Ltd., Macclesfield, Cheshire, UK*
CLEMENS WITTWEHR • *Systems Toxicology Unit, Joint Research Centre, European Commission, Ispra, Italy*
BERND WOLLSCHEID • *Department of Health Sciences and Technology, ETH Zurich, Zurich, Switzerland; Institute of Molecular Systems Biology, ETH Zurich, Zurich, Switzerland*
MARC ZIMMERMANN • *Fraunhofer Institute SCAI, Sankt Augustin, Germany*

Chapter 1

The Adverse Outcome Pathway: A Conceptual Framework to Support Toxicity Testing in the Twenty-First Century

Edward Perkins, Natàlia Garcia-Reyero, Stephen Edwards, Clemens Wittwehr, Daniel Villeneuve, David Lyons, and Gerald Ankley

Abstract

The need to rapidly characterize the risk of large numbers of chemicals has moved the traditional toxicological paradigm from animal testing to a pathway-based approach using in vitro assay systems and modeling where possible. Adverse Outcome Pathways (AOPs) provide a conceptual framework that can be used to link in vitro assay results to whole animal effects in a pathway context. AOPs are defined and examples are provided to demonstrate key characteristics of AOPs. To support development and application of AOPs, a knowledge base has been developed containing a Wiki site designed to permit documentation of AOPs in a crowd-sourced manner. Both empirical and computational methods are demonstrated to play a significant role in AOP development. The combination of computational approaches, including different modeling efforts, together with apical end points within the pathway-based framework will allow for a better understanding of the linkage of events from a molecular initiating event to a potential adverse outcome, therefore defining key events, AOPs, and even networks of AOPS. While these approaches are indeed very promising, the ability to understand and define key events and key event relationships will remain one of the more complex and challenging efforts within AOP development. In order to make AOPs useful for risk assessment these challenges need to be understood and overcome. An interdisciplinary approach including apical and molecular measurements, computational, and modeling efforts is currently being one of the most promising approaches to ensure AOPs become the useful framework they were designed to be.

Key words Predictive toxicology, Ecotoxicology, Key events, Computational biology, Network science, Risk assessment

1 Toxicity Testing: A Changing Paradigm

Traditionally, toxicology has focused on understanding the hazards of chemicals. Much of this work has relied on examining how chemicals and other materials affect animals such as mice, rats, fish, and other species. Overall, however, relatively few of the chemicals actively used in commerce and released directly or indirectly into the environment have been tested for toxicity. For example, in the

Julia Hoeng and Manuel C. Peitsch (eds.), *Computational Systems Toxicology*, Methods in Pharmacology and Toxicology, DOI 10.1007/978-1-4939-2778-4_1,

USA, extensive, legislatively mandated toxicity testing is generally limited to pesticide active ingredients regulated under the Federal Insecticide, Fungicide and Rodenticide Act [1], some high production volume chemicals, and contaminants for which water quality guidelines need to be developed. Little or no testing has been required for the majority of chemicals, which number into the tens of thousands [2] and are regulated principally under the US Toxic Substances Control Act ([3]; e.g., industrial and commercial chemicals). There is a strong interest globally in more comprehensive chemical characterization to support safety evaluation and risk-based decision-making. However, when using standard toxicity testing approaches, both the costs associated with testing and the time required to conduct the assays, generate, and analyze the data make it impossible to extensively test large numbers (thousands) of chemicals [4]. Additionally, in many parts of the world there are mounting legal, regulatory, economic, and social pressures that demand a reduction in animal testing and the development of suitable alternatives such as in vitro and/or in silico testing [5–7].

As a result, it has become clear that new approaches are needed to determine the potential hazards of chemicals that come in contact with humans and environmental receptors. A landmark report by the US National Research Council (NRC) considered a vision and strategy for improving and modernizing the practice of regulatory toxicology using advances in systems biology, computational methods, bioinformatics, and in vitro screening to provide more accurate and rapid assessments of chemical hazards to humans [8]. The report envisioned "a new toxicity-testing system that relies mainly on understanding "toxicity pathways"—the cellular response pathways that can result in adverse health effects when sufficiently perturbed (Fig. 1a). Such a system would evaluate biologically significant alterations without relying on studies of whole animals." The principal recommendations were to develop approaches that: (1) replace animal testing with in vitro assays that assess perturbation of toxicity pathways, which are defined as "Cellular response pathways that, when sufficiently perturbed in an intact animal, are expected to result in adverse health effects" [8]; (2) develop methods to assess chemical effects in a high throughput manner amenable to automation; and (3) use human cells or cell lines rather than animal cell lines so that testing results are more relevant to human populations [8]. Computational toxicology, systems biology, and bioinformatics play major roles in interpreting and applying this approach, especially when extrapolating to humans and population level effects [10, 11]. While the report focused exclusively on human health toxicology, many aspects of the recommendations are also applicable to ecological risk assessment concerning chemical impacts on other species [12].

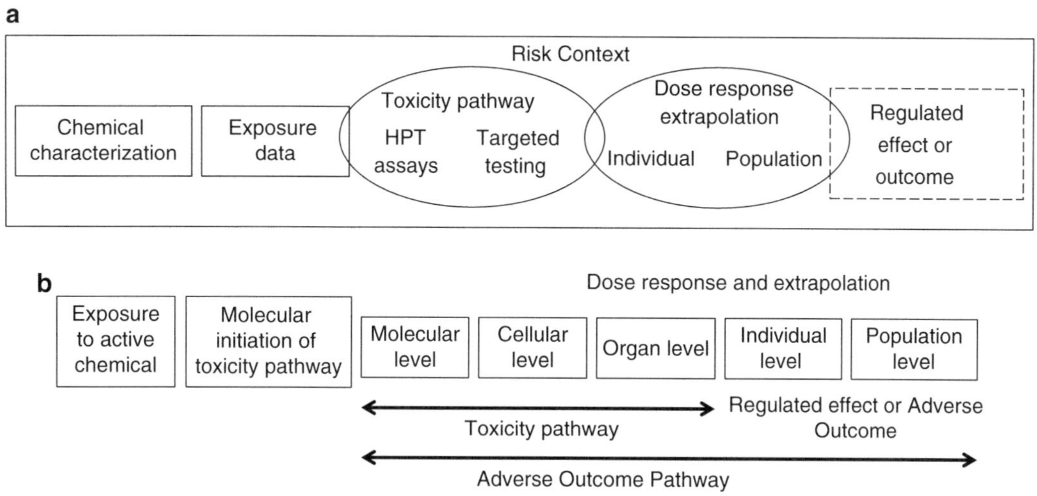

Fig. 1 Organizing framework for toxicity testing in the twenty-first century. (**a**) The twenty-first century toxicity testing recommended by the US National Research Council [8] integrated with elements such as chemical characterization, exposure assessment, population effects, regulated end point, and risk context required for chemical risk assessment. (Adapted from [8]). (**b**) The Adverse Outcome Pathway (AOP): A pathway-based framework integrating toxicological pathways and adverse outcomes (Adapted from ref. [9]

2 Organizing Framework for Implementation of the NRC Vision

The NRC vision was implemented in the Toxicology in the 21st Century (Tox21) program, a multi-agency effort involving the US National Toxicology Program of the National Institutes of Environmental Health Sciences, the US National Institutes of Health Chemical Genomics Center, the US Environmental Protection Agency (EPA), and the US Food and Drug Administration [13]. The Tox21 program focused on addressing the recommendations put forward by the NRC report. As a result, significant advances have been made in using human-relevant high-throughput (HTP) assays to assess chemical effects. Our understanding of the pathways by which chemicals cause effects has also greatly advanced. Valuable assays for key toxicological events such as receptor activation or inhibition, and activation of pathways and cellular events have been developed; however, it is recognized that these measurements alone are not sufficient to predict toxicity of a chemical. As data accumulated on chemical hazards using approaches recommended in the NRC report, it was recognized that a logical, biologically plausible pathway-based framework was needed to facilitate data collection and interpretation, as well as to enable transparent communication of what chemical testing results mean relative to a regulated outcome. To capture these concepts, a framework was needed that could connect elements in the NRC vision to outcomes that were relevant to both risk management decision-making and regulatory agencies such as the US EPA (Fig. 1a).

One such approach organizes knowledge about how chemicals interact with biological molecules to initiate causally linked events in toxicological pathways that lead to adverse outcomes (Fig. 1b). This general approach has been formalized for toxicological assessment for both human health and ecological end points as the adverse outcome pathway (AOP) framework [9]. The AOP framework focuses on defining measurable key events that mark the progression from chemical interaction with a biological molecule that initiates a perturbation (molecular initiating event; MIE) to an adverse outcome considered relevant to risk assessment. An important difference between toxicological pathways and an AOP is that an AOP is intentionally focused on regulated outcomes. Data aligned with this framework can therefore be interpreted in a way that is useful in assessing chemical hazards. For example, using the AOP framework, HTP screening assay results can be related to an adverse outcome such as reproductive impairment by describing key events that connect the two and by providing evidence for causal or statistically robust correlative linkages between them.

Consequently, the AOP framework provides a logical context in which in vitro data such as HTP receptor binding or activation assays can be linked to downstream events measured in animal assays. Inherent to the application of this framework is the need to develop complementary computational and modeling tools including bioinformatics, systems biology, network inference, pathway based-models, and computational biology models, as indicated in the NRC report. These tools are essential for linking results from different assays and end points across different levels of biological organization, including the population level, and extrapolating from in vitro effects to in vivo impacts.

3 Development of the AOP Concept

An AOP has been defined as "a conceptual construct that portrays existing knowledge concerning the linkage between a direct MIE and an adverse outcome at a biological level of organization relevant to risk assessment" [9]. MIEs are linked to adverse outcomes via key events that consist of measurable and essential changes in biological state that represent progression from a molecular level biological perturbation to an apical outcome. The AOP concept has its roots in early considerations of how to effectively apply biomarkers and quantitative structure activity relationships in ecological risk assessment (e.g., [14, 15]). The basic concept, which had long been considered in ecotoxicology, came to the fore with the publication of NRC's vision and strategy for toxicity testing in the twenty-first century, which is built on the assumption that measures of the initiation or early progression of toxicity can be understood and interpreted in a manner that is meaningful for risk-based regulatory decision-making. However, previously development of the linkages that could be considered an AOP was an informal process that was used mainly to guide thinking and experimental design. To apply this new paradigm in toxicity testing, it was critical to formally describe and support these linkages in a scientifically robust, transparent, and defensible manner.

One of the most seminal aspects of the evolution of this concept came from the convergent development of a largely analogous framework within the human health toxicology community. The World Health Organization (WHO) International Program on Chemical Safety (IPCS) developed a mode of action (MOA) framework to evaluate the relevance of experimental animal data for predicting human health outcomes [16–19]. The description of the MOA framework is similar to the AOP concept in that both aimed to describe a series of measureable and essential key events that reflected the progression of biological changes (perturbations) through which exposure to a chemical could result in adverse health outcomes. A critical aspect of the MOA framework is that it includes a weight of evidence (WoE) evaluation using modified Bradford Hill considerations on causality [20, 21]. Thus, the IPCS MOA framework offered a more systematic approach to evaluation and documentation of the linkages making up an MOA or AOP than had been broadly considered or adopted in the ecotoxicological community.

Integration of the AOP concept described by Ankley et al. [9] with the critical documentation and evaluation aspects of the MOA framework was facilitated by the Organization for Economic Cooperation and Development (OECD; [22]). The OECD recognized the value of the AOP concept for: (1) supporting test guideline development, particularly as it relates to developing alternatives to whole organism tests focused on apical end points; (2) supporting the use of quantitative structure–activity relationship (QSAR) models for grouping chemicals; and (3) developing hypothesis-driven testing strategies, including tiered approaches and "integrated approaches to testing and assessment (IATA)". The first examples of a template for documenting AOPs, in this case to support the application of QSARs and/or expert systems for chemical categorization and hazard prediction, came from a 2010 OECD workshop on "Using Mechanistic Information in Forming Chemical Categories" [23]. In 2012, the OECD formally launched its AOP Development Program under the auspices of the Extended Advisory Group on Molecular Screening and Toxicogenomics [22]. One of the first activities of the program was the development of a guidance document on developing and assessing AOPs [24]. The guidance document represented the first attempt to describe the desired contents of a formal AOP description and to integrate key elements of the AOP and MOA frameworks, including components of the MOA framework's WoE evaluation. Subsequent to the

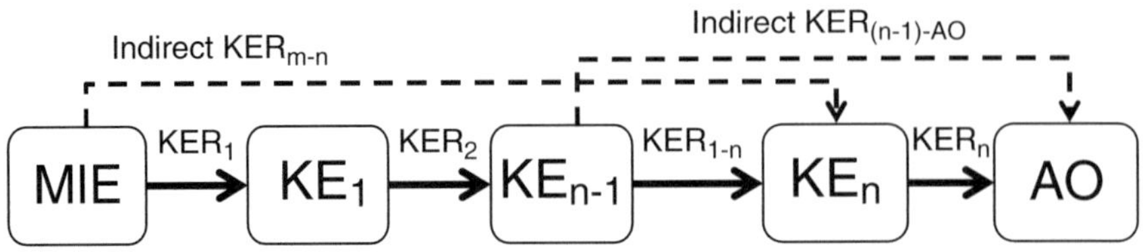

Fig. 2 Components of an Adverse Outcome Pathway (AOP). An AOP begins with a molecular initiating event (MIE) that initiates key events (KEs) causally linked to an adverse outcome (AO) of regulatory concern. MIEs, KEs, and AOs are connected by key event relationships (KERs). Indirect KERs represent empirical or statistical relationships between nonadjacent events. Here, an indirect KER going from KE_{n-1} to the AO, is labeled indirect $KER_{(n-1)-AO}$ (Adapted from ref. [24])

publication of the original guidance document, experiences while concurrently developing formal AOP descriptions and a knowledge base to facilitate the collaborative development and dissemination of the descriptions, led to the evolution of a revised template for AOP descriptions outlined in the publication of the "User's Handbook Supplement to the Guidance Document for Developing and Assessing AOPs" [24] which was implemented through the AOP Knowledge Base ([25], see Sect. 4 of this chapter).

Using the current template, AOPs are described through two primary elements, key event (KE) descriptions and key event relationship (KER) descriptions (Fig. 2; [26, 27]). The KE descriptions detail the essential and measurable changes in biological state that can be used to document progression of a particular type of toxicity. As such, a KE description includes a description of what change in biological state is being considered, the methods that can be used to measure that state, and the biological domain in which that event has relevance (e.g., taxa, life-stage, sex; [24, 27, 28]). KER descriptions detail the biological plausibility, WoE, and quantitative understanding that can support predictive inference from one KE in the AOP to the next [24, 27]. Independent descriptions of KEs and KERs, linked into a specific sequence connecting a MIE to an adverse outcome, represent the primary content of an AOP description (Fig. 2). However, to facilitate their regulatory application, AOP descriptions are complemented by an evaluation of the WoE supporting the AOP. The WoE for the AOP is based on the strength of KERs connecting its KEs. These relationships are evaluated based on modified Bradford Hill considerations [20, 21, 29]. This evaluation considers: (1) biological plausibility/concordance: whether the sequence of events and their relationships are consistent with established biological knowledge; (2) essentiality of the KEs: whether preventing an upstream KE from occurring blocks all subsequent downstream KEs; and (3) concordance of empirical observations, including dose–response concordance (i.e., whether upstream KEs are affected at lower concentrations than downstream KEs), temporal concordance (i.e., whether upstream KEs precede downstream KEs), incidence (i.e., whether upstream KEs occur more frequently than downstream KEs), and consistency (i.e., whether a KE or KER has been observed in multiple independent experiments employing relevant, orthogonal, test systems). The "User's Handbook" and AOP-KB provide a systematic and consistent approach for assembling formal AOP descriptions that include all of these elements [24, 25]

4 Examples of Ecological Health and Human Health-Oriented AOPs

A growing number of potential AOPs have been described in the literature, and several of these have been or will be added to the AOP-KB described below (e.g., [9, 28, 30–35]). Simple linear pathways are often used to represent AOPs. However, AOPs function in the context of other pathways with greater or lesser interactions depending on the level of biological organization. To illustrate these concepts, we present two examples of well-characterized AOPs with different types of information they contain and different interactions with other AOPs through common KEs. The first example, aromatase inhibition leading to reproductive dysfunction, is a very well understood linear pathway supported by extensive data, statistical and mathematical models that permit prediction of population level effects based on changes that occur at early key events. In contrast, the second example, AOPs related to thyroid disruption, demonstrates the complexity of AOPs in real life. In real life, AOPs occur as overlapping and interacting networks. These examples also provide insight into how computational toxicology methods might support development, discovery and application of AOPs to toxicology.

4.1 AOP Depicting Aromatase Inhibition Leading to Reproductive Dysfunction in Fish

The AOP for aromatase inhibition leading to reproductive dysfunction in fish is a well-characterized AOP that depicts the relationship between inhibition of cytochrome P450 aromatase (CYP19) and reproductive impacts in fish at both the individual and population levels [9, 25]. Experimental work with the fathead minnow (*Pimephales promelas*), a model fish species used widely in regulatory testing and ecotoxicology research [36], served as the foundation for development of this AOP. However, based on knowledge of the cross-species conservation of CYP19 structure and function in the context of reproductive endocrinology, this AOP likely has broad biological applicability that could include most fish species as well as, perhaps, other oviparous vertebrates [37, 38]. Aromatase catalyzes the conversion of testosterone to 17β-estradiol during normal biological function. Estradiol (E2) is involved in different reproductive processes in fish, including stimulation of the production of vitellogenin (VTG; an egg yolk precursor protein) in the liver of females, through activation of the estrogen receptor.

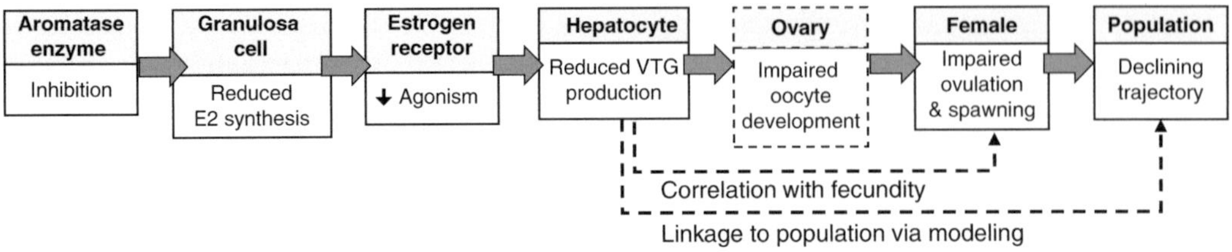

Fig. 3 Adverse Outcome Pathway (AOP) for aromatase inhibition leading to reproductive dysfunction in fish. *Dashed lines* represent indirect key event relationships (KERs). The *dashed box* represents a key event (KE) that is difficult to measure (Adapted from [9]

Vitellogenin subsequently enters the bloodstream and is taken up by the ovary, where it undergoes structural modifications and is incorporated into developing oocytes. A number of environmental contaminants, including pesticides such as the conazole fungicides, can bind to and inhibit fish CYP19 [39, 40]. Exposure of reproductively active female fish to aromatase inhibitors decreases ovarian aromatase activity, resulting in a series of measurable downstream responses (i.e., KEs), including a depression in plasma E2 concentrations, a decrease in plasma VTG, reduced deposition of VTG into developing oocytes, and depressed egg production (fecundity) that can be translated, via modeling, into potential population-level impacts (Fig. 3).

There are seven pairs of KEs (and associated KERs) in the aromatase inhibition-reproductive dysfunction AOP for fish (Fig. 3; [25]): inhibition of CYP19 activity (the MIE) resulting in decreased ovarian production of E2 (KE1); reduced plasma concentration of E2 (KE2); depressed VTG production in the liver (KE3); decreased plasma VTG concentrations (KE4); impaired oocyte development (KE5); reduced fecundity (KE6); and decreased recruitment of new individuals to the population (KE7). Reductions in VTG have also been demonstrated to be significantly correlated with fecundity, which can be used to predict impacts on reproduction [41] resulting in an indirect KER linkage between KE4 and KE6. Indirect KERs are a special class of relationships between events where significant observable or statistical relationships exist between two nonadjacent events. Indirect KER are generally predictive of a downstream event.

The aromatase inhibition-reproductive dysfunction AOP is archived in the OECD-sponsored AOP-KB and has undergone a formal WoE evaluation based on modified Bradford Hill considerations to assess its robustness [25]. Overall, based on WoE rankings of the individual KERs indicated above, and a WoE evaluation of the entire pathway, the aromatase inhibition reproductive-dysfunction AOP is rated as strong. The well-characterized and highly correlated concentration–response relationships between KEs have been useful in predicting impacts at the population level [41]. This AOP demonstrates several critical concepts: (1) AOPs represent a plausible series of biological events that describe the effects of perturbation from the molecular level to an impact of regulatory concern, here population viability; (2) KEs represent major measurable events within an AOP that are necessary for the adverse outcome to occur rather than all potential contributing events; and (3) indirect KERs are composed of statistical or observed relationships between distant KEs that are predictive effects on KEs from upstream events. Because this AOP is ranked as strong, it is useful for many applications. For example, because in vitro assays for the MIE, aromatase inhibition, are strongly linked to effects on fish reproduction, in vitro assays measuring inhibition of CYP19 could be used with some confidence to predict potential of a chemical to cause reproductive impairment.

4.2 Network of AOPs Related to Disruption of Thyroid Function

While the previous AOP provided a well-characterized linear example of a pathway, AOPs exist as interacting pathways where one can influence another. Indeed, the activation of several different MIEs can lead to the same adverse outcome of reproductive dysfunction in fish [9]. A well-developed example of the interacting and overlapping nature of AOPs is the AOPs that describe disruption of thyroid dysfunction (reviewed in [42]). These separate AOPs provide an informative example of how AOPs with common KEs can be merged to form a network of AOPs (Fig. 4).

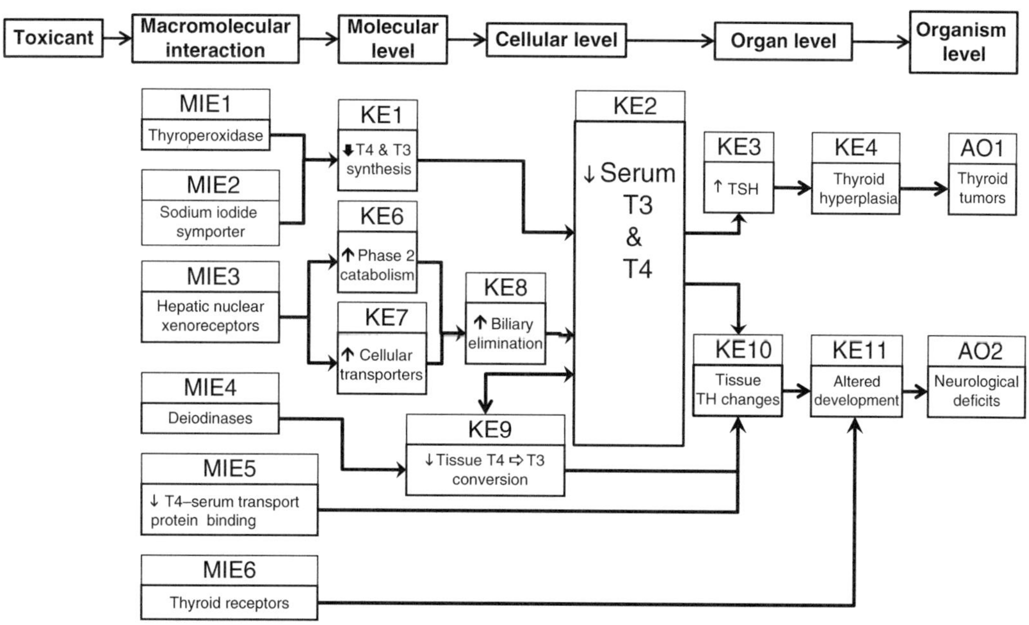

Fig. 4 Major Adverse Outcome Pathways (AOPs) for thyroid disruption. Multiple molecular initiating events (MIEs) converge on the key event (KE) of reduced levels of thyroid hormones in blood serum (KE2). Different species experience different adverse outcomes (AO) as a result of thyroid hormone (TH) reduction in serum: in rats, reduced hormone levels can lead to thyroid tumors (AO1); in both humans and rats, reduced hormone levels can lead to neurological deficits (AO2)

In this network, interspecies differences in downstream KEs can lead to different adverse outcomes. For example, unlike the previous example AOP, activation of one MIE such as thyroperoxidase inhibition can lead to two different outcomes, thyroid tumor formation (AO1), was observed in rats but not humans, while a second potential adverse outcome, neurological deficits due to altered development (AO2), was observed in both humans and rats.

Thyroid function is highly conserved across vertebrate species and is important in vertebrate development [43]. As a result, chemical disruption of thyroid hormone (TH)-dependent pathways can have significant adverse impacts [44]. A number of different chemicals have been found to disrupt thyroid pathway functions including agrochemicals, polychlorinated biphenyls, perchlorate, and bisphenols [42]. Interference with normal TH function can result from these chemicals interacting with several different MIEs that converge at the KE of reduced serum TH (KE2 in Fig. 4). Numerous studies have shown that inhibition of thyroid peroxidase (MIE1) and sodium iodide symporter (MIE2), both of which are involved in the synthesis of TH, leads to a decrease in the thyroid hormone triiodothyronine (T3) and the prohormone of T3, thyroxine (T4) (KE1) (e.g., [45–48]). The same two chemicals inhibit both thyroid peroxidase and sodium iodide symporters from numerous species, thereby allowing the prediction of chemical effects on circulating TH across species, though downstream effects may be species specific [42]. Decreased T3 and T4 synthesis leads to lower circulating serum levels of TH (KE2). Activation of hepatic nuclear receptors (MIE3) can also lead to a reduction in serum T3 and T4 levels by increasing phase 2 catabolism activity (KE6) or cellular transporter activity (KE7), which results in the increased elimination of THs through the biliary system (KE8) [49–52]. Inhibition of deiodinases (MIE4) can lead to the common KE of reduction in serum T3 and T4 because deiodinases catalyze the synthesis of T3 from T4 in peripheral tissues (KE9), thereby contributing both to overall circulating TH levels and to tissue TH levels (KE10) [45]. Vertebrates respond to low levels of TH by releasing thyroid stimulating hormone from the pituitary through a negative feedback loop to upregulate TH synthesis in the thyroid gland (KE3) [53]. In rats, when this feedback loop can no longer compensate for the inhibition of TH synthesis, increased thyroid-stimulating hormone (TSH) levels lead to thyroid hyperplasia (KE4) and ultimately thyroid follicular tumors (AO1); however, KE4 and AO1 are not relevant to the MOA for thyroid adverse effects in humans [45, 54].

A second significant adverse effect of reduced TH levels can be developmental deficits (AO2; [55]). TH deficiency has been found to result in developmental defects in humans and animals. Decreased TH tissue levels (KE10) have been shown to disrupt normal development (KE11) in a wide range of species including frogs [56, 57], flatfish [58], salmonids [59], and humans [55]. Because of the similarity in KE11 across species, disruption of metamorphosis in the frog (*Xenopus laevis*) has been used as a measure of the potential of a chemical to disrupt thyroid function and possibly affect neurological development [60]. Disruption of thyroid pathways during development can lead to dramatic neurological deficits (AO2), including impacts on visual and motor skills, or language and memory skills in humans. An alteration in the binding of T4 to serum transport proteins (MIE5) can also lead to changes in tissue TH levels, resulting in altered development and a second adverse outcome of neurological deficits. A sixth event leading to AO2 involves chemicals that act as TH receptor agonists or antagonists (MIE6) resulting in altered TH-dependent gene activation, abnormal development, and neurological impacts [42, 55]. The AOPs for thyroid dysfunction demonstrate several concepts not demonstrated by the linear AOP for aromatase inhibition leading to reproductive dysfunction: (1) KE can be shared by multiple AOPs; (2) AOPs exist as networks; (3) Thresholds, such as exceeding the ability of feedback loops to compensate for the inhibition of TH synthesis, can dictate which adverse outcome is realized after activation of a MIE; (4) Integration of AOPs from multiple species can assist in development of screening models such as *X. laevis*. Finally, these example AOPs demonstrate the utility of developing AOPs both across and within species.

5 AOP Knowledge Management and a New Paradigm for AOP Development

The OECD program on the development of AOPs [22] has been highly active, with 25 ongoing AOP development projects as of June 2014. In addition to AOP development, there are seven ongoing AOP case studies where an AOP is evaluated for its ability to inform regulatory decision-making in the context of specific chemicals. In 2013, the OECD recognized that international efforts to develop and use AOPs would be limited without a structured mechanism for capturing and disseminating the collective AOP knowledge. In response, the OECD endorsed an international effort between the US and the European Commission to jointly develop an AOP-KB to house this information [25].

The goal of the AOP-KB is to allow the scientific community, in one central location, to develop, share, and discuss their AOP-related knowledge. By doing this, the AOP-KB overcomes the following challenges to AOP development:

- Knowledge about the KEs, which include the MIE and adverse outcome as the two anchors for the AOP, is often distributed among many experts within the scientific community;
- Knowledge about KEs is not organized nor easily accessible;

- Many AOPs share common KEs resulting in duplication of effort and lack of coherence;
- Purely text-based descriptions of KEs are difficult to use in a systematic decision-making context such as the use of computational modeling to link HTP screening toxicity data to potential adverse outcomes at the organism or population level.

The AOP-KB allows all stakeholders to build AOPs by entering and then linking information about MIEs, KEs, adverse outcomes, and chemical initiators. The AOP-KB is designed to facilitate the reuse of MIE, KE, and adverse outcome information across AOPs, which prevents redundancy and makes the collective knowledge about those entities available in all the AOPs in which they appear. The AOP-KB has been developed in conjunction with the "User's Handbook" [24], so that the structure of the knowledge base matches the established best practices for assembling AOP information. In particular, the evidence supporting an AOP is captured in a consistent and organized manner, making it easier for regulatory decision makers to access, understand, and use the information in an efficient and transparent manner.

The AOP-KB project brings together four independently developed platforms, the AOP-Wiki, Effectopedia, AOP-Xplorer, and the Intermediate Effects Database [25], so that stakeholders can benefit from having a shared base of knowledge. The four modules share, exchange, and synchronize information via the AOP-KB Hub. This system allows bidirectional flow of the core AOP information shared among the different modules and provides seamless integration for the user. In addition, the application programming interface implemented in the AOP-KB Hub will be open to third party developers allowing extension of the capabilities beyond the four core modules.

The four components of the AOP-KB were designed to complement each other in the context of their focus on distinct types of AOP-related information. The AOP-Wiki [25] is focused on capturing a basic description of the AOP along with supporting evidence. This is accomplished using a mix of form-based entry fields for structured information coupled with free-text entry of supporting documentation. Effectopedia [61] provides a graphical user interface for entering AOP information and captures information that is much more structured than the information in AOP-Wiki with an emphasis on promoting computational modeling. It uses a visually expressed modular structure that captures semantically annotated knowledge, computational models, and algorithms along with the supporting evidence. AOP-Xplorer [62] is a computational tool that permits the automated graphical representation of AOPs in a network context. This module allows integration of external information to enhance the knowledge captured in the AOP-KB and thereby enhances both the development and use of

AOPs. The Intermediate Effects Database hosts chemical-related data derived from nonclassical methods, and contains information about how individual compounds trigger MIEs and/or KEs. Cross-referencing these empirical observations with the AOP descriptions in the other tools make it possible to predict the extent to which a chemical compound is involved in a certain AOP.

The four individual modules together form the comprehensive AOP-KB, which was designed to allow scientists and regulators develop and use of AOPs no matter where they are in the world. Detailed evaluation of the evidence supporting an AOP can be coupled with computational predictions derived and interpreted based on the broader biological context. Regulatory decision-making requires that an AOP is considered in the broader biological context in which it acts; therefore, having a mechanism for this level of aggregation is considered essential for the ultimate use of AOPs captured in the knowledge base. While the AOPs are designed to be chemical independent [26], the use of AOPs requires information about the potential for chemicals to perturb the pathways. For this reason, the AOP-KB incorporates chemical information and links out to additional chemical-specific resources such as the US EPA Aggregated Computational Toxicology Resource [63] and the European Chemicals Agency's International Uniform Chemical Information Database [64]. The AOP-Wiki is currently the only module online in the AOP-KB. Two of the three other modules (AOP-Xplorer, Effectopedia) are already functional at their respective host sites (awaiting compatibility testing and establishment of data exchange protocols with the AOP-Wiki), while the Intermediate Effects Database will rely on a reporting template currently in the adoption process at OECD.

The AOP-KB is intended to accelerate AOP development while facilitating information exchange between the scientific and regulatory communities. Accordingly, the AOP-KB is expected to provide a new paradigm for AOP development. Previously, the development of AOPs could be a time-consuming and labor-intensive endeavor, which, in ideal cases, would be equivalent to the effort required to write a comprehensive review article (e.g., see [34]). However, because an AOP can span a broad range of biological space, AOP development often requires scientists to tackle topics outside of their area of expertise, thereby requiring substantially more effort than is generally required for a standard review article. The AOP-KB has been designed specifically to encourage crowd-sourcing to develop the AOPs so as to leverage the collective expertise of the scientific community.

Quality control in this crowd-sourced environment will be maintained via a mix of complementary approaches. First, the ability to author or edit AOPs will be restricted to scientists who have demonstrated a clear understanding of existing guidance on AOP development from the OECD AOP Development Program [22].

As AOPs mature, authors will be encouraged to submit their AOPs to the OECD through the AOP Development Program for formal review. New authors will be given the opportunity to work with a mentor who has previously authored an AOP that has successfully undergone an OECD review. This step ensures that a minimum of acceptable information is available to begin an AOP. Second, AOPs are presented in two layers: one where AOPs are under development and a second where AOPs are ready for comment and citation. In the development layer, direct commenting and editing is limited to participants in that AOP development project. Interested parties can contact the corresponding author listed for the AOP to contribute information and ask to participate in the AOP. At the second, commenting level, each AOP is open for commenting from the general public so that anyone has the opportunity to raise concerns with the AOP. AOPs that have been submitted to the OECD for formal review will be identifiable, and the frozen version of the AOP reviewed by the OECD will be accessible from the AOP-Wiki. Third, volunteer "gardeners" that are subject matter experts will prune, edit and comment on AOPs to provide quality control for the crowd sourced AOPs. It is hoped that this process will improve the quality of the AOP information and, at the same time, increase the speed with which new AOPs are developed. Because the AOP-KB has been designed with the needs of regulators in mind, it should also facilitate information exchange between the scientific experts who are developing the AOPs and the regulators who will use them to inform their decisions.

6 Methods Used to Derive the AOPs

AOP development involves the identification of a sequence of essential and measurable biological changes (KEs) that reflect a causal progression from a MIE to an adverse outcome of regulatory significance. Predictive value is derived from the KERs that form the basis for scientifically supported inference along the identified sequence of events. Consideration of the modified Bradford Hill criteria that were used to evaluate the confidence in AOPs [24] provides some important insights into the empirical approaches and strategies that can be used to derive AOPs.

6.1 Hypothesis-Based Approaches

Hypothesis-based approached that have been used to derive AOPs stem from an understanding of normal, unperturbed biology pathways, many of which are well-studied and thoroughly described in the public literature. The strongest consideration in the WoE evaluation is the biological plausibility of the relationship based primarily on a consideration of existing biological data [24]. For example, does a functional relationship between two KEs suggest that if one is perturbed the other is also likely to be perturbed?

An approach that has been employed successfully for AOP development involves the organization of knowledge of unperturbed or normal biology underlying critical functions (e.g., reproduction, development, or immune function) into conceptual models that can be used to guide consideration of how the perturbation of various elements of a system could impact function. For example, Ankley et al. [9] employed a conceptual model of the fish reproductive axis [65] to develop testable hypotheses regarding molecular perturbations that could potentially lead to reproductive impairment in fish. In a similar effort, as part of a workshop focused on developing AOPs related to fish early life stage toxicity, conceptual models of important morphological processes during normal fish development were used to identify KEs whose disruption could plausibly lead to adverse outcomes such as impaired growth or survival [28]. Known molecular signaling pathways involved in regulating the development of specific tissues were also used to develop hypotheses regarding the type(s) of molecular perturbation that could lead to altered organogenesis, as an intermediate KE [28]. Similar approaches could be used to hypothesize potential relationships between specific molecular, biochemical, or physiological perturbations and a broad range of health outcomes in humans and/or other organisms. Thus, first and foremost, an understanding of normal biology is the critical foundation on which AOP development is based.

Hypothesis-based AOPs rely heavily on evidence for essentiality, an important consideration in the WoE evaluation of an AOP [24]. A KE in an AOP is viewed as "essential" if preventing that event also prevents all subsequent downstream KEs (including the AO) from occurring. Consequently, evidence that blocking a KE of interest can rescue an organism from the consequence(s) of downstream effects is considered very powerful support for a causal role of that KE in an AOP. Thus, certain types of experimental designs and the resultant empirical data can be particularly useful for establishing confidence in an AOP. For example, genetic knockout or knockdown studies are often well suited for evaluating the essentiality of a specific gene or protein interaction for triggering downstream events in a pathway. Wilbanks et al. [35] used this type of approach to investigate the role of the antagonism of peroxisome proliferator-activated receptor alpha (PPARα) as an MIE in an AOP where the antagonism of PPARα signaling led to decreases in long-chain fatty acid metabolism and ultimately to decreases in the endurance of mice in swim challenge experiments. Knockout of the PPARα gene partially rescued mice from the reduction in swimming endurance, demonstrating the essentiality of PPARα in this AOP. In some cases, targeted pharmacological or biochemical interventions (for example providing an exogenous source of a key metabolite that is depleted as part of a KE) can be used to establish essentiality. Recovery experiments in which an animal is initially exposed to a chemical that elicits a series of KEs along the AOP, then the chemical is removed and the events are observed to return to normal in the same sequence in which they were elicited [66, 67] can also provide evidence of essentiality.

AOPs represent generalizable and predictable motifs of biological failure that are specific to a particular type of biological perturbation (i.e., a MIE), rather than a particular chemical [26]. This is one reason why evidence establishing biological plausibility and/or essentiality is weighted more heavily than direct observation of empirical responses to specific stressors. Toxicology studies with certain chemicals (i.e., those associated with a given MIE) can contribute substantially to the overall WoE supporting a given AOP. However, many toxicology studies are rather poorly suited for AOP development. This is generally because AOP development is focused on establishing predictive linkages between observations made at different levels of biological organization that represent progression of a toxic insult to an ultimate adverse outcome. Most standard guideline toxicity tests used to generate empirical data focused solely on direct observation of an apical outcome without collecting any mechanistic information. Conversely, many mechanistically oriented studies measure only one or a few KEs without demonstrating direct connection to the apical outcome.

Toxicology experiments that incorporate multiple measurements across levels of biological organization, across a concentration range, and over a time-course are the best suited for the purpose of AOP development. Time-course studies provide the capacity to evaluate temporal concordance among KEs; i.e., whether the first KEs in a hypothesized pathway sequence occur before subsequent KEs. Testing a range of concentrations also makes it possible to evaluate dose–response concordance, namely whether upstream KEs are observed at concentrations equal to or lower than those that elicit downstream KEs, including the adverse outcome. There can be technical reasons for detecting a downstream KE at lower concentrations than the concentrations shown to elicit an upstream KE, e.g., if the measurement technique for the downstream event is much more precise than the technique used to measure an upstream event. However, in general, if the causal sequence represented by the KEs is accurate, events that lead to an adverse outcome should occur at concentrations equal to or less than the concentrations that cause the apical outcome. Finally, the incorporation of multiple measurements representing KEs into studies that generate the empirical data can help establish the co-occurrence of the KEs without confounding variables associated with study design and implementation. The more frequently two responses to various stressors are observed together, the more likely it is that they are related KEs along an AOP.

While the ideal toxicology experiment for AOP development should consider multiple measurement end points, at multiple scales, across multiple doses, and at multiple time-points, such experiments can be technically challenging, too costly, and too time-consuming to implement on a regular basis. Furthermore, even if the ideal design can be employed for a given chemical, establishing confidence in an AOP still requires establishing that other chemicals that cause the same perturbation (MIE) can elicit the same series of responses and that the similar response motif is conserved across the species for which the AOP is thought to be relevant. Consequently, to develop empirical support for an AOP, it is almost always necessary to collect empirical data from a variety of studies. Furthermore, it is helpful to begin organizing information into an AOP framework as early in the process as possible. For example, starting with a conceptual model can help in the development of a hypothetical but biologically plausible AOP that can inform the selection of the multiple end points to measure across biological levels of organization in a toxicological study. Likewise, assembling data from published literature into an AOP framework can help identify critical gaps in the information needed to establish concordance among the KEs. The identification of such gaps can help inform the design of experiments that are most likely to enhance confidence in an AOP. Overall, beginning the development of an AOP with knowledge about the type of evidence that will help to evaluate the strength of or confidence in an AOP can be used to more effectively design experiments aimed at empirically establishing AOPs.

6.2 Discovery-Based Approaches

Most of the empirical approaches and strategies to derive AOPs discussed above take advantage of known information or information from sequential testing. New approaches for the discovery of MIEs, KEs, and their relationships with adverse outcomes are required when the available information is insufficient to use hypothesis-based approaches.

When linked to adverse outcomes, omics (e.g., transcriptomics, metabolomics, proteomics) measurements can provide a unique understanding on how chemicals might cause an adverse outcome. For example, omics data have already been used to identify chemical MIEs, MOAs, KEs, and potential adverse outcomes (e.g., [68–72]). Many omics experiments focus on a limited exposure series or single time points and rely on knowing which differentially expressed genes or other molecules belong to which known biological pathways to assign biological functions (e.g., enzyme type or disease pathway). The functional information can then be used to interpret the impact of perturbing a receptor or other MIE on an animal. This approach is useful in identifying the effects on known functions and relationships between genes, proteins, and metabolites especially if direct manipulation, such as gene knockouts or targeted drug application to block the function of enzymes or other key events, is used to test hypothesized relationships arising from bioinformatics analysis.

The discovery and investigation of new KEs and relationships between KEs and adverse outcomes where limited information exists can be accomplished using a network biology approach. Network biology examines interactions or statistical relationships between large numbers of measureable events to determine specific functions within a network. Measurable events can range from gene expression levels, protein levels, metabolite levels, enzymatic responses, and histological measures to reproductive or mortality measures. For example, statistical relationships between gene expression levels can be used to infer genes that control different functions in a gene regulatory network. Essentially this approach identifies genes that consistently change expression prior to a second gene changing expression. A significant correlation between the two genes is taken as an indication that the expression of the first gene is related to the second [73–75].

New relationships between KEs can also be discovered by analyzing the effects of perturbing several different MIEs and examining what changes in events could lead to the adverse outcome of interest. Network biology approaches can be used to infer relationships between measured events (potential KEs) and apical outcomes (adverse outcome of interest). The network topology can then be used to infer relationships and potentially discover new biological relationships and knowledge that were not known previously. For example, network inference methods have been used to detect interactions among multiple different levels of biological organization, such as between genes and hormones [76], genes and metabolites [77], or even genes, microsatellites, chemistry, and phenotypes [78].

To be useful for AOP development, discovery-based approaches should incorporate the modified Bradford Hill criteria for causation to judge the strength of AOPs based on the: biological plausibility of each of the KEs and KERs, support for the KEs being necessary for the adverse outcome to occur, the extent of empirical support for each KER and the overall AOP, time concordance (i.e., the KEs occur in the expected order and before the outcome), and dose concordance (i.e., the KE should occur at a dose at or below the dose associated with the adverse outcome). With these criteria in mind, discovery approaches, such as network inference, should incorporate multiple events and a specific adverse outcome in order to discover new KEs and KERs. This approach can be especially powerful when perturbations associated with multiple concentrations or doses of chemicals are used to examine quantitative relationships between fixed time exposures (steady state or static) and biological events. Several algorithms have been proposed to analyze steady state data [75].

Valuable information for AOP development can also be gained using network approaches that examine time series data. These approaches generally examine time dependent data to build relationships between events consistent with the criteria for AOPs. Biological systems are highly dynamic and continuously respond to environmental or genetic changes [79]. Understanding the dynamics is especially important when feedback mechanisms may be involved in governing dose or concentration response thresholds, as was found in the endocrine system and synthesis of estradiol in fish [66]. Time series analyses are also useful in that they can identify sequences of events, directions of influence, or potential causality (where the occurrence of one event leads to another event). Algorithms that have been used in time series analyses include mutual information approaches, ordinary differential equations, and dynamic Bayesian networks: e.g., Time-Delay ARACNE [80]; Time-Lagged Context Likelihood of Relatedness (tlCLR) [81, 82], an extension of the CLR algorithm; the Inferelator, which is based on ordinary differential equations [81, 82]; and Bayesian Network Inference with Java Objects (BANJO; [83]).

A major challenge in establishing a quantitative understanding of KERs is how much stimulation is needed to go from one KE to another. Recent studies have focused on identifying network modules that are sets of molecules and interactions that participate in a particular biological function (reviewed by [84]) that could provide support for the biological plausibility of different parts of an AOP. Network motifs or modules have also been suggested to provide a basis for understanding the thresholds for cellular responses or KERs [85].

Another important area where discovery-based approaches could be used in AOP development is in understanding how applicable a pathway is to other species. To do this, approaches need to determine whether significant changes have occurred that would affect the function of an AOP in another species; e.g., changes in signaling or sensitivity. Some progress has been made in this area. Comparisons of genetic interactomes (sets of molecular interactions) between different organisms have previously been restricted to small sets of genes but, more recently, Ryan et al. [86] compared the conservation of biological functions in related networks in different species of yeast. Their comparison of a set of functional network modules for *Schizosaccharomyces pombe* to a set from *Saccharomyces cerevisiae* revealed a hierarchical model for the evolution of genetic interactions, with the highest conservation within protein complexes, lower conservation among biological processes, and the lowest between distinct biological processes. They also found that, despite the extensive rewiring of interactions and the evolutionary distance between the two yeast species, both networks retained conserved features and displayed similar levels of functional cross talk between equivalent biological processes. Approaches of this type may be useful in examining cross-species conservation of the complex pathways that underlay AOPs.

7 Practical Risk/Regulatory Applications

The AOP framework evolved from efforts to relate biomarker and QSAR results to apical effects in order to support regulatory applications. The AOP framework has established a scientifically rigorous context for defining and documenting how the perturbation of a single MIE causally leads to an adverse outcome. An AOP can be useful for a wide range of regulatory needs depending on how much uncertainty or how many data gaps exist within the AOP (Fig. 5). In this context, Villeneuve et al. [26] suggested three operationally defined stages of AOP development (putative, qualitative, and quantitative) that broadly reflect the gradient of development toward increasing levels of scientific confidence and predictive sophistication [26].

Putative AOPs, where a hypothesized set of KEs and KERs are supported primarily through biological plausibility and/or statistical inference, are the most uncertain. Putative AOPs often have significant data gaps in describing the KEs necessary for linking an MIE to the adverse outcome and generally lack a formal WoE evaluation. Despite these limitations, putative AOPs can be useful frameworks from which potential links between molecular/cellular level perturbations and apical hazards can be characterized; however, the

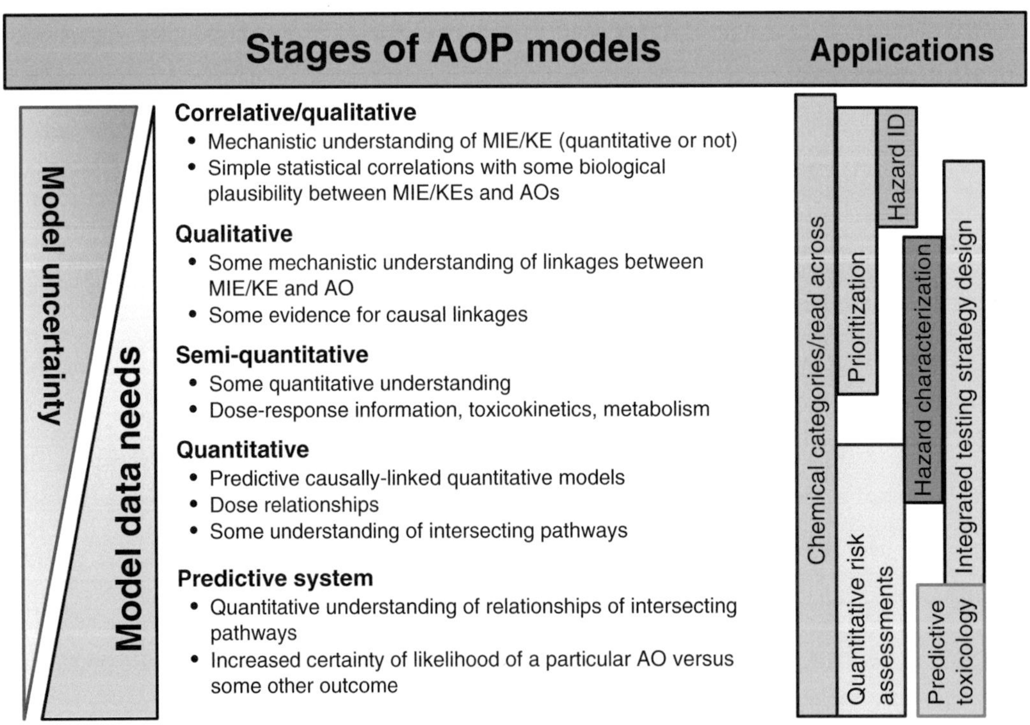

Fig. 5 Stages of AOP models and the risk or hazard assessment applications that they support

data gaps or uncertainties that exist need to be carefully taken into account. For example, if a chemical is found to cause a KE that is causally or statistically linked to an adverse outcome in a putative AOP, this could be enough evidence to support chemical categorization, hazard identification, and hazard assessment, or for prioritization for further testing.

At the qualitative AOP stage, the KEs have been assembled with descriptions of how they can be measured and KERs supported by empirical evidence along with plausibility or statistical inference and a qualitative evaluation of the overall WoE supporting the AOP have been developed. At this stage, the supporting information has been assembled and described in a manner consistent with OECD guidelines [24]. In qualitative AOPs the uncertainties lie in the lack of information available for developing quantitative relationships; e.g., an insufficient number of doses to establish dose–response relationships. In general, qualitative AOPs are suitable for applications where a quantitative evaluation of the ability of particular chemicals to cause a KE or adverse outcome is not required, e.g., chemical categorization, prioritization, hazard identification, hazard characterization, and integrated approaches to testing and assessment [87].

At the quantitative AOP stage, the KEs are supported by descriptions of how each KE can be measured and the accuracy and precision with which the measurements are made have been recorded. Additionally, a quantitative understanding exists of the magnitude and/or duration of change in the upstream KE needed to evoke some magnitude of change in the downstream KE. Quantitative AOPs are especially suitable for applications where outcomes are predicted based on limited testing data. These AOPs can support quantitative estimations of the probability or severity of an adverse outcome, based on one or more measurements of upstream KEs linked in the AOP.

8 Anticipated Impacts and Challenges

The AOP concept has already had significant impacts on efforts to characterize chemical hazards. It has served as a focal point and created a common, scientifically rigorous framework for harmonizing efforts in chemical hazard screening and assessment. The AOP concept is poised to become indispensable to hazard and risk assessment through its acceptance as a common approach for documenting different mechanisms and biological pathways that cause impacts relevant to regulatory concern. While the AOP framework has great promise, several challenges remain for it to be fully accepted for regulatory applications. For instance, while AOP development can benefit from multidisciplinary efforts, one of the main challenges on succeeding will be the ability to communicate across fields of expertise. Scientists with different backgrounds, developers, and regulators need to communicate how data are handled, the critical features needed for different applications, and how AOPs can be used in different contexts so that the AOP framework can truly advance. Another potential challenge is that while the AOP-KB is structured to allow for conceptualization and visualization of AOP networks, it does not allow for analysis or exploration of those networks. Therefore, there is a critical need for approaches, such as the AOP-Xplorer [62] that allow the exploration and analyses of AOP networks to be able to predict the results of multiple perturbations to truly understand AOP networks and mixture effects. Another challenge is that, unlike risk assessments, AOPs are not chemical specific. Thus, there is a need to merge the non-chemical specific MOA description provided by AOPs with the chemical-specific information required for a chemical risk assessment. Here, new approaches for assessing exposure, absorption, distribution, metabolism, and elimination will need to be used in conjunction with AOPs. Another important challenge is that AOPs can be modulated by many different factors including exposure history, habitat, life history, genetic susceptibility, life stages, and non-chemical stressors. These factors will need to be incorporated into the AOP framework to make truly accurate predictions.

AOP developers also face difficulties in measuring relationships among different levels of organization that may often be conceptual and hard to quantify. Development of tools to link results from different assays and end points across levels of biological organization would greatly facilitate extrapolation from in vitro effects to in vivo impacts, especially at the population level. Concurrent with the development of new tools will be development of databases with large numbers of AOP descriptions such as the AOP-KB that include all possible biological pathways and adverse outcomes that could be impacted by chemicals. Toxicological testing should focus more on incorporating dose responses over multiple KEs or on the adverse outcomes in order to sufficiently describe the KERs necessary to develop quantitative AOPs. This focus should also include identifying the chemical thresholds that lead to an adverse outcome, because not all changes within a specific level of biological organization will necessarily be translated into adverse effects at a higher level [85]. Computational approaches can be extremely helpful in developing a framework that can be applied to both ecological and human health risk assessment. Clearly, it is imperative that multidisciplinary teams work together to develop the framework. Thus, it is crucial that researchers from different disciplines, risk assessors, and government agencies learn to speak the same language, as communication will be crucial for AOP development. Furthermore, the use of the AOPs should be independent of AOP development, but both should inform each other and allow the development of a framework that is truly predictive and useful for risk assessment.

References

1. United States Code (2012) Federal Insecticide, Fungicide, and Rodenticide Act. United States Senate. http://www.ag.senate.gov/download/fifra. Accessed 20 Jan 2015
2. United States Environmental Protection Agency (2014) TSCA chemical substance inventory. United States Environmental Protection Agency. http://www.epa.gov/oppt/existingchemicals/pubs/tscainventory. Accessed 14 March 2014
3. United States Public Law (1976) Toxic Substances Control Act. United States Senate Public Law 94-469. http://www.gpo.gov/fdsys/pkg/STATUTE-90/pdf/STATUTE-90-Pg2003.pdf. Accessed 20 Jan 2015
4. Rovida C, Hartung T (2009) Re-evaluation of animal numbers and costs for in vivo tests to accomplish REACH legislation requirements for chemicals – a report by the transatlantic think tank for toxicology (t(4)). ALTEX 26:187–208
5. European Commission (2003) Directive 2003/15/EC (7th Amendment to the European Cosmetics Directive 76/768/EEC). http://ec.europa.eu/consumers/sectors/cosmetics/files/doc/antest/%282%29_executive_summary_en.pdf. Accessed 14 Mar 2014
6. European Parliament (2010) Directive 2010/63/EU of the European Parliament and of the Council of 22 September 2010 on the protection of animals used for scientific purpose. Official J Eur Union L276:33–79
7. Schiffelers M-JWA, Blaauboer BJ, Hendriksen CFM et al (2012) Regulatory acceptance and use of 3R models: a multilevel perspective. ALTEX 29:287–300
8. US National Research Council Committee on Toxicity (2007) Toxicity testing in the 21st century. The National Academies Press, Washington, DC
9. Ankley GT, Bennett RS, Erickson RJ et al (2010) Adverse outcome pathways: a conceptual framework to support ecotoxicology research and risk assessment. Environ Toxicol Chem 29:730–741
10. Andersen ME, Krewski D (2008) Toxicity testing in the 21st century: bringing the vision to life. Toxicol Sci 107:324–330
11. Cote I, Anastas PT, Birnbaum LS et al (2012) Advancing the next generation of health risk assessment. Environ Health Perspect 120:1499–1502
12. Villeneuve D, Garcia-Reyero N (2010) Vision and strategy: predictive ecotoxicology in the 21st century. Environ Toxicol Chem 30:1–8
13. National Institutes for Health (2014) Tox21 program. National Center for Advancing Translational Sciences. http://www.ncats.nih.gov/research/reengineering/tox21/tox21.html. Accessed 20 Jan 2015
14. Bradbury SP, Feijtel TC, Van Leeuwen CJ (2004) Meeting the scientific needs of ecological risk assessment in a regulatory context. Environ Sci Tecnol 38:463A–470A
15. Haux C, Forlin L (1988) Biochemical methods for detecting effects of contaminants on fish. Ambio 17:376–380
16. Boobis AR, Cohen SM, Dellarco V et al (2006) IPCS framework for analyzing the relevance of a cancer mode of action for humans. Crit Rev Toxicol 36:781–792
17. Cohen SM, Boobis AR, Meek ME et al (2006) 4-Aminobiphenyl and DNA reactivity: case study within the context of the 2006 IPCS Human Relevance Framework for Analysis of a cancer mode of action for humans. Crit Rev Toxicol 36:803–819
18. Dellarco VL, McGregor D, Berry SC et al (2006) Thiazopyr and thyroid disruption: case study within the context of the 2006 IPCS Human Relevance Framework for analysis of a cancer mode of action. Crit Rev Toxicol 36:793–801
19. Boobis AR, Doe JE, Heinrich-Hirsch B et al (2008) IPCS framework for analyzing the relevance of a noncancer mode of action for humans. Crit Rev Toxicol 38:87–96
20. Hill AB (1965) The environment and disease: association or causation? Proc R Soc Med 58:295–300
21. Meek ME, Boobis A, Cote I et al (2014) New developments in the evolution and application of the WHO/IPCS framework on mode of action/species concordance analysis. J Appl Toxicol 34:1–18
22. Organization for Economic Co-operation and Development (2011) Testing of chemicals: adverse outcome pathways, molecular screening and toxicogenomics. http://www.oecd.org/chemicalsafety/testing/adverse-outcome-pathways-molecular-screening-and-toxicogenomics.htm. Accessed 20 Jan 2015
23. OECD, Organization for Economic Co-operation and Development (2011) Report of the workshop on using mechanistic information in forming chemical categories, Series on testing and assessment, No. 138 ENV/JM/MONO(2011)8. Organization for Economic Cooperation and Development, Environment Directorate, Paris, France, p 45
24. Adverse Outcome Knowledge Base (2014) Users handbook supplement to the guidance for developing and assessing AOPs. https://aopkb.org/common/AOP_Handbook.pdf. Accessed 20 Jan 2015

25. Adverse Outcome Knowledge Base (2014) Adverse outcome pathway knowledge base (AOP-KB). www.aopkb.org . Accessed 20 Jan 2015
26. Villeneuve DL, Crump D, Garcia-Reyero N et al (2014) Adverse outcome pathway (AOP) development I: strategies and principles. Toxicol Sci 142:312–320
27. Villeneuve DL, Crump D, Garcia-Reyero N et al (2014) Adverse outcome pathway development II: best practices. Toxicol Sci 142: 321–330
28. Villeneuve D, Volz DC, Embry MR et al (2014) Investigating alternatives to the fish early-life stage test: a strategy for discovering and annotating adverse outcome pathways for early fish development. Environ Toxicol Chem 33:158–169
29. OECD, Organization for Economic Co-operation and Development (2013) Guidance document on developing and assessing adverse outcome pathways, Series on testing and assessment, No. 184 ENV/JM/MONO(2013)6. Organization for Economic Cooperation and Development, Environment Directorate, Paris, France, p 45
30. Volz DC, Belanger S, Embry M et al (2011) Adverse outcome pathways during early fish development: a conceptual framework for identification of chemical screening and prioritization strategies. Toxicol Sci 123:349–358
31. MacKay C, Davies M, Summerfield V et al (2013) From pathways to people: applying the adverse outcome pathway (AOP) for skin sensitization to risk assessment. ALTEX 30: 473–486
32. Vinken M, Landesmann B, Goumenou M et al (2013) Development of an adverse outcome pathway from drug-mediated bile salt export pump inhibition to cholestatic liver injury. Toxicol Sci 136:97–106
33. Rattner BA, Lazarus RS, Elliott JE et al (2014) Adverse outcome pathway and risks of anticoagulant rodenticides to predatory wildlife. Environ Sci Technol 48:8433–8445
34. Russom CL, LaLone CA, Villeneuve DL et al (2014) Development of an adverse outcome pathway for acetylcholinesterase inhibition leading to acute mortality. Environ Toxicol Chem 33:2157–2169
35. Wilbanks MS, Gust KA, Atwa S et al (2014) Validation of a genomics-based hypothetical adverse outcome pathway: 2,4-dinitrotoluene perturbs PPAR signaling thus impairing energy metabolism and exercise endurance. Toxicol Sci 141:44–58
36. Ankley GT, Villeneuve DL (2006) The fathead minnow in aquatic toxicology: past, present and future. Aquat Toxicol 78:91–102
37. Celander MC, Goldstone JV, Denslow ND et al (2011) Species extrapolation for the 21st century. Environ Toxicol Chem 30:52–63
38. Norris DO, Carr JA (2013) Vertebrate endocrinology, 5th edn. Academic, Waltham, MA
39. Ankley GT, Jensen KM, Durhan EJ et al (2005) Effects of two fungicides with multiple modes of action on reproductive endocrine function in the fathead minnow (Pimephales promelas). Toxicol Sci 86:300–308
40. Skolness SY, Blanksma CA, Cavallin JE et al (2013) Propiconazole inhibits steriodogenesis and reproduction in the fathead minnow (*Pimephales promelas*). Toxicol Sci 132:284–297
41. Miller DH, Jensen KM, Villeneuve DL et al (2007) Linkage of biochemical responses to population-level effects: a case study with vitellogenin in the fathead minnow (*Pimephales promelas*). Environ Toxicol Chem 26:521–527
42. Crofton KM (2008) Thyroid disrupting chemicals: mechanisms and mixtures. Int J Androl 31:209–223
43. Paris MM, Laudet VV (2008) The history of a developmental stage: metamorphosis in chordates. Genesis 46:657–672
44. Brucker-Davis F (1998) Effects of environmental synthetic chemicals on thyroid function. Thyroid 8:827–856
45. Capen CC (1997) Mechanistic data and risk assessment of selected toxic end points of the thyroid gland. Toxicol Pathol 25:39–48
46. Hurley PM (1998) Mode of carcinogenic action of pesticides inducing thyroid follicular cell tumors in rodents. Environ Health Perspect 106:437–445
47. Wolff J (1998) Perchlorate and the thyroid gland. Pharm Rev 50:89–105
48. Van Sande J, Massart C, Beauwens R et al (2003) Anion selectivity by the sodium iodide symporter. Endocrinology 144:247–252
49. Oppenheimer JH, Bernstein G, Surks MI (1968) Increased thyroxine turnover and thyroidal function after stimulation of hepatocellular binding of thyroxine by phenobarbital. J Clin Invest 47:1399–1406
50. McClain RM, Levin AA, Posch R et al (1989) The effect of phenobarbital on the metabolism and excretion of thyroxine in rats. Toxicol Appl Pharmacol 99:216–228
51. Visser TJ, Kaptein E, Gijzel AL et al (1993) Glucuronidation of thyroid hormone by human bilirubin and phenol UDP-glucuronyltransferase isoenzymes. FEBS Lett 324:358–360
52. Hood A, Klaassen CD (2000) Differential effects of microsomal enzyme inducers on in vitro thyroxine (T(4)) and triiodothyronine (T(3)) glucuronidation. Toxicol Sci 55:78–84

53. Chiamolera MI, Wondisford FE (2009) Thyrotropin-releasing hormone and the thyroid hormone feedback mechanism. Endocrinology 150:1091–1096
54. Hill RN, Crisp TM, Hurley PM et al (1998) Risk assessment of thyroid follicular cell tumors. Environ Health Perspect 106:447–457
55. Zoeller RT (2007) Environmental chemicals impacting the thyroid: targets and consequences. Thyroid 17:811–817
56. Dodd MHI (1976) The biology of metamorphosis. In: Lofts B (ed) Physiology of amphibia. Academic, New York, NY, pp 467–599
57. Leloup J, Buscaglia M (1977) La triiodothyronine, hormone de la metamorphose des amphibiens. C R Acad Sci 284:2261–2263
58. Power DM, Llewellyn L, Faustino M et al (2001) Thyroid hormones in growth and development of fish. Comp Biochem Physiol C Toxicol Pharmacol 130:447–459
59. Dickhoff WW, Folmar LC, Gorbman A (1978) Changes in plasma thyroxine during smoltification of coho salmon, *Oncorhynchus kisutch*. Gen Comp Endocrinol 36:229–232
60. Degitz SJ, Holcombe GW, Flynn KM et al (2005) Progress towards development of an amphibian-based thyroid screening assay using *Xenopus laevis*. Organismal and thyroidal responses to the model compounds 6-propylthiouracil, methimazole, and thyroxine. Toxicol Sci 87:353–364. doi:10.1093/toxsci/kfi246
61. International QSAR Foundation (2014) Effectopedia. http://www.effectopedia.org. Accessed 20 Jan 2015
62. Perkins EJ (2014) AOP-Xplorer. http://aopxplorer.org. Accessed 20 Jan 2015
63. United States Environmental Protection Agency (2014) Aggregated Computational Toxicology Resource (ACToR). http://actor.epa.gov. Accessed 15 Jan 2015
64. European Chemicals Agency (2014) International Uniform Chemical Information Database (IUCLID) http://iuclid.eu. Accessed 20 Jan 2015
65. Villeneuve DL, Larkin P, Knoebl I et al (2007) A graphical systems model to facilitate hypothesis-driven ecotoxicogenomics research on the teleost brain-pituitary-gonadal axis. Environ Sci Technol 41(1):321–330
66. Villeneuve D, Mueller ND, Martinović D et al (2009) Direct effects, compensation, and recovery in female fathead minnows exposed to a model aromatase inhibitor. Environ Health Perspect 117:624–631
67. Ankley GT, Bencic D, Cavallin JE et al (2009) Dynamic nature of alterations in the endocrine system of fathead minnows exposed to the fungicide prochloraz. Toxicol Sci 112(2): 344–353. doi:10.1093/toxsci/kfp227
68. Garcia-Reyero N, Escalon BL, Prats E et al (2014) Effects of BDE-209 contaminated sediments on zebrafish development and potential implications to human health. Environ Int 63:216–223
69. Garcia-Reyero N, Kennedy AJ, Escalon BL et al (2014) Differential effects and potential adverse outcomes of ionic silver and silver nanoparticles in vivo and in vitro. Environ Sci Technol 48:4546–4555
70. Hallén K, Björkegren J, Tegnér J (2006) Detection of compound mode of action by computational integration of whole-genome measurements and genetic perturbations. BMC Bioinformatics 7:51
71. Wang R-L, Bencic D, Villeneuve DL et al (2010) A transcriptomics-based biological framework for studying mechanisms of endocrine disruption in small fish species. Aquat Toxicol 98:230–244
72. Warner CM, Gust KA, Stanley JK et al (2012) A systems toxicology approach to elucidate the mechanisms involved in RDX species-specific sensitivity. Environ Sci Technol 46:7790–7798
73. Basso K, Margolin AA, Stolovitzky G et al (2005) Reverse engineering of regulatory networks in human B cells. Nat Genet 37: 382–390
74. Della Gatta G, Palomero T, Perez-Garcia A et al (2012) Reverse engineering of TLX oncogenic transcriptional networks identifies RUNX1 as tumor suppressor in T-ALL. Nat Med 18:436–440
75. Lingeman J (2012) Network inference in molecular biology. Springer, New York, NY, ISSN: 2191-8112
76. Perkins EJ, Chipman JK, Edwards SW et al (2010) Reverse engineering adverse outcome pathways. Environ Toxicol Chem 30:22–38. doi:10.1002/etc.374
77. Garcia-Reyero N, Ekman DR, Habib T et al (2014) Integrated approach to explore the mechanisms of aromatase inhibition and recovery in fathead minnows (*Pimephales promelas*). Gen Comp Endocrinol 203:193–202
78. Williams TD, Turan N, Diab AM et al (2011) Towards a system level understanding of non-model organisms sampled from the environment: a network biology approach. PLoS Comput Biol 7:e1002126. doi:10.1371/journal.pcbi.1002126.t004

79. Ideker T, Krogan NJ (2012) Differential network biology. Mol Syst Biol 8:1–9
80. Zoppoli P, Morganella S, Ceccarelli M (2010) TimeDelay-ARACNE: Reverse engineering of gene networks from time-course data by an information theoretic approach. BMC Bioinformatics 11:154
81. Greenfield A, Madar A, Ostrer H et al (2010) DREAM4: combining genetic and dynamic information to identify biological networks and dynamical models. PLoS One 5:e13397
82. Madar A, Greenfield A, Vanden-Eijnden E et al (2010) DREAM3: network inference using dynamic context likelihood of relatedness and the inferelator. PLoS One 5:e9803
83. Yu J, Smith VA, Wang PP et al (2004) Advances to Bayesian network inference for generating causal networks from observational biological data. Bioinformatics 20:3594–3603
84. Mitra K, Carvunis AR, Ramesh SK et al (2013) Integrative approaches for finding modular structure in biological networks. Nat Rev Genet 14:719–732
85. Zhang Q, Bhattacharya S, Conolly RB et al (2014) Molecular signaling network motifs provide a mechanistic basis for cellular threshold responses. Environ Health Perspect 122: 1261–1270
86. Ryan CJ, Roguev A, Patrick K et al (2012) Hierarchical modularity and the evolution of genetic interactomes across species. Mol Cell 46:691–704
87. Tollefsen KE, Scholz S, Cronin MT et al (2014) Applying Adverse Outcome Pathways (AOPs) to support Integrated Approaches to Testing and Assessment (IATA). Regul Toxicol Pharmacol. doi:10.1016/j.yrtph.2014.09.009, Epub ahead of print

Chapter 2

Metabolic Pathway Databases: A Word of Caution

Miranda D. Stobbe

Abstract

Metabolic pathway databases collect and organize knowledge on metabolism that has been gathered over the course of decades of research. It is, however, far from trivial to accurately represent all this knowledge in a format suitable for a wide range of computational analyses. Nevertheless, many research groups have taken up this challenge, inspired by the great value and potential of pathway databases. Nowadays, these databases are routinely used in the interpretation of omics data. They have also proven to be powerful resources for various other types of analyses, including the identification of potential drug targets and in silico phenotype prediction. The number of metabolic pathway databases continues to grow, describing the metabolic network for an increasing number of organisms. For *Homo sapiens* and several other organisms, such as *S. cerevisiae* and *A. thaliana*, there are even multiple databases that describe their metabolic network.

A word of caution is, however, warranted when working with metabolic pathway databases, because no two are alike. Databases have different ways of representing a metabolic network in a digital format and they define concepts like a pathway differently. Moreover, the breadth and depth with which a network is covered varies. Several comparisons have also shown that the consensus between databases that describe the metabolic network of the same organism is limited. Furthermore, each database offers users a different way to extract the knowledge represented. At the root of these differences lies the main intended use of the database, the strategy used to reconstruct the metabolic network and the degree of curation.

Taken together, these differences should be borne in mind when working with metabolic pathway databases, especially when one needs to choose between several databases that describe the metabolic network of the same organism. The choice of a particular database could influence (the interpretation of) the results of the analysis at hand. In this chapter, we provide further insights into metabolic pathway databases to understand their individual strengths and limitations, which is important to be aware of for any essential input of an analysis. Ultimately, these databases should be integrated and consolidated to arrive at a more complete and biologically accurate description of the metabolic network. This will, however, require a broad community effort.

Key words Metabolic network, Network reconstruction, Curation, Data extraction, Community effort

1 Introduction

The first scientific paper on metabolism can be traced back to the 17th century, when Santorio Sanctorius published his results on body weight fluctuations during the course of a day [1]. Two

Julia Hoeng and Manuel C. Peitsch (eds.), *Computational Systems Toxicology*, Methods in Pharmacology and Toxicology, DOI 10.1007/978-1-4939-2778-4_2, © Springer Science+Business Media New York 2015

centuries later, the realization that the reactions within a cell are the same as those studied in chemistry marked a new era in the field of metabolism. Another important milestone was the discovery in 1897 that enzymes catalyze metabolic reactions [2]. The first complete metabolic pathways were described in the 1930s, among which was the tricarboxylic acid (TCA) cycle [3]. Although they are frequently studied in isolation, pathways interact and together constitute what is referred to as the metabolic network. To understand metabolism in all its detail in healthy and in diseased states, it is fundamental to study it at a systems level. This is analogous to the fact that it is impossible to understand the principles of powered flight and the working details of a modern aircraft by only considering the components of an airplane laid out on a hanger floor [4]. The importance of a systems level approach has been recognized in several research fields, including pharmacology [5, 6].

To be able to study the metabolic network as a whole, metabolic pathway databases are essential. They collect and organize our knowledge on metabolism, which is scattered across a multitude of scientific publications and other types of databases. A metabolic pathway database serves as a digital encyclopedia on metabolism, and often provides powerful tools to visualize experimental data. Another key objective is to accurately represent the metabolic network in a format suitable for computational analyses. One of the first metabolic pathway databases to be made publicly available was the Kyoto Encyclopedia of Genes and Genomes (KEGG) in 1995 [7]. Since then, numerous metabolic pathway databases have been constructed, covering an increasing number of organisms [8]. This growth has been fueled by the tremendous progress in the assembly and functional annotation of whole genomes over the last 20 years. Nowadays, how to capture the increasing amount of knowledge on metabolism has become a firmly established topic of research in the relatively short history of bioinformatics. Furthermore, metabolic pathway databases are now used routinely for the analysis of different types of omics data.

However, a word of caution is warranted when working with metabolic pathway databases, because no two of them are alike. First of all, the breadth and depth with which a metabolic network is covered varies. Several comparisons have also shown that there is a surprisingly limited consensus between pathway databases that describe the metabolic network of the same organism [9–14]. A striking example of this lack of consensus is given by the results of our comparison of ten descriptions of one of the most classical pathways, the

TCA cycle [15] (Fig. 1). The ten human pathway databases agree on only three reactions. Metabolic pathway databases also differ in how they represent the metabolic network in a digital format, to what detail and how concepts, like a pathway, are defined. Moreover, the databases each offer a different way of accessing the data. Explanations for these differences include that the main intended use of the databases differs, the manner in which the network has been reconstructed varies and the networks are curated to different degrees.

It is important to be aware of the differences between databases, as this will enable one to make a well-informed decision on which one to use if more than one is available for the organism of interest. The choice of a particular database is likely to influence the (interpretation of the) results of the analyses, as shown by the diverging outcomes of methods that have been applied to multiple metabolic pathway databases [26–31]. Furthermore, as is true for any essential input of an analysis, it is important to familiarize oneself with the metabolic pathway database(s) one is working with to understand its strengths and limitations. As a guide to gain a deeper understanding of the database(s) of interest, we propose the following set of questions:

- What is the main objective of the database?
- How is the metabolic network reconstructed?
- What is the manner and degree of curation of the metabolic network?
- What is the breadth and depth with which the metabolic network is covered?
- How is knowledge represented in the database?
- How can knowledge be extracted from the database?

We will first go into what the main differences are between metabolic pathway databases. Next, we will discuss the different possible answers to the above stated questions and how the varying answers could influence the analysis of interest. Illustrative examples are taken from our previously published analyses of five human metabolic pathway databases [14, 32]: the Edinburgh Human Metabolic Network (EHMN) [17], *H. sapiens* Recon 1 [18], HumanCyc [19], KEGG [21], and Reactome [23]. In the last section of this chapter we will discuss the importance of a broad community effort to further improve the description of the metabolic network. The multiple descriptions of the network of the same organism provide a unique way to reveal controversial biological knowledge by providing different, possibly contradictory, views. Moreover, they form complementary pieces of the puzzle, i.e., the complete metabolic network. The integration of the descriptions will provide a more complete and accurate network. This will require joining forces across multiple scientific disciplines, including biochemistry and systems biology.

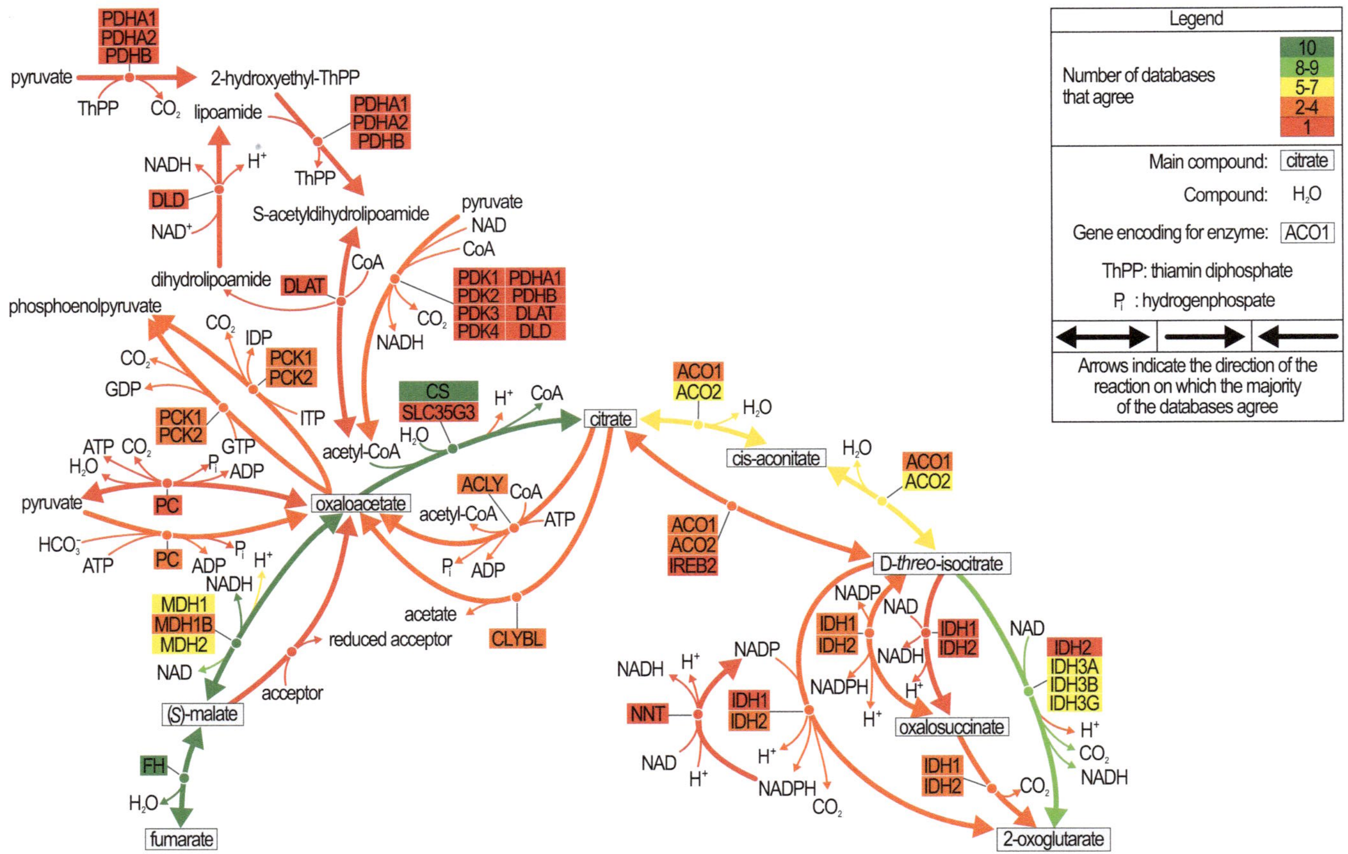

Fig. 1 Union of the descriptions of the TCA cycle given by ten pathway databases. This overview shows the reactions and genes annotated to play a role in the TCA cycle by the following ten databases: BioCarta [16], EHMN [17], *H. sapiens* Recon 1 [18], HumanCyc [19], INOH [20], KEGG [21], Panther [22], Reactome [23], SMPDB [24], and UniPathway [25]. *Grey rectangles* indicate the main metabolites of the TCA cycle. All genes linked to a reaction by any of the databases are grouped together. *Colors* indicate the level of agreement on a reaction and on the gene(s). The direction of an *arrow* is determined by what is indicated by the majority of the databases. In cases where there is no majority, irreversible was chosen. This figure and its caption have been adapted from Stobbe et al., "Improving the description of metabolic networks: the TCA cycle as example". The FASEB Journal, 2012, Sep;26(9):3625–36 [15] with the permission of the Journal of the Federation of American Societies for Experimental Biology

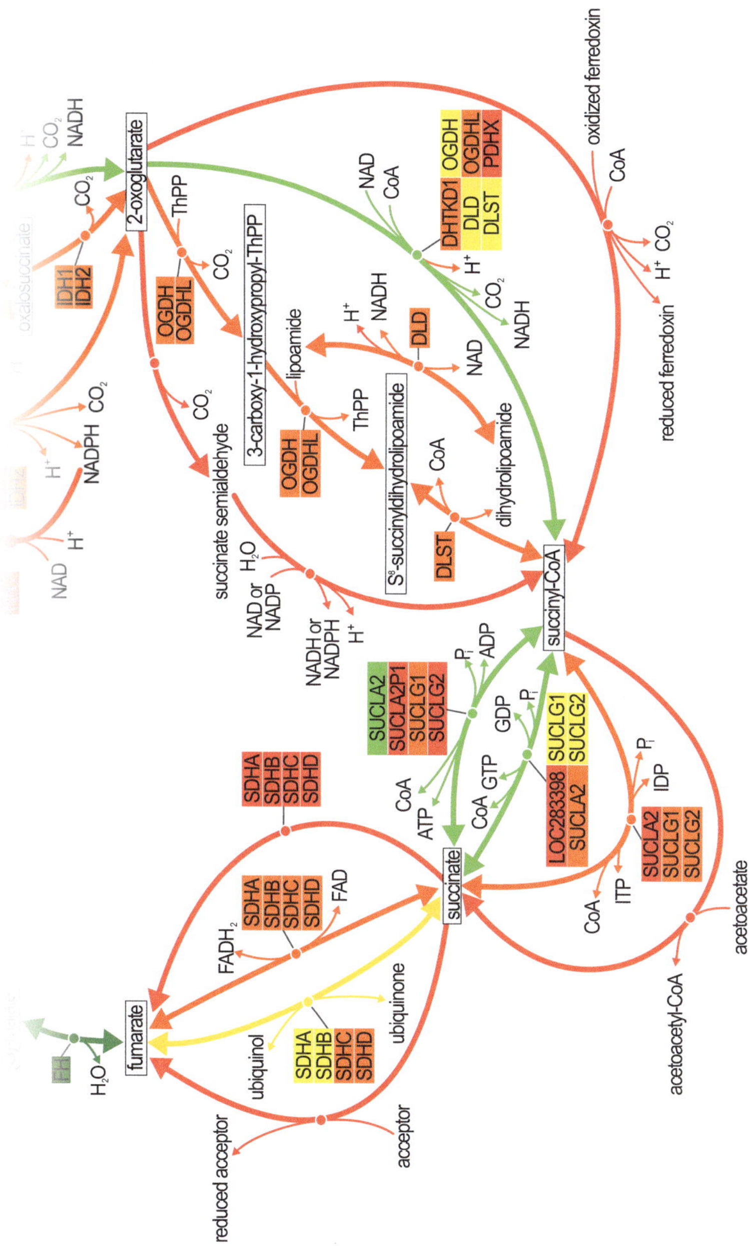

Fig. 1 (continued)

2 What Are the Main Differences between Databases?

We can divide the differences between databases into those related to the description of the metabolic network and those related to the representation of the network. The differences result from, among other things, the main intended use of the database, how the network was reconstructed, and how it was curated, which we will discuss in further detail in Sect. 3 of this chapter.

2.1 Differences in Content

Several comparisons of databases that describe the network of the same organism have shown that the consensus on the network is limited [9–14]. For example, in the comparison of two metabolic networks of *S. cerevisiae*, the consensus on the reaction level was only around one third [11]. A similar result was found for a comparison of two metabolic networks of *A. thaliana* [10], where only about one third of the reactions could be matched unambiguously. The overlap between the five human pathway databases that we have analyzed was about 3 % of the almost 7000 reactions they have combined [14]. If only two networks were compared, the consensus varied from 11 % (EHMN versus Reactome) to 28 % (EHMN versus KEGG). Furthermore, there are large differences in the breadth and depth with which the metabolic network is covered. This is depended on, among other things, to what detail the metabolism of the particular organism has been unraveled and the degree to which the network has been curated. About 70 % of the total number of reactions that the five human pathway databases had combined could only be found in one of the databases.

There are various explanations for the lack of consensus between databases and difference in the breadth and depth, which we will discuss throughout this chapter. One important explanation to keep in mind is that curators may have interpreted the literature differently [11]. In our comparison of ten descriptions of the TCA cycle, we discovered that none of these descriptions accurately reflected what is written in the literature [15]. In some cases, this could be explained by a lack of conclusive evidence or even conflicting information in the literature, which may have led to curators drawing different conclusions. The same will likely be true for many other pathways. Aung et al. [33], for example, revised the pathway "fatty acid, glycerolipid, and glycerophospholipid metabolism" in the consensus model Yeast v6.0 [34]. They found several inconsistencies with respect to the literature, including "*incorrect gene-reaction associations*," "*inappropriate assignments of reaction directionality*," and "*missing reactions relevant to the synthesis and degradation of triglycerides*." This further emphasizes that reconstructions of metabolic networks continue to be work in progress.

2.2 Differences in Knowledge Representation

Another important ongoing challenge is how to represent every detail of the reconstructed metabolic network in a way that is suitable for a wide range of computational analyses. Numerous decisions have to be made as to how to represent and connect the different elements of the metabolic network. Furthermore, the different aspects of the network, such as a pathway, have to be defined. Figure 2 shows a selection of elements that a database would need to represent to capture the various details of the gluconeogenesis pathway. At the same time, a level of abstraction is required to provide the user with a more general overview. Each database has its own approach for representing the metabolic network, which the databases continue to further refine, even long standing databases like KEGG, as evidenced by a recent publication [21]. Moreover, as more detailed knowledge is gained on, for example, tissue- and cell-specificity, the representation will need to be even further expanded. We will describe the main differences in knowledge representation between databases on the level of the entire database, network, pathway and reaction. Note that the design decisions taken on one level may also influence the representation on other levels. Interestingly, Tieri and Nardini [36] also observed some of these differences in knowledge representation for signaling pathway databases.

2.2.1 Database Level

One difference on this most global level is that some databases describe the network of a single organism while others describe the network for multiple organisms. Those that describe the network for multiple organisms may differ in how tightly the different networks are linked. In KEGG, the exact same representation is used for each organism, including a single pathway diagram for all organisms taken together for the majority of the pathways. Some organism-specific data are lost because of this, as we will discuss on the pathway and reaction level further on. In other database collections the individual networks are more separated with organism-specific pathway diagrams or even with their own dedicated website, using the same main framework to represent the network. Curation of each network in a database collection is not necessarily performed by a single research group. If changes to the main framework are not followed by each individual research group, some differences in representation of the networks may arise over time. This is, for example, the case for YeastCyc [31], part of the BioCyc collection, which has been adopted by a research group different to the one that designed the framework BioCyc [32].

A second difference on the database level is that some databases are limited to the metabolic network, e.g., *H. sapiens* Recon 1, while other databases, e.g., KEGG and Reactome, also incorporate other types of networks, such as the signaling network. These different types of biological processes are inherently intertwined. Therefore, describing them in a single database provides a more complete and biologically more accurate overview. KEGG describes the different types of networks, but they stand largely

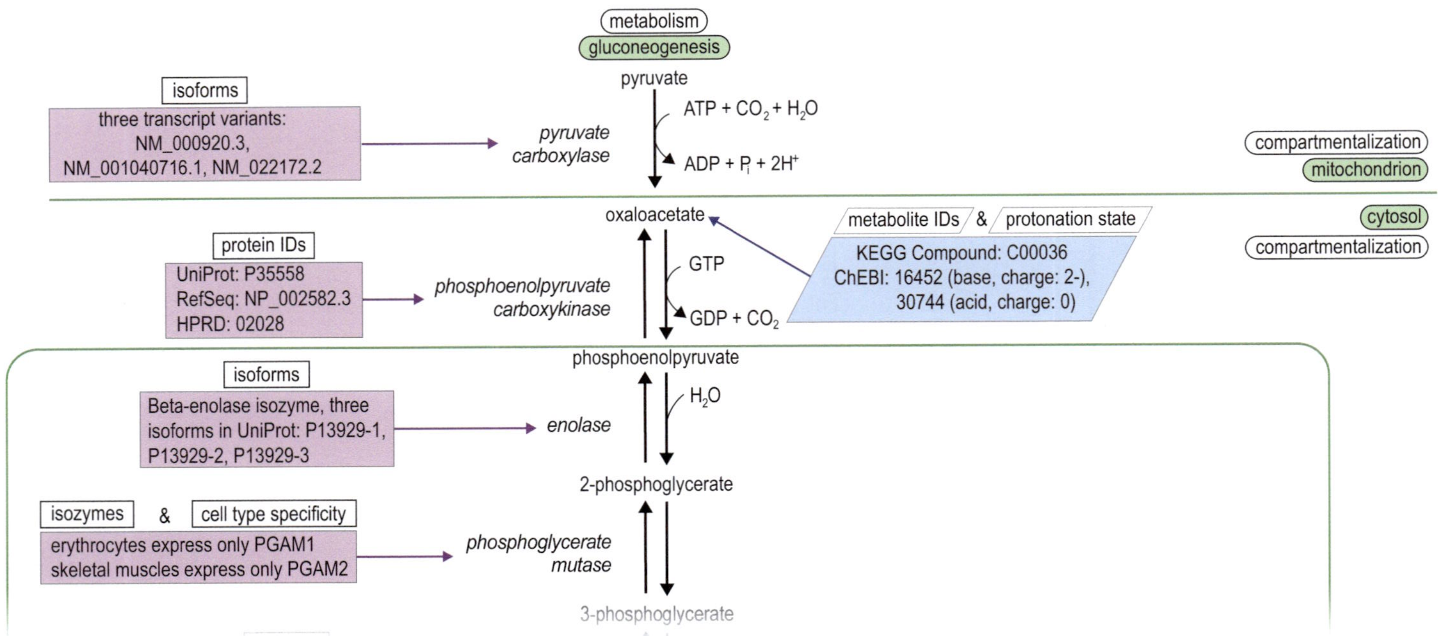

Fig. 2 Gluconeogenesis. A selection of the types of information that would need to be stored to accurately represent the various details of the gluconeogenesis pathway [35]. This figure and its caption have been adapted from Stobbe et al., "Knowledge representation in metabolic pathway databases". Brief Bioinform, 2014, 15 (3):455–470 [32] by permission of Oxford University Press

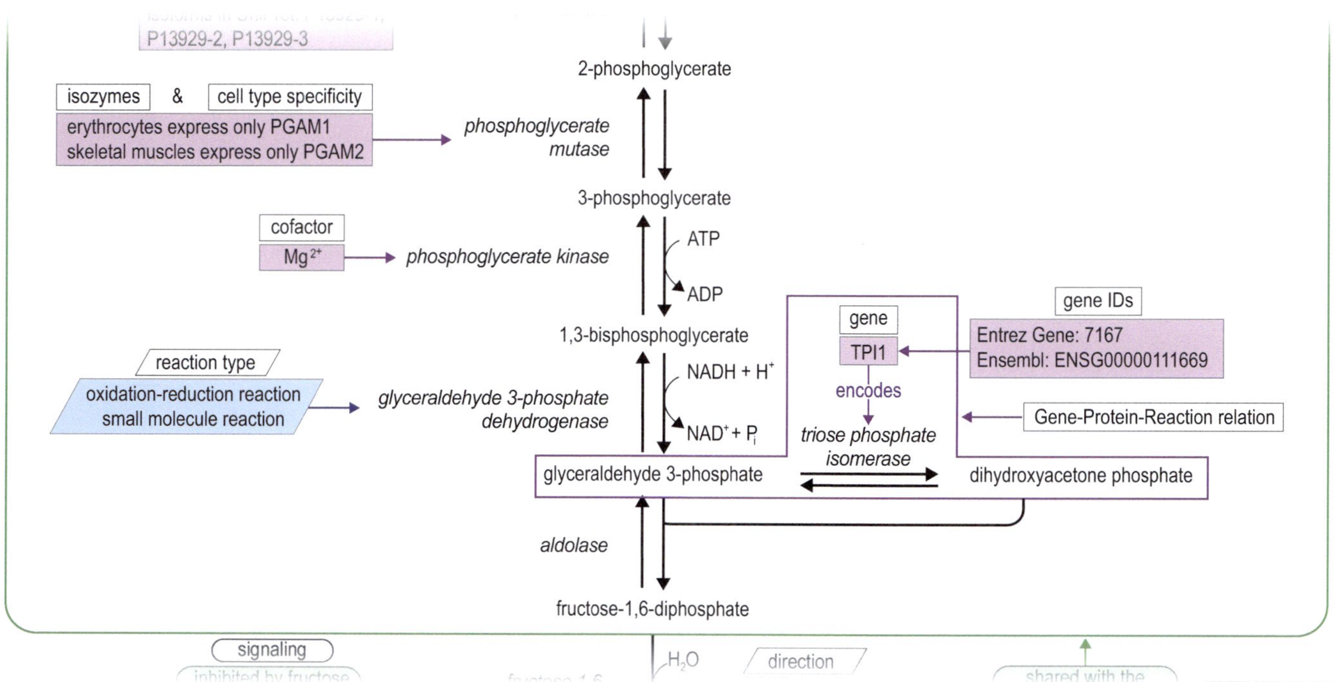

Fig. 2 (continued)

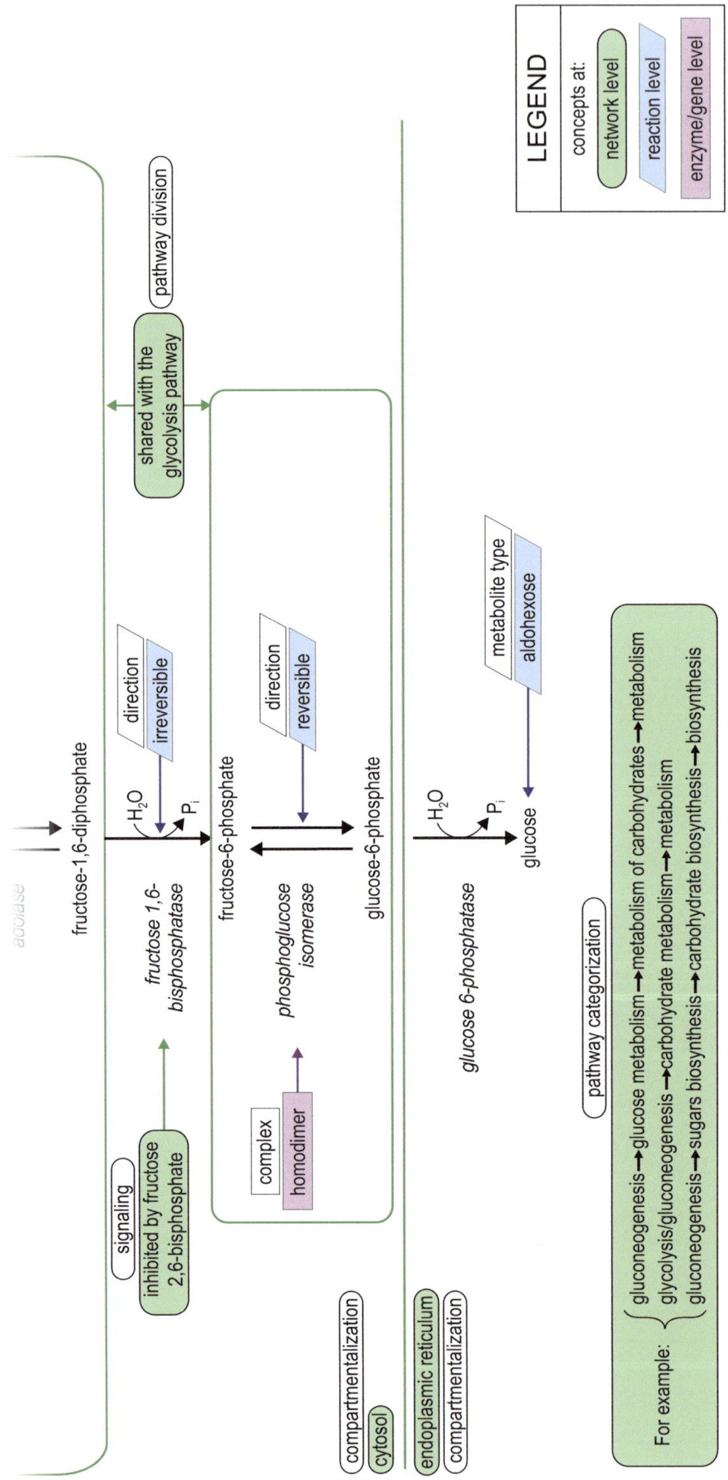

Fig. 2 (continued)

on their own and are modeled in a different way. In Reactome, the different types of process are represented in the same way and are linked directly to each other. It is not always trivial to (automatically) filter out only the metabolic reactions. The decisions that developers make on the database level also influence the knowledge representation on network level.

2.2.2 Network Level

At the network level, one difference between databases is the criteria used to decide which reactions, and consequently the enzymes and genes linked to them, are part of the metabolic network and which are not. The genes that were only found in one of the five human pathway databases in our comparison were significantly enriched for processes that could be considered peripheral to metabolism proper. Examples are processes related to ion transport, protein metabolism (such as proteolysis), and RNA metabolism (such as tRNA processing). This is one explanation for the high percentage of reactions (~70 %) that could only be found in a single database [14]. Note though that even when we compared only core metabolic processes, such as amino acid metabolism and carbohydrate metabolism, we did not observe a major increase in the percentage of consensus [14]. The percentage on the reaction level increased only by 1 % to about 4 % of the slightly more than 3700 reactions.

Another difference on the network level is that not all metabolic networks are compartmentalized, nor are all networks fully mass and charge balanced. In KEGG, compartmentalization is not represented, which is possibly because this type of information is organism-specific. A side-effect of this is that hardly any transport reactions are included. Furthermore, whether a network is charge balanced or not also has an effect on the metabolite level, as we will discuss further on. One thing that all pathway databases do have in common is that they divide the network into pathways.

2.2.3 Pathway Level

The division into pathways provides useful insights into the functional organization of the network, despite this being an artificial division. There is, however, no widely accepted definition of a pathway. All five human databases that we have analyzed have different definitions, each based on different biological criteria (Table 1). Consequently, there are substantial differences in the average size of a pathway and the amount of overlap between them (Table 1). The granularity with which a biological process is described is an additional factor that may influence the size of a pathway; we will discuss this in more detail in the next paragraph. Note that the pathways in KEGG are not human-specific, which is a consequence of their decision on the database level to represent the network of multiple organisms in the same framework. To illustrate the effect of this, if we look at the pathway "lysine biosynthesis" it seems to be part of the human network; however, humans cannot produce lysine. The genes and reactions that are indicated in the diagram to be part of this pathway in human are

Table 1
Pathway definition and statistics

Database	Definition	Number of pathways	Average number of reactions per pathway	% of reactions occurring in >1 pathway*
EHMN	Redivision of the pathways of KEGG and EMP: – less overlap between pathways – small functionally related pathways are grouped – human-specific	69	52	1 %
Recon 1	Definition of KEGG, but human-specific	96	30	9 %
HumanCyc	Guidelines used: – a single biological process – evolutionary conserved – regulated as a unit – boundaries at stable and high-connectivity metabolites	257	4	13 %
KEGG	Centered on the synthesis and/or degradation of one or more related substrates	84	23	10 %
Reactome	A series of reactions, connected by their participants, leading to a biological outcome	171	8	12 %

Statistics are based upon the data used for a comparison of the content of the five databases [14]. Only the metabolic pathways of HumanCyc, KEGG and Reactome are considered. For Reactome and HumanCyc, the lowest level in the hierarchy was used when counting the number of pathways. If reactions only differ in direction and/or compartments, they are counted as one.

*With respect to the total number of reactions assigned to at least one pathway.

EMP: Enzymes and Metabolic Pathways database.

This table and its caption have been adapted from Stobbe et al., "Knowledge representation in metabolic pathway databases". Brief Bioinform, 2014, 15 (3):455–470 [32] by permission of Oxford University Press

the ones in common with the degradation pathway. Another difference between databases with respect to pathways is that in some databases reactions of the same pathway are not necessarily all of the same type. In Reactome, for example, the activation of an enzyme may be linked to the next step in the pathway in which it is the catalyst of a metabolic reaction.

2.2.4 Reaction Level

A metabolic reaction can be defined as the synthesis or degradation of chemical compounds, which may or may not be a reversible process. The granularity with which a particular biological process is described frequently varies across databases and even within a database. In one database, a conversion may be represented as a single step, while in another, two or more steps are used. An example of this is the conversion of 2-oxoglutarate into succinyl-CoA in the

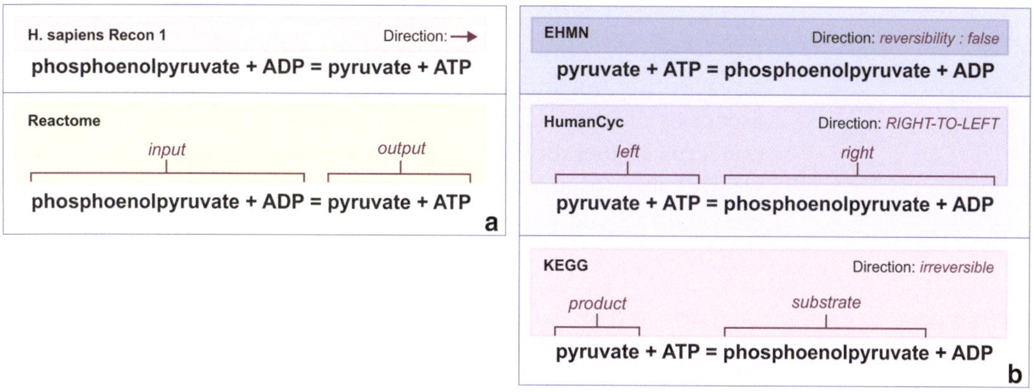

Fig. 3 Last, irreversible, step of glycolysis. Each database indicates the direction of the reaction differently. Words in *purple* and *italics* indicate reserved terms in a database. (**a**) *H. sapiens* Recon 1 and Reactome store the reaction in the physiological direction. (**b**) EHMN, KEGG, and HumanCyc store the reaction in the direction defined by NC-IUBMB. This figure and its caption have been adapted from Stobbe et al., "Knowledge representation in metabolic pathway databases". Brief Bioinform, 2014, 15 (3):455–470 [32] by permission of Oxford University Press

TCA cycle (Fig. 1). Another example is the representation of the fatty acid β-oxidation pathway, which is a repetitive process consisting out of cycles of four steps. In each cycle the fatty acid is shortened by two carbons. In *H. sapiens* Recon 1 four of these cycles are summarized into a single step, while KEGG describes all 16 steps. In Reactome, each step of the mitochondrial part of the fatty acid β-oxidation pathway is described. However, for the peroxisomal part, only the first cycle is described step-by-step, and the eight subsequent cycles are captured in a single reaction. Leaving out the intermediate steps prevents the network from being able to simulate the buildup of intermediate products of a cycle if there is an enzyme deficiency [37–40]. There are various reasons why the granularity with which a biological process is described may differ. A database may decide to leave out intermediate products that remain bound to the catalyzing enzyme or that cannot be measured. Another database may prefer to represent each step to describe the biological mechanism in full detail. It could also be that there is no conclusive evidence available to be able to describe the intermediate steps. Finally, especially for highly repetitive processes, curators may simply prefer to summarize it in just a few steps. It is not possible to tell in an automated way if a difference in granularity is a disagreement on the biology or merely a difference in representation.

All five human pathway databases also use different terminology to represent each individual reaction (Fig. 3). Furthermore, two of the databases store reactions in the physiological direction in humans (Fig. 3a), while the other three chose the direction as defined by NC-IUBMB (Fig. 3b), which is not organism-specific. Note that EHMN does not represent the correct direction in the example given in Fig. 3. The reaction is stored in the non-physiological direction and it is only stated that the reaction is irreversible. KEGG has the correct direction in this case, but the reversibility of a reaction is not indicated for each organism separately. Another aspect of the representation of a reaction in which databases differ concerns its metabolites, which we will discuss next. Furthermore, we will show that the relation between a reaction, its catalyzing enzyme and the encoding gene is represented differently across databases.

Metabolite Level

One way in which pathway databases differ on the metabolite level is the amount of information they provide about a metabolite. This may range from only the name to detailed information like structural information and the Gibbs free energy of formation. If a link to an external metabolite database is provided, which is not always the case, lack of information can potentially be easily remedied. Another difference on the metabolite level is that in charge balanced networks the chemical formula and charge of the metabolites are based on their ionization state at a specific pH level instead of the neutral form. Note that the two different forms of the same metabolite do not have the same identifier in a metabolite database like ChEBI [37]. A final example of a difference on this level is that some databases represent the broad substrate specificity of an enzyme using generic metabolites such as "alcohol". HumanCyc provides a metabolite ontology for this purpose. However, it is not always possible to (automatically) deduce for each of the generic reactions what the specific reactions are. A more generic metabolite may also be used in cases where the substrate specificity of an enzyme is unknown.

Reaction–Enzyme–Gene Relation

In general, a metabolic reaction is catalyzed by one or more enzymes that are encoded by genes. An enzyme can be a single protein or a complex comprising multiple components, which may be the same proteins (homomers) or different proteins (heteromer). How this relationship between a reaction, the catalyzing enzyme and the encoding gene is captured differs among databases (Fig. 4). In the representation of KEGG and EHMN, one cannot distinguish multiple isozymes that can each independently catalyze a reaction from multiple proteins that together form a complex and are all required to catalyze the reaction. This could lead to incorrect conclusions, when one is, for example, interested in what happens if a protein is deficient. If this protein is part of a complex that is inactive when the protein is lacking, the reaction may still seem feasible in EHMN and KEGG. The difference between isozymes and complexes is captured in *H. sapiens* Recon 1, but not on the protein level, as one might expect, but on the gene level instead.

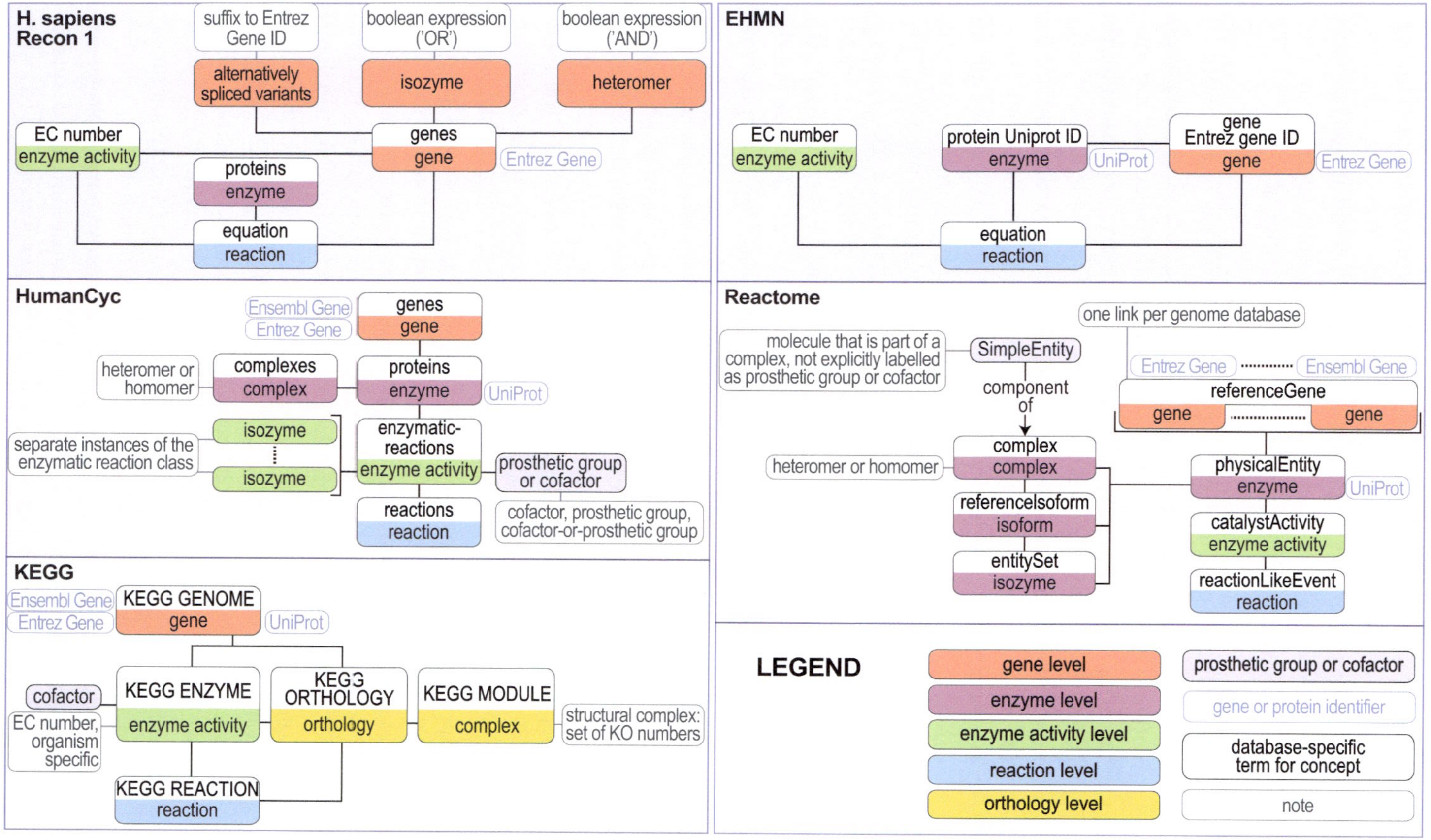

Fig. 4 Differences in representation of the Reaction–Enzyme–Gene Relation. The *boxes* with Entrez Gene, Ensembl Gene, and UniProt indicate where in the data model the specific identifier is provided. KEGG: The KEGG MODULE database describes four types of modules, among which are a limited set of structural complexes. This figure and its caption have been adapted from Stobbe et al., "Knowledge representation in metabolic pathway databases". Brief Bioinform, 2014, 15(3):455–470 [32] by permission of Oxford University Press

3 Gaining Further Insight into Metabolic Pathway Databases

The following questions, which were set forward in the introduction, can serve as a guide to gain a deeper understanding of the database(s) one is interested in using for an analysis. They could also aid in deciding which database to select if there are multiple databases available that describe the network of the organism of interest.

3.1 What Is the Main Objective of the Database?

The general goal for every pathway database is the same: to provide an accurate and complete description of the metabolic network in a digital format. The more specific objectives and main intended uses of databases, however, may vary. For example, the main objective of *H. sapiens* Recon 1 is to provide a quantitative mathematical model. This makes this database particularly suitable for applications such as phenotype prediction. For Reactome, however, this is not one of its main purposes. Its focus is on, among other things, serving as a knowledge base of (human) biological processes, which can also be used for educational purposes. This role as a digital encyclopedia is of lesser importance for *H. sapiens* Recon 1, which is, for example reflected in the limited textual description of pathways and limited features of their website. It is important to know the specific focus of a database, because it influences the decisions the developers make on, for example, what information to represent.

3.2 How Is the Metabolic Network Reconstructed?

Research groups have taken different approaches as to how to reconstruct the metabolic network (Table 2). One approach is to construct the metabolic network pathway by pathway, until every metabolic process is described. Initially, the coverage of the metabolic network will be low, which will improve over time as new pathways are added. Examples of databases that are being constructed in this way are Reactome and WikiPathways [49]. A second approach that is frequently used is to construct an initial draft of the metabolic network based on the functional annotation of the genome of the organism of interest. Based on this genome annotation, reactions are retrieved from other databases, such as the Nomenclature Committee of the International Union of Biochemistry and Molecular Biology (NC-IUBMB), KEGG LIGAND [50], and MetaCyc [51]. This initial network will need to be further refined using expert knowledge, scientific literature, existing reconstructions, and other resources. Examples of databases using this approach are KEGG and BioCyc [52]. A possible additional step, which is mostly taken by research groups that focus on systems biology, is to convert the genome-based network into a constraint-based, quantitative, mathematical model [53]. *H. sapiens* Recon 1 [18] is an example of this. A third approach to construct the metabolic network uses gene orthology and an already existing network of another organism as a template. This method is based on the assumption that if a gene product in one organism catalyzes a certain reaction, then the ortholog in another organism will likely encode a protein that catalyzes the same reaction. Similar to the second approach we have discussed, further refinement is required after the construction of an initial draft. For most organisms in Reactome, this approach was taken using the human network as a template [54]. KEGG also uses orthology as part of the method to construct the metabolic network for the majority of the organisms [21].

3.3 What Is the Manner and Degree of Curation of the Metabolic Network?

Curation of the initial draft of the metabolic network, whether generated by a genome- or orthology-based approach, is essential. For networks that are built incrementally, the curation is done per pathway before it is added to the database. The degree of curation is an important factor to take into account, especially when choosing between multiple databases, because it directly relates to the quality of the reconstruction. In addition, it is also important to take into consideration whether or not the network is kept up-to-date with new discoveries. The manner and degree of curation differs substantially among databases (Table 2).

3.3.1 Manner of Curation

The main source of information used to curate a metabolic network is the literature. Extracting knowledge from the extensive amount of available literature is a very labor-intensive and time-consuming task. The curator needs to evaluate all available evidence and determine its strength. Furthermore, there may also be controversy in the literature and the curator needs to decide how to reflect this or which view to incorporate in the database. Curators for Reactome are asked to choose the current consensus and to not incorporate multiple, opposing views [4]. Furthermore, for parts of the metabolic network, no conclusive evidence may be available. HumanCyc prefers to include multiple pathway variants if there is no conclusive evidence available to permit a choice between them. This is done to bring them to the attention of the scientific community [19]. In addition to the literature, for the curation of a mathematical model, such as *H. sapiens* Recon 1 [18], experimental data are used to verify the models ability to simulate known metabolic functions and, in an iterative process, to further refine the reconstruction.

There are different ways in which the curation of a metabolic network is organized. A single research group is involved in the curation of KEGG [55]. Reactome invites independent researchers to curate the biological process on which they are experts [56]. Their contribution is peer reviewed before it is added to the database. Once a process is added, it will, after a period of time, be revisited to refine it based on new information. For the curation of *H. sapiens* Recon 2 [57] a community-driven approach was taken in the form of jamboree meetings. A jamboree is a focused meeting in which experts from different research groups and various

Table 2
Selection of publicly available pathway databases

Database	Description
BioCyc	Collection of Pathway/Genome Databases, each describing the genome and metabolic pathways of a single organism
BiGG	*B*iochemical *G*enetic and *G*enomic knowledgebase of large scale metabolic networks
EHMN	*E*dinburgh *H*uman *M*etabolic *N*etwork
KEGG	*K*yoto *E*ncyclopedia of *G*enes and *G*enomes: resource for understanding high-level functions and utilities of the biological system from molecular-level information
Panther	*P*rotein *A*nalysis *TH*rough *E*volutionary *R*elationships: protein sequences are grouped into functional subfamilies, which allows for a more accurate association with biological pathways
Reactome	Open-source, open access, manually curated and peer-reviewed pathway database
UniPathway	A resource for the exploration and annotation of metabolic pathways
WikiPathways	Open, public platform dedicated to the curation of biological pathways by and for the scientific community

Construction strategy	Curation	Number of organisms
Genome-based	Three levels of curation: Tier 1—highly curated, tens of man-years of literature-based curation Tier 2—moderately curated, i.e., partly curated and partly automatically derived Tier 3—automatically derived, not curated	>3500
Genome-based	Curation is based on primary literature articles, reviews, biochemical textbooks and by performing functional validation tests	8[a]
Genome-based	Curated is based on literature and the literature-based *E*nzymes and *M*etabolic *P*athways database	1 (human)
Genome-based or orthology[b]	The 15 main databases are manually curated. Several auxiliary databases that contain genomic information, such as the one for draft genomes, are automatically generated	>3000
Incremental	Manually curated by external experts with an option for community pathway curation	85
Incremental or orthology[b]	Human network is curated and four spin-offs, e.g., FlyReactome and Gallus Reactome, maintained by other research groups Curation is done by invited biological experts and each module is peer-reviewed	20
Starting from reactions, which are linked to proteins	Curated is based on literature and existing metabolic resources such as KEGG and MetaCyc	>2800
Incremental	Curated by the community	29

The descriptions are taken from the respective websites of the databases [41–48]
[a]Ten networks are described, three of which are from *Escherichia coli*
[b]Depends on the specific organism

disciplines, including systems biology, chemistry and molecular biology, come together to curate the network. The jamboree meetings were followed by further curation by a smaller group of experts. Another community-based method of curation is used in WikiPathways, where, similar to Wikipedia, the entire community can contribute by adding a new pathway or by editing an existing one. One key advantage is that pathways can be constantly updated if new information is uncovered. The pathways of Reactome are also available in the WikiPathways format at a dedicated portal [58]. The curation team of Reactome uses the input from the users of their portal to further refine the main version of the network. Finally, the degree to which curation has been done varies per database and per organism.

3.3.2 Degree of Curation

For most organisms, the network in database collections such as BioCyc has been generated automatically without any subsequent manual curation [52, 59]. Only for a limited set of organisms the time that researchers have put into the curation combined sums up to years, or even decades in a few cases such as EcoCyc [60]. The level of curation can also vary within one network reconstruction depending on the metabolic process. Research groups may have a specific interest in particular parts of the metabolic network and are therefore able to describe them in more detail. For example, in *H. sapiens* Recon 2 [57] a model of acylcarnitine and fatty-acid oxidation [61] was included. This model was constructed for a separate project by a subgroup of the researchers who contributed to the complete model.

Literature references linked to each element of the metabolic network are the main way in which pathway databases indicate the evidence on which it has been based. It is, however, often not clear what type of evidence is described in the articles referred to and how strong this evidence is, without actually reading the articles. It could be that the article describes biochemical evidence in the particular organism at hand, but also that it has solely been postulated as a hypothesis or proven for another organism. Only some databases explicitly indicate what type of evidence there is. In *H. sapiens* Recon 1, curators give a cumulative confidence score to reflect what kind of evidence there is available [53]. A score of 1, for example, indicates that the reaction is included only because it improved the performance of the model. A score of 4 indicates biochemical evidence is available or a combination of physiological and sequence evidence, i.e., "*evidence for gene function*" [53]. In BioCyc, an evidence ontology is used to specify what type of evidence is available [62]. Examples of evidence codes are "*Inferred from experiment*" and "*Inferred from computation*". The indication of the type of evidence could be used to filter out parts of the network for which for example experimental evidence is lacking, if so required for the analysis of interest.

3.4 What Is the Breadth and Depth with Which the Metabolic Network Is Covered?

The extent of our knowledge on metabolism varies from organism to organism and, in general, it is far from complete. The reconstructions of the metabolic network continue to require further expansion and refinement. The breadth and depth with which a metabolic network is covered varies substantially as the various comparisons of databases that we discussed have shown. The completeness of the metabolic network and the level of curation are important factors to consider when choosing between multiple databases. To get an idea of the breadth and depth with which the metabolic network is covered, one could look at which metabolic processes have been covered and the statistics given on the number of components. However, it is important to bear the following things in mind, especially when comparing these statistics across databases. If a network is compartmentalized, then often a large percentage of the reactions are related to transport. Furthermore, one specific type of metabolic processes may account for a large percentage of the reactions. For example, the curators of EHMN explicitly focused on lipid metabolism, which together with transport reactions, explained 23 % of the reactions that could only be found in one of the five databases that we compared [63]. Another factor that influences the number of reactions is the extent to which alternative substrates are described for one main reaction. To illustrate this, Kuffner et al. [63] compared the reactions from BRENDA, ENZYME and KEGG. The overlap between the reactions of these three databases was 21 %. However, this overlap increased to 67 % when only the main reactions of BRENDA, as defined by NC-IUBMB, were compared, without considering reactions derived from these with alternative substrates. To some extent, the granularity with which processes are described may also influence the number of reactions. Also note that networks that were initially constructed in an automated way may contain a large set of isolated components.

If multiple, highly curated databases are available, it may be difficult to choose between them given the lack of consensus and the complementary knowledge they contain. Instead of choosing a single pathway database, another possible option would be to apply the analysis of interest to multiple databases. This also permits the assessment of the robustness of the results. Carbonetto and Stephens [64] chose another option: they applied their pathway enrichment method to the pathways of eight databases combined. A disadvantage of both options is that it requires extracting the data from each of the databases, which may be challenging, as we will discuss further on. Moreover, the differences in the outcome of the analysis may be caused both by differences in the biology described and conceptual differences between the databases. Ideally, we should strive towards a single reconstruction of the metabolic network for each organism that combines all the knowledge captured by the individual databases.

3.5 How Is Knowledge Represented in the Database?

There are two main ways in which knowledge representation plays a role when working with metabolic pathway databases. One way is the definition used for various concepts, which can influence the (interpretation of) results of an analysis. The definition of a pathway is an example of this [29, 30]. Mora and Donaldson [29] tested on three pathway databases (KEGG, PID [65] and Reactome) a pathway-based metric to predict which proteins are likely drug targets. The metric is defined as the number of pathways in which a protein plays a role. Proteins with a large number of associated pathways are assumed to be important for the cell and are therefore likely drug targets. The metric had the highest predictive power in KEGG, which is probably related to its definition of a pathway. Another example of an analysis in which the definition of a pathway plays an important role is the, frequently used, gene set enrichment analysis (GSEA) [66]. The metabolic processes that are enriched in a gene expression dataset may depend on which pathway database was used to define the gene sets, and it is hard to tell which definition will give the most useful results. Moreover, the names of pathways are not consistent between databases [13]. Consequently, it could be that the results only appear to be different, while in reality, the enriched pathways contain a similar set of genes linked to the same reactions. The opposite may also occur: the names of a pathway may be similar, but this does not necessarily mean that the pathways contain the same set of genes. Note that it is potentially problematic if the majority of the pathways are sparse with respect to the number of genes, such as in HumanCyc (Table 1). Sets with fewer than 15 genes are left out in GSEA by default, because the scores used to determine enrichment are inflated for these small sets. In the case of HumanCyc, one could use their pathway hierarchy to select superpathways in which multiple processes are grouped to avoid this problem.

The second way in which knowledge representation plays a role is that it may not capture every detail of the metabolic network. Table 3 provides an overview of the questions one could consider to evaluate the strengths and limitations of the knowledge representation used by the database(s) under consideration. These questions need to be evaluated in context of the analysis at hand, because it depends on the requirements of the analysis what is truly a limitation and which strong points of a database are relevant. One example of this is that for accurate in silico simulations compartmentalization is essential, while for analyses like GSEA it is not. Note, however, that the lack of this type of knowledge may still be of influence, as we have observed for the TCA cycle [15]. Four gene products that are located in the cytosol were indicated to play a role in this pathway, while this pathway takes place entirely in the mitochondria. A second example of what does not necessarily need to be a limitation for the analysis at hand is that some databases only describe metabolism. There are various applications that are specific to metabolic networks, such as flux balance analyses [67, 68]. This type of analysis requires that only metabolic reactions can be

Table 3
Questions to use as a guideline

Level	Questions to ask
Database	Does the database describe the network of a single or multiple organisms? Is only metabolism described or are also other processes described?
Network	How are the boundaries of the metabolic network defined? Is the network compartmentalized? Is the network charge and mass balanced?
Pathway	What is the definition of a pathway?
Reaction	How is a reaction and its direction represented? What is the granularity with which conversions are described? What information is provided on the metabolites? Are generic metabolites used to represent for example the broad substrate specificity of an enzyme? How is the relation between reaction, enzyme, and gene represented? Can isozymes be distinguished from protein complexes?

These questions can serve as a guideline to determine the strengths and limitations of the knowledge representation used by the databases

extracted from the database, which may be a challenge in databases where the different types of reactions are intertwined. Additional examples of what is and what is not captured by the databases are given in the previous section and are further detailed in our article on knowledge representation [32].

A possible third way in which knowledge representation could affect the analysis is if multiple pathway databases are required that have been built by different research groups. An example of this type of analysis is the study of the evolution of the metabolic network across several organisms. In this case, the difficulty is to avoid detecting differences in representation rather than differences in biology.

Finally, another practical aspect of knowledge representation to consider is whether the network components, i.e., metabolites, proteins, and genes, are unambiguously identified. We will discuss this in more detail below, as well as how databases deal with the representation of the lack of knowledge.

3.5.1 Are the Network Components Identified Unambiguously?

To map experimental data onto the metabolic network it is essential that the components, i.e., metabolites, proteins and genes, can be identified unambiguously. Names often do not suffice, because they may not be unique and many synonyms may exist. Therefore, pathway databases use identifiers from external databases for unique identification, for example, Entrez Gene [58] for genes, UniProt [59] for proteins, and ChEBI for metabolites. For each of the three types of components, multiple external databases are available, which differ in the information they provide and how they assign identifiers. An example of the latter is that in ChEBI, different identifiers are assigned to the acid and base form of a metabolite, while in KEGG Compound, both forms have the same identifier. There is also not always a one-to-one relation between the identifiers of genome databases. For this reason, a simple question like how many genes are included in each database, is more difficult than one might think, because it will depend on which gene identifier one uses to determine this. Metabolic pathway databases also have different preferences for the main external database they use to identify and indirectly define the components. KEGG follows the definition of a gene that NCBI Entrez Gene uses, while HumanCyc has its own definition of a gene, which is closer to Ensembl. Both also provide gene identifiers from other genome databases. In Reactome, a gene is represented as a collection of identifiers from multiple genome databases. Finally, the unambiguous annotation of metabolites is especially challenging, because none of the metabolite databases is yet complete and they are to some extent complementary to each other. Often pathway databases deal with the lack of coverage of metabolite databases by annotating metabolites with identifiers from more than one metabolite database. This is very labor-intensive, especially because the identifiers also need to be kept up-to-date [69]. It can be that not every metabolite of the network has an identifier from the same metabolite database. Moreover, many metabolites most likely even lack an identifier.

3.5.2 How Is a Lack of Knowledge Indicated?

It is important to be aware that a lack of knowledge in a database is rarely indicated in a systematic way. *H. sapiens* Recon 1, for example, indicates "*cytosol*" if the compartment of the reaction is unknown. Consequently, one cannot distinguish a reaction that truly takes place in the cytosol from a reaction for which the compartment is unknown. In the five databases we have analyzed, only in HumanCyc spontaneous reactions could be distinguished from reactions for which the enzyme is unknown. In KEGG, this information is also available, but only in an unstructured comment field. Moreover, as KEGG contains the metabolic network of multiple organisms, the information on the genes is needed to retrieve the reactions from the organism of interest. Consequently, spontaneous reactions cannot be retrieved easily.

3.6 How Can Knowledge Be Extracted from the Database?

After familiarizing oneself with how knowledge is represented by a database, the design decisions taken, and the definitions used, the next step is to learn how to extract the data needed for the analysis of interest. As Tieri and Narini state in the first sentence of the abstract of their article on the usability of signaling pathway databases [36] "*issues and limitations related to accessibility, understandability and ease of use of signalling pathway databases may hamper or divert research workflow, leading, in the worst case, to the generation of confusing reference frameworks and misinterpretation of experimental results*". A similar observation could be made for metabolic pathway databases.

Data can be retrieved manually by browsing the website of the database. For our comparison of ten descriptions of the TCA cycle, we retrieved the data we needed from each database by copying it from their websites. The benefit of doing this in an automated way instead, did not outweigh the effort we would have had to put in to be able to do this for all ten databases given that they each offer a different way of retrieving this data. However, in general, for computational analyses, it is not feasible to manually extract the required data and a more automated method is needed. This may, depending on the data required, be quite a challenge. As Fearnley et al. [70] described for signaling databases, you soon realize that you have opened "*Pandora's box*" when it comes to extracting data from pathway databases. Various data warehouses have been built to aid in this by offering unified access and retrieval of (part of) the data from multiple pathway databases. We will first go into the different ways pathway databases offer to retrieve the data, followed by a brief discussion on data warehouses.

3.6.1 Manual Data Extraction

The level of sophistication of the website varies across databases. In some databases, the Web interface allows only for basic search options, such as the name of a metabolite or a gene. This is the case for *H. sapiens* Recon 1, which is in line with its main objective to provide a mathematical model and not necessarily serve as a digital encyclopedia on metabolism. This is in contrast to databases like HumanCyc and Reactome, which do provide a wide range of search options. Moreover, both databases provide additional tools to, for example, perform an online pathway enrichment analysis.

It is not always clear how the search functions offered by a database exactly work, as Tieri and Narini [36] described for signaling pathway databases. An example they give is that when a user searches for a specific gene, this may return only the pathway the gene plays a role in; however, it could also return any pathway in which the gene is mentioned in supporting texts. Tieri and Narini recommend challenging the database to ensure that one retrieves all relevant information by using, for example, multiple identifiers to search for the same gene or protein. Finally, websites evolve over time to further improve the user experience and accommodate changes in data representation. Sometimes the changes made are substantial, requiring users to reacquaint themselves with how the website works. Occasionally, the old interface may still be provided. All databases also provide a way to extract data from their website in a more automated manner, which we will discuss next.

3.6.2 Automated Data Extraction

The manner in which data can be extracted in an automated manner differs across databases (Table 4). Databases may even provide multiple options and in some cases more than one will be needed to extract all the information required. In general, the common denominator for any of the available ways to retrieve data is that one needs to understand the knowledge representation and

Table 4
Formats and the associated challenges to extract the data

Format		Challenges
Downloadable files	Flat text/Excel	– requires parsing – often a long list of small issues to resolve, e.g., inconsistent formatting – multiple files may need to be combined – database specific – may change in new release
	Database-specific XML-based	– requires parsing – multiple files may need to be combined – each pathway in a separate file – database specific – may change in new release
	Standard exchange formats (e.g., BioPAX, SBML)	– requires parsing – no single exchange format is provided by every database – different interpretations of the same exchange format – changes with version of standard (infrequent)
Relational database		– requires writing queries – a large number of tables is needed to extract information[a] – database specific – database scheme may change in new release
Application programming interface		– requires programming in e.g., Java – not all data may be provided – possibly limited access – database specific – may change in new release

Note that each database has its own format(s) and that it partly depends on the database if all challenges will need to be met or whether there may be additional database-specific ones

[a]in the case of a normalized database schema

terminology used by the database. Moreover, it is not always trivial to correctly connect the different components of a network, i.e., pathway, reaction, catalyzing enzyme, and encoding gene. Furthermore, how challenging it is to extract data depends on the format in which it is offered and what data are required for the analysis of interest. An example of the latter is that it is easier to extract only all genes for each pathway than to extract a reaction in a certain pathway along with the catalyzing enzyme and the encoding gene. Additionally, how data is represented and can be extracted may change in a new release of the database. Finally, there are also various challenges that are specific to each format, which we will discuss in more detail below.

Downloadable Files

Sometimes the most convenient or only available option is to download a set of flat text files and then extract the required data. Often the data are spread across a number of files, which may need to be combined, depending on what one wants to retrieve from the database. How the data are structured in these files is specific to each database. For example, a reaction may be presented as a single string, such as "*phosphoenolpyruvate+ADP->pyruvate+ATP*". This string will need to be parsed to get its direction and the individual metabolites to be able to connect them to the provided external metabolite identifiers. In another database, the input and output of a reaction and its direction are provided separately, which makes it easier to parse. The challenge of working with flat text files is that often a long list of small issues need to be resolved, which taken together may take quite some time. Examples include inconsistent formatting and badly structured text fields, such as when punctuation marks are used within a single term and also to separate multiple terms.

To provide the data in a more formally defined format, KEGG and WikiPathways both have their own XML-based file formats, KGML [71] and GPML [49], respectively. The structure of this file format is provided by the set of rules defined by the XML. As for the flat text files, parsing is still required. Furthermore, Zhou et al. [13] observed inconsistencies in the GPML files; for example, some XML tags were in upper case and others in lower case, and there were random insertions of whitespace characters. In the KGML files of KEGG not all information is contained, but only what is needed for the pathway diagram. For example, only the main metabolites of a reaction are described. The complete reaction is available in a separate flat text file or can be retrieved via the KEGG API.

To improve the exchange of knowledge between databases, several XML-based standards have been proposed such as BioPAX [72] and SBML [73]. Here, the terminology is predefined and the file is highly structured. If every database provided their data in such a format, a single script should, in principle, be sufficient to obtain the information one needs from any database. However, not all databases offer their data in such a format and the ones that do, have not all chosen the same standard exchange format. Moreover, databases interpret these standards in their own way [55, 70] and the differences we discussed related to knowledge representation are also visible in these formats [32]. For example, for BioPAX we observed that one database stored a reversible reaction in both directions separately, while in another database, the reaction was stored only once with its direction indicated as "*reversible*".

Relational Database

A (downloadable) relational database allows fast programmatic access to retrieve data in the form of queries, without requiring a lot of parsing. Reactome is an example of a database that provides the data in this format. The database schema used is quite extensive and requires some time to fully understand. In version 48 of Reactome, 13 tables were needed to extract a reaction from the human glycolysis pathway along with the catalysts, encoding gene, and the external gene and metabolite identifiers (Fig. 5).

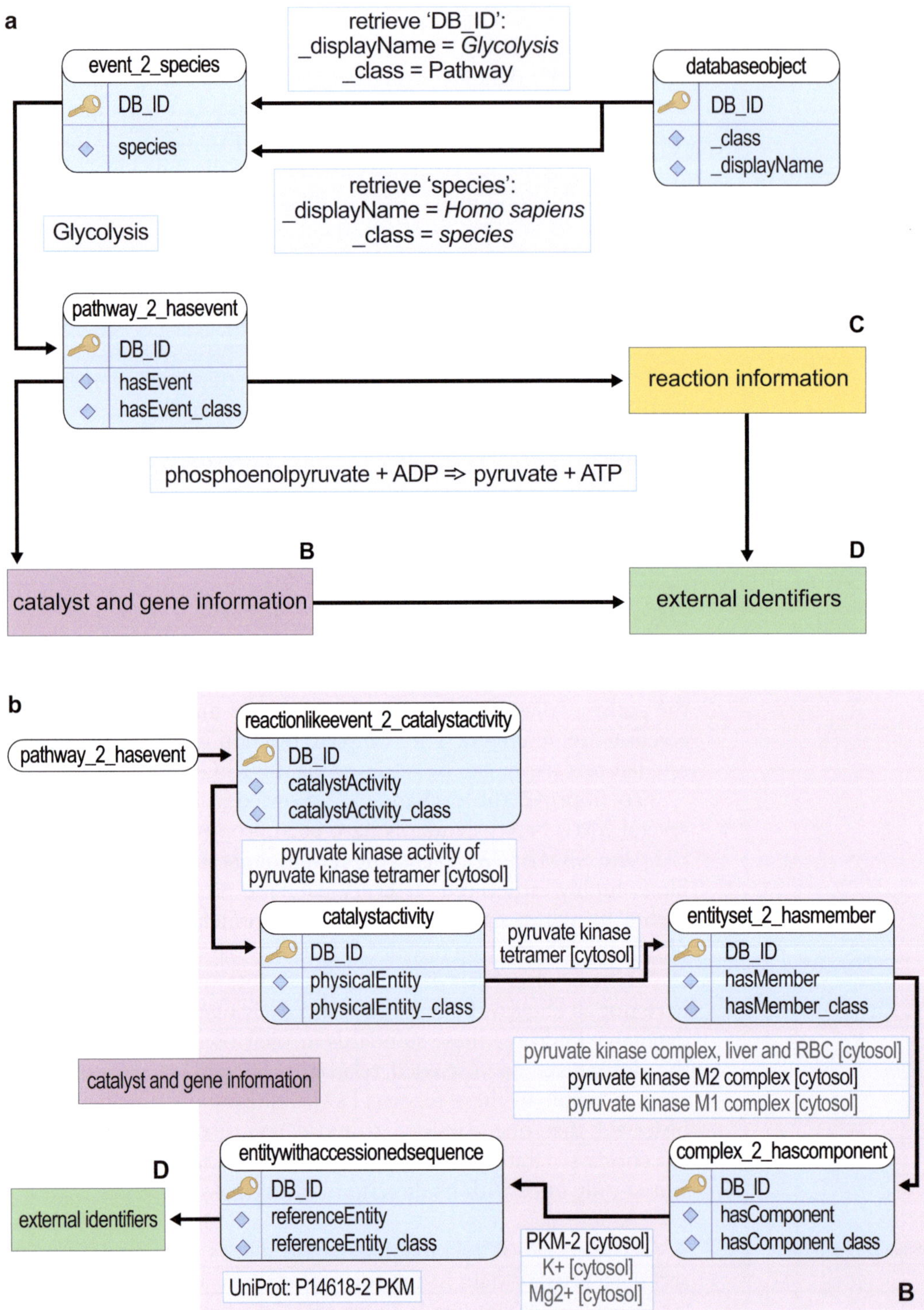

Fig. 5 Overview of tables needed to extract data from Reactome (v48). A simplified overview of the tables needed to extract the indicated reaction of the human glycolysis pathway, along with the catalysts, encoding genes, and the external identifiers. The names in the *grey rectangles* were retrieved from the *databaseobject* table. For the two catalysts in *grey font*, the same tables are needed as the one in *black font*. The two metabolites that are part

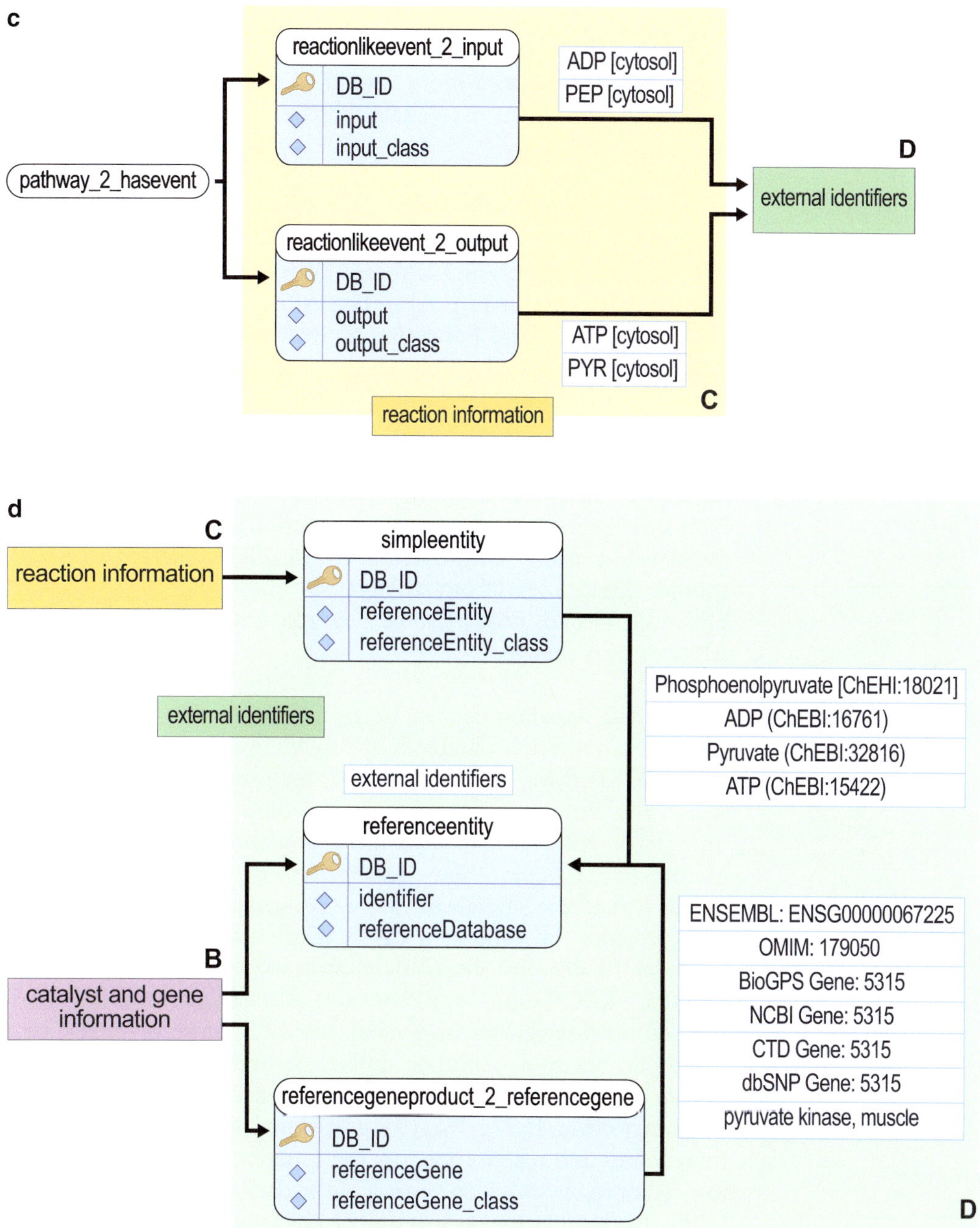

Fig. 5 (continued) of the catalyst were not further considered for the overview. Note that it depends on the value indicated in the fields ending with "_class" as to which tables are needed, which may differ depending on the reaction and the catalyst. (**a**) Global overview of the different sets of tables needed to extract the various elements. (**b**) Overview of the tables needed to retrieve the catalyst and gene information. (**c**) Overview of the tables needed to retrieve the input and output of a reaction. (**d**) Overview of the table needed to retrieve identifiers for the various elements

This relatively high number of tables is inherent to the fact that the schema has been normalized to reduce redundancy. Specifically for pathway analyses, a more simplified relational database is also available. Note that the database schema may change in new releases.

Application Programming Interface (API)

A number of databases provide an API to extract data, e.g., KEGG, BioCyc, and Reactome. Different forms of APIs are used; for example, Reactome offers a SOAP-based Web Services API and BioCyc provides a Common Lisp API. The API does not always provide access to all the data. For example, in KEGG, only pairwise relations are available, depending on the analysis this may not be sufficient. If this is the case, then one needs to be careful how to connect the different elements of a network to each other to prevent incorrect combinations. Furthermore, there may be restrictions on the amount of data one can retrieve via the API. BioCyc asks users to limit their queries to, on average, one per second. This is to prevent the database from being overloaded. Finally, in principle, an API should shield users from changes in the underlying database. However, despite this, in some new releases of the database, the API may also change.

3.6.3 Data Warehouses

Several data warehouses have been built to simplify the extraction of data from multiple databases by providing a single point of access (Table 5). Pathway Commons [74], for example, allows users to search the entire content of six (metabolic) pathway databases at once, to download the data all in the same format, and to have programmatic access to it. Some data warehouses are focused on a particular part of the (metabolic) network. IntPath [13], for example, only provides the gene-to-pathway relation, and can be used as a platform to perform enrichment analyses across three databases (BioCyc, KEGG and WikiPathways). ConsensusPathDB [75] is focused on interactions in general, not only those described by a metabolic reaction. Another difference between the data warehouses is whether they standardized the data derived from the databases they cover. In MetRxn [9], great care has been taken to match metabolites and reactions across the different sources. Finally, it is important to check which version of the database has been included in the data warehouse, as it is quite a challenge to keep up with the new releases of the metabolic pathway databases.

4 Future Outlook: The Importance of a Community Effort

Metabolism plays an important role in several prevalent diseases such as cancer, cardiovascular disease, and diabetes. To gain further insights into the molecular mechanisms involved, a biologically accurate and complete model of the metabolic network is of the utmost

Table 5
Examples of data warehouses

Name	Description	URL
BioModels Database	Depository for qualitative models	http://www.ebi.ac.uk/biomodels-main/
BioWarehouse	Toolkit for constructing database warehouses	http://biowarehouse.ai.sri.com/
ConsensusPathDB	Interaction database	http://cpdb.molgen.mpg.de/
IntPath	Integrated pathway gene relationship database	http://compbio.ddns.comp.nus.edu.sg:8080/IntPath/
MetRxn	Knowledgebase of metabolites and reactions	http://metrxn.che.psu.edu/
PathwayAPI	Aggregated database combining and unifying databases from WikiPathways, Ingenuity and KEG	http://www.pathwayapi.com/
PathCase	Software system for storing, managing, analyzing, and querying pathways	http://nashua.cwru.edu/pathwaysweb/
Pathway Commons	Collection of pathway information from multiple organisms	http://www.pathwaycommons.org/about/

The descriptions are derived from the respective websites of the databases

importance. Furthermore, the reliance on databases will continue to increase, given the deluge of high-throughput data. Pathway databases are increasingly used as a primary knowledge resource by biologists. One of the ultimate objectives is to construct a mathematical model that can be used to generate experimentally verifiable hypotheses, to identify potential drug targets and to simulate the effect of network perturbations, such as loss of function [76].

The currently available human metabolic pathway databases contain complementary information, as evidenced by the difference in the breadth and depth of their coverage of the metabolic network. Similar observations hold for various other organisms for which multiple databases are available, such as *S. cerevisiae* and *A. thaliana*. By combining the knowledge contained in each of these databases and reconciling the conflicting information, we will arrive at a more complete and more accurate description. Moreover, this would ensure that the maximum benefit is derived from all the time, effort, and money that have already been put into these pathway databases. At the same time, it will be nearly impossible to represent the network in a way that is suitable for all possible analyses.

One option, similar to what was proposed by Kitano et al. [77], would be to have a central database with the description of the metabolic network for which different representations are available.

4.1 Integration and Consolidation

Multiple research groups have realized the importance of integrating and consolidating the knowledge contained in the multiple metabolic pathway databases. It is, however, a huge challenge, given the numerous differences between the databases. Various initiatives have been started in this respect, including exchange formats, algorithms to (semi-)automatically integrate networks, data warehouses, and our own application called C_2Cards [78]. Each of these examples has its strengths and limitations. The standards for exchanging pathway information provide a single terminology to be used to represent the metabolic network. However, these exchange formats are not being used in the recommended way, as pointed out by Fearnley et al. [70] and Soh et al. [55]. There is too much room for different interpretations of the standards, which is defeating their purpose. It may be tempting to develop yet another standard that is better in one way or another than the existing ones. However, this will not solve the current issues if database developers do not precisely follow the recommendations of the standard. A second type of initiative is the development of algorithms that semi-automatically combine multiple networks into one. Examples of this include the algorithm of Radrich et al. [10] and the MetaMerge algorithm of Chindelevitch et al. [79]. The core of the resulting integrated network, in both cases, comprises the reactions found in all databases that exactly match each other. Note that, especially if more than two databases are combined, this core may be quite small. The integrated network further contains reactions that could not be exactly matched, which are marked as such. This is helpful for the subsequent manual curation that will still be necessary, because the algorithms do not resolve conflicting information nor filter out incorrect information. Moreover, several conceptual differences were not taken into account, such as differences in the number of steps used to describe a biological process. The third type of initiative is the construction of data warehouses such as Pathway Commons and MetRxn, which are excellent resources for constructing an integrated network based on data from multiple databases. However, unifying the databases into a single resource will not be enough, as this does not resolve the conflicting and erroneous information. Finally, we have built a Web application to aid in the integration of multiple human pathway databases, which is called C_2Cards [78]. One important difference compared with data warehouses is that the primary aim is not to provide a union of the databases, but to point out the differences between them. For the application to be successful, it will, however, require the involvement of a large group of experts.

Ultimately, to truly integrate the available descriptions of the metabolic network, a broad community effort will be needed, or as Mons et al. [80] phrased it, we need to call upon "*a million minds*". WikiPathways is an example of implementing this idea of involving the scientific community to curate the metabolic network. Following the approach of Wikipedia, everyone can curate an existing pathway or add a new one. *H. sapiens* Recon 2 is one of the success stories in which a jamboree was used to build an improved version of Recon 1. After the jamboree, additional curation was performed by a smaller group of experts, which resulted in the current state-of-the-art network for human metabolism. Although successful, it has been proposed by Monk et al. [81] that this way of involving the community could be further improved by involving experts in a more structured way. They propose to use crowdsourcing, which allows individual researchers to each contribute to a single reconstruction of the metabolic network of a particular organism. Another important challenge is to keep up with the latest discoveries.

4.2 Keeping Up with New Discoveries

Even for a pathway like the TCA cycle, new articles continue to be published. From the growing number of articles on metabolism, knowledge needs to be extracted and incorporated into the database, which requires large amounts of time, manpower and funding. As Amos Bairoch [82] stated: "*It's quite depressive to think that we are spending millions in grants for people to perform experiments, produce new knowledge, hide this knowledge in often badly written text and then spend some more millions trying to second guess what the authors really did and found.*" Further refinement of the description of the metabolic network will require the continuous commitment of experts from various scientific disciplines. This may be quite challenging, because many experts may not be familiar with the reconstruction and curation of a metabolic network [81]. Moreover, for experts who do not work with metabolic pathway databases it may not be directly evident what the benefit is of contributing to such an endeavor. The time and efforts required should, therefore, be kept to a minimum and the contributions of each expert should be clearly acknowledged.

Various approaches have been suggested to obtain new pieces of knowledge in a machine-readable format directly from the authors of articles. This will also prevent misinterpretations of what is described. One proposed solution is to let authors semantically annotate their key findings and allow readers to further improve this annotation [83]. This will facilitate automatic retrieval of this knowledge. Another possible solution is the concept of nanopublications [84], which are traceable and citable author statements comprising a single piece of knowledge formulated as a triplet. A simple example would be: "a *statement*, e.g., reaction X (subject) is catalyzed by (predicate) enzyme Y (object), *conditions* under which the statement holds, e.g., a specific compartment, and *provenance* of the statement, e.g., author and literature." In this way, the authors of the statement are explicitly acknowledged for their contribution. Journals could play a role by encouraging authors to contribute the knowledge published in formats like the nanopublication or as an addition to WikiPathways. For mathematical models, a requirement could be that it needs to be deposited in the BioModels database [85]. This is not unlike what is already required for microarray data. Moreover, funding agencies and tenure track committees currently measure contributions to science mainly in terms of publications. A cultural shift is needed to give appropriate recognition to these alternative ways of contributing to science. Initiatives to collaborate and improve an existing description of the metabolic network should be encouraged. The construction of yet another database that describes the network of a particular organism for which a reconstruction already exists, is not the solution.

In summary, it is of the utmost importance to join forces across multiple scientific fields to further improve the description of the metabolic network for a wide range of organisms.

Acknowledgements

Hereby I would like to acknowledge that this chapter has been based on and single sentences have been used from two previously published articles, i.e., the article titled "*Critical assessment of human metabolic pathway databases: a stepping stone for future integration*" [14] published in BMC Systems Biology and the article titled "*Knowledge representation in metabolic pathway databases*" [32] published in Briefings in Bioinformatics. The same holds for the general introduction and discussion of my thesis titled "*The road to knowledge: from biology to databases and back again*" [86]. Also examples and conclusions were used from the previously published article titled "*Improving the description of metabolic networks: the TCA cycle as example*" [15], which has been explicitly referenced where applicable.

I would like to thank the reviewers for their insightful comments and suggestions for improving the content, comprehensibility, structure and grammar of the chapter. I also thank Samantha Elmhurst for remaking the figures and tables to give them a more uniform design and Julia Hoeng for answering my questions during the process of writing this chapter. Furthermore, I would like to thank Perry Moerland for his suggestions on the content and structure of this chapter and also his contributions to the articles that either formed the base of this chapter or provided inspiration for it. Finally, I would like to thank also my other coauthors of the articles that either formed the base of this chapter or provided inspiration for it; these coauthors are Sander Houten, Gerbert Jansen, Antoine van Kampen, Trebor Rengaw, Morris Swertz, Ines Thiele, and Ronald Wanders.

References

1. Sanctorius S (1614) Ars de statica medicina
2. Buchner E (1897) Alkoholische gärung ohne hefezellen (vorläufige mitteilung). Ber Dtsch Chem Ges 30:117–124
3. Krebs HA, Johnson WA (1937) The role of citric acid in intermediate metabolism in animal tissues. Enzymologia 4:148–156
4. Vastrik I, D'Eustachio P, Schmidt E et al (2007) Reactome: a knowledge base of biologic pathways and processes. Genome Biol 8:R39
5. Kell DB, Goodacre R (2014) Metabolomics and systems pharmacology: why and how to model the human metabolic network for drug discovery. Drug Discov Today 19:171–182
6. Tang J, Aittokallio T (2014) Network pharmacology strategies toward multi-target anticancer therapies: from computational models to experimental design principles. Curr Pharm Des 20:23–26
7. Kanehisa M (1996) Toward pathway engineering: a new database of genetic and molecular pathways. Sci Technol Jpn 34–38
8. Pathguide. www.pathguide.org Accessed 10 Oct 2014
9. Kumar A, Suthers P, Maranas C (2012) MetRxn: a knowledgebase of metabolites and reactions spanning metabolic models and databases. BMC Bioinformatics 13:6
10. Radrich K, Tsuruoka Y, Dobson P et al (2010) Integration of metabolic databases for the reconstruction of genome-scale metabolic networks. BMC Syst Biol 4:114
11. Herrgård MJ, Swainston N, Dobson P et al (2008) A consensus yeast metabolic network reconstruction obtained from a community approach to systems biology. Nat Biotechnol 26:1155–1160
12. Altman T, Travers M, Kothari A et al (2013) A systematic comparison of the MetaCyc and KEGG pathway databases. BMC Bioinformatics 14:112
13. Zhou H, Jin J, Zhang H et al (2012) IntPath – an integrated pathway gene relationship database for model organisms and important pathogens. BMC Syst Biol 6:S2
14. Stobbe MD, Houten SM, Jansen GA et al (2011) Critical assessment of human metabolic pathway databases: a stepping stone for future integration. BMC Syst Biol 5:165
15. Stobbe MD, Houten SM, van Kampen AHC et al (2012) Improving the description of metabolic networks: the TCA cycle as example. FASEB J 26:3625–3636
16. Dharuri H, Henneman P, Demirkan A et al (2013) Automated workflow-based exploitation of pathway databases provides new insights into genetic associations of metabolite profiles. BMC Genomics 14:865
17. Lee DS, Park J, Kay KA et al (2008) The implications of human metabolic network topology for disease comorbidity. Proc Natl Acad Sci U S A 105:9880–9885
18. Zelezniak A, Pers TH, Soares S et al (2010) Metabolic network topology reveals transcriptional regulatory signatures of type 2 diabetes. PLoS Comput Biol 6, e1000729
19. Mora A, Donaldson I (2012) Effects of protein interaction data integration, representation and reliability on the use of network properties for drug target prediction. BMC Bioinformatics 13:294
20. Green ML, Karp PD (2006) The outcomes of pathway database computations depend on pathway ontology. Nucleic Acids Res 34:3687–3697
21. Elbers CC, van Eijk KR, Franke L et al (2009) Using genome-wide pathway analysis to unravel the etiology of complex diseases. Genet Epidemiol 33:419–431
22. Stobbe MD, Jansen GA, Moerland PD et al (2014) Knowledge representation in metabolic pathway databases. Brief Bioinform 15:455–470
23. Hao T, Ma HW, Zhao XM et al (2010) Compartmentalization of the Edinburgh Human Metabolic Network. BMC Bioinformatics 11:393
24. Duarte NC, Becker SA, Jamshidi N et al (2007) Global reconstruction of the human metabolic network based on genomic and bibliomic data. Proc Natl Acad Sci U S A 104: 1777–1782
25. Romero P, Wagg J, Green ML et al (2004) Computational prediction of human metabolic pathways from the complete human genome. Genome Biol 6:R2
26. Kanehisa M, Goto S, Sato Y et al (2014) Data, information, knowledge and principle: back to metabolism in KEGG. Nucleic Acids Res 42:D199–D205

27. Croft D, O'Kelly G, Wu G et al (2011) Reactome: a database of reactions, pathways and biological processes. Nucleic Acids Res 39:D691–D697
28. Aung HW, Henry SA, Walker LP (2013) Revising the representation of fatty acid, glycerolipid, and glycerophospholipid metabolism in the consensus model of yeast metabolism. Ind Biotechnol 9:215–228
29. Heavner BD, Smallbone K, Price ND et al (2013) Version 6 of the consensus yeast metabolic network refines biochemical coverage and improves model performance. Database. doi:10.1093/database/bat059
30. Tieri P, Nardini C (2013) Signalling pathway database usability: lessons learned. Mol BioSyst 9:2401–2407
31. Yeast Pathways Database. http://pathway.yeastgenome.org/ Accessed 10 Oct 2014
32. Ron Caspi, Kate Dreher, Peter D. Karp (2013) The challenge of constructing, classifying and representing metabolic pathways. FEMS Microbiol Lett 345(2):85–93
33. Das AM, Illsinger S, Lücke T et al (2006) Isolated mitochondrial long-chain ketoacyl-CoA thiolase deficiency resulting from mutations in the HADHB gene. Clin Chem 52:530–534
34. Molven A, Matre GE, Duran M et al (2004) Familial hyperinsulinemic hypoglycemia caused by a defect in the SCHAD enzyme of mitochondrial fatty acid oxidation. Diabetes 53:221–227
35. Wanders RJA, Ijlst L, Poggi F et al (1992) Human trifunctional protein deficiency: a new disorder of mitochondrial fatty acid β-oxidation. Biochem Biophys Res Commun 188:1139–1145
36. Wanders RJA, Ijlst L, van Gennip AH et al (1990) Long-chain 3-hydroxyacyl-CoA dehydrogenase deficiency: identification of a new inborn error of mitochondrial fatty acid β-oxidation. J Inherit Metab Dis 13:311–314
37. Chemical Entities of Biological Interest (ChEBI). https://www.ebi.ac.uk/chebi/ Accessed 10 Oct 2014
38. Pico AR, Kelder T, van Iersel MP et al (2008) WikiPathways: pathway editing for the people. PLoS Biol 6, e184
39. Goto S, Okuno Y, Hattori M et al (2002) LIGAND: database of chemical compounds and reactions in biological pathways. Nucleic Acids Res 30:402–404
40. Caspi R, Altman T, Dreher K et al (2012) The MetaCyc database of metabolic pathways and enzymes and the BioCyc collection of pathway/genome databases. Nucleic Acids Res 40:D742–D753
41. Karp PD, Ouzounis CA, Moore-Kochlacs C et al (2005) Expansion of the BioCyc collection of pathway/genome databases to 160 genomes. Nucleic Acids Res 33:6083–6089
42. Thiele I, Palsson BØ (2010) A protocol for generating a high-quality genome-scale metabolic reconstruction. Nat Protoc 5:93–121
43. Matthews L, Gopinath G, Gillespie M et al (2009) Reactome knowledgebase of human biological pathways and processes. Nucleic Acids Res 37:D619–D622
44. Soh D, Dong D, Guo Y et al (2010) Consistency, comprehensiveness, and compatibility of pathway databases. BMC Bioinformatics 11:449
45. Croft D, Mundo AF, Haw R et al (2014) The reactome pathway knowledgebase. Nucleic Acids Res 42:D472–D477
46. Thiele I, Swainston N, Fleming RMT et al (2013) A community-driven global reconstruction of human metabolism. Nat Biotechnol 31:419–425
47. Reactome portal at WikiPathways. www.wikipathways.org/index.php/Portal:Reactome Accessed 10 Oct 2014
48. List of BioCyc Pathway/Genome Databases. http://biocyc.org/biocyc-pgdb-list.shtml Accessed 10 Oct 2014
49. Keseler IM, Mackie A, Peralta-Gil M et al (2013) EcoCyc: fusing model organism databases with systems biology. Nucleic Acids Res 41:D605–D612
50. Sahoo S, Franzson L, Jonsson JJ et al (2012) A compendium of inborn errors of metabolism mapped onto the human metabolic network. Mol BioSyst 8:2545–2558
51. Karp PD, Paley S, Krieger CJ et al (2004) An evidence ontology for use in pathway/genome databases. In: Altman RB, Dunker AK, Hunter L, Jung TA, Klein TE (eds) Pac symp biocomput. World Scientific Publishing Co-Pte. Ltd., New York, NY, pp 190–201
52. Kuffner R, Zimmer R, Lengauer T (2000) Pathway analysis in metabolic databases via differential metabolic display (DMD). Bioinformatics 16:825–836
53. Carbonetto P, Stephens M (2013) Integrated enrichment analysis of variants and pathways in genome-wide association studies indicates central role for IL-2 signaling genes in type 1 diabetes, and cytokine signaling genes in Crohn's disease. PLoS Genet 9, e1003770
54. Schaefer CF, Anthony K, Krupa S et al (2009) PID: the pathway interaction database. Nucleic Acids Res 37:D674–D679
55. Subramanian A, Tamayo P, Mootha VK et al (2005) Gene set enrichment analysis: a knowledge-based approach for interpreting genome-wide expression profiles. Proc Natl Acad Sci U S A 102:15545–15550

56. Latendresse M, Krummenacker M, Trupp M et al (2012) Construction and completion of flux balance models from pathway databases. Bioinformatics 28:388–396
57. Orth JD, Thiele I, Palsson BØ (2010) What is flux balance analysis? Nat Biotechnol 28: 245–248
58. NCBI Entrez Gene. http://www.ncbi.nlm.nih.gov/gene/ Accessed 10 Oct 2014
59. UniProt. http://www.uniprot.org/ Last accessed 10 Oct 2014
60. Haraldsdóttir H, Thiele I, Fleming RMT (2014) Comparative evaluation of open source software for mapping between metabolite identifiers in metabolic network reconstructions: application to Recon 2. J Cheminform 6:2
61. Fearnley LG, Davis MJ, Ragan MA et al (2013) Extracting reaction networks from databases-opening Pandora's box. Brief Bioinform. doi:10.1093/bib/bbt058
62. Kanehisa M, Goto S, Kawashima S et al (2004) The KEGG resource for deciphering the genome. Nucleic Acids Res 32:D277–D280
63. Demir E, Cary MP, Paley S et al (2010) The BioPAX community standard for pathway data sharing. Nat Biotechnol 28:935–942
64. Hucka M, Finney A, Sauro HM et al (2003) The systems biology markup language (SBML): a medium for representation and exchange of biochemical network models. Bioinformatics 19:524–531
65. Cerami EG, Gross BE, Demir E et al (2010) Pathway commons, a web resource for biological pathway data. Nucleic Acids Res 39:D685–D690
66. Kamburov A, Stelzl U, Lehrach H et al (2013) The consensusPathDB interaction database: 2013 update. Nucleic Acids Res 41:D793–D800
67. Oberhardt MA, Palsson BØ, Papin JA (2009) Applications of genome-scale metabolic reconstructions. Mol Syst Biol 5:320
68. Kitano H, Ghosh S, Matsuoka Y (2011) Social engineering for virtual 'big science' in systems biology. Nat Chem Biol 7:323–326
69. Stobbe MD, Swertz M, Thiele I et al (2013) Consensus and conflict cards for metabolic pathway databases. BMC Syst Biol 7:50
70. Chindelevitch L, Stanley S, Hung D et al (2012) MetaMerge: scaling up genome-scale metabolic reconstructions, with application to Mycobacterium tuberculosis. Genome Biol 13:R6
71. Mons B, Ashburner M, Chichester C et al (2008) Calling on a million minds for community annotation in WikiProteins. Genome Biol 9:R89
72. Monk J, Nogales J, Palsson BØ (2014) Optimizing genome-scale network reconstructions. Nat Biotechnol 32:447–452
73. Bairoch A (2009) The future of annotation/biocuration. Nat Preced. doi:10.1038/npre.2009.3092.1
74. Jensen LJ, Bork P (2010) Ontologies in quantitative biology: a basis for comparison, integration, and discovery. PLoS Biol 8, e1000374
75. Groth P, Gibson A, Velterop J (2010) The anatomy of a nanopublication. Inform Serv Use 30:51–56
76. Li C, Donizelli M, Rodriguez N et al (2010) BioModels database: an enhanced, curated and annotated resource for published quantitative kinetic models. BMC Syst Biol 4:92
77. Stobbe MD (2012) The road to knowledge: from biology to databases and back again. Ph.D. thesis. http://dare.uva.nl/record/1/385827
78. BioCarta. http://www.biocarta.com/ Accessed 10 Oct 2014
79. Yamamoto S, Sakai N, Nakamura H et al (2011) INOH: ontology-based highly structured database of signal transduction pathways. Database. doi:10.1093/database/bar052
80. Mi H, Muruganujan A, Thomas PD (2013) PANTHER in 2013: modeling the evolution of gene function, and other gene attributes, in the context of phylogenetic trees. Nucleic Acids Res 41:D377–D386
81. Jewison T, Su Y, Disfany FM et al (2014) SMPDB 2.0: big improvements to the small molecule pathway database. Nucleic Acids Res 42:D478–D484
82. Morgat A, Coissac E, Coudert E et al (2012) UniPathway: a resource for the exploration and annotation of metabolic pathways. Nucleic Acids Res 40:D761–D769
83. Berg JM, Tymoczko JL, Stryer L (2012) Glycolysis and Gluconeogenesis. In: Biochemistry. W.H. Freeman and Company, New York, Palgrave MacMillan, 7th edition, p. 469–513
84. BioCyc. www.biocyc.org Accessed 10 Oct 2014
85. BiGG Database http://bigg.ucsd.edu/. Accessed 10 Oct 2014
86. EHMN. http://www.ehmn.bioinformatics.ed.ac.uk/. Accessed 10 Oct 2014
87. KEGG. http://www.genome.jp/kegg/. Accessed 10 Oct 2014
88. Panther. http://pantherdb.org/ Accessed 10 Oct 2014
89. Reactome. http://www.reactome.org/ Accessed 10 Oct 2014
90. UniPathway. http://www.grenoble.prabi.fr/obiwarehouse/unipathway. Accessed 10 Oct 2014
91. WikiPathways. http://www.wikipathways.org/index.php/WikiPathways. Accessed 10 Oct 2014

Chapter 3

Causal Biological Network Database: A Comprehensive Platform of Causal Biological Network Models Focused on the Pulmonary and Vascular Systems

Marja Talikka, Stephanie Boue, and Walter K. Schlage

Abstract

While the ever-increasing amounts of scientific data provide a more detailed description of toxic effects, it is not trivial to extract information that will contribute to a better biological understanding. Sophisticated computational methods have been developed to separate mathematically the biological signal from the noise in high-throughput datasets; however, visualizing and putting a signal into a relevant biological context using a priori knowledge is equally important. The Causal Biological Network (CBN) Database addresses these pressing needs in pulmonary and vascular biology. Each of the network models deposited in the CBN Database is scripted in the Biological Expression Language (BEL), a semantic language that represents scientific findings in a computable format. The biological areas covered by the CBN models include cell fate, response to cell stress, cell proliferation, inflammation, tissue repair, and angiogenesis, all in the context of the pulmonary and vascular systems. With specific biological boundaries, multiple types of gene-centric entities, literature support at edge level, and interactive visualization, the CBN Database offers a coherent illustration of important biological processes. Moreover, the computability of the CBN models provides the possibility of data-driven enhancement that delivers an efficient combination of literature knowledge and high-throughput data in a single model. The CBN Database can be applied in computational toxicology and could be extended to drug discovery, biomarker identification, and personalized medicine.

Key words Biological expression language, Computational toxicology, Causal biological network models, Database

1 Background

State-of-the-art toxicological risk assessment and applications in drug discovery demand powerful computational approaches to distinguish biological signals from noise in high-throughput data. Furthermore, selected findings from the scientific literature need to be presented in a format that can be used to interpret a specific signal in the context of a priori knowledge. A frequent example is the mining of (genome-wide) gene expression changes observed

Julia Hoeng and Manuel C. Peitsch (eds.), *Computational Systems Toxicology*, Methods in Pharmacology and Toxicology, DOI 10.1007/978-1-4939-2778-4_3,

between treated and control samples, and a number of tools with unique features have been developed to extract significant biological mechanisms from a list of differentially expressed genes. In Web applications dedicated to the analysis of differentially expressed genes, such as the DAVID Bioinformatics Database [1] or the Gene Set Enrichment Analysis (GSEA) method [2], users can choose from a number of pathway repositories, such as Reactome [3], Biocarta [4], and the Kyoto Encyclopedia of Genes and Genomes (KEGG) [5] pathways to interpret the lists of differentially expressed genes. With several export formats, Reactome provides a hierarchical representation of biological pathways, subpathways, and pathway steps (reactions). These pathways are curated to reflect human biology; however, orthologous reactions have been inferred for more than 20 other species [6]. Biocarta harvests the knowledge of the scientific community and emphasizes the graphical representation of its 354 pathways; it could be considered an "open source graphics database." The May 2014 version of KEGG Pathways, one of the first freely available databases of biological signaling pathways [7], contained over 400 pathway maps. Designed initially for metabolic pathways, the KEGG collection now includes general signaling and disease pathways referencing a set of published articles that support the biology described in each pathway.

The Signaling Pathway Integrated Knowledge Engine (SPIKE) contains relationships drawn from other large signaling databases including KEGG, Reactome, and several protein–protein interaction networks [8, 9]. SPIKE offers a dynamic visualization package and, in addition to the prearranged pathways made available, users can create and export their own networks from the content available in the SPIKE database and save networks in XGMML format.

Other initiatives have tried to model known biology in the form of pathways and networks. For example, each of the UCSD Signaling Gateway Molecule Pages [10] focuses on a single protein and provides essential information on its involvement in cellular signaling. The descriptions of the biological activity, regulation, and localization of the proteins are written by subject matter experts. The pages are regularly updated from highly structured datasets that are retrieved from publicly available data sources (e.g., UniProt and Genbank) [10, 11].

Other pathway tools include the National Cancer Institute Interaction Database [12] with 137 human pathways with 9,248 interactions plus 322 human pathways with 7,575 interactions imported from BioCarta and Reactome.

While all these resources have many useful features, very few of the available pathway tools combine the formal language used to record the literature support at the edge level, interactive visualization, and computability with the possibility of data-driven enhancement, all at the same time. Moreover, these tools tend to operate without specific biological boundaries, such as tissue type or species.

In this chapter, we describe the Causal Biological Network (CBN) Database that combines all these features into a comprehensive representation of biological processes relevant to pulmonary and cardiovascular tissues. More specifically, we introduce the Biological Expression Language (BEL) that was used to build the CBN models, describe the model structure as well as the biological areas covered by the models. The CBN database that hosts the models and offers their use to scientific community is also described. Finally, we discuss some future prospects for CBN models and their utility in biomarker discovery.

2 Structure and Properties of the CBN Models

2.1 Biological Expression Language (BEL)

The CBN models were curated using the Biological Expression Language (BEL), a standardized language that uses controlled vocabulary. BEL represents scientific information as semantic triplets where causal or non-causal relationships between biological entities can be presented in a computable format. In BEL, biological *entities* are defined as official symbols within specified *name spaces/ontologies* and described as abundances, modifications, activities, processes, or transformations, called the BEL *functions* (Fig. 1a). The BEL triplet consists of a subject, a predicate, and an object where, in the example shown, CCND1 (subject) is represented as increasing (predicate) the kinase activity of CDK4 (object) (Fig. 1b). OpenBEL (http://www.openbel.org) is an open consortium for the BEL framework that provides tools and resources for scripting and editing with BEL language (BEL editor) and for visualizing BEL-encoded network models (Knowledge Assembly Model navigator).

2.2 Structure of the CBN Models

A CBN model consists of a backbone, which is a graph assembly of BEL triplets that together form a coherent illustration of the biological process of interest (Fig. 2). Unlike protein–protein interaction networks where all the nodes are proteins, the BEL-encoded CBN models can capture heterogeneous functional entities, including biological molecules such as proteins, DNA variants, coding and noncoding RNA, chemicals, lipids, methylation states, and/or other modifications (e.g., phosphorylation), as well as biological processes, phenotypic, and/or clinical observations. The biological entities, i.e., nodes in the network, are connected by edges. Non-causal edges link different forms of a biological entity, such as its mRNA and protein abundance, and causal edges represent cause-effect relationships between biological entities as described in the scientific literature. Hence, most causal edges in the network model are supported by at least one (often multiple) literature reference that can be easily traced to the originally published scientific findings.

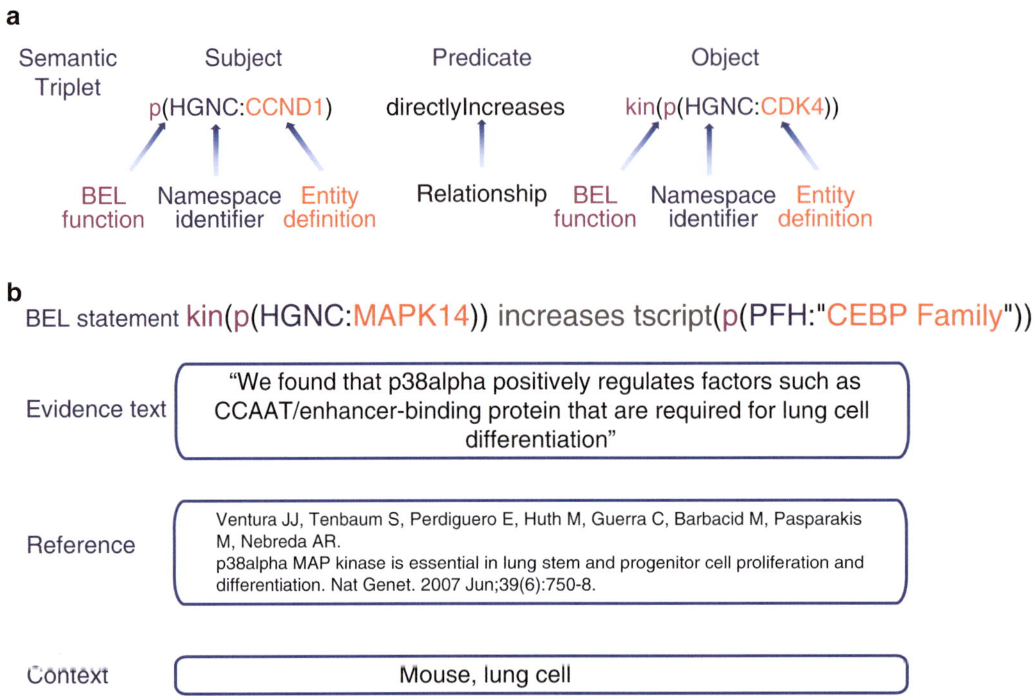

Fig. 1 Biological Expression Language. (**a**) Example of a BEL statement. Biological entities are described as BEL functions and assigned a name space identifier. The relationships between the biological entities describe the impact that one entity has on another entity in a specific context. (**b**) BEL document containing the BEL statement, the evidence text that was extracted from scientific literature, the reference to the source of the information, and the context. *p* protein abundance, *kin* kinase activity, *HGNC* Hugo Gene Nomenclature Committee

In addition to the model backbone, each CBN model can have a measurable layer that consists of information on the ability of certain entities depicted as nodes in the model backbone to regulate the expression of particular genes. These genes are associated with their upstream controller node with the appropriate sign indicating upregulation or downregulation by the upstream controller. For example, many genes are known to be regulated by the transcriptional activity of the Nuclear Factor (Erythroid-derived 2)-like 2 (NFE2L2) as shown in Fig. 2. The number of downstream gene expression entities can vary from a few to several hundred depending on the node. Similar to the BEL documents that describe the model backbone, the causal relationships between the nodes and their downstream gene expression layer are maintained in a knowledgebase. These gene expression entities can be used by the Reverse Causal Reasoning (RCR) algorithm to predict the activation or inhibition of the upstream controller node [13]. To compute these predictions, RCR employs gene expression changes observed in the data and the knowledgebase, which contains a priori information on gene expression changes resulting in the activity of a given biological entity. RCR relies on two metrics, richness and concordance.

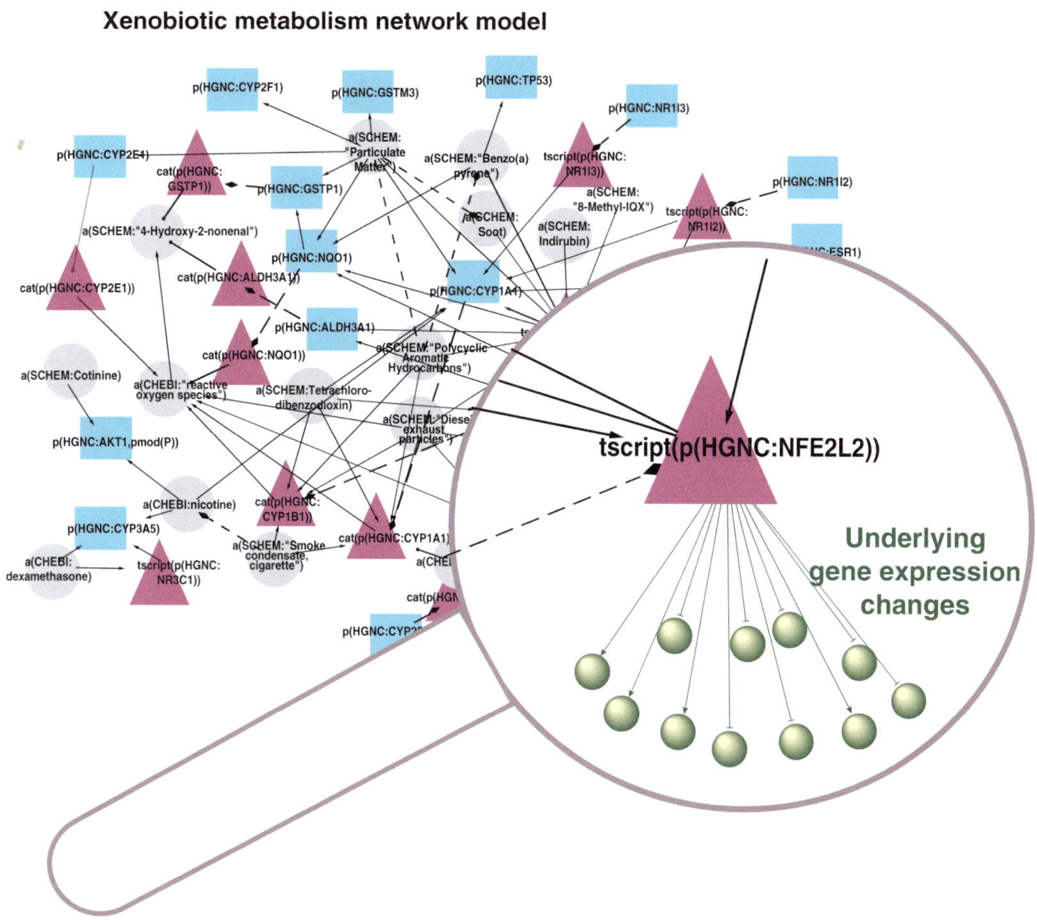

Fig. 2 Structure of the models in the CBN Database. The biological network models in the CBN Database contain two layers. The model backbone depicts the causal relationships between biological entities, where types of entities are illustrated with different symbols (abundances as *grey circles* and *blue squares*, activity nodes as *purple triangles*). Each backbone node is labeled with a BEL function, the name space identifier, and the entity definition. The network models also consist of a measureable layer that is used to infer the activities of the backbone entities based on transcriptomics data. *p* protein abundance, a chemical abundance, *tscript* transcriptional activity, *cat* catalytic activity, *HGNC* Hugo Gene Nomenclature Committee, *NFE2L2* Nuclear factor (erythroid-derived 2)-like 2

Richness is calculated as a p-value based on a hypergeometric probability distribution and indicates how significant the signal is (e.g., gene expression changes). Concordance is calculated as a *p*-value based on a binomial distribution and indicates confidence/consistency in the directionality of the prediction (e.g., is the predicted activity of the backbone node increased or decreased) [13].

2.3 Properties of the CBN Models

Unlike other common approaches for building pathway or connectivity maps, the CBN models were built within defined biological and context (species, tissues, cell types, experiment types) boundaries. The majority of causal edges have been annotated with specific information regarding species, tissue and cell type, or any other context-specific information.

The causality of the network edges permits inference and computation on the network as a whole. This computation does not rely on the commonly used "forward assumption", where gene expression changes are used as proxies for the abundance or activity of the encoded protein. Instead, RCR leverages the a priori knowledge in the knowledgebase and predicts the activity of a backbone node based on the differential expression of genes that it regulates, thereby allowing inferences from the expression of genes whose biological function is entirely unknown without any additional assumption. In addition to providing tools to create and handle BEL documents, the OpenBEL framework contains computational tools that can be used to analyze network models using a RCR approach (Whistle) [13]. Moreover, novel algorithms have been developed to compute a quantitative measure for the impact of gene expression changes on a given biological process by utilizing the network models composition and structure (Chap. 15 in this book) and [14–18].

The CBN models are dynamic and can be updated as new knowledge becomes available. Hosted on the CBN Web site (www.causalbionet.com), the models can be browsed, visualized, and downloaded in a portable network visualization format for further analysis possible with freely available tools, including Cytoscape [19].

A comparison of the CBN Database with other pathway databases is shown in Fig. 3.

3 Biological Areas Covered by the CBN Models

The CBN models have been built over many years and mainly contain processes relevant to pulmonary and vascular biology (Fig. 4a). Each model comprises canonical signaling pathways (e.g., MAPK signaling, Fig. 4b) or biological signaling pathways that are relevant to specific cell types (e.g., certain immune cells in inflammatory processes, Fig. 4c).

3.1 Cell Proliferation

Cellular proliferation in the lung was the first biological process that was modeled. The model represents the cell cycle module as the core unit, which consists of the elements that regulate entry to and exit from the cell cycle [20]. The cell cycle model is coupled with models that represent pathways that are known to regulate proliferation, e.g., growth factor signaling, cell interaction, intra- and extra-cellular signaling, and epigenetics [20].

3.2 Cellular Stress Response

Cellular stress is represented in specific pathways in the lung and vascular tissues that are activated in response to different stressors [21]. The oxidative stress model describes management of

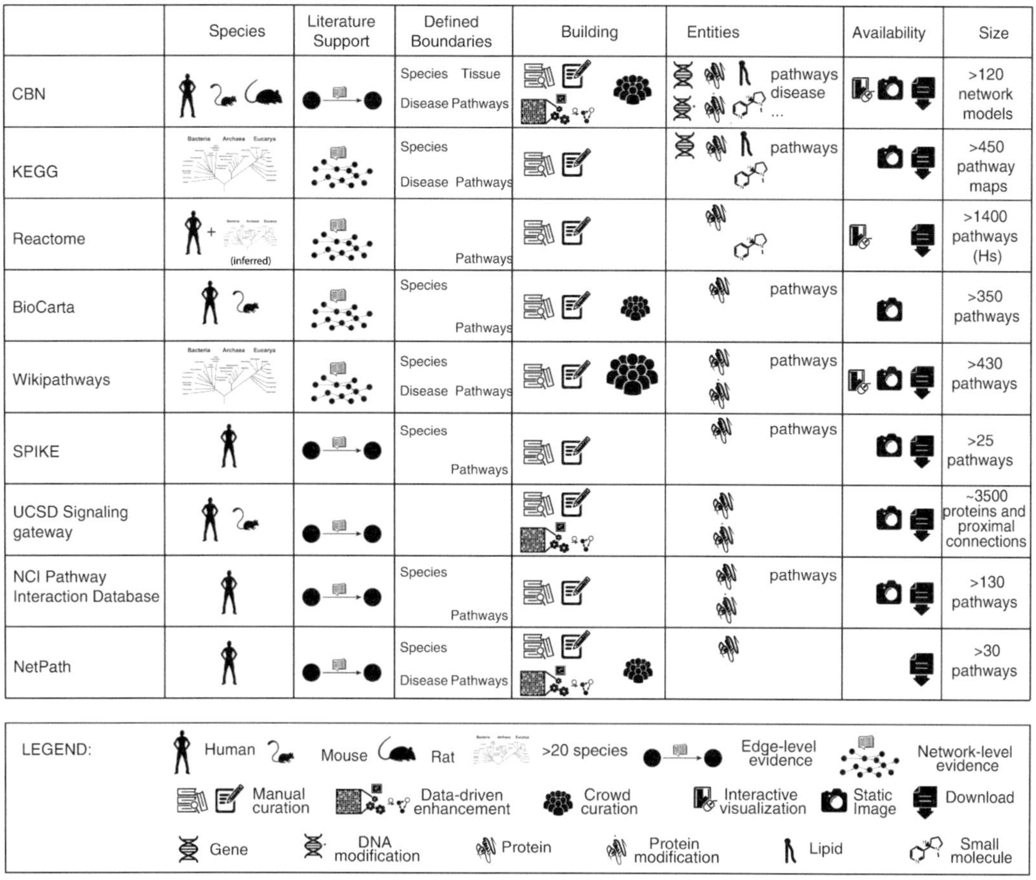

Fig. 3 Comparison of the CBN Database to other pathway tools

intracellular free radicals, glutathione metabolism, and response to endogenous and exogenous oxidants and anti-oxidants. For example, nuclear factor NFE2L2 signaling has been modeled in detail as an essential mediator of oxidative stress and the xenobiotic metabolism network model was built around the aryl hydrocarbon receptor (AHR) node. AHR is activated by xenobiotic compounds, which results in the regulation of the expression of several target genes, including the cytochrome P450 genes *CYP1A1* and *CYP1B1* [22, 23]. Various environmental stressors, such as "soot", "particulate matter", and "diesel exhaust particles" were also included in the network backbone [24–26]. The original xenobiotic metabolism network model was updated to reflect the processes in the lung tissue more comprehensively [27] and this model has been used successfully to assess the response of different experimental systems to cigarette smoke exposure [15, 27]. Hypoxic stress was modeled with a focus on the hypoxia inducible factor (HIF1A) and osmotic stress was modeled with a focus on the nuclear factor of activated T cells (NFAT5), aquaporin, and cystic

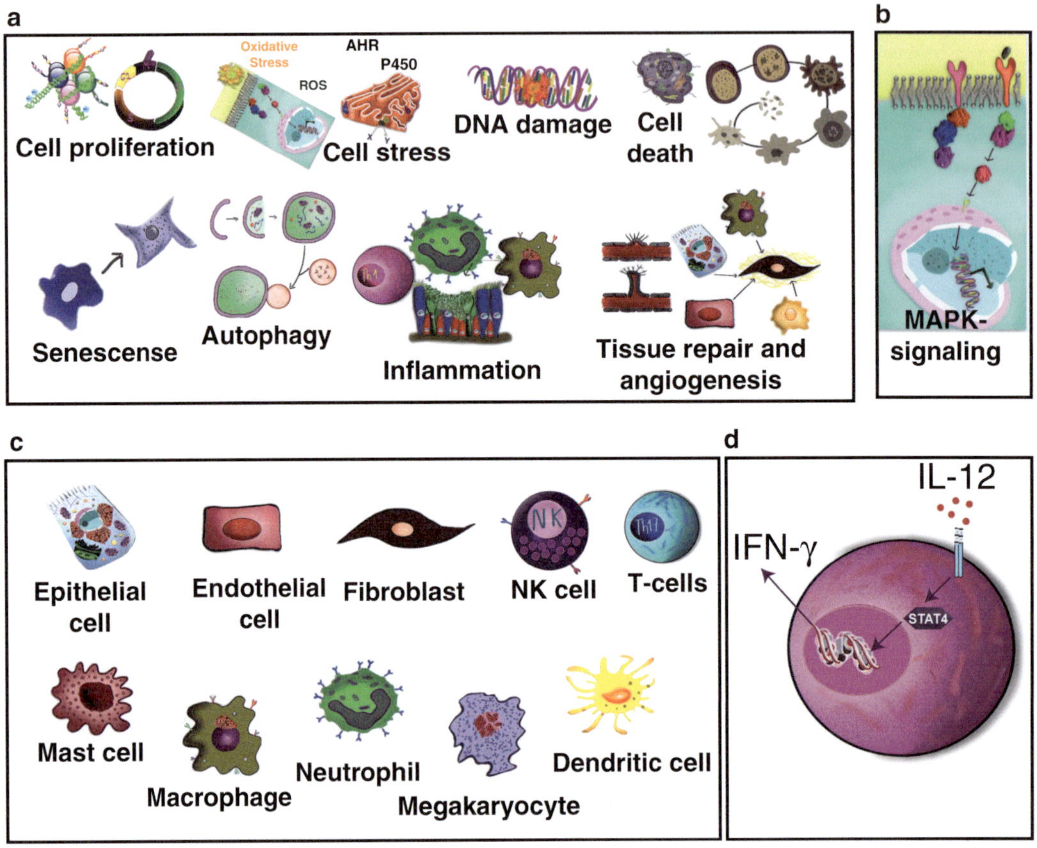

Fig. 4 Types of pulmonary network models in the CBN Database. (**a**) Lung-specific network models. (**b**) MAPK signaling as an example of the several canonical signaling pathways. (**c**) Signaling that is specific to certain cell types are modeled within the main biological processes shown in (**a**). (**d**) Cell-specific models that were built following an input–output design

fibrosis transmembrane conductance regulator (CFTR) pathways. The endoplasmic reticulum (ER) stress model represents the unfolded protein response and pathways downstream of the main mediators of stress [21]. The endothelial shear stress model is related to cardiovascular biology and describes the impact of atheroprotective (laminar) and atherogenic (turbulent) shear stress on monocyte adhesion.

3.3 Pulmonary Inflammation

Several CBN models were built to represent the main cell types involved in acute pulmonary inflammation induced by environmental stressors and included the signaling characteristic for pulmonary epithelial cells, macrophages, neutrophils, T-cell subsets (cytotoxic T cells, T-helper cells, and regulatory T-cells), natural killer cells, dendritic cells, megakaryocytes, and mast cells (Fig. 4c). The cell-specific models were built initially following an input–output design, where inputs were signaling ligands/triggers that induce or suppress an intracellular signaling cascade, and outputs were the cellular/physiological products of these signaling pathways, i.e., mostly secreted cytokines or biological processes (Fig. 4d) [28].

3.4 DNA Damage, Autophagy, Cell Death, and Senescence

When a tissue cannot deal efficiently with stress, the existence of the cell is jeopardized. The DNA damage, autophagy, cell death, and senescence models (Fig. 4a) describe the DNA damage response and the different fates that a cell may adopt on exposure to a stress that overwhelms the compensatory stress response mechanisms of the cell. The response includes the different processes that lead to cell death, i.e., apoptosis and necroptosis, as well as processes that do not lead to death but qualitatively change the phenotype or function of the cell. In the autophagic response, intracellular organelles and proteins are degraded and recycled to maintain nutrient and energy homeostasis. Irreversible growth arrest in response to various external stimuli was modeled in the senescence network models [29].

3.5 Tissue Repair and Angiogenesis

CBN models were built to describe tissue repair and angiogenesis in the lung tissue that, when unresolved, may progress eventually to irreversible disease [30]. The angiogenesis network models represent key angiogenic processes of sprouting and tubulogenesis mediated via Notch-Delta signaling as well as the regulation of the HIF1A transcription factor resulting in angiogenesis [30]. The effects of growth factor signaling on various angiogenic processes such as migration, proliferation and vascular permeability were also modeled. The fibrosis network model consists of causal relationships between molecules involved in epithelial-to-mesenchymal cell transition in lung tissue repair, as well as characteristics of lung fibrosis such as the transforming growth factor TGFB1, collagen, and the matrix metalloproteinases. This network also contains signaling components of the Wnt, Hedgehog, and epidermal growth factor receptor (EGFR) pathways that promote fibrosis and epithelial-to-mesenchymal cell transition.

The tissue repair and angiogenesis network models were both accompanied by modules that portrayed their regulation by the immune system, e.g., the activation of macrophages, dendritic cells, neutrophils, and monocytes.

3.6 Vascular Inflammation

The essential pathobiological events that contribute to vascular disease were covered in six distinct CBN models [31]. The first five models, endothelial cell activation, endothelial cell–monocyte interaction, foam cell formation, smooth muscle cell activation, and platelet activation represent central atherogenic mechanisms that are activated during vascular inflammatory responses in distinct cellular populations. The plaque destabilization model describes molecular events that occur in advanced, unstable atherosclerotic lesions.

4 Construction of the CBN Models

4.1 Boundary Definition

Network model construction has been described in detail in several publications [20, 21, 28–30]. The first step in building the CBN models was defining the model boundaries, including the selection of the appropriate tissue/cell context with the aim of confining the models to cell types that are found in the lung and vascular system, i.e., fibroblasts, epithelial and endothelial cells, and cells of the immune system. Preferably, all causal relationships in a CBN model should be traceable back to scientific literature that describes experiments in a non-diseased pulmonary (e.g., lung tissue, bronchial epithelial cells, alveolar type II cells) or cardiovascular context (e.g., coronary artery endothelial cells). Immortalized cell lines were generally excluded; however, an exception was made for experiments where cell lines were used as tools to investigate molecular relationships in overexpression studies. In addition, well-established canonical pathways without tissue-specific evidence were included in the CBN models. For example, the glutathione synthesis pathway was included in the cellular stress network model under the assumption that this machinery intended to battle metabolically generated reactive oxygen species (ROS) can operate in most mammalian cell types, even though explicit evidence was not identified in a pulmonary or vascular context [21].

Boundaries were also set based on the biological processes that were included in the model. Care was taken to ensure maximum coverage of the biology without excessive overlap among the CBN models. For example, inflammatory cytokine signaling and the p53-dependent DNA damage response were not included in the cell proliferation network [20], because the plan was to model them in detail in specific networks focused on inflammation [28] and DNA damage [29]. Similarly, the pro-apoptotic arm of the ER stress response was excluded from the cell stress network model [21] and included in a separate CBN model for cell death-related processes [29]. However, despite the specific boundary criteria for each network to reflect the biological relationships involved in the specific processes, the CBNs can still contain shared elements. One such an element is NFκB signaling, which (because it is involved in a number of cellular processes) is modeled in several different networks to maintain the entirety of the biological process.

4.2 Two-Step Model Building

The CBN models were built by first constructing a scaffold from causal relationships extracted from scientific literature (Fig. 5a). Relationships derived from human tissue contexts were prioritized, and statements in mouse and rat contexts were used only to complete a model when human information was not available. Causal relationships that were not already in the knowledgebase with the evidences that obeyed the boundaries were curated from scientific literature and added to the model.

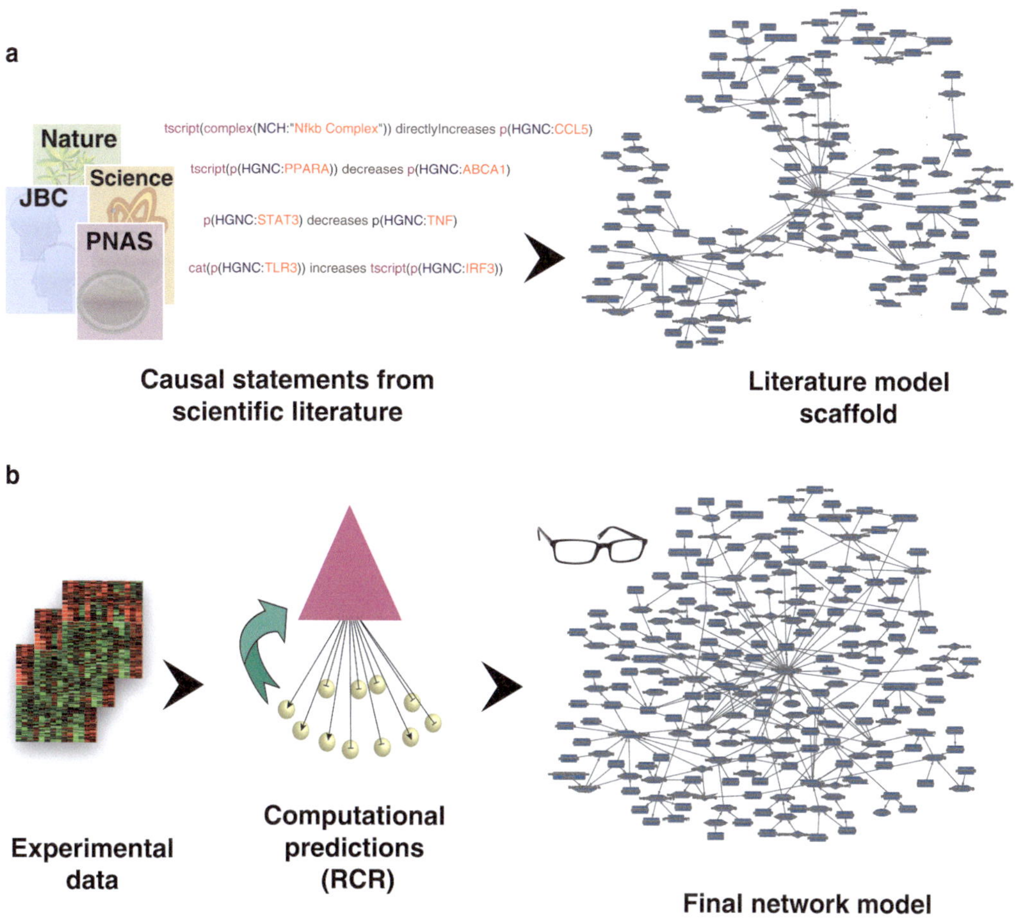

Fig. 5 Construction of the models in the CBN Database. The two main steps in the CBN model construction process are presented using a hypothetical example. (**a**) BEL statements extracted from scientific literature form the model scaffold. (**b**) The model scaffold is augmented with biological entities predicted from relevant transcriptomics data. The resulting integrated model is verified manually. *RCR* Reverse Causal Reasoning

Elements crucial to the biology that may have been missed during the thorough literature review were captured using RCR on relevant molecular profiling datasets. For this data-driven model enhancement step, several gene expression datasets from boundary-relevant experiments were chosen and scored against the knowledgebase (Fig. 5b). Any entity whose activity was predicted to be changed based on gene expression data was evaluated for its presence in the literature-derived backbone. If it had been missed during the construction of the literature model and if sufficient evidence was available in the context of the experiment under investigation, it was included in the relevant network and linked to existing nodes in the literature model. As many datasets as possible were used in this model enhancement step to increase the comprehensiveness of the biology modeled.

An example of model enhancement in building the tissue repair and angiogenesis network models where two datasets were used is given here. The gene expression data from mouse lung after bleomycin instillation (GSE25640), which is a known inducer of fibrosis and thus triggers appropriate processes involved in the tissue repair networks [32], and the gene expression data from human lung microvascular endothelial cells in vitro subjected to hypoxia for 48 h (GSE11341) were used [33]. As a result, the literature model gained additional 140 new biologically relevant nodes. The resulting integrated model illustrates the value of the combined use of molecular profiling data and prior biological knowledge of cause-and-effect relationships.

4.3 Model Validation

Each integrated CBN model was reviewed by a group of scientists and the majority of the resulting final models were validated using at least one dataset that was not used in the building process. Model validation using public gene expression data aimed to demonstrate that the CBN models captured the expected biology by linking the differential gene expression with observed phenotypes and biological functions. In essence, the nodes that were predicted to change by RCR based on the transcriptomics data were mapped onto the models and signaling pathways relevant to the phenotypic/functional observations in the experiment were highlighted. The apoptosis, necroptosis, autophagy, and angiogenesis networks and the megakaryocyte subnetwork could not yet be validated in this way because of the lack of suitable datasets that fulfill the boundary and technical criteria.

To overcome the lack of suitable public datasets, the cell cycle module within the proliferation network model was validated experimentally using transcriptomics data from a dedicated experiment designed to follow cells arrested in S phase with subsequent release from arrest and progression of the cell cycle [34]. Normal human bronchial epithelial cells (NHBE) in culture were arrested in G1 phase with a small molecule cyclin-dependent kinase (CDK4/6) inhibitor, PD-0332991. The inhibition of CDK4 and CDK6 has been shown to specifically reduce the phosphorylation of retinoblastoma (Rb) protein at the Ser780/Ser795 residues, thereby inducing an exclusive G1 arrest [35]. The cultures were subsequently released from cell cycle arrest to reenter the cell cycle, confirmed by labeling the S-phase cells at 2, 4, 6, and 8 h after inhibitor washout. To validate the cell proliferation network, RCR was applied to the transcriptomics dataset derived from these first arrested and subsequently released NHBE cells (E-MTAB-1272). Within the cell cycle module, 26 nodes were predicted to be regulated following release compared with cells that remained arrested in G1 phase. In agreement with the specific impact of PD-0332991 on CDK4, the corresponding node was predicted to be activated upon inhibitor washout. Moreover, several other major players in the entry to G1 phase were identified. This example also highlighted the value of the CBN models in mode of action studies; in this case, a chemical proxy node representing flavopiridol, a cyclin-dependent kinase inhibitor under clinical development [36], was inhibited as the cells progressed to G2 phase [34].

In addition to serving as network validation datasets, the scoring of transcriptomics data against the CBN model by RCR may reveal mechanisms that were previously not known to be involved in the biological processes being modeled. New information of this kind can help generate new hypotheses that could be verified experimentally in the appropriate context to gain new biological insights.

5 Evolution of the CBN Models

5.1 From Non-diseased to COPD-Relevant Network Models

The network models are not static entities because they can be updated when new knowledge becomes available and can be modified to reflect biology relevant to a specific disease. For example, the original 87 non-disease network models relevant to lung biology [20, 21, 28, 29, 37] were supplemented with biological processes relevant to chronic obstructive pulmonary disease (COPD). Several transcriptomics datasets from the small airways, lungs, and alveolar macrophages of asymptomatic smokers and COPD patients were used to infer processes that were impacted in COPD subjects compared with asymptomatic smokers. In addition to data-driven identification of missing biological entities, COPD-relevant processes were extracted from literature. Existing extracellular matrix degradation and efferocytosis models were heavily modified and new CBN models for B-cell activation and T-cell recruitment and activation were reassembled and augmented to comprehensively represent COPD-relevant biology (Fig. 6a). Additionally, correlative relationships were curated from literature describing the differential abundance of blood markers and microRNAs found in COPD patients.

5.2 Crowd-Sourced Network Verification

To ensure that the biology captured in the CBN models is accurate and up-to-date, the CBNs are subjected to continuous scrutiny by the scientific community. The open review of the networks was initiated in a challenge run as part of the systems biology verification (sbv) IMPROVER (Industrial Methodology for Process Verification in Research) project [38]. Ninety networks that were augmented with COPD-relevant processes were consolidated into 50 main biological areas by manually combining 56 models to yield 16 agglomerated networks, leaving 34 standalone networks. Additional fine-tuning allowed redundant biology to be eliminated and harmonized canonical signaling within the different CBNs.

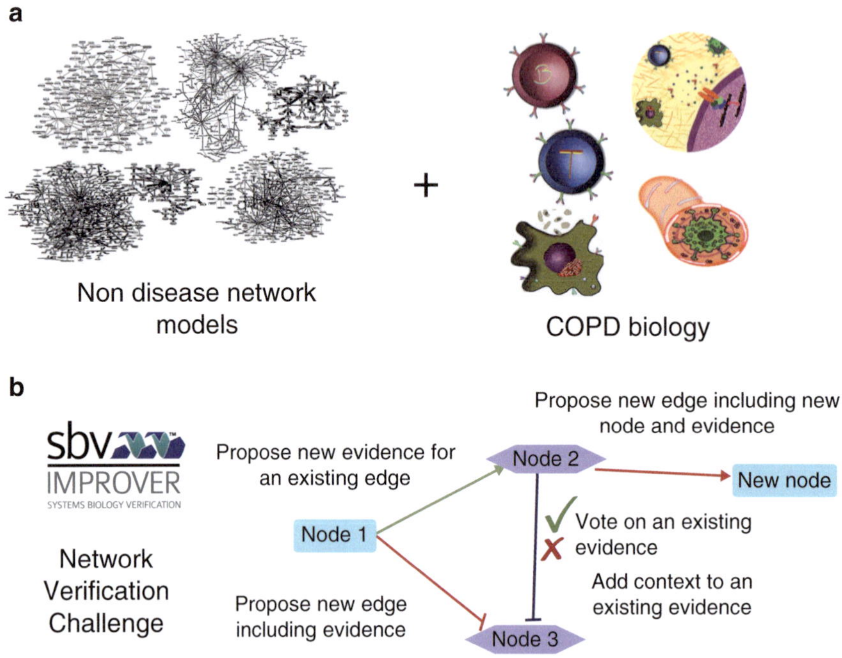

Fig. 6 Model evolution and crowd verification. (**a**) The non-disease network models were augmented with COPD relevant biology. (**b**) Principles of the crowd verification process. During the open phase, participants could add edges or nodes to the networks as well as add, approve, or reject evidence supporting an edge. *COPD* chronic obstructive pulmonary disease

The challenge was designed to collect the knowledge of crowds and attract biologists from across multiple fields of biology to review the networks. A proven social networking approach was used to generate high-quality curation results [39]. The challenge was hosted on the Network Verification Challenge (NVC) Web site, which displayed and supported a full set of actions to review and modify the networks [40]. During the open phase, participants could add edges or nodes in the network as well as add, approve, or reject evidence supporting the edges (Fig. 6b). Additions to and modifications of the biological networks during the online verification process were evaluated further during an in-person scientific 'jamboree' session where the networks were discussed and the input consolidated [41]. These activities resulted in significant modification of some of the CBN models and are still ongoing.

For example, the neutrophil signaling model was affected the most by the verification process with the refinement of the role of cytoskeleton and adhesion in chemotaxis as well as neutrophil activation. The partially verified network versions are subjected to continuous verification via the NVC platform. The crowd-sourced network verification process is described in detail in Chap. 8.

All the current CBN models are listed in Table 1 along with a short description of the biology of each model.

Table 1
Network models available in the CBN database (August 2014 release)

Biological process	Network model	Version	Short description	Number of nodes	Number of edges
Cell proliferation	Cell cycle	1.2	Core machinery regulating entry and exit from the mammalian cell cycle	185	303
	Growth factor	1.2	Extracellular growth factors involved in regulating lung cell proliferation	277	443
	Cell interaction	1.1	Signal transduction pathways leading to cell proliferation that originate from the interactions between common cell adhesion molecules and extracellular matrix components	77	100
	Epigenetics	1.1	Epigenetic modulators of lung cell proliferation and connections from these epigenetic mediators to the core cell cycle components	36	32
	Calcium	1.1	Calcium signaling in the context of cell proliferation, increases in calcineurin complex activity as a result of Wnt-mediated calcium signaling and vascular endothelial growth factor (VEGF) signaling, and the induction of nuclear factor of activated T cells (NFATC1) transcriptional activity	19	20
	Clock	1.1	Mechanisms that regulate circadian clock signaling leading to cell proliferation	77	147
	Hedgehog	1.1	Hedgehog signaling in cell proliferation and branching morphogenesis in the developing mammalian lung	44	56
	Hox	1.1	Hox (homeobox) signaling pathway that leads to cell proliferation as well as downstream targets of members of the HOX family	32	33
	Jak-Stat	1.1	Canonical Jak-Stat signaling leading to cell proliferation	64	88
	Mapk	1.1	Causal mechanisms that regulate different families of MAPK signaling (p38, JNK, ERK, MAP2K) in the context of cell proliferation	44	52
	mTor	1.1	Causal mechanisms that are involved in the canonical mammalian target of rapamycin (mTOR) signaling pathway, including AMPK, TSC1/2, RHEB, and RPS6KB1	44	59
	Notch	1.1	Causal mechanisms that regulate Notch signaling that controls lung cell proliferation as well as differentiation through DLK1 and JAG1/2	19	19

(continued)

Table 1
(continued)

Biological process	Network model	Version	Short description	Number of nodes	Number of edges
	Nuclear receptors	1.2	Nuclear receptor signaling pathways that lead to cell proliferation and cell cycle progression	37	43
	PGE2	1.1	Regulatory control of prostaglandin E2 (PGE2) in the context of cellular proliferation, cAMP-mediated activation of PKA-, Epac/Rap1-, and Tiam1/Vav2-dependent pathways of Rac1 activation	68	79
	Wnt	1.1	Wnt signaling primarily in the context of heightened cell proliferation in response to lung injury, multiple pathways leading to the degradation of beta-catenin	57	90
	Oxidative stress	1.2	Intracellular free radical management, cellular responses to endogenous/exogenous oxidants and antioxidants and the glutathione metabolism	337	511
	NFE2L2 signaling	1.2	Activation of the nuclear factor NFE2L2, its interaction with MAF proteins and its downstream target genes such as ATF4, GCLC, NQO1, and PRDX1	97	144
Response to cellular stress	Xenobiotic metabolism response	1.2	Response to various environmental stressors and signaling components that regulate aryl hydrocarbon receptor (AHR) and cytochrome p450 enzymes	71	126
	Hypoxic stress	1.2	Activation of the hypoxia inducible factor HIF1A and its targets, control of transcription, protein synthesis, and cross talk with oxidative stress, ER stress, and osmotic stress response pathways.	125	190
	Osmotic stress	1.1	Causal mechanisms that regulate osmotic stress including NFAT5, aquaporins, and CFTR pathways downstream of the hyperosmotic response	75	89
	Endoplasmic Reticulum Stress	1.1	Unfolded protein response and the pathways downstream of the three key stress mediators: PERK (EIF2AK3), ATF6, and IRE1alpha (ERN1).	99	143
	Endothelial Shear Stress	1.0	Effects of laminar (atheroprotective) and turbulent (atherogenic) shear stress on monocyte adhesion, including NFKB and nitric oxide pathways.	92	141

(continued)

Table 1 (continued)

Biological process	Network model	Version	Short description	Number of nodes	Number of edges
Pulmonary inflammation	Cytotoxic T-cell signaling	1.1	Causal mechanisms that are activated in CD8+ cytotoxic T-cells following T-cell receptor (TCR) ligation. The chemokines secreted by macrophages and dendritic cells, as well as the cognate T-cell receptors involved in mediating T-cell recruitment to compromised lung tissue during COPD development are included	44	46
	Th1_Th2 Signaling	1.2	Causal mechanisms that are activated in T-helper 1 (Th1) cells and in T-helper 2 (Th2) cells following engagement with macrophages and B-cells, respectively, The chemokine pathways involved in mediating T-cell recruitment to compromised lung tissue during COPD development and Th1-produced cytokines that activate cytotoxic T-cells, natural killer cells, and macrophages, and Th2-produced cytokines that activate B cells and eosinophils and suppress macrophage activation are highlighted	118	167
	Th17 signaling	1.1	Causal mechanisms that are activated in interleukin IL17-producing T-helper (Th17) cells following T-cell receptor (TCR) ligation. The chemokines secreted by macrophages and dendritic cells, as well as the cognate TCRs, involved in mediating T-cell recruitment to compromised lung tissue during COPD development are included	71	75
	Treg signaling	1.1	Causal mechanisms that are activated in CD4+ Foxp3+ regulatory T-cells (Treg) following T-cell receptor (TCR) ligation. The chemokines secreted by macrophages and dendritic cells, as well as the cognate T-cell receptors involved in mediating T-cell recruitment to compromised lung tissue during COPD development are included	46	55

(continued)

Table 1
(continued)

Biological process	Network model	Version	Short description	Number of nodes	Number of edges
	Dendritic cell signaling	1.2	Key cytokine pathways leading to the activation of dendritic cells and their migration to both lung tissue and lymph nodes. Chemokines and surface receptors expressed by dendritic cells that serve to promote T-cell migration and subsequent differentiation in the lung during COPD are included	184	241
	B-cell signaling	1.1	Causal mechanisms linking the surface receptors expressed on B-cells to a variety of downstream signaling pathways leading to B-cell activation and differentiation. Key components involved in the class switching process, which may contribute to the generation of an autoimmune component in COPD, are included	59	81
	Macrophage signaling	1.2	Pathways related to the recruitment of macrophages to sites of tissue injury and the differentiation of macrophages from hematopoietic progenitors. Cell surface receptors and signaling pathways involved in macrophage efferocytosis, a process known to be largely inhibited during COPD development, are included	222	292
	Efferocytosis	1.0	The efferocytosis model was derived from the Plaque Destabilization submodel, which was refined to focus specifically on surface receptors and key components involved in efferocytosis, and to exclude general phagocytosis-related processes	28	29
	Mast cell activation	1.1	Causal mechanisms that occur in mast cells following exposure to antigen and IgE/Fc receptor ligation and stimulation with IL4 and the KIT ligand (KITLG). Elements involved in downstream signal transductions that lead to mast cell activation and subsequent release of several cytokines and other factors are included	55	66
	Megakaryocyte differentiation	1.1	Causal mechanisms that are involved in megakaryocyte differentiation in response to upstream signals (e.g., IL11, CXCL12, and THPO)	122	158

(continued)

Table 1 (continued)

Biological process	Network model	Version	Short description	Number of nodes	Number of edges
	Neutrophil signaling	1.2	Causal mechanisms involved in various cellular functions in neutrophils, such as polarization, extravasation, respiratory burst, response to stimuli, and chemotaxis, in response to upstream signals	227	372
	NK signaling	1.1	Causal mechanisms involved in natural killer (NK) cell activation and the induction of target cell cytolysis in response to upstream signals, including interleukins, TGFB1, IFNs, and ITGB2	63	77
	Epithelial innate immune activation	1.1	Aspects of the innate immune system that are activated initially in response to cigarette smoke exposure and remain in a state of constitutive activation during COPD development. These processes include Toll-like receptor activation and respective pathways leading to subsequent proinflammatory cytokine production by lung epithelial cells	132	176
	Tissue damage	1.1	Causal mechanisms leading to the secretion of damage-associated molecular patterns (DAMPs) and pathogen associated molecular patterns (PAMPs) following tissue injury. Pathways related to the detection of DAMPs, which serve to perpetuate heightened immune responses to noninfectious tissue insult, are included	86	111
	Epithelial mucus hypersecretion	1.1	Causal mechanisms that lead to increased expression of the mucin MUC5AC and mucus hypersecretion in pulmonary epithelial cells in response to pro-inflammatory signals and environmental insults. COPD-specific signaling components such as mucins and mucociliary clearance processes are included	131	206

(continued)

Table 1
(continued)

Biological process	Network model	Version	Short description	Number of nodes	Number of edges
Cell fate	Apoptosis	1.2	Signaling pathways involved in the induction of apoptosis in response to environmental stimuli. The pathways include ER stress signaling, MAPK components that regulate the pro-apoptotic proteins, NFkB signaling, PKC-mediated apoptosis, and activation of TNFR1 and Fas receptor leading to downstream signaling of caspase-8 as well as the transcriptional activity of TP53 in apoptosis through signaling to its downstream components	270	421
	Autophagy	1.1	Induction of autophagy following pulmonary tissue damage as a result of exposure to environmental insults including the regulation of mTOR, ATG signaling, and regulation of protein translation via S6K and AMPK.	138	202
	Mechanisms of cellular senescence	1.2	Induction of oncogene-induced senescence, stress-induced senescence, and replicative senescence in response to a variety of external stimuli. Mechanisms that lead to manifestation of the senescence-associated secretory phenotype (SASP) are included	139	186
	Necroptosis	1.1	Activation of the Fas receptor and TNFR1 receptor, leading to RIPK/ROS (reactive oxygen species) signaling and the subsequent induction of necroptosis as a result of exposure to environmental insults	68	90
	Response to DNA damage	1.1	Multiple DNA damage response pathways as a result of exposure to environmental insults including components that regulate TP53, TP63, and TP73 activities, the impact of G1/S and G2/M transitions of the cell cycle downstream of DNA damage, and DNA repair processes following single-strand and double-strand DNA damage. The model describes inhibition of DNA repair by various exogenous compounds and components that function in nucleotide excision repair	253	369

(continued)

Table 1 (continued)

Biological process	Network model	Version	Short description	Number of nodes	Number of edges
Tissue repair and angiogenesis	Angiogenesis	1.1	Causal mechanisms leading to angiogenic processes of migration, proliferation, and vascular permeability as well as key angiogenic processes of sprouting and tubulogenesis. The model also contains causal mechanisms involved in the regulation of HIF1A transcriptional activity leading to angiogenesis	235	322
	Wound healing	1.1	Causal mechanisms involved in the wound healing process including cell migration as a result of matrix metalloproteinase (MMP) degradation of matrix components, and progenitor cell differentiation mediated by NOTCH1, SOX2, KRT14, and CTNNB1	193	317
	Fibrosis	1.1	Fibrosis and epithelial-to-mesenchymal cell transition (EMT) in lung tissue repair. Signaling components of the Wnt, Hedgehog, and EGFR pathways that promote fibrosis and EMT are included	178	352
	Endothelial innate immune activation	1.1	Acute response of healthy pulmonary endothelium to pro-inflammatory stimuli; production of inflammatory chemokines, adhesion molecules, and other mediators during endothelial cells activation leading to lung microvascular endothelial cells dysfunction. Angiogenic regulation by immune cells is included	133	194
	ECM degradation	1.1	Key events leading to degradation of the extracellular matrix (ECM) in the context of COPD, including the release of proteases from inflammatory cells resulting in destruction of the alveolar walls	67	125

(continued)

Table 1 (continued)

Biological process	Network model	Version	Short description	Number of nodes	Number of edges
Vascular inflammation	Endothelial cell activation	1.0	Causal mechanisms involved in cell adhesion, lipid oxidation, endothelial permeability, and proinflammatory signaling through release of inflammatory mediators	133	407
	Platelet activation	1.0	Causal mechanisms involved in the activation of platelets following exposure to damaged endothelial cell components. Mechanisms leading to platelet activation and the release of alpha granule contents are included	134	179
	Endothelial cell–monocyte interaction	1.0	Causal mechanisms involved in the activation of endothelial cells in the arterial tree at sites that are prone to atherosclerotic lesions, including the upregulation of cell adhesion molecules and chemokines that mediate the recruitment of circulating monocytes. Mechanisms of endothelial cell (EC) stimulation by low-density lipoprotein (LDL), oxidized LDL, cell-derived microparticles and leukotrienes, leading to expression of several proinflammatory chemokines and adhesion molecules are included. This further leads to interaction of EC adhesion molecules with corresponding monocytes' receptors/ligands and results in increased monocytes chemotaxis, adhesion and transmigration	86	112
	Foam cell formation	1.0	Pathways that lead to activation of transcription factors involved in the transformation of macrophages into lipid-laden foam cells	192	285
	Smooth muscle cell activation	1.0	Causal mechanisms involved in the main branches of signaling that lead to smooth muscle cell proliferation, which is a major component of smooth muscle cell activation. The low-density lipoprotein receptor-related LRP1 and LRP6 play suppressive roles in smooth muscle cell activation by inhibiting NFKB and PDGFB signaling, respectively	137	179

(continued)

Table 1 (continued)

Biological process	Network model	Version	Short description	Number of nodes	Number of edges
	Plaque destabilization	1.0	Causal mechanisms leading to plaque rupture, focused on mechanisms related to fibrous cap thinning and decreased cellularity of the plaque core; matrix degradation, inhibition of matrix synthesis by smooth muscle cells, macrophage apoptosis, defective efferocytosis leading to necrosis, and intra-plaque neovascularization induced by hypoxia leading to intra-plaque hemorrhage	246	494

6 The CBN Database

The CBN Database Web site (http://causalbionet.com/) offers an interface that accesses a Mongo database that contains different versions of the network models, ranging from the 98 original non-diseased models (version 1.0, 98) to the consolidated COPD-relevant models (1.1), and to the crowd-verified network models (1.2) (Fig. 7); i.e., the history of the networks is maintained and one may go back to previous versions of the models that may have been used for scientific publications [42]. Since a Mongo database provides native JSON storage, it works very well for application stores that are Web or mobile-based. Most of the data required for an application operation is encapsulated by one "document" or record in the Mongo database. Table 1 shows the different network models that are available in the CBN Database.

The CBN Database Web site can be viewed on any device. The Web site allows network visualization using a Web-based network visualization engine powered by D3.js (www.d3js.org). All available versions of the networks can be accessed through the Web site. Each network model is represented by a unique icon that provides a rapid visual indication of the biological processes modeled in the network.

The CBN Database will be updated and maintained regularly. The CBN Database Web site is associated with the Bionet Web site [43] that hosts the continuing verification of the networks, thereby allowing frequent updates in the CBN models. Table 2 shows the current CBN Database statistics.

All the network models in the CBN Database, including supporting evidence, can be downloaded and imported into Cytoscape

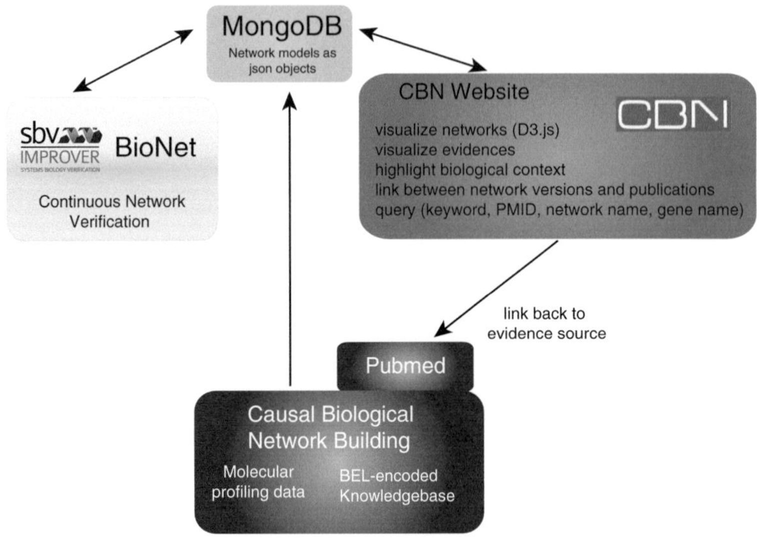

Fig. 7 CBN Database and Web site architecture. The CBN Database Web site connects to a Mongo repository, where different versions of the network models are stored as json objects. The networks can be fed from the Mongo repository into the BioNet site where the networks are subjected to crowd verification. Among its many features, by connecting back to the original scientific literature the CBN interphase allows the underlying evidences associated with each network edge to be browsed

Table 2
CBN database statistics (2015)

Number of CBN models	148
Total number of nodes	12,260
Total number of edges	17,347
Total number of evidences	491,233 (73 % Human, 19 % Mouse, 8 % Rat context)
Unique nodes	3,847
Unique edges	9,597 (65 % causal, 35 % non-causal)

for further visualization and evaluation by individual users who will also be able to submit their own BEL models to the CBN Database and have then subjected to the crowd verification process.

The CBN Database provides help and training pages and offers features such as a faceted search that allows users to search for networks by keyword, gene, protein, or small molecule name, biological process, PubMed ID, and digital object identifier (doi) of supporting evidence. The CBN Database also allows filtering of the nodes and edges and access to all pieces of supporting evidence linked to the original articles in PubMed (Fig. 8).

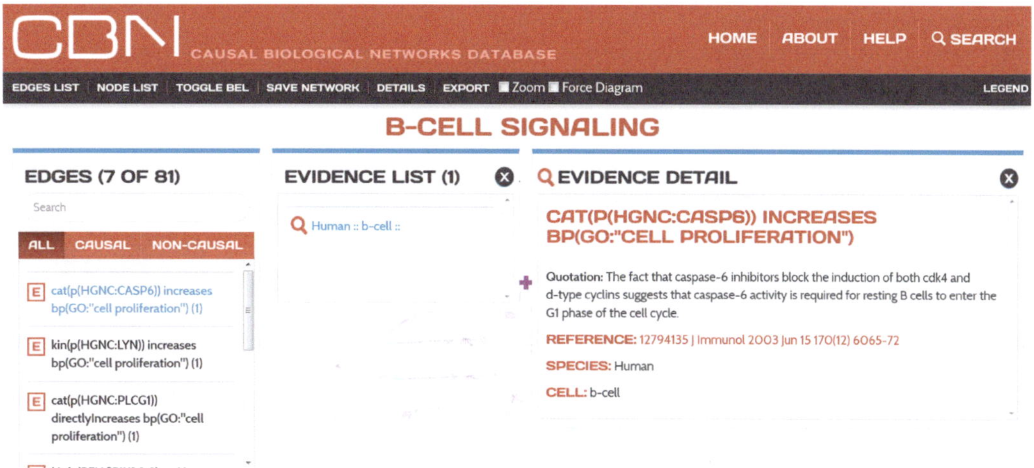

Fig. 8 Visualization of network edges in CBN models. Screenshot of the CBN Database platform in which a chosen edge with cell-type relevant evidence is shown. All the edges in the B-cell singling network model are shown on the *left*; evidences for a selected edge are shown in the *middle*; and literature reference details, including an evidence line (quotation), publication details, species and tissue/cell context, are on the *right*

7 Future Prospects

7.1 Semi-automated Statement Curation

Scientific knowledge increases exponentially, and for the CBN models to stay up-to-date and to evolve further, the knowledgebase needs to be regularly updated. While manual curation is considered the most accurate way of capturing scientific knowledge, it is extremely time consuming. Furthermore, the knowledge can be subject to human interpretation and as a result the curation results from individual scientists may not always be aligned. Text mining has been shown to be a promising addition to purely manual curation. Automated entity recognition and the BEL statement creator can help produce a high number of causal statements that may otherwise be missed by a human reader. A pipeline that involves a manual curation step has been proposed, where extracted BEL statements are checked for accuracy at the assembled knowledge level instead of at the more common entity recognition level and corrected before the final BEL document is compiled [43]. This step will also remove false positive statements, an inevitable side effect of machine-based curation. A semi-automated approach of this kind was found to result in high recall combined with high accuracy and is undoubtedly the future of network building. Text mining methods and the results are described in detail in Chap. 7. The semi-automated text mining workflow can be linked to the BEL framework, which will allow the generation of knowledge-based network models for any biological areas of interest (Fig. 9). The openBEL framework contains a small knowledgebase that can be used by scientists to begin creating their own biological networks.

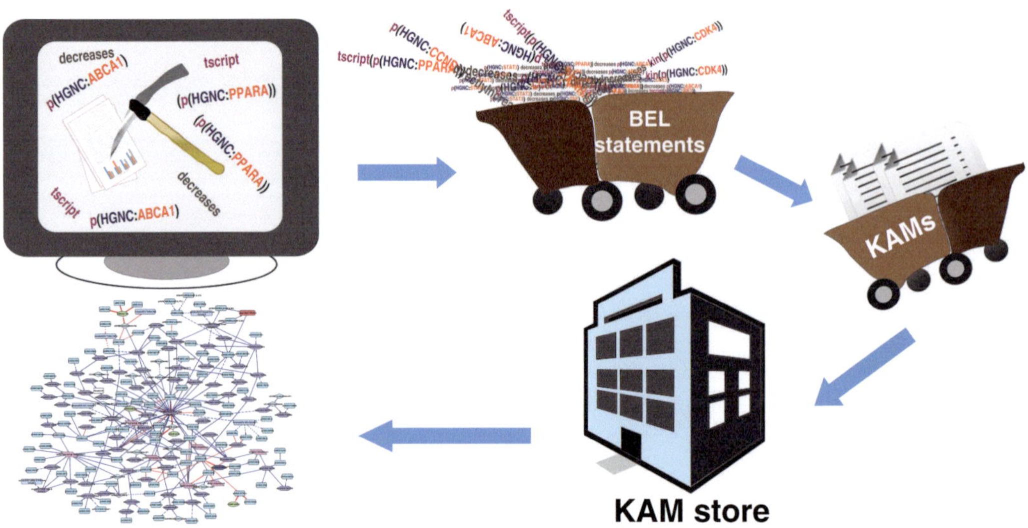

Fig. 9 Integration of automated text mining into the BEL framework. An automated text mining pipeline extracts data from the scientific literature and proposes causal statements in BEL. The automated process is followed by manual verification by a subjects matter expert. The curated BEL statements are processed using the BEL framework toolset, which includes a compiler to generate a knowledge assembly model (KAM) for each BEL document. A Cytoscape plugin allows the conversion of the BEL document format into XGMML format for visualization in Cytoscape

7.2 Integration of Additional Biological Entities

The current CBN models contain some biological entities related to epigenetic modifications; e.g., the histone deacetylase (HDAC) family and a DNA methyltransferase (DMT) family member DNMT1 [20]. The BEL community is working to further develop the language to capture various DNA bound endpoints, such as single nucleotide polymorphisms and methylated DNA loci. It would be extremely helpful to include entities that represent gene dosage effects (heterozygosity, amplification), activating and inactivating mutations, and an interface to genomics platforms. Because lipidomics studies generate knowledge on the biological relevance of specific lipid species, they have also been integrated in BEL.

In addition to gene expression entities, other high-throughput methods could be used to generate other measurable entities. For example, when causal connections are modeled from the backbone entities to changes in protein abundances in response to treatment, such data modalities could be used as the measurable entity in network scoring. Alternatively, protein data can be mapped onto the network model backbone to give insights into the potentially impacted pathways. The same argument will apply to any high-throughput measurement such as lipidomics or genomics endpoints.

As mentioned previously for the microRNA and blood modules in the COPD CBN models, a correlative measurement could be added to the models to link biological mechanisms to clinical outcomes and biomarker measurements.

7.3 Lung Tumorigenesis Network Model

Similar to the approach that was used to augment the CBN models with COPD biology, the models could be adapted for any other pulmonary or cardiovascular disease by expanding and redefining the scope and the boundaries. Because COPD shares many characteristics with lung cancer, the CBN models could serve as a basis for future efforts to model biological processes involved in cellular transformations in the human lung. The resulting models would be representative of pre-neoplastic lesions and tumor biology, the causal relationships that accurately reflect the biology that describes early (preneoplastic) lung lesions and early-stage lung tumors. A lung tumorigenesis network model would contain several processes and signaling pathways that were previously excluded from the non-disease model.

7.4 CBN Models in Biomarker Discovery

The assessment of the impact of toxic compounds by correlations between symptoms and clinical findings has very limited use in establishing biomarkers that could predict adverse outcomes at an early stage. It has been recognized that gene expression changes in an exposed tissue can reflect pathology, such as drug-induced liver injury [44, 45], ulcerative colitis [46], and formaldehyde-induced tumorigenesis in the nasal epithelium [47, 48]. Expanding from the sole use of apical endpoints, drug discovery has set out to use gene expression to predict a toxic response, to determine modes of action, and for the development of new drugs [49, 50]. Given the fact that the CBN models capture biological perturbations in a well-defined context, they will be useful for many investigators who work with high-throughput datasets derived from pulmonary or vascular tissues for either mapping the biological pathways that are impacted by a given exposure or for the comparison of results from experimental datasets to other datasets. The ability of scientists to be able to submit their own models for crowd verification and the provision of openBEL framework tools are particularly important for the development of a larger set of applications, and databases, such as the CBN, set the stage for broad scientific collaboration. In addition to its original orientation towards systems toxicology, the CBN Database can thus provide essential tools for biomarker discovery, target identification, prediction of adverse outcome, and mode of action studies, as well as patient stratification in the area of personalized medicine.

Acknowledgements

The research described in this chapter of the book was funded by Philip Morris International.

References

1. Dennis G Jr, Sherman BT, Hosack DA et al (2003) DAVID: database for annotation, visualization, and integrated discovery. Genome Biol 4:P3
2. Subramanian A, Tamayo P, Mootha VK et al (2005) Gene set enrichment analysis: a knowledge-based approach for interpreting genome-wide expression profiles. Proc Natl Acad Sci U S A 102:15545–15550
3. Joshi-Tope G, Gillespie M, Vastrik I et al (2005) Reactome: a knowledgebase of biological pathways. Nucleic Acids Res 33: D428–D432
4. Nishimura D (2001) BioCarta. Biotech software & internet report. Comput Software J Sci 2:117–120
5. Kanehisa M, Goto S (2000) KEGG: Kyoto Encyclopedia of Genes and Genomes. Nucleic Acids Res 28:27–30
6. Croft D, Mundo AF, Haw R et al (2014) The reactome pathway knowledgebase. Nucleic Acids Res 42:D472–D477
7. Kanehisa M, Goto S, Sato Y et al (2014) Data, information, knowledge and principle: back to metabolism in KEGG. Nucleic Acids Res 42:D199–D205
8. Elkon R, Vesterman R, Amit N et al (2008) SPIKE–a database, visualization and analysis tool of cellular signaling pathways. BMC Bioinformatics 9:110
9. Paz A, Brownstein Z, Ber Y et al (2010) SPIKE: a database of highly curated human signaling pathways. Nucleic Acids Res 39(Database issue):793–799
10. Li J, Ning Y, Hedley W et al (2002) The molecule pages database. Nature 420:716–717
11. Saunders B, Lyon S, Day M et al (2008) The molecule pages database. Nucleic Acids Res 36:D700–D706
12. Schaefer CF, Anthony K, Krupa S et al (2009) PID: the pathway interaction database. Nucleic Acids Res 37:D674–D679
13. Catlett NL, Bargnesi AJ, Ungerer S et al (2013) Reverse causal reasoning: applying qualitative causal knowledge to the interpretation of high-throughput data. BMC Bioinformatics 14:340
14. Hoeng J, Deehan R, Pratt D et al (2012) A network-based approach to quantifying the impact of biologically active substances. Drug Discov Today 17:413–418
15. Hoeng J, Talikka M, Martin F et al (2013) Case study: the role of mechanistic network models in systems toxicology. Drug Discov Today 19:183–192
16. Martin F, Thomson TM, Sewer A et al (2012) Assessment of network perturbation amplitude by applying high-throughput data to causal biological networks. BMC Syst Biol 6:54
17. Thomson TM, Sewer A, Martin F et al (2013) Quantitative assessment of biological impact using transcriptomic data and mechanistic network models. Toxicol Appl Pharmacol 272: 863–878
18. Martin F, Sewer A, Talikka M et al (2014) Quantification of biological network perturbations for mechanistic insight and diagnostics using two-layer causal models. BMC Bioinformatics 15:238
19. Shannon P, Markiel A, Ozier O et al (2003) Cytoscape: a software environment for integrated models of biomolecular interaction networks. Genome Res 13:2498–2504
20. Westra JW, Schlage WK, Frushour BP et al (2011) Construction of a computable cell proliferation network focused on non-diseased lung cells. BMC Syst Biol 5:105
21. Schlage WK, Westra JW, Gebel S et al (2011) A computable cellular stress network model for non-diseased pulmonary and cardiovascular tissue. BMC Syst Biol 5:168
22. Fujii-Kuriyama Y, Mimura J (2005) Molecular mechanisms of AhR functions in the regulation of cytochrome P450 genes. Biochem Biophys Res Commun 338:311–317
23. Sagredo C, Øvrebø S, Haugen A et al (2006) Quantitative analysis of benzo-a-pyrene biotransformation and adduct formation in Ahr knockout mice. Toxicol Lett 167:173–182
24. Baulig A, Garlatti M, Bonvallot V et al (2003) Involvement of reactive oxygen species in the metabolic pathways triggered by diesel exhaust particles in human airway epithelial cells. Am J Physiol Lung Cell Mol Physiol 285: L671–L679
25. Ferecatu I, Borot M-C, Bossard C et al (2010) Polycyclic aromatic hydrocarbon components contribute to the mitochondria-antiapoptotic effect of fine particulate matter on human bronchial epithelial cells via the aryl hydrocarbon receptor. Part Fibre Toxicol 7:18–32
26. Rouse RL, Murphy G, Boudreaux MJ et al (2008) Soot nanoparticles promote biotransformation, oxidative stress, and inflammation in murine lungs. Am J Respir Cell Mol Biol 39:198–207
27. Iskandar AR, Martin F, Talikka M et al (2013) Systems approaches evaluating the perturbation of xenobiotic metabolism in response to cigarette smoke exposure in nasal and bron-

chial tissues. BioMed Res Int 2013:512086, doi:10.1155/2013/512086
28. Westra JW, Schlage WK, Hengstermann A et al (2013) A modular cell-type focused inflammatory process network model for non-diseased pulmonary tissue. Bioinform Biol Insights 7:167–192
29. Gebel S, Lichtner RB, Frushour B et al (2013) Construction of a computable network model for DNA damage, autophagy, cell death, and senescence. Bioinform Biol Insights 7:97–117
30. Park J, Schlage W, Frushour B et al (2013) Construction of a computable network model of tissue repair and angiogenesis in the lung. J Clin Toxicol S12:2161-0495
31. De Leon H, Boue S, Schlage WK et al (2014) A vascular biology network model focused on inflammatory processes to investigate atherogenesis and plaque instability. J Transl Med 12:185
32. Liu T, Baek HA, Yu H et al (2011) FIZZ2/RELM-beta induction and role in pulmonary fibrosis. J Immunol 187:450–461
33. Costello CM, Howell K, Cahill E et al (2008) Lung-selective gene responses to alveolar hypoxia: potential role for the bone morphogenetic antagonist gremlin in pulmonary hypertension. Am J Physiol Lung Cell Mol Physiol 295:L272–L284
34. Belcastro V, Poussin C, Gebel S et al (2013) Systematic verification of upstream regulators of a computable cellular proliferation network model on non-diseased lung cells using a dedicated dataset. Bioinform Biol Insights 7:217
35. Fry DW, Harvey PJ, Keller PR et al (2004) Specific inhibition of cyclin-dependent kinase 4/6 by PD 0332991 and associated antitumor activity in human tumor xenografts. Mol Cancer Ther 3:1427–1438
36. Fornier M, Rathkopf D, Shah M et al (2007) Phase I dose-finding study of weekly docetaxel followed by flavopiridol for patients with advanced solid tumors. Clin Cancer Res 13: 5841–5846
37. Park WJ, Kothapalli KS, Reardon HT et al (2012) A novel FADS1 isoform potentiates FADS2-mediated production of eicosanoid precursor fatty acids. J Lipid Res 53:1502–1512
38. Sbv Improver Project Team (2013) On crowd-verification of biological networks. Bioinform Biol Insights 7:307
39. Ansari S, Binder J, Boue S, Di Fabio A, Hayes W, sbv Ipt et al (2013) On crowd-verification of biological networks. Bioinform Biol Insights 7:307–25, Pubmed Central PMCID: 3798292. Epub 2013/10/24. eng
40. sbv Improver Network Verification Challenge. https://bionet.sbvimprover.com
41. sbv Improver Project Team, Boue S, FieldsB, Hoeng J, Park J, Peitsch MC, Schlage WK, et al. (2015) Enhancement of COPD biological networks using a web-based collaboration interface.F1000Res 4
42. Boue S, Talikka M, Westra JW, Hayes W, Di Fabio A, Park J et al (2015) Causal biological network database: a comprehensive platform of causal biological network models focused on the pulmonary and vascular systems. Database 2015, PubMed
43. Szostak, J, Ansari, S, Madan S et al. (2015) Construction of biological networks from unstructured information based on a semiautomated curation workflow. Database. In press
44. Fielden MR, Brennan R, Gollub J (2007) A gene expression biomarker provides early prediction and mechanistic assessment of hepatic tumor induction by nongenotoxic chemicals. Toxicol Sci 99:90–100
45. Huang J, Shi W, Zhang J et al (2010) Genomic indicators in the blood predict drug-induced liver injury. Pharmacogenomics J 10:267–277
46. Toedter G, Li K, Sague S et al (2012) Genes associated with intestinal permeability in ulcerative colitis: changes in expression following infliximab therapy. Inflamm Bowel Dis 18: 1399–1410
47. Andersen ME, Clewell HJ 3rd, Bermudez E et al (2010) Formaldehyde: integrating dosimetry, cytotoxicity, and genomics to understand dose-dependent transitions for an endogenous compound. Toxicol Sci 118:716–731
48. Monticello TM, Swenberg JA, Gross EA et al (1996) Correlation of regional and nonlinear formaldehyde-induced nasal cancer with proliferating populations of cells. Cancer Res 56: 1012–1022
49. Hamadeh HK, Bushel PR, Jayadev S et al (2002) Gene expression analysis reveals chemical-specific profiles. Toxicol Sci 67: 219–231
50. Waring JF, Jolly RA, Ciurlionis R et al (2001) Clustering of hepatotoxins based on mechanism of toxicity using gene expression profiles. Toxicol Appl Pharmacol 175:28–42

Chapter 4

Data Sources for Signature Discovery in Toxicology

Dragana Mitic Potkrajac, Branka Rakic, Gordana Apic, and Robert B. Russell

Abstract

From a systems biology point of view, signature of a chemical can be defined as a collection of data or a measure of cellular response to a certain chemical, where biomarkers are often the best characteristics objectively measured and evaluated as an indicator of the biological processes. Chemical profiles (signatures) related to chemical-induced gene expression and gene changes coupled to particular pathologies require additional interpretation and processing to identify true biomarker candidates and together with the mechanistic hypothesis underlying biological effects of that chemical enable comprehensive knowledge about chemical toxic effect. Knowledge bases such as ToxWiz capture a broad spectrum of mechanistic hypothesis and pathways for toxic effects derived from precise expert analysis of millions of scientific articles, establishing connections between mechanism of disease, pathology, and toxic endpoints, and representing them in a form of biological pathways for underlying toxic endpoints. A unique module connected to ToxWiz knowledge base represents a hand-curated database of gene expression signatures. This module contains a good size collection of 1000 unique toxicity signatures related to chemical-induced toxicities in human, rat, mouse, and several nonmammalian species covering studies with 297 compounds, known to induce major toxicities in liver, kidney, and most other major organ systems. Implemented software tools based on systems biology principles allow analysis of novel compounds for toxic effects and allows the analysis of your own data in order to identify new biomarker candidates by interrogating your -omics data with gene signatures module. By describing biological pathways underlying the toxic effects, and discovering and exploring related biomarkers, these tools promise to help design safer chemicals. Furthermore, we describe here an example exercise on annotated public data, using ToxWiz knowledge base and tools, which confirm and expand the science on the hazards of tobacco smoke exposure via this approach and we demonstrate its respectable power to make accurate predictions of possible toxicity and generation of mechanistic hypothesis of its effect, as an indication of the cellular response in vascular endothelium upon exposure to tobacco smoke components.

Key words Systems biology, ToxWiz, Biomarkers, Molecular mechanism, Toxic endpoint, Chemical signature, Predictive algorithm, Data extraction, Manual curation, Gene expression data

1 Introduction

There is a constant concern in the past decades about general chemical safety for humans and the environment which raises a need to better understand the mechanisms underlying toxic effects

Julia Hoeng and Manuel C. Peitsch (eds.), *Computational Systems Toxicology*, Methods in Pharmacology and Toxicology,
DOI 10.1007/978-1-4939-2778-4_4, © Springer Science+Business Media New York 2015

and safety issues. Mechanisms can often identify biomarkers of toxicity, which makes prediction of toxic effects much easier. From a systems biology point of view, signature of a chemical can be defined as a collection of data or a measure of cellular response to a certain chemical, where biomarkers are often the best characteristics objectively measured and evaluated as an indicator of the biological processes. Biomarkers as a source of information about mechanism of action of chemicals, their safety, and metabolic profile are becoming indispensable in reporting the toxicological effect of a chemical on an in vitro or in vivo system, thus demonstrating practical benefit for predicting pathological events and toxic endpoints. Toxicity biomarkers have been growing in attention in the pharmaceutical and other industrial sectors when assessing the impact of new chemicals on human physiology or the environment. Regulatory requirements for drugs, agrochemicals, cosmetics, and other chemicals are changing and becoming more focused on the mechanisms related to the potential safety issues. Toxicology was not traditionally meant to explain the molecular basis of observed toxicity; little was known about the mechanism but new molecular biology data make it possible to try to improve our understanding of underlying mechanisms. Knowledge bases such as ToxWiz capture a broad spectrum of mechanistic hypothesis and pathways for toxic effects derived from precise expert analysis of millions of scientific articles, establishing connections between mechanism of disease, pathology, and toxic endpoints, and representing them in a form of biological pathways for underlying toxic endpoints. ToxWiz also contains over 2500 organ-, tissue-, and pathology-specific toxic endpoints and over four million of expert-curated literature data points to support the underlying mechanistic hypothesis. A unique module connected to ToxWiz represents a hand-curated database of gene expression signatures. This module contains a good size collection of 1000 unique toxicity signatures related to chemical-induced toxicities in human, rat, mouse, and several nonmammalian species covering studies with 297 compounds, known to induce major toxicities in liver, kidney, and most other major organ systems.

Chemical profiles (signatures) related to chemical-induced gene expression and gene changes coupled to particular pathologies, together with the mechanistic hypothesis underlying biological effects of that chemical, enable comprehensive knowledge about chemical toxic effect. Additionally, software tools based on systems biology principles allow analyses of novel compounds for toxic effects. Software tools, such as ToxWiz, provide a unique insight into toxicology and safety of novel compounds and perceiving the molecular mechanism behind toxicity (Fig. 1).

Additional tool allows you analysis of your own data in order to identify new biomarker candidates by interrogating your -omics data with gene signatures module, implemented in ToxWiz

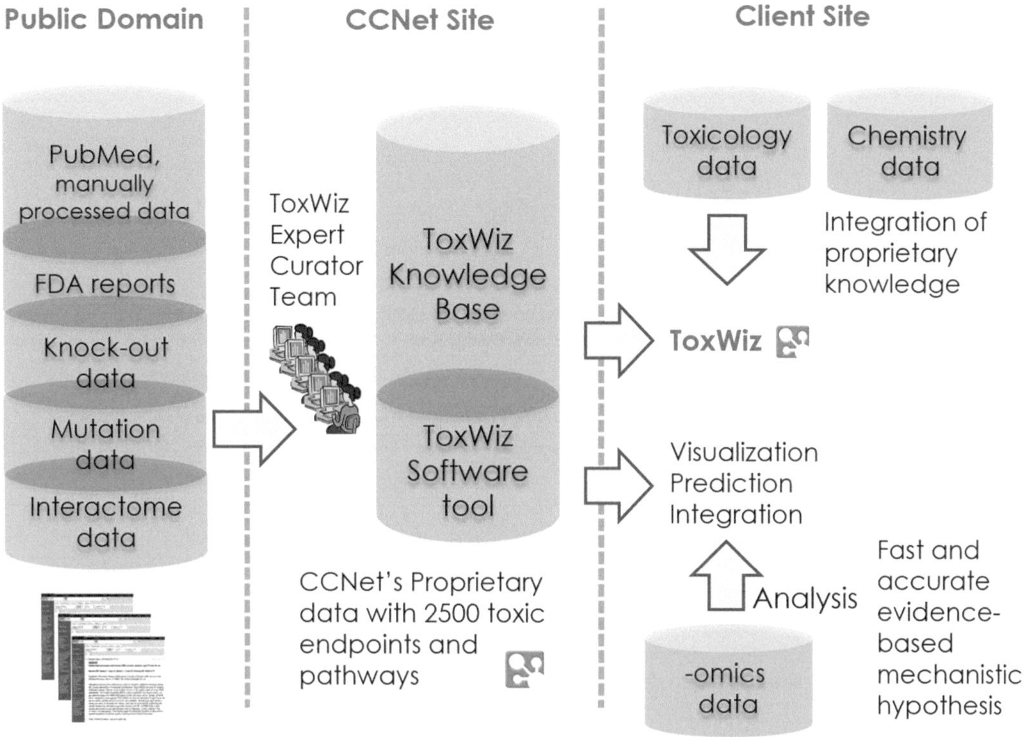

Fig. 1 The process of elucidating mechanisms of toxicity, showed from *left* to *right*. ToxWiz knowledge base gathers over 150 years of manual expert curation from over 17 million scientific articles from PubMed and process all relevant data including FDA post-marketing reports, the entire human interactome (protein–protein interaction), and knockout and mutation data. The curators use text mining software with over 500,000 ontology terms and other systems biology software tools such as chemical structure searches in order to help them identify the relevant information faster. These data are then manually processed, interpreted, and integrated via a ToxWiz software tool in a unique ToxWiz knowledge base describing over 2500 toxic endpoints with their mechanistic hypothesis and with a direct links to evidence with over four million expert-curated articles supporting the hypothesis (pathways and pathologies). On the client side, researcher can easily assess the integrated data from a ToxWiz knowledge base in order to explain the mechanisms of observed toxicity. Additionally, the ToxWiz software tool allows generating mechanistic hypothesis based on the chemical structure of the compound, as well as corresponding to the analysis of time- and dose-dependent -omics experiments and seeing them in the light of hypothesized mechanism. The systems biology principles make it possible to use ToxWiz software tool to collate client's proprietary data on chemical compounds, toxicity reports, or clinical data, in order to bring the proprietary knowledge accumulated in the client's site into predictive context

database (Fig. 2). By describing biological pathways underlying the toxic effects, and discovering and exploring related biomarkers, these tools promise to help design safer chemicals.

We will provide insights into how knowledge about biological pathways and pathway analysis tools, together with reported biomarkers, can be used in both retrospective (explaining mechanisms of observed toxicity) and prospective ways (predicting toxicity) for better safety assessment and understanding of the molecular mechanisms of toxic effects.

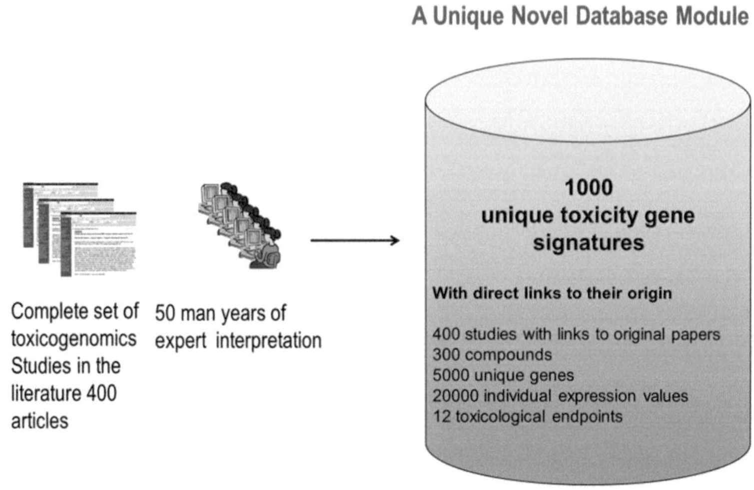

Fig. 2 Hand-curated database of gene expression signatures represents additional module connected with ToxWiz database which allows analysis of gene expression data. It contains complete set (all to date available) of gene expression signatures proposed by study authors from the literature related to chemical-induced toxicities in human, rat, mouse, and several nonmammalian species covering studies with 297 compounds in around 400 publications. The set includes more than 5000 unique genes, and nearly 20,000 individual expression values gathered into over 1000 toxicity gene signatures. These data are accessible via ToxWiz database and tools. The unique analysis tool allows you to study particular pathologies for gene or protein biomarker candidates and to analyze your own data (gene expression, RNAseq, proteomics), for overlaps with more than a thousand manually curated pathology signatures. In order to define biomarker candidate from your own gene expression dataset, a novel method for biomarker identifications was developed, which uses a combination of different scores

2 Signatures in Toxicology: Toward Understanding Mechanisms of Toxicity and Predicting Toxicities for Chemicals

There is a clear and continuous effort in discovering and evaluating the most reliable indicators of chemical exposure leading to changes in humans and our environment. Biomarkers are of growing importance in the pharmaceutical and industrial sectors for assessing cellular responses to various disease states and impact of drugs and new chemicals on human physiology or the environment. Sufficient data from confident sources can help toxicologists to rapidly employ the best efforts to reduce or eliminate the exposure and protect vulnerable population. Microarrays are considered as a powerful analysis to deduce potential sets of genes indicating a particular pathology, but require additional interpretation and processing to identify true biomarker candidates. Different "omics" biomarkers (genomic, proteomic, metabolomic) represent a collection of cellular responses to drug or chemical exposures and provide association to mechanisms of cell injury/cell death [1].

Employment of modern classification algorithms onto the gene and protein expression patterns has already demonstrated practical benefits for predicting pathological events and toxic endpoints. A multidisciplinary team of molecular toxicologist, molecular biologists, biochemists, chemists, and pharmacologists has developed a hand-curated module of gene expression signatures, implemented as part of ToxWiz knowledge base (Fig. 2), which allows analysis of these signatures. From the public domain a complete set (all to date available) of gene expression signatures proposed by study authors from the literature related to chemical-induced toxicities in human, rat, mouse, and several nonmammalian species covering studies with 297 compounds in around 400 publications compounds has been extracted. The set comprises of more than 5000 unique genes, and nearly 20,000 individual expression values gathered into over 1000 toxicity gene signatures. The tool allows you to upload and analyze your own data (e.g., gene expression, RNAseq, proteomics) to identify new biomarker candidates from overlaps with more than a thousand manually curated pathology signatures.

Knowing that similar chemical structures have a tendency to induce, bind, and be metabolized by similar macromolecules in the cell, the efforts have been brought forward to identify toxicity biomarkers, which have been a major focus in predictive toxicology. Biomarkers can represent changes in the expression of genes as measured by toxicogenomics [2], can be in the form of metabolites produced possibly due to effects of a chemical on normal cellular metabolism [3], or in some cases can be proteins indicative of particular toxic responses, such as cytokines in the case of inflammation [4]. Biomarkers can be described as a result of the overall biological response to a xenobiotic, which is nearly always the consequence of the chemical, or one of its metabolites binding to a number of proteins in the cell.

During the last 50–60 years of molecular biology research, it has been reported that many chemicals, used as tool compounds, can bind to certain receptor or cause changes in the gene expression and some of these molecules have been already identified as toxicity biomarkers. Molecular biology also developed different means to perturb a molecular target and to observe a produced phenotype (often pathology) in ways similar to a toxic compound affecting a target or other off-target proteins. Even so, not much is known and reported about mechanisms of toxicity. Mechanistic knowledge buried in the scientific literature requires expert analysis in order to be related to molecular biology and catalogued systematically in a database. In the case of the ToxWiz knowledge base, a multidisciplinary team accumulated over 150 man-years of expert literature curation associated with toxicity biomarkers, biomarker candidates, and molecules associated with toxic endpoints (Fig. 1).

A systems biology approach made it possible for the first time to assemble information on the basic set of biological pathways and to catalogue literature-supported mechanistic hypothesis for over 2500 toxic endpoints. All the manually expert-curated information was put in systems biology framework within ToxWiz tool (Fig. 3). This system now enables systematic analysis of toxic effects, -omics data, and predictions of toxic endpoints or biomarker identification.

3 Systems Biology Approach Enables Unprecedented Insights into Molecular Mechanisms of Toxicology

New technologies for biological research and their application in understanding molecular mechanism of toxicology resulted in accumulation of vast amount of complex data. New challenge for scientists lays in handling, processing, and interpreting this information. Systems biology grants a mean for not only handling and processing but more importantly interpreting these data in a meaningful manner. This approach enables a comprehensive look at hundreds or thousands of interconnected molecules and how these combine to reveal definable phenotypes, such as disease or toxic responses.

One of the approaches in systems biology is building models/databases to integrate present knowledge and to analyze new experimental datasets. ToxWiz is such a database built for understanding and analyzing molecular mechanisms. This rich database contains over four million of expert-curated data points that form a holistic "biological context network" organized in pathways and clusters. Pathways are represented as series of interactions within a cell and laid out as a network of nodes (i.e., genes, proteins, chemicals, or effects) connected by edges (i.e., the lines joining molecules or effects together). On the other hand clusters are catalogued as groups of proteins/genes and chemicals associated to toxic endpoints, disease, or compound type. One of the advantages of ToxWiz database is that it fully integrates various data about chemicals, genes, drug targets, metabolites, and pathologies describing their relationships or interactions. Moreover all these data are coupled with specific algorithms that enable prediction of biological and toxicological endpoints, ultimately suggesting underlying molecular mechanisms of action and toxicity. Exploiting such an approach and database enables analysis of collections of completely sequenced genomes and genomics information and direct comparison of gene isoforms and fast cross-species validation.

This unprecedented method in biological research focuses on understanding the connections between chemicals, genes, proteins, and toxic or biological endpoints and ultimately leads to making a more informed assessment of risk or benefit of a drug/chemical. Combining annotated and systemized data with new datasets (e.g.,

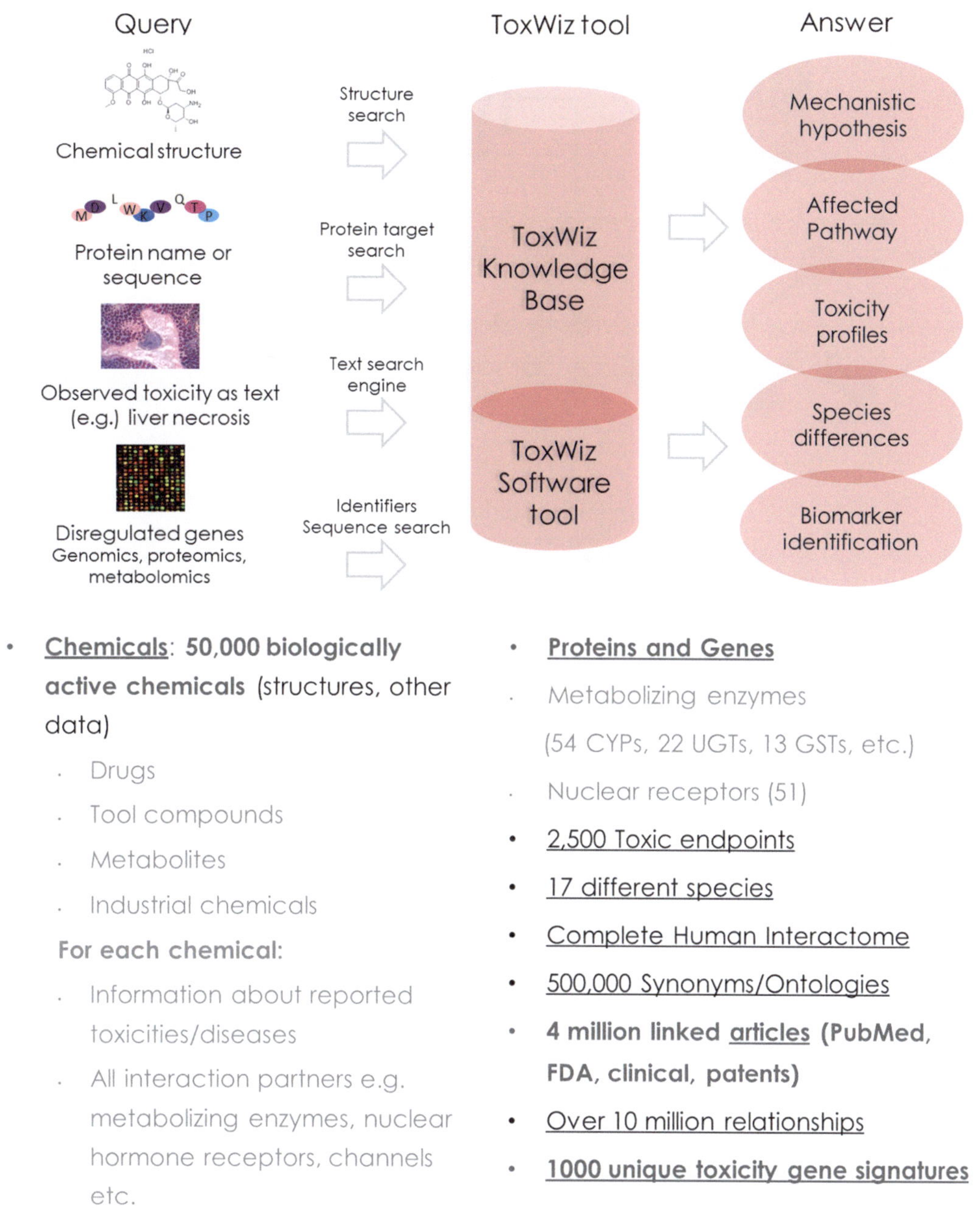

Fig. 3 Predictions of toxic endpoints and mechanistic hypothesis, together with biomarker identification, can be generated by querying with chemical structures only, or combined with text searches for toxicities, pathologies, or -omics data. The different ways of searching the ToxWiz knowledge base (structure search, text search with synonyms, identifier, or sequence search) are integrated into the ToxWiz software tools. Relevant content of the database is shown below the graph

deep sequencing, proteomics, metabolomics) enables one to make an accurate predictions of toxic endpoints, biomarkers, or other aspects of pharmacology, and possibly associated mechanistic explanations (Fig. 1).

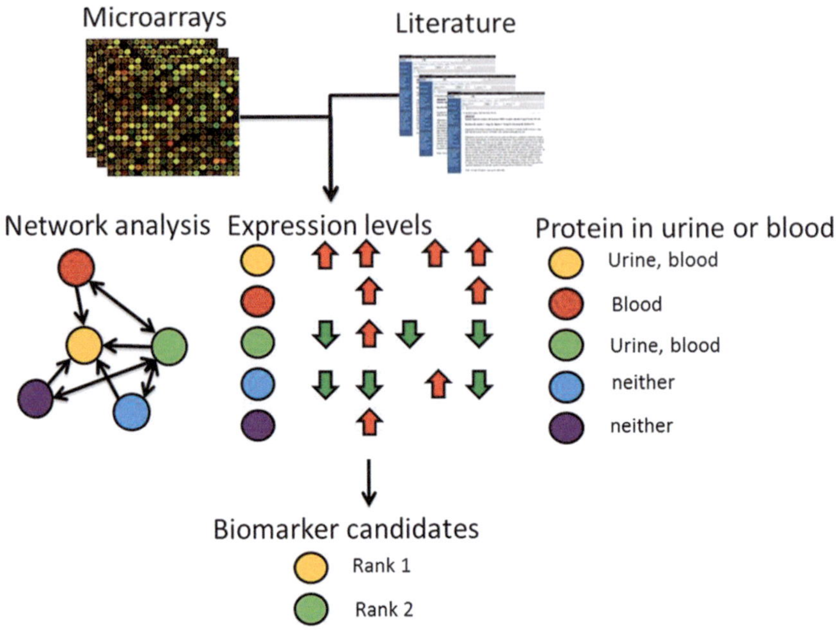

Fig. 4 Workflow for predicting biomarkers from gene expression datasets. Gene expression data from the literature related to chemical-induced toxicities were used to create module database of signatures. In order to define biomarker candidate from your own gene expression dataset, a novel method for biomarker identifications was developed, which uses a combination of scores. Biomarkers are ranked by a combination of scores: (1) presence of the datasets where biomarkers were sought (e.g., cardiotoxicity); (2) absence in control datasets (e.g., other toxicities); (3) network centrality within the dataset (by exploiting curated dataset of interactions between protein, gene, and chemicals, which considers the nature of the interaction and its confidence); and (4) previously observed presence of the protein coded by the gene in urine or blood samples [11]

The ultimate goal is to make toxicology more a predictive science than a diagnostic as it used to be. Toxicology is an ideal place to apply this kind of technology, since there is a pressing need to integrate a lot of information in order to improve efficiency.

Knowledge bases and tools such as ToxWiz allow one to rationalize toxicity findings by understanding mechanisms and their significance. Moreover these tools allow identification and prediction of biomarkers of downstream cellular effects of a chemical.

In order to define biomarker candidate from your own dataset, a novel method for biomarker identifications was developed, which uses a combination of scores (Fig. 4). Biomarkers are ranked by a combination of scores: (1) presence of the datasets where biomarkers were sought (e.g., cardiotoxicity); (2) absence in control datasets (e.g., other toxicities); (3) network centrality within the dataset (by exploiting curated dataset of interactions between protein, gene, and chemicals, which considers the nature of the interaction and its confidence); and (4) previously observed presence of the protein in urine or blood samples.

By applying this algorithm to ToxWiz database one can effectively identify potential toxicities and biomarkers. Advantage of using the database is that one have an access to a high-quality information, while the software allows querying the database from several starting points. The starting point in database querying can be one or a list of proteins, genes, chemicals, or text. Regardless the starting point as a result, one can get specific answers about the mechanism, affected pathways, and biomarkers (Fig. 3).

Another challenge is that xenobiotics can cause diverse toxicities affecting several organs which make it hard to rank predictions of toxicities by their significance and to form a mechanistic hypothesis. The ToxWiz algorithm allows one to rank possible affected pathways or toxicities. The algorithm derives a score for each predicted pathway/toxicity by considering xenobiotic directly and indirectly associated genes/proteins, type, and amount of evidence.

A case study performed with ToxWiz, with well-established tobacco smoke toxicities, is given below, and shows how these techniques are able to predict nonobvious associations that might be missed by more conventional means.

4 Cellular Response to Secondhand Smoke: Systems Biology Approach to Signature Discovery

There is a constant interest in science and public health expert groups toward understanding the underlying mechanisms of the diseases caused by tobacco smoke and the complex role of tobacco smoke constituents in the development of various diseases including cardiovascular diseases. Secondhand smoke contains around 90 known toxic compounds emitted from tobacco combustion. We focused on eight compounds labeled as harmful and potentially harmful constituents (HPHCs) by FDA (Docket No. FDA–2012–N–0143) [5] and present mainly in the sidestream smoke with sidestream-to-mainstream smoke ratio of two [6]. Those are arsenic, benzanthracene, benzene, benzo[b]fluoranthene, cobalt, hydrazine, lead, and propiolaldehyde (Fig. 5).

We addressed the four broad data sources: datasets from high-throughput analysis relevant for tobacco constituents toxicity, publicly available epidemiological studies focused on measuring risks associated with smoke, peer-reviewed published literature from both in vivo and in vitro studies, and publicly available knowledge bases comprehending notification on compounds, proteins, functional information, clinical studies, etc. We have processed and curated this data from diverse sources and integrated them into our existing knowledge base ToxWiz.

For each curated molecule, being chemical or a protein, available toxicology data were harmonized with the unique toxicology ontology so that each endpoint is presented on a cellular, tissue, and organ level. Additionally, we have identified and curated all the proteins associated with a chemical compound allowing possible investigation of mechanistic hypothesis (Fig. 6). On that way, we

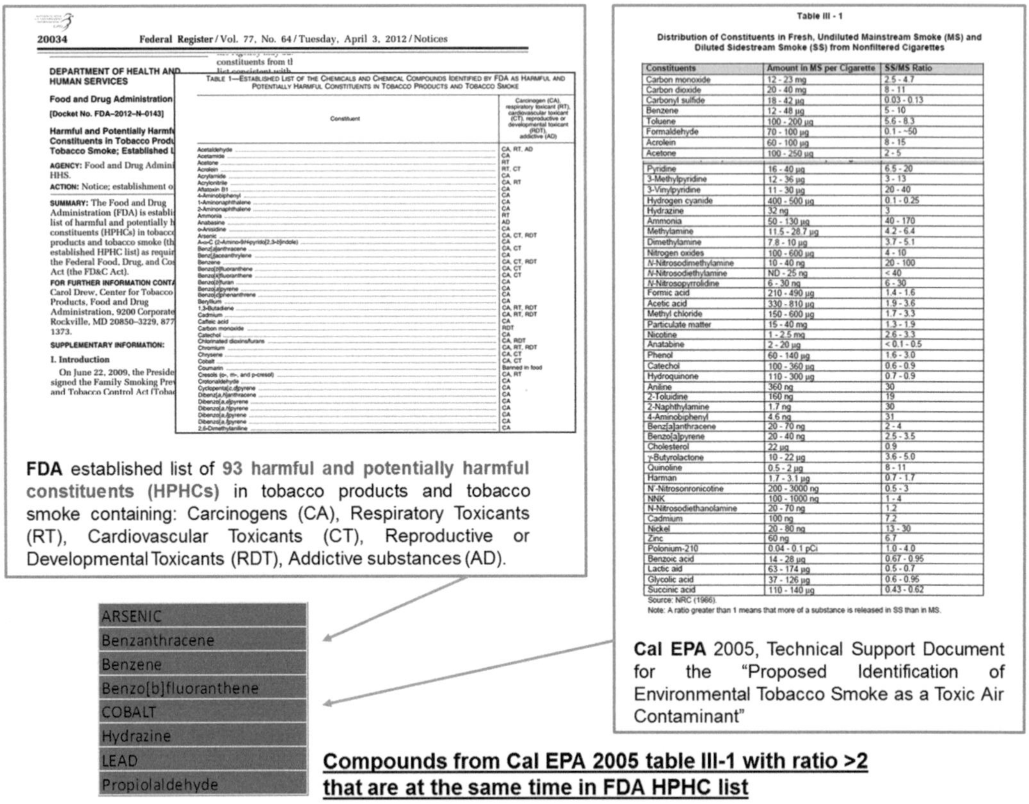

Fig. 5 Compounds from a tobacco smoke chosen to benchmark the databases and tools for cellular response. We focused on eight compounds labeled as harmful and potentially harmful constituents (HPHCs) by FDA (Docket No. FDA–2012–N–0143) present mainly in the sidestream smoke with sidestream-to-mainstream smoke ratio of two (according to Cal EPA 2005). Those were arsenic, benzanthracene, benzene, benzo[b]fluoranthene, cobalt, hydrazine, lead, and propiolaldehyde

identified a collection of molecules comprising cellular response upon exposure to these eight compounds from tobacco smoke, and were able to provide association to mechanisms of cell injury/cell death.

The chemicals in secondhand smoke poison the heart muscle, interfere with the ability of blood vessels to adjust themselves to control blood pressure and flow, increase the buildup of blockages of blood vessels (which lead to heart attacks), and make blood stickier. As a benchmark we considered eight compounds labeled as harmful and potentially harmful tobacco constituents (HPHCs) according to FDA and Cal EPA Cal 2005, and looked at the effects of these eight compounds on cardiovascular system. Our systems report a number of different cardiovascular toxic endpoints associated with the individual tobacco smoke compound, listed in Table 1.

Vast amount of processed data presents the potential for generation of valuable mechanistic hypothesis. On the example of vascular endothelial cell apoptosis, analysis of the network of interactions between the molecules (chemicals, genes, and proteins)

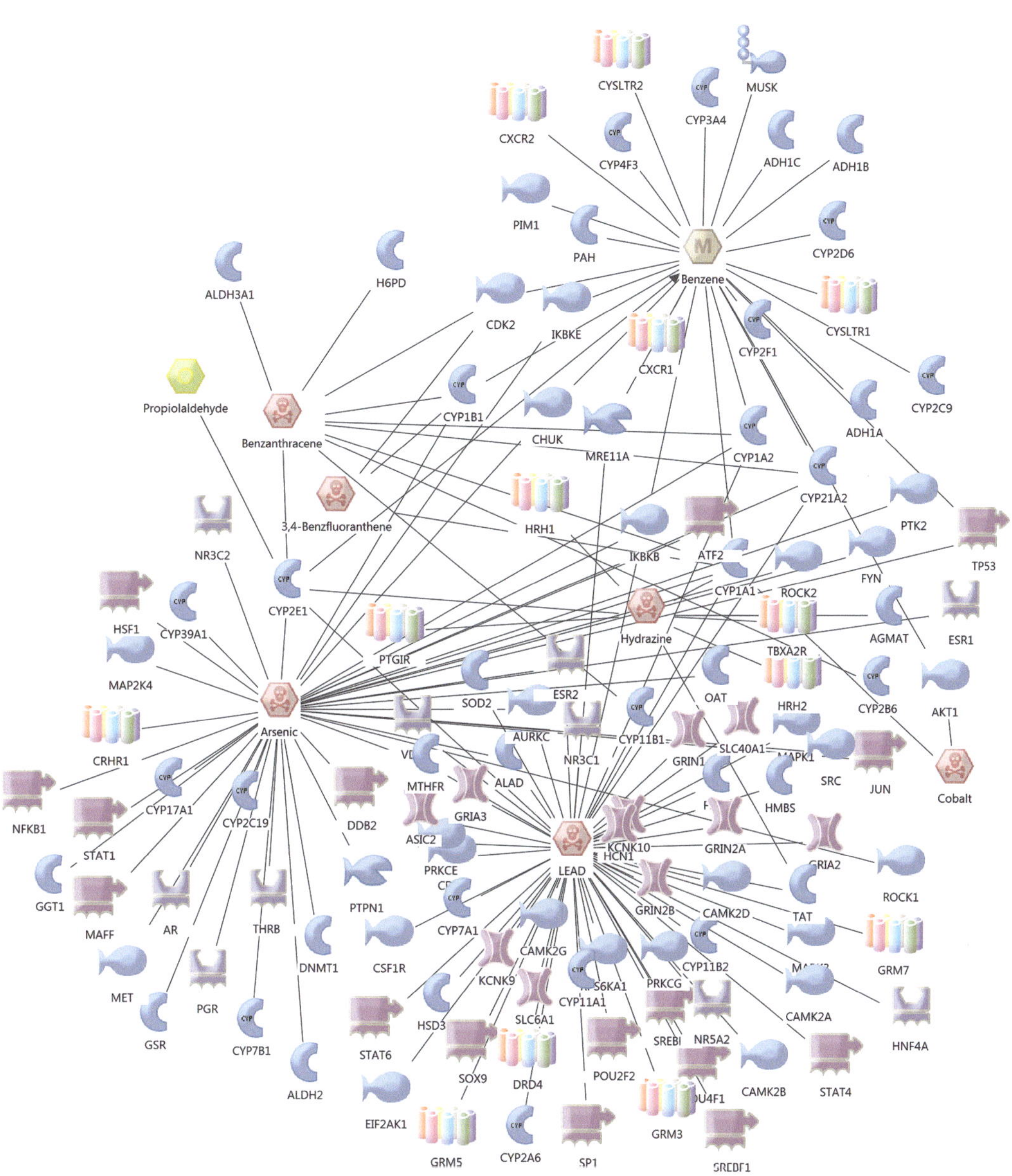

Fig. 6 Some of the downstream molecules (kinases, phosphatases, nuclear receptors, CYPs, channels, transcription factors, GPCR, receptor kinases, metabolizing enzymes, presented as shapes of different colors) as possible targets of eight compounds (*hexagon shapes*) from tobacco smoke suggested by ToxWiz, with connecting lines having expert-curated literature references linked to them, allowing fast access to literature supporting the interactions

and toxicities, diseases, and biological pathways enabled correlation of selected group of tobacco smoke components with endothelial cell toxicities, thus providing a mechanistic view of the effects (Fig. 7). The mechanisms behind these associations are mostly related to complex alternations in gene expression following exposure to tobacco smoke components.

In more details, there are three main routes of tobacco smoke components leading to the vascular endothelial cell apoptosis:

Table 1
A sample of cardiovascular endpoints reported by ToxWiz in silico tool for a selection of eight HPHC tobacco compounds, with names of endpoints in the first column, types of the system they belong to in the second one, molecules associated with a toxic endpoint in the third column, and molecules interacting or influencing a given toxicity endpoint in the last column

Toxic endpoint clusters

...	Molecular Mechanism	Type	Molecules Inside	Molecules Next To
1	Blood toxicities cluster	Circulatory system	2: NR3C1, IGF1R	73: PTGS2, NOS3, STAT1, NFKB1, CYP1A1, STAT5B, C...
2	Peripheral blood apoptosis cluster	Circulatory system	2: NR3C1, IGF1R	70: PTGS2, NOS3, CYP1A1, STAT5B, NFKB1, STAT1, C...
3	Cardiac hypertrophy association cluster	Circulatory system		69: NOS3, PTGS2, CYP1A2, CYP1A1, IGF1R, CYP2E1, ...
4	Peripheral blood toxicities cluster	Circulatory system	2: NR3C1, IGF1R	70: PTGS2, NOS3, CYP1A1, STAT5B, NFKB1, STAT1, C...
5	Peripheral blood toxicities association cluster	Circulatory system	1: NFKB1	67: PTGS2, JUN, IGF1R, NR3C1, NOS3, STAT1, STAT5...
6	Peripheral blood apoptosis association cluster	Circulatory system	1: NFKB1	67: PTGS2, JUN, IGF1R, NR3C1, NOS3, STAT1, STAT5...
7	Cardiotoxicity association cluster	Circulatory system		63: PTGS2, CYP2E1, CYP1A2, CYP21A2, CYP1A1, CYP...
8	Vascular smooth muscle toxicities cluster	Circulatory system	7: RUNX2, ATF3, NFKB1, STA...	74: CYP1A2, CYP2E1, CYP1A1, CYP21A2, STAT5B, EG...
9	Cardiotoxicity cluster	Circulatory system	9: MAPK3, CYP21A2, CYP1B1...	75: CYP2E1, CYP2B6, HPRT, IGF1R, STAT1, NFKB1, E...
10	Acute toxic toxicities association cluster	Circulatory system		41: CYP2B6, MAPK9, ATF3, CYP2E1, THRA, JUN, NR3...
11	Aorta toxicities cluster	Circulatory system	1: NOS3	74: PTGS2, IGF1R, CYP21A2, CYP1A1, CYP2E1, STAT...
12	Endothelial cell necrosis cluster	Circulatory system	4: ATF3, E2F1, NFKB1, NOS3	75: PTGS2, IGF1R, STAT1, CYP2E1, EGFR, CYP1A1, C...
13	Heart apoptosis association cluster	Circulatory system	1: NFKB1	65: PTGS2, NOS3, EGFR, JUN, IGF1R, MAPK3, NR3C1,...
14	Heart apoptosis cluster	Circulatory system	3: MAP2K5, NOS3, IGF1R	72: PTGS2, NFKB1, STAT1, CYP2E1, CYP1A2, EGFR, M...
15	Heart ventricle hypertrophy cluster	Circulatory system		69: NOS3, PTGS2, IGF1R, CYP21A2, CYP1A2, NR3C1, ...
16	Heart ventricle toxicities cluster	Circulatory system		69: NOS3, PTGS2, IGF1R, CYP21A2, CYP1A2, NR3C1, ...
17	Heart necrosis cluster	Circulatory system	1: NOS3	74: PTGS2, IGF1R, CYP2E1, NFKB1, STAT1, CYP21A2,...
18	Heart ischaemia cluster	Circulatory system		69: NOS3, PTGS2, CYP1A2, CYP2E1, CYP21A2, CYP1A...
19	Heart toxicities cluster	Circulatory system	20: LDHB, THRA, MAP2K5, LD...	62: CYP1A2, NFKB1, MAPK3, CYP21A2, CYP2E1, PTAF...
20	Vascular smooth muscle inflammation cluster	Circulatory system	5: ATF3, NFKB1, STAT1, NOS...	76: CYP2E1, STAT5B, IGF1R, CYP1A1, CYP21A2, CYP...
21	Myocardial fibrosis association cluster	Circulatory system		49: NOS3, CYP2B6, PTGS2, CYP1A2, NR3C1, JUN, IGF...
22	Artery calcification cluster	Circulatory system	1: RUNX2	71: PTGS2, NOS3, CYP2E1, IGF1R, STAT1, STAT5B, C...
23	Herg current association cluster	Circulatory system		52: CYP1A2, CYP2E1, CYP2B6, PTGS2, CYP21A2, CYP...
24	Cardiac myocyte apoptosis cluster	Circulatory system	1: MAPKAPK2	68: NOS3, PTGS2, IGF1R, STAT1, STAT5B, NFKB1, EG...
25	Artery toxicities cluster	Circulatory system	2: RUNX2, NOS3	74: PTGS2, IGF1R, CYP2E1, STAT1, CYP1A1, CYP1A2,...
26	Vascular smooth muscle hypertrophy cluster	Circulatory system	2: NOS3, IGF1R	73: PTGS2, CYP1A2, STAT1, CYP1A1, EGFR, PTAFR, ...
27	Heart hypertrophy cluster	Circulatory system	18: LDHB, THRA, MAP2K5, LD...	64: MAPK3, CYP1A2, NFKB1, CYP2E1, CYP21A2, JUN, ...
28	Endothelial cell apoptosis cluster	Circulatory system	8: MAPK3, FGFR2, E2F1, EGF...	73: NFKB1, CYP1A1, CYP2E1, STAT5B, CYP1A2, CYP2...
29	Heart apoptosis inhibition cluster	Circulatory system		49: NOS3, PTGS2, CYP1A2, JUN, IGF1R, CYP1A1, STA...
30	Vascular endothelial cell apoptosis association cluster	Circulatory system		62: NFKB1, JUN, MAPKAPK2, NOS3, PTGS2, STAT1, E...
31	Vascular smooth muscle fibrosis cluster	Circulatory system		59: PTGS2, NOS3, EGFR, MAPK3, IGF1R, STAT1, NR3...
32	Pulmonary toxicities association cluster	Circulatory system		60: NOS3, PTGS2, PTAFR, CYP1A2, CYP21A2, CYP2E1...
33	Serum fibrosis cluster	Circulatory system		61: PTGS2, NOS3, STAT1, CYP2E1, CYP1A2, STAT5B, ...
34	Cardiac myocyte apoptosis association cluster	Circulatory system		64: NFKB1, MAPK3, JUN, IGF1R, EGFR, MAPKAPK2, PT...
35	Aortic endothelium apoptosis cluster	Circulatory system		63: JUN, NR4A1, IGF1R, NOS3, CCR1, PTGS2, MAPKA...

- FAS-mediated apoptotic pathway by decreased expression of NR4A1 mRNA [7] and increased expression and phosphorylation of STAT5B proteins [8]
- Endothelial NOS3 pathway in which lead and arsenic influence increased phosphorylation and activity of NOS3 protein [9], thus leading (via TGF beta, FGF2, TNF) to endothelial cell apoptosis
- Benzene deregulation of vascular endothelial cell gene expression by downstream NFkB-regulated genes (TGFB1, MAPK14, TNF) [10].

Deregulation of these genes/proteins and affected signaling pathways they are part of indicate the overall cellular response in vascular endothelium upon exposure to tobacco smoke components.

An understanding of molecular mechanisms is not only important when developing new drugs but also for safety of chemicals in a workplace, industrial pollutants, and naturally occurring hazardous compounds found in food and drinking water. Significant progress in genomics has provided tools for the investigation and interpreta-

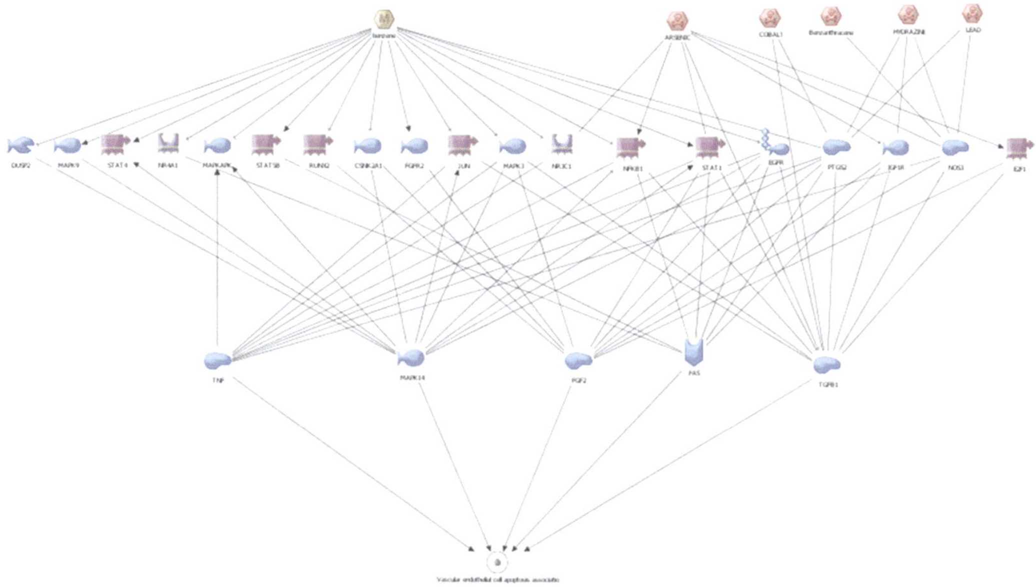

Fig. 7 Mechanistic view of vascular endothelial cell apoptosis association shows connected genes/proteins from curated data. The connecting lines have expert-curated literature references linked to them, thus allowing fast access to literature supporting the mechanistic hypothesis

tion of important biochemical events. By combining in vitro and in silico systems, we should be better able to define biomarkers of toxicity, which will in turn allow better translation of effects across species and add specificity to predictions of toxic response.

Exercise on annotated public data confirm and expand the science on the hazards of tobacco smoke exposure via this approach. We have benchmarked ToxWiz database and tools with eight compounds labeled as harmful and potentially harmful tobacco constituents (HPHCs) according to FDA and Cal EPA 2005, as shown in this study for cardiovascular toxic effects. Further, we demonstrated its respectable power to make accurate predictions of possible toxicity and to generate mechanistic hypothesis of its effect, as an indication of the cellular response in vascular endothelium upon exposure to tobacco smoke components.

5 Future Prospects

Biology is entering a period in which our knowledge of biological systems is producing new biologically based technologies. These in turn lead to new discoveries and vast amounts of data, further enhancing our need for fast processing and interpretation. This demand for more capable methodology will further develop systems biology approach and development of specialized databases and algorithms. Already now manually curated databases, such as the ToxWiz, are exponentially growing in size which is improving their reliability and value in research process, especially in the first phases of toxicity profiling. Needless to say those algorithms are following the trend of technology and knowledge development. This will provide a better understanding of mechanisms underlying safety issues reducing time and costs of research not only of a drug discovery process but also in the assurance of safety of a wide variety of chemicals such as pesticides, herbicides, and detergents, which we are surrounded with in everyday life.

6 About CCNet

Cambridge Cell Networks (CCNet) based in Cambridge, UK, supplies a range of industry-leading content on biological pathways, gene expression data, chemistry, and toxicology, combined with integrated pathway visualization and exploration tools to the pharmaceutical and biotechnology industries. Using cutting-edge biological and computational methods combined with knowledge management techniques, CCNet offers a novel approach to pathway analysis, providing effective target validation and predictive toxicology data, which will ensure the production of safer drugs. CCNet has facilities in three countries and is staffed by a team of expert biochemists, pharmacologists, bioinformaticians, biologists, chemists, and industrial toxicologists.

References

1. Fowler BA (2012) Biomarkers in toxicology and risk assessment. EXS 101:459–470. doi:10.1007/978-3-7643-8340-4_16
2. Fielden MR, Kolaja KL (2006) The state-of-the-art in predictive toxicogenomics. Curr Opin Drug Discov Dev 9(1):84–91
3. Heijne WH, Kienhuis AS, van Ommen B, Stierum RH, Groten JP (2005) Systems toxicology: applications of toxicogenomics, transcriptomics, proteomics and metabolomics in toxicology. Expert Rev Proteomics 2(5):767–780
4. Tsutsui H, Matsui K, Okamura H, Nakanishi K (2000) Pathophysiological roles of interleukin-18 in inflammatory liver diseases. Immunol Rev 174:192–209
5. Food and Drug Administration (2012) Department of Health and Human Services. Harmful and potentially harmful constituents in tobacco products and tobacco smoke; Established list (DocketNo.FDA–2012–N–0143), (http://www.fda.gov/downloads/TobaccoProducts/GuidanceComplianceRegulatoryInformation/UCM297981.pdf)
6. Cal EPA (2005) Technical support document for the "Proposed Identification of Environmental Tobacco Smoke as a Toxic Air Contaminant" (http://www.arb.ca.gov/toxics/id/summary/etspt_a.pdf)
7. Faiola B, Fuller ES, Wong VA, Recio L (2004) Gene expression profile in bone marrow and hematopoietic stem cells in mice exposed to inhaled benzene. Mutat Res 549(1-2): 195–212
8. Pontillo CA, García MA, Peña D, Cocca C, Chiappini F, Alvarez L, Kleiman de Pisarev D, Randi AS (2011) Activation of c-Src/HER1/STAT5b and HER1/ERK1/2 signaling pathways and cell migration by hexachlorobenzene in MDA-MB-231 human breast cancer cell line. Toxicol Sci 120(2):284–296. doi:10.1093/toxsci/kfq390, Epub 2010 Dec 30
9. Vaziri ND, Ding Y (2001) Effect of lead on nitric oxide synthase expression in coronary endothelial cells: role of superoxide. Hypertension 37(2):223–226
10. Joyce DE, Gelbert L, Ciaccia A, DeHoff B, Grinnell BW (2001) Gene expression profile of antithrombotic protein c defines new mechanisms modulating inflammation and apoptosis. J Biol Chem 276(14):11199–11203, Epub 2001 Feb 5
11. Cambridge Cell Networks (2013) Scientific webinar: SABIDA – New biomarker identification tool. http://camcellnet.com/wp-content/uploads/2013/11/SABIDA-Scientific-webinar.pdf, page 9. Accessed 10 Nov 2014

Chapter 5

Data Management and Processing in Toxicoinformatics: From Chemical Databases to Automatic Extraction of Unstructured Resources

Juliane Fluck and Marc Zimmermann

Abstract

Toxicoinformatics needs electronic access to all available information about toxicological substances and their influences on biological systems. Compound information such as physicochemical properties and pharmacological data is of utmost importance. In the past, mainly commercial information providers supplied access to this chemical information. Now, increasing amounts of this type of information are becoming available in public databases. A number of representation standards have been developed for the unique representation of chemical compounds, and an increasing number of databases have collected different chemical and toxicological information. Despite these efforts, most information about toxic substances and their relationships is electronically available only in unstructured text resources. This makes it necessary to extract the knowledge from those resources to make them available in an electronic readable format for further processing. Thus, in addition to an overview of chemical representation and resources, short overviews of the current status of information extraction techniques are given and a suitable network representation for compound-target interactions is described. Examples are provided to illustrate the capability of these methods.

Key words Toxicoinformatics, Chemoinformatics, Knowledge representation, Database systems, Text mining

1 Introduction

Toxicology assessment plays a major role in human health and environmental risk management. Historically, toxicity tests were based on testing high doses of chemicals in large numbers of animals. Over the last decade, animal tests have been increasingly replaced by tests on cell and on genomic studies. New programs have been started to reform risk assessment and toxicity testing methods—for example, the "*Toxicity testing in the 21st Century*" of the National Research Council [1]. Components of this vision include the use of human cell-based, high-throughput assays

Julia Hoeng and Manuel C. Peitsch (eds.), *Computational Systems Toxicology*, Methods in Pharmacology and Toxicology,
DOI 10.1007/978-1-4939-2778-4_5, © Springer Science+Business Media New York 2015

that cover the diversity of toxicity pathways, targeted animal, dose-response, and extrapolation modeling. Population-based and human exposure data have been proposed for use in the interpretation of toxicity tests.

The US Environmental Protection Agency started the ToxCast or "*toxicity forecaster*," program for large-scale testing of hundreds of chemicals against various biological assays. To analyze the data, decision support systems and predictive models are required. For example, in one of the studies in the ToxCast program, 976 chemicals (including failed pharmaceuticals, alternative plasticizers, food additives, and pesticides) were analyzed across 331 cell-free enzymatic and ligand-binding high-throughput screening assays [2, 3]. Another project is the US Food and Drug Administration (FDA) National Center for Toxicological Research that was set up to study drug-induced liver injury (DILI). DILI is one of the key areas in the FDA's Critical Path Initiative. Hepatotoxicity is the most common cause for the withdrawal of drugs from the market. The FDA has developed clinical guidelines for industry and has set up a liver toxicity knowledge base. The Agency for Toxic Substances and Disease Registry (ATSDR), based in Atlanta, Georgia, is a federal public health agency of the US Department of Health and Human Services. It provides information and fact sheets for toxic substances to prevent harmful exposures and diseases related to toxic substances.

In a cooperative research project between national institutes and private companies, Japan started the Toxicogenomics Project (TGP) to create a toxicology database. Gene expression data were generated from liver and kidney organs, or from cultured cells exposed to chemicals to identify toxicological markers. The TGP gathered data on gene expression and toxicity in rats and the primary cultured hepatocytes of rats and humans following exposure to 150 compounds [4]. In addition, it identified more than 30 biomarkers in a second phase. In Europe, regulation concerning the Registration, Evaluation, Authorization and Restriction of Chemicals (REACH) has been established. The European Chemical Agency (ECHA) is responsible for the REACH system and all chemicals imported or produced in Europe have to be registered. Substance Information Exchange Forms (SIEFs) have been introduced to deal with the same substance by different manufacturers and to simplify the registration of chemicals. Material safety data sheets are necessary for every chemical. In addition, for all substances being supplied over one ton, chemical safety reports containing risk assessments are required. These reports can be very extensive. Cadmium sulfide is one of the chemicals in the list of substances that is of very high concern for authorization and the corresponding data sheet is a PDF file of 22 pages (cf http://echa.europa.eu/documents/10162/a53b1eae-ee39-43f6-9428-668d9a381100). Another long-term European Union strategy is the Safety Evaluation Ultimately Replacing Animal Testing (SEURAT). In SEURAT-1 projects, knowledge and technology building blocks are being developed for the assessment of human safety to replace the currently repeated dose systemic toxicity testing in vivo.

All these initiatives generate new and relevant toxicity data that are provided to the public through web pages, or specified databases and web portals. Some of the information in the different specialized databases is also incorporated into larger chemical repositories. An overview of all these resources is given in Sect. 2, including relevant chemical and toxicology information. The corresponding links are summarized in Table 1 (without any claims of completeness).

Table 1
Websites and their corresponding links

Website	URL
Public programs	
US ToxCast or "toxicity forecaster" program	http://www.epa.gov/ncct/toxcast/
FDA's Critical Path Initiative	http://www.fda.gov/ScienceResearch/SpecialTopics/CriticalPathInitiative/CriticalPathOpportunitiesReports/
FDA's clinical guidelines for industry	http://www.fda.gov/downloads/Drugs/.../Guidances/UCM174090.pdf
Agency for Toxic Substances and Disease Registry (ATSDR), Atlanta, Georgia, USA	http://www.atsdr.cdc.gov/
SEURAT 1	http://www.seurat-1.eu/
Representation of chemical compounds	
IUPAC	http://www.iupac.org/home/about.html
SMILES tutorial	http://www.epa.gov/med/Prods_Pubs/smiles.htm
Mol file format description	http://infochim.u-strasbg.fr/recherche/Download/Fragmentor/MDL_SDF.pdf
HELM	http://www.openhelm.org/
Chemical and drug databases	
CAS Registry	http://www.cas.org/content/chemical-substances
REAXYS	https://www.reaxys.com/reaxys/session.do
PubChem	http://pubchem.ncbi.nlm.nih.gov/
ChEBI	www.ebi.ac.uk/chebi/init.do
ChEMBL	https://www.ebi.ac.uk/chembl/
Chemspider	http://www.chemspider.com/

(continued)

Table 1
(continued)

Website	URL
World Drug Index	http://thomsonreuters.com/en/products-services/pharma-life-sciences/life-science-research/world-drug-index.html
DrugBank	http://drugbank.ca/
KEGG	http://www.genome.jp/kegg/
TTD	http://bidd.nus.edu.sg/group/ttd/ttd.asp
HMDB	http://hmdb.ca/
STICH	http://stitch.embl.de/
Open PHACTS	http://www.openphacts.org/open-phacts-discovery-platform
Open PHACTS integration guideline	http://www.openphacts.org/specs/2013/WD-datadesc-20130912/
Toxicological databases	
ToxNet	http://toxnet.nlm.nih.gov/
CTD	http://ctdbase.org/
ACToR	http://actor.epa.gov/actor/faces/ACToRHome.jsp
ToxBank	http://toxbank.net/
Open TG-GATEs	http://toxico.nibio.go.jp/english/index.html
Toxygate	http://toxygates.nibio.go.jp/toxygates/
CEBS	http://www.niehs.nih.gov/research/resources/databases/cebs/index.cfm
LTKB	http://www.fda.gov/ScienceResearch/BioinformaticsTools/LiverToxicityKnowledgeBase/
SIGMA-ALDRICH Material Safety data	http://www.sigmaaldrich.com/safety-center.html
Chemical modeling environments	
Collector	http://phi.imim.es/collector/
OCHEM	https://ochem.eu/home/show.do
Modeling languages for systems biology	
SBML	http://sbml.org
KEGG SMBL model	http://www.systems-biology.org/001/001.html
Reactome SMBL model	http://www.reactome.org/download/
BioPax	http://www.biopax.org/
BEL	http://www.openbel.org/

(continued)

Table 1
(continued)

Website	URL
Text resources	
PubMed	http://www.ncbi.nlm.nih.gov/pubmed
PubMed toxicology subset	http://www.nlm.nih.gov/bsd/pubmed_subsets/tox_strategy.html
TOXLINE	http://toxnet.nlm.nih.gov/cgi-bin/sis/htmlgen?TOXLINE
Named entity recognition tools	
GNAT	http://gnat.sourceforge.net/
Peregrine	https://trac.nbic.nl/data-mining/
ProMiner	http://www.scai.fraunhofer.de/en/business-research-areas/bioinformatics/products/prominer.html
OSCAR	https://bitbucket.org/wwmm/oscar4/wiki/Home
Chemspot	https://www.informatik.hu-berlin.de/forschung/gebiete/wbi/ resources/chemspot/chemspot?set_language = en&cl = en
Leadmine	http://www.nextmovesoftware.com/leadmine.html
Text mining assessments and corpora	
TREC	http://trec.nist.gov/overview.html
BioCreative	http://www.biocreative.org/tasks/biocreative-iv/chemdner/
SemEval2013	http://www.cs.york.ac.uk/semeval-2013/
BioNLP shared tasks	http://2013.bionlp-st.org/
PPI benchmark set	http://informatik.hu-berlin.de/forschung/gebiete/wbi/ppi-benchmark/ppi-benchmark?set_language = en&cl = en
Text mining frameworks and tools	
UIMA	http://uima.apache.org/
IBM Watson	http://www.enterpriseirregulars.com/32391/beyond-jeopardy-with-ibm-watson--quick-analysis/
SCAIView	http://www.scaiview.com/scaiview-academia.html
GOPubMed	http://www.gopubmed.com/web/gopubmed/
TEES	https://github.com/jbjorne/TEES/wiki/TEES-2.1
UIMA-HPC	http://www.uima-hpc.de/en/about-uima-hpc.html

The website description can be found in the text
All links were last accessed in August 2014

Although structured toxicity data are generated by the initiatives described above, additional important information is provided as unstructured text and this is especially true for safety reports or risk assessments. Asbestos, for example, can be found in 83 National Priorities List sites of the Environmental Protection Agency and a PDF document of 441 pages is provided by ATSDR. Even when newly generated toxicogenomic data are available in databases, corresponding relationships and information about molecular pathways are often available only in published literature. Toxicoinformatics aims to integrate the molecular and cellular pathways that are the targets of chemical interactions to understand their influences on health or diseases. This information is necessary to identify biomarkers and significant susceptibility factors that may influence dose-response curves. Hence, comprehensive chemical–protein networks are necessary resources for toxicoinformatics. A short overview of network representation is given in Sect. 3, where BEL is introduced as a modeling language that is suitable for encoding causal toxicology information.

Most of the toxicology information required to create these networks is still communicated mainly through text in toxicology reports, risk assessments, or scientific publications. In particular, important information for new developed chemicals can usually only be found in safety reports or in patent records. Hence, methods for automatic information retrieval and extraction are needed in this area.

In Sect. 4, basic aspects of information retrieval and information extraction processes are described. These processes are not established yet in the field of toxicoinformatics. Nevertheless, without those methods it is difficult to extract valuable information from the large numbers of textual chemical and toxicology resources that are available. Two application examples that illustrate the relevance of these methods are given in Sect. 5.

2 Chemical Information Resources

In chemistry, there has been a long-standing tradition of compiling all available information about chemicals and how they are synthesized into large compendia. Examples are *Handbuch der Organischen Chemie* (Friedrich Konrad Beilstein, 1st edition 1880–1882) and *Houben-Weyl – Methoden der Organischen Chemie* (1st edition 1909–1911). The first compound databases and registration systems evolved from these early book series; however, it remains a challenge to represent, index, and retrieve chemicals from books or databases. In Sect. 2.1, different chemical representations are described followed by an overview of the main chemical databases.

2.1 Representation of Chemical Compounds

The unique identification of the molecules of interest is necessary for all computational methods that aim to integrate different data resources. Eighty years ago, the International Union of Pure and Applied Chemistry (IUPAC) developed a rule-based nomenclature system for the naming and description of small molecules. However, despite the fact that chemistry follows simple physical rules, the representation of complex molecules using IUPAC nomenclature can result in complicated regular expressions that may be ambiguous [5]. Moreover, in many publications and databases, trivial names and/or ambiguous abbreviations have been used, especially when the focus of the scientific reports is on the biological effects of drugs [6]. Furthermore, combinations of different types of names, production identifiers, as well as brand names or short names are all still in use.

The preferred representation of chemical entities is the chemical structure of the molecule. Molecules are most often represented as atoms and bonds between the atoms. This representation is a universal language of chemistry that can be understood by both humans and computers. However, some representations (e.g., depictions and drawings) can be interpreted more easily by humans while others (e.g., connection tables) are better suited to the machine [7]. Numerous tools for drawing and structure-based comparisons of molecules are available. An overview of the available techniques and tools is presented by Li et al. [8]. The most common interchange format for these structure representations is the MDL molfile format, which is a coordinate-based atom description with a connection table for the bonds [9]. The current de facto standard version is molfile V2000, although, more recently, the V3000 format has been circulated. InChI [10] and SMILES [11] have been developed to describe chemical structures using a unique textual compound identifier (also called line notations). In addition, the largest commercial chemical database CAS provides unique CAS Registry numbers (e.g., 92786-62-4 for mercury) for the whole of its chemical compound content. Although these registry numbers are proprietary and contain no structural information, they are often used for normalization and mapping.

Chuckles was developed by Siani et al. to represent peptides and peptoid sequences [12]. These authors also developed Chortles to represent oligomeric and template-based mixtures [13]. The notations in both these formats are similar to those used in SMILES, where monomer IDs are concatenated into a string with special characters to indicate branches, cycles, and mixtures.

A unique representation of macromolecules that are too complex or large to draw (e.g., molecules such as peptides, nucleotides, and antibody–drug conjugates) has been developed by the HELM (Hierarchical Editing Language for Macromolecules) project. HELM was developed as an open standard by the Pistoia Alliance, a nonprofit precompetitive alliance of life science technology partners. HELM solves the complexity problem through a hierarchical approach [14] and the notation is constructed using definitions at different levels (*Complex Polymer*, *Simple Polymer*, *Monomer*, and *Atom*) where higher level components are defined as a combination of lower level components. In HELM, small-molecule structure formats (such as SMILES) that represent monomers and Chuckles-type notations that represent simple polymers can be combined to form a complex polymer notation that can represent the connections between simple polymers. Hierarchy information is built into the HELM notation language itself, and this allows for different representation levels depending on the users' requirements. As an alternative, a molecule-like representation of macromolecules can be produced by SCSR (Self-Contained Sequence Representation) [15] that uses the enhanced v3000 molfile format in which connection tables describe large molecule structures and pseudo-atom structures. These macromolecular representations have not yet been incorporated into the current databases. It remains to be seen which of these representations becomes accepted and used in the community as standard.

2.2 Chemical and Toxicological Databases

Repositories that describe the available chemical compounds are the basis for all data processing steps in chemoinformatics and toxicoinformatics. Computational (in silico) models are used widely in biomedical research to predict the diverse biological properties of chemical compounds. The advantages of in silico methods over experimental techniques are particularly obvious in drug safety evaluation, where they are cheap, are fast, do not expend compound, and can be carried out even before the compound is synthesized chemically. In many cases, however, the predictive quality of such methods is far from perfect. Among the diverse issues that can hamper the quality of in silico models, probably the most important one is related to the availability of appropriate data to train the models. Statistically derived models require training data sets that contain large collections of structurally diverse compounds for which the biological endpoints have been measured with enough precision, using consistent experimental methods. In academic environments, this is seldom the case and in the vast majority of studies, the models were built using published data collected from the literature. The size of training data sets is limited by difficulties in extracting the data because the data sources are often of heterogeneous quality and contain data obtained using heterogeneous methods. Despite these difficulties, many databases have integrated various chemical knowledge resources into a single resource with cross-links to the information in the original databases. Other resources have gone even further by adding knowledge about compounds and their protein targets or relevant biomarkers. Here, we give a short overview of publicly available databases and projects that collect toxicological information.

The most important and largest resources for chemical compounds are the CAS Registry, and REAXYS. The CAS Registry provides information about more than 88 million organic and inorganic substances as well as over 65 million sequences. SciFinder is a research discovery application based on the CAS Registry that provides access to substances and reactions in chemistry and related sciences. REAXYS (formerly CrossFire Beilstein database) is a large repository of information about the bioactivity and physical properties, environmental fates, and reactions of more than ten million organic compounds. Besides structural information, the entities are associated with chemical and physical facts, bioactivity data, and literature references.

The PubChem, Chemical Entities of Biological Interest (ChEBI), ChEMBL, and ChemSpider databases represent sources within a broad chemical space. PubChem consists of three linked databases, PubChem Substance, PubChem Compound, and PubChem BioAssay [16], and is part of the NCBI Entrez information retrieval system (http://www.ncbi.nlm.nih.gov/). PubChem Compound contains almost 50 million entries of pure and characterized chemical compounds, and PubChem Substance contains over 130 million entries with information about mixtures, extracts, complexes, and uncharacterized substances or proteins. In PubChem Substance, often the same compound can be found more than once; for example, sertraline is recorded 69 times and the records contain different partly overlapping information from different resources. PubChem Bioassay contains several million entries with more than 6000 high-throughput screening results. ChEMBL contains bioactive molecules and is hosted and supported by the European Bioinformatics Institute (EBI) [17]. It currently contains 1,359,508 distinct compounds, with 2,419,715 activities, and 9414 targets collected from more than 50,000 publications. Another EBI resource, ChEBI, is a freely available controlled vocabulary of small molecular entities [18] that are organized in an ontological classification and are grouped by their chemical structure and functional properties.

Other more specialized databases in the area of drugs are available (e.g., the commercial World Drug Index, and the open accessible databases DrugBank and Kyoto Encyclopedia of Genes and Genomes (KEGG) Drug). The World Drug Index contains chemical and biomedical data for over 80,000 marketed and development drugs with internationally recognized drug names, synonyms, trade names, and trivial names. Each record consists of a chemical structure that is classified by drug activity, mechanism of action, treatment, manufacturer, synonyms, and medical information. DrugBank currently contains around 7600 drugs and is a specific database about pharmaceuticals, which combines detailed chemical, pharmacological, and pharmaceutical information with drug target information [19]. It provides trivial, brand, and brand mixture names, IUPAC nomenclature, and a structure for almost every entity in SMILES or InChI notations. KEGG is a composite database that integrates genomic, chemical, and systemic functional information that contains two sub-databases: KEGG COMPOUND and KEGG DRUG [20]. KEGG DRUG covers all drugs approved in the USA and in Japan. In both databases, every entry is linked to a unique chemical structure and to standard generic names that could be an IUPAC type and a trivial name. The Therapeutic Target Database (TTD) currently contains more than 20,000 drugs and more than 2300 targets with information about drug efficacy, target validation information, clinical trials, and investigative drugs [21]. The database also contains information about almost 1800 biomarkers for 300 disease conditions. KEGG COMPOUND and the Human Metabolome Database (HMDB) [22] focus on metabolic substances. The types of compounds in KEGG COMPOUND span from single ions (e.g., Mg^{2+}) to simple compounds (e.g., different sugars or cofactors of enzymes, metabolites, products of microorganisms, or nuclear receptor compounds like GW 6471), to peptides and basic RNAs, and then to all the essential endogenous molecules of cells. HMDB covers more than 40,000 metabolites.

ChemSpider is a chemistry-focused Wiki provided by the Royal Society of Chemistry (RSC). ChemSpider is a free, online chemical database that offers access to physical and chemical properties, molecular structure, spectral data, synthetic methods, safety information, and nomenclature for almost 25 million unique chemical compounds, which have been sourced from and linked to almost 400 separate data sources on the Internet [23]. Wikipedia has also become relevant for the representation of chemical information after ChemBoxes were introduced. The boxes list information on database identifiers, physicochemical properties, and hazards (e.g., the antidepressant sertraline). For example, for sertraline, the information content in Wikipedia varies in different language versions of the entry (c.f. Figs. 1 and 2).

Database resources that focus more on toxicology are provided by the National Library of Medicine (NLM) Environmental Health and Toxicology Portal, the US Environmental Protection Agency, and the European Union's SEURAT data sources.

ToxNet is an integrated database system of hazardous chemicals, toxic releases, and environmental health on the NLM Environmental Health and Toxicology Portal. One of the components of ToxNet is the Hazardous Substances Data Bank (HSDB) that provides comprehensive, peer-reviewed toxicology data for more than 5,000 chemicals. TOXLINE is the NLM toxicology reference database that is another component of ToxNet. TOXLINE provides bibliographic information that covers the biochemical, pharmacological, physiological, and toxicological effects of drugs and other chemicals. It contains over four million

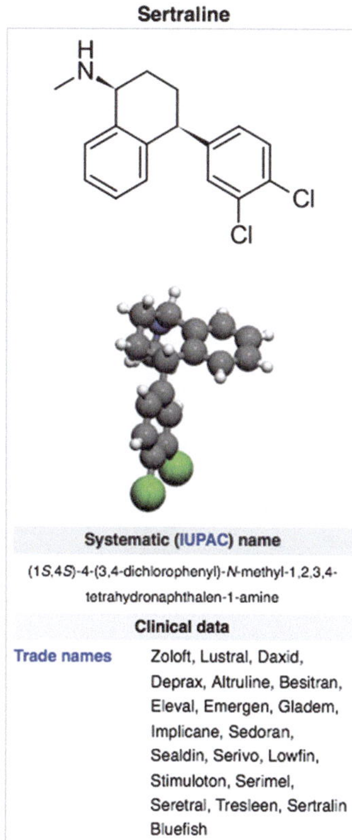

Fig. 1 The ChemBox for sertraline in the English version of Wikipedia (http://en.wikipedia.org/wiki/Sertraline), which provides the two- and three-dimensional structures, the IUPAC name, and trade names (accessed August 2014)

references, most with abstracts, indexing terms, and CAS Registry Numbers. The toxicology subset of MEDLINE/PubMed is part of TOXLINE. The Comparative Toxicogenomic database (CTD) is a resource that provides information about the mechanisms by which drugs and environmental chemicals influence the function of biological processes and human health. CTD contains chemical–gene/protein interactions, chemical–disease relationships, and gene–disease relationships [24]. In collaboration with Pfizer, a large amount of scientific literature information (88,000 articles) has been incorporated into the CTD [25]. ACToR (Aggregated Computational Toxicology Resource) and ToxRefDB (Toxicity Reference Database) are hosted by the US Environmental Protection Agency. ACToR is the Agency's online warehouse of all publicly available chemical toxicity data. ToxRefDB captures thousands of in vivo animal toxicity studies on hundreds of chemicals.

ToxBank provides a dedicated data warehouse for toxicity data management and modeling. It has been described as a "gold standard" compound database and repository of selected test compounds, and is a reference resource for cells, cell lines, and tissues of

Sicherheitshinweise	
Bitte die eingeschränkte Gültigkeit der Gefahrstoffkennzeichnung bei Arzneimitteln beachten	
GHS-Gefahrstoffkennzeichnung [3]	
Hydrochlorid	
keine GHS-Piktogramme	
H- und P-Sätze	H: *keine H-Sätze*
	P: *keine P-Sätze*
Toxikologische Daten	• 336 mg·kg^{-1} (LD_{Lo}, Maus, oral, Hydrochlorid)[4] • 500 mg·kg^{-1} (LD_{50}, Maus, oral, Hydrochlorid)[2] • 840 mg·kg^{-1} (LD_{Lo}, Ratte, oral, Hydrochlorid)[4] • 2,857 mg·kg^{-1} (TD_{Lo}, Mann, oral, Hydrochlorid)[5] • 7 mg·kg^{-1} (TD_{Lo}, Frau, oral, Hydrochlorid)[6]

Fig. 2 The ChemBox for sertraline in the German version of Wikipedia (http://de.wikipedia.org/wiki/Sertralin), which provides toxicological data (i.e., LD50 measurements) (accessed August 2014)

relevance for in vitro systemic toxicity research [26]. Currently, ToxBank contains gold standards for hepatotoxic and cardiotoxic compounds. The Open TG-GATEs website is the public website of the Japanese TGP collection of toxicogenomic data. Toxygate is an analysis platform that was developed for this database. It combines microarray data sets with semantically linked data such as pathway information or compound–protein interactions [27].

CEBS (Chemical Effects in Biological Systems) is an integrated public repository for toxicogenomics data, including the study design and timeline, clinical chemistry and histopathology findings, and microarray and proteomics data [28]. Currently CEBS holds 22 studies of rats, 4 studies of mice, and 1 study of *Caenorhabditis elegans*. It contains more than 4000 microarray hybridizations, and 75 two-dimensional gel images annotated with protein identifications that were obtained by matrix-assisted laser desorption/ionization (MALDI) and tandem mass spectrometry (MS/MS). The data derived by the ToxCast program are also included in the CEBS database.

The LTKB (Liver Toxicology Knowledge Base) provides a benchmark data set with 137 most-DILI-concern drugs that have either been withdrawn/discontinued from markets, 85 less-DILI-concern drugs for which DILI events have been highlighted either in the "Adverse Reactions" or "Warnings and Precautions" sections with mild DILI content, and 65 no-DILI-concern drugs with labels that contained no DILI indication.

STITCH and Open PHACTS are two databases that focus on the pharmacological data of chemicals. STITCH (Search Tool for Interactions of Chemicals) [29] contains a consolidated set of chemicals derived from PubChem. Relationships between the chemicals are inferred either from similar bioassay activity profiles in PubChem or from pharmacological actions assigned to the chemicals in the Medical Subject Headings (MeSH) [30] and associations between chemicals, or between proteins and chemicals, are integrated from pathway or experimental databases. In addition, text-mining methods have been integrated to further enhance the content of the chemical associations and relationships. Currently, STITCH contains interactions between 300,000 small molecules and 2.6 million proteins from 1133 organisms. Open PHACTS is a unique initiative that has brought major industry and nonprofit groups together to develop a shared platform for integration and knowledge discovery. The project aims to deliver on multiple fronts, namely enhancing the quality of relevant Resource Description Framework (RDF) data, addressing key scientific bottlenecks, developing and promoting open standards, and creating a lasting infrastructure for cross-sector collaboration [31]. At its core, the Open PHACTS platform is a data integration platform for pharmacological data (i.e., compound and target information) (Fig. 3). Collector, developed at the Parc de Salut Mar in Barcelona, allows collections of compounds that have been annotated with experimental data to be extracted from the Open PHACTS Discovery platform. These compound collections are used directly as training series for building quantitative structure–activity relationship (QSAR) predictive models. Collector implements a flexible set of filters to select compounds according to the annotated target, the biological data ranges, and the diverse physicochemical properties of the compounds.

Even in the age of semantic web and fast-growing resources of large interlinked open data (LOD) one of the largest challenges is the normalization of chemical compounds (concepts). In order to refer to a certain compound a fixed URI or registry number is needed. For that purpose, concept stores like concept wiki (http://www.conceptwiki.org/) have been established. Mapping services allow mapping between different concept stores using different

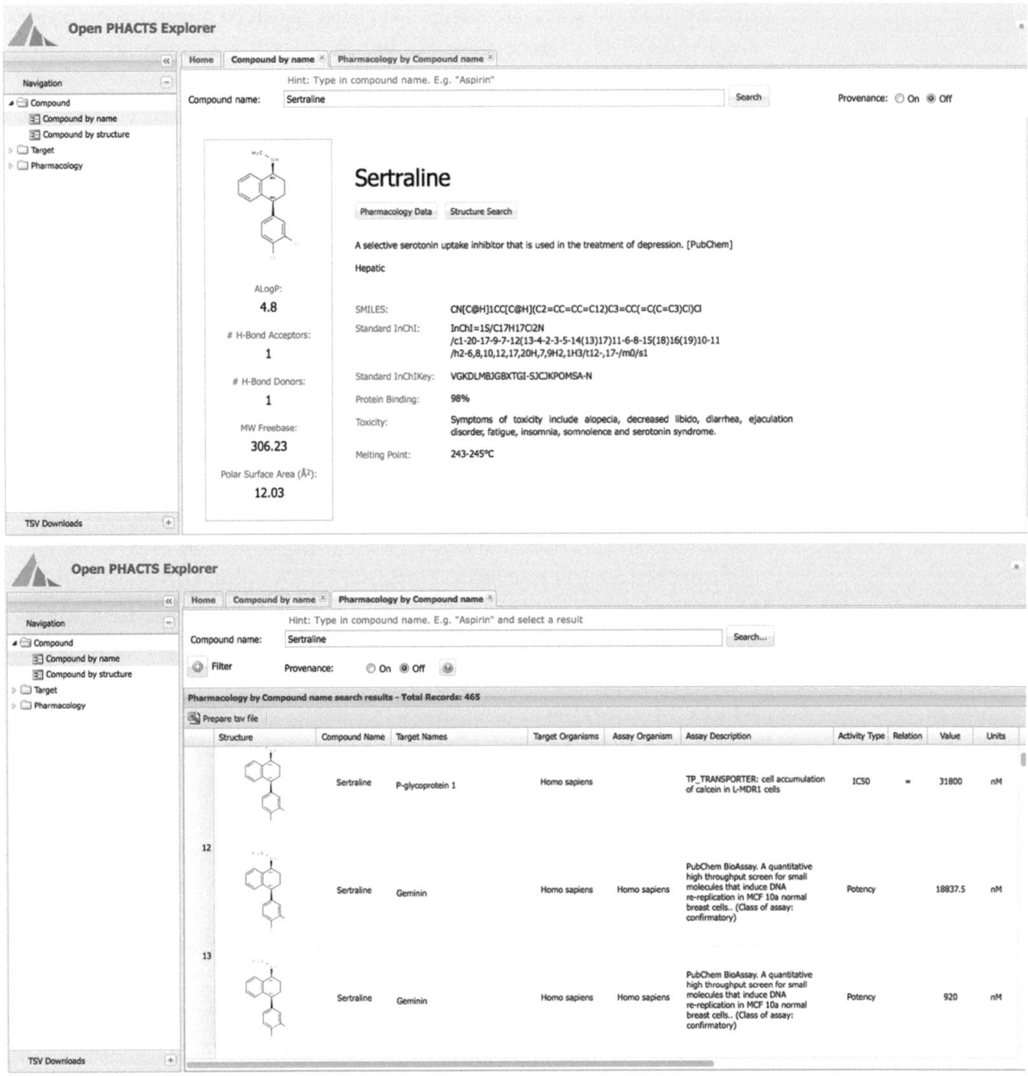

Fig. 3 Screenshot of the Open PHACTS Explorer, the compound browser of the Open PHACTS platform. The system contains 465 records of pharmacology data for sertraline (i.e., activity, selectivity, and potency measurements) for different species (accessed August 2014)

identifiers and namespaces. Depending on the level of detail, such as stereo centers, tautomeric states, protonation states, and mixtures, it is still hard to decide if two structurally similar compounds are identical or not. Depending on the level of detail needed it is important to discriminate closely related concepts or to merge the information into a broader context. Therefore, in Open PHACTS so-called lenses have been introduced to switch to different detail levels. Normalization is not only relevant for chemical concepts but also for associated properties and measurements. In Open PHACTS a lot of effort has been put into the normalization of units to make experimental data interoperable and comparable. Open PHACTS has released guidelines helping in automatic data set discovery by adding metadata to each new data set. The best way to describe the properties of the data set itself is the use of Vocabulary of Interlinked Datasets (VoID). This will allow us in the future to search for new QSAR data automatically and combine it with the already linked data.

So far, we have described a variety of online tools that are available to store information on chemical compounds (i.e., online databases). For example, DrugBank, ChemSpider, and PubChem store chemical information but no QSAR/QSPR (quantitative structure property relationship) modeling tools to predict toxicology properties have been implemented in these databases [32]. The Online Chemical Modeling Environment (OCHEM) is a Web-based platform that aims to automate and simplify the typical steps required for QSAR modeling. The platform consists of two major subsystems: a database of experimental measurements for compounds and a modeling framework (Fig. 4). OCHEM contains a set of tools for easy upload of user-contributed data, a search engine to search chemical and biological data, and the option to modify thousands of records. The OCHEM database is based on the Wiki principle and is focused primarily on the quality and verifiability of the data.

We have introduced a variety of chemical databases and tools that link chemical structures to property values, which are defined by the biological system (i.e., the effect of a compound as measured on an organism). Information about the targeted molecular and cellular pathways is missing from these resources. In the next section, common network representations in the area of systems biology are described.

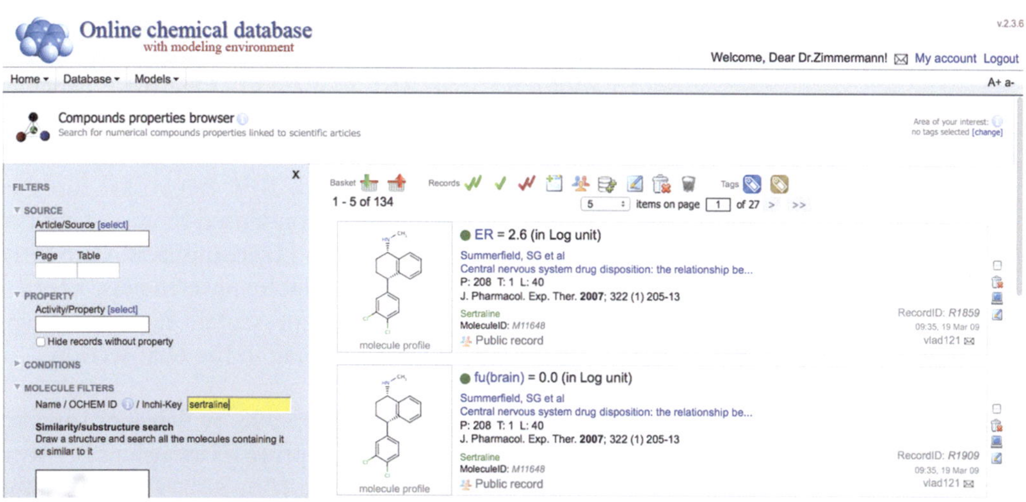

Fig. 4 Screenshot of the OCHEM platform, which combines information on compounds such as sertraline with QSAR, QSPR, and toxicology models (accessed August 2014)

3 Network Representations for Computational Processes

Currently, no formal knowledge representation (i.e., a special language or a defined data model) is available for toxicology information. Toxicology information can be represented at the molecular level in a similar way to pathways modeled in biological systems. For pathways, two machine-readable representations have been used: Biological Pathway Exchange (BioPAX) [33] and Systems Biology Markup Language (SBML) [34]. BioPAX is an RDF/OWL-based standard language that allows the integration, exchange, visualization, and analysis of biological pathway data. It is used by all the major pathway databases as the common format for data exchange. SBML is an XML-based data exchange format that supports a formal mathematical representation of chemical reactions including kinetic parameters. SBML is well established in the biomedical community for discrete mathematical modeling and many tools and packages are already available. A number of biological networks are available in the SBML and BioPax formats, for example, KEGG and Reactome, a pathway knowledge base. SBML qualitative models that allow logical and causal modeling have been developed recently [35]; however, how far these SBML models will be used for logic modeling remains to be seen. A disadvantage is that while BioPAX and SBML representations have been designed to be computer readable, for humans they are difficult to understand. As a result, network visualization and analysis tools such as database front ends and network viewers are necessary to visualize the information for humans. A disadvantage for toxicoinformatics is that toxicology information is usually based at cellular, tissue, organ, or organism levels. Hence, in many cases, a more general causal representation is necessary to model complex toxicological relationships. The OpenBEL framework, which uses the Biological Expression Language (BEL), provides such a representation. In addition to the BEL framework, OpenBEL provides a BEL-to-RDF converter that allows the storage facilities and reasoning methods developed for the semantic web (e.g., triple stores and SPARQL queries) to be used with BEL networks. Another advantage is that BEL provides a very condensed representation of the data that can be read by humans and by computers. Currently, not many toxicology networks are available; nevertheless, here we will consider BEL as a format for the representation of toxicological networks and describe it in more detail in the next section.

3.1 BEL as a Suitable Representation for Toxicological Networks

BEL was designed and used initially in 2003 by Selventa (operating as Genstruct® Inc. at the time) to capture relationships between biological entities in scientific literature [36, 37]. BEL allows the representation of causal relationships across a wide range of mechanistic details and between different levels of molecular events, cellular processes, and organism-scale phenotypes. It is flexible enough to store toxicological information from multiple knowledge layers and a broad range of analytical and decision-supporting applications. Discrete toxicology information and the relevant contextual information can be represented as qualitative causal relationships that can drive knowledge-based analytics.

BEL has been used in more than 80 commercial pharmaceutical and life science projects but is very new to the academic community. In 2012, BEL was released to the public as OpenBEL. The OpenBEL portal contains information about BEL, its formatted content, as well as compatible tools for research, and has an active community for the expansion of BEL to different use case scenarios. Currently, the various networks that were built in BEL have been focused mainly on disease mechanisms [38, 39]. The BEL framework exploits the BEL networks for hypothesis generation and causal reasoning by combining network data with large experimental data [40, 41].

BEL represents scientific findings by capturing causal and correlative relationships in a given context, including information about the biological system and experimental conditions, and supporting evidence is captured and linked to the publication references. BEL is designed specifically to adopt external vocabularies and ontologies, and therefore can represent life science knowledge in a formal language and schema that is already known by the community. Entities in BEL statements are mapped to accepted predefined namespaces that specify a set of domain entities (e.g., CHEBI). An overview of currently used namespaces is available on the OpenBEL portal. BEL defines semantic triples (BEL statements) that are stored in structured human-readable BEL document files. A BEL statement is defined as a subject–predicate–object triple, where the subject is always a BEL term, the object can be either a BEL term or a BEL statement (recursive nature of BEL), and the predicate is one of the BEL relationship types. A BEL term is composed of a BEL function, a corresponding entity, and a referencing namespace. The two main classes of BEL terms define the abundance of an entity (e.g., chemical compound) and a biological process (e.g., disease). Optionally, statements can be enriched by context information annotations including evidence sentences, tissue type, species, or cell line. Two annotation types are reserved, i.e., citation and evidence. Evidence should hold the exact evidence sentence, while Citation is the source of this evidence. A complete definition of BEL is available in the BEL Language documentation and an example statement coded in BEL together with the relevant context information is shown in Fig. 5. Namespaces and annotations are defined at the beginning of the document followed by the citation information. In this example, two BEL statements together with the corresponding evidence sentences and context information are shown. The first piece of evidence describes a relationship

```
DEFINE NAMESPACE CHEBI AS URL
http://resource.belframework.org/belframework/1.0/namespace/chebi-names.belns
DEFINE NAMESPACE HGNC AS URL
http://resource.belframework.org/belframework/1.0/namespace/hgnc-approved-symbols.belns
DEFINE NAMESPACE MESHBP AS URL
http://resource.belframework.org/belframework/1.0/namespace/mesh-biological-processes.belns
DEFINE ANNOTATION Cell AS  URL
 http://resource.belframework.org/belframework/1.0/annotation/mesh-cell.belanno
DEFINE ANNOTATION BodyRegion AS  URL
http://resource.belframework.org/belframework/1.0/annotation/mesh-body-region.belanno
DEFINE ANNOTATION Species AS  URL
http://resource.belframework.org/belframework/1.0/annotation/species-taxonomy-id.belanno


Set Citation {"PubMed","Sertraline, an antidepressant, induces apoptosis in hepatic cells through
the mitogen-activated protein kinase pathway.","Toxicological sciences Feb;137(2):404-
15",“Chen, Si|Xuan, Jiekun|Wan, Liqing|Lin, Haixia|Couch, Letha|Mei, Nan|Dobrovolsky, Vasily
N|Guo, Lei",“24194395"}

Set Evidence „Sertraline, an antidepressant, induces apoptosis in hepatic cells through the
mitogen-activated protein kinase pathway.“
SET Body region = „Hepatocytes“
a(CHEBI:sertraline) increase bp(MESHPP:Apoptosis)

Set evidence „ In the current study, we investigated and characterized molecular mechanisms of
sertraline toxicity in human hepatoma HepG2 cells. Sertraline activated the intrinsic checkpoint
protein caspase-9.“
Set Species = Homo sapiens
Set Cell = Hep G2 Cells
a(CHEBI:sertraline) increase p(HGNC:CASP9)
```

Fig. 5 BEL document with namespace and annotation definitions and two BEL statements with corresponding evidence sentences and annotations

between a chemical substance sertraline and the biological process apoptosis in the context of hepatocytes. The second piece of evidence is a relationship between two molecular entities, sertraline and the human protein CASP9, in the context of the species human and the cell line Hep G2.

As this example shows, BEL is not only a computer representation of causal networks but is also an intuitive language for its human users. BEL is aligned well to the way information is communicated in publications. The condensed representation of BEL statements and their human and machine readability are great advantages that make it easy for curators to convert published literature knowledge into structured BEL statements. A number of BEL networks that encode the effect of chemical agents on biological networks are already available; however, networks that encode toxicity information are still rare, although this is changing because the toxicological information in structured databases is growing constantly. As can be seen from the examples above, most information in these databases is still hidden in free text forms or is attached to them as unstructured text documents. Manually extracting and structuring this information will exceed the capacity of all toxicology programs. Clearly, automatic methods are needed to support information retrieval and extraction. Some such methods and their applications are summarized in Sect. 4.

4 Text Mining: Information Retrieval and Information Extraction Technologies

The electronic availability of toxicological publications has increased rapidly in recent years. Currently, Toxline contains over four million references, and other large resources such as material safety data sheets are now available online on supplier websites (such as SiGMA-ALRICH material safety data sheets website, cf link Table 1). Alone, these resources do not necessarily lead to a better access to this information and methods to automatically support data mining are required to help:

- Retrieve relevant documents
- Discriminate between relevant and irrelevant information
- Extract information from unstructured text in a human- and computer-readable structured format

The BioNLP community is an active biological research group that has developed a number of information retrieval and information extraction methods for use in the life sciences. In particular, a number of open assessments have provided training and test data that have driven tool development. Some recent assessments have been focused on the chemistry domain. Here, a short overview of the main text mining techniques and relevant assessments in the area of chemistry is given.

4.1 Information Retrieval

Information retrieval involves the retrieval of documents or information from a large document collection that is relevant to a user's query [42]. In the simplest case, the query can be a simple text search. In more advanced cases, the query can be keywords or semantic searches, where keywords or hierarchical ordered vocabularies are used to index every document in the collection. PubMed is a biomedical search engine that combines a simple text search with a semantic search through MeSH annotations that are linked to every abstract. The PubMed retrieval system works well, but is restricted by a limited annotated vocabulary and a delay in the manual annotation of new abstracts. Semantic search engines that use automatic annotation of terminology (i.e., named entity recognition (NER)) can overcome this restriction. SCAIView and GOPubMed in the biomedical domain support NER searches. For example, SCAIVIEW provides document visualization along with overviews of the found entities together with links to respective databases.

These search systems can be extended to question answering tools, where users can pose natural language questions to a system. A very popular example is IBM Watson, which won the Jeopardy quiz show, and IBM now collaborates with several health units and universities to adapt Watson to the clinical domain. The Text REtrieval Conference (TREC) is an academic platform that was set up to encourage research on information retrieval from large test collections. In chemistry, the TRECChem track between 2009 and 2011 provided more than one million full-text patent documents and an additional article collection of more than 30 different chemistry-related full-text articles for different retrieval tasks [43]. In the Technical Survey task, chemistry-related questions were asked and a ranked list of relevant documents was given back to the users. For this task, the terms in the question had to be semantically expanded to retrieve relevant documents. A number of different ranking functions have been developed to order the retrieved documents. An example of such a ranking function is the term frequency-inverse document frequency (TF-IDF), which is a term weighting scheme that is used often for document indexing and retrieval purposes [44]. TF-IDF is defined as a statistical measure of how frequently a term is found in a general versus a selected document collection.

Other relevant information retrieval tasks are based on similarity comparisons of text or whole documents and are mainly applicable to the clustering of large text collections or the categorization of a collection into predefined classes. In TRECCHEM, these techniques were applied to solve the Prior Art Search task in which a set of text patents were given and the task was to retrieve sets of documents that could potentially invalidate the given test patents. TRECCHEM employed a retrieval approach that used term annotations combined with a consideration of co-citations that produced a remarkable performance in extracting relevant articles [45].

4.2 Named Entity Recognition Systems

The recognition of relevant terminology is one of the most important processes in information extraction. As mentioned above, the annotation of relevant concepts in a text has been used in search engines to retrieve documents of interest and to highlight the terminology that is found. Moreover, the annotation of terminology can improve categorization and clustering. NER is a necessary preprocess for relationship extraction in most established systems.

In the biomedical domain, NER systems have been developed for a number of biological and medical entities and different systems are available for the classification of different terminology classes such as genes, chemicals, and diseases. Another application of NER is the normalization of different synonyms to well-defined concepts, for example, assigning chemical names to a chemical reference structure or CAS-RN. For curators in particular, the automatic normalization of gene names to their corresponding sequence database entries can save time in the manual extraction of information from the literature [46], and normalization is essential for the automated extraction of molecular toxicology networks from text.

A number of critical assessments of text mining have been established over recent years, and these have provided a common platform for the comparison and exchange of methods and the standardization of input and output formats. The recognition of gene and protein names has reached F-measures of around 0.87 for gene mention recognition [47] and 0.81 for recognition and normalization of human genes and proteins [48]. The F-measure described the relevance of an extraction or retrieval result (1.0 indicates a perfect answer).

A number of open-source and commercial NER normalization tools are available. GNAT [49] and Peregrine [50] are open-source gene and protein NER normalization tools that have been found to perform well. ProMiner [51] is a commercial entity recognition system that performed very well and it is now being used by curators of databases such as the Mouse Genome informatics (MGI) database [52]. The usability of these systems depends on their recognition performance, the availability of the tool either as open source or as a commercial version, their user friendliness, and the availability of regular dictionary updates, error curation, and user general support.

In the last few years, text mining has been expanded to include the chemical domain. Chemical names can be even more challenging than gene and protein names because they span from single elements or ions (e.g., Ca^{2+} and Mg^{2+}) via the plethora of metabolites to chemical compounds and toxins. Currently, the most used open-source tool is OSCAR [53], which consists of methods for the recognition of chemical names and other chemical-related terms. Similarly, Chemspot uses a combined machine learning and dictionary approach for the recognition of chemical names [54]. Another example is a tagger, which is focused on the recognition of IUPAC names [55].

In 2013, the BioCreative team initiated BioCreative IV together with the CTD database curators to support the curation workflow at CTD. The CTD has already implemented a text-mining pipeline to help rank and prioritize the articles to be curated [56]. Two tasks have been set up to further support the CTD curators: the ChemNER (CEM) task and the chemical document indexing (CDI) task. The CEM task focused on the recognition of chemical compounds [57]. The manual annotated training corpus consists of 10,000 abstracts from the medical chemistry, biochemistry, organic chemistry, and toxicology subdisciplines. It contains annotations for IUPAC names, and trivial or family names. The performance of CEM task systems was similar to the gene mention recognition with the best systems reaching F-scores up to 0.87 for the chemical mention recognition and up to 0.88 for the indexing task. Although most of the successful methods used combinations of machine learning methods such as [58], the commercial tool LeadMine performed very well with a combination of grammar rules and dictionaries [59]. In these tasks, no normalization was evaluated, and only the recognition of chemical names was evaluated. It may be easier to close the gap to normalized entities with the chemical names already recognized, but speculation will remain until training corpora with normalized chemical names are publicly available. In the CDI task, participants were asked to provide web services to recognize gene/protein, chemical, disease, and chemical/gene-specific action term mentions, each within the context of the CTD controlled vocabulary structure. The web services could be integrated directly into the CTD text mining workflow to further enhance document ranking. The CTD aims to collaborate with participants who have developed high-performing tools to further extend the impact of the curation workflow.

The support of database providers such as CTD via the definition of tasks and the annotation of training corpora in open assessments is an important way to accelerate method development. These assessments lead to transparent comparisons of the different approaches, fast uptake of successful methods, and enhancement of the availability of tools. Yet it remains to be seen how many of these tools will become publicly available for the integration into new workflows.

4.3 Relationship Extraction

In the life sciences, most of the available applications for relationship extraction are focused on protein–protein relationships. This focus is due mainly to the high interest in biomedical information but is also due to the availability of training corpora. The simplest way to annotate and classify a relationship is a sentence-based pairwise annotation of concepts. If more than one entity exists in a sentence, all binary relationships between the occurring entities are annotated (Fig. 6). The direction of the relationship as well as the relationship type are not annotated. Five different small corpora have been published for training and testing protein–protein interactions. Despite the different sizes and application focus of these corpora, they have been combined to provide a larger training base [60]. A number of binary classification methods spanning co-occurrence and pattern-based or machine learning methods have been developed. Machine learning approaches have been shown to perform well in the biomedical domain,

Sertraline treatment resulted in up-regulation of brain BDNF, phospho-ERK and Bcl-2 that may be involved in the procognitive effect of sertraline.

Binary relation information:

Sertraline	-	BDNF	Yes
Sertraline	-	ERK	Yes
Sertraline	-	Bcl-2	Yes
BDNF	-	ERK	No
BDNF	-	Bcl-2	No
ERK	-	Bcl-2	No

BioNLP relation information

T1	Protein	Sertraline		
T2	Protein	BDNF		
T3	Protein	ERK		
T4	Protein	Bcl-2		
T5:	Positive_regulation	up-regulation		
E1	Positive_regulation: T5		Theme:T2	Cause:T1
E1:	Positive_regulation: T5		Theme:T3	Cause:T1
E1:	Positive_regulation: T5		Theme:T4	Cause:T1

Fig. 6 Example of relationship extraction from a textual statement about the effect of sertraline on gene regulation within the brain. The relationships can be coded either as interaction pairs or in the treelike BioNLP format

although their performances differ between the different corpora. For example, the application implemented by Tikk et al. [61] produced F-scores from 0.42 to 0.66, depending on the corpus that was used for training.

In 2011, a similar classification task was started for drug–drug interactions (DDI) and it was repeated in 2013 as part of the SemEval assessment. The DDIExtraction 2013 task relied on a DDI corpus that consisted of 341 documents from the DrugBank database and from Medline abstracts [62]. In addition to the DDI relationship classification, four different types of relationship classes had to be found (advice, effect, mechanism, interaction). Overall, the relationship classification was quite high with F-scores between 0.70 and 0.80 for the DrugBank set, but only 0.35 to 0.42 for the Medline set. The further classification into the four subtypes led to decrease in performance of around 15 %.

Opposed to this simple classification (Fig. 6) are the more fine-grained annotations in the training corpora for the BioNLP shared tasks. These annotations are closer to the needs of the user but more effort is needed to set up proper guidelines and for consistent pre-annotation. The BioNLP shared task started in 2009 and took place for the third time in 2013. Detailed information about the different tasks and the results can be found at the BioNLP website. The training corpus of the Genia (GE) task was built on the GENIA corpus by adding fine-grained information for relationship extraction that was close to the needs of systems biology [63]. This BioNLP GE task addressed protein–protein regulation relationships as well as protein modification events. The top-scoring systems achieved F-scores between 0.50 and 0.56 for the extraction of simple events from a combination of abstracts and full text. When the focus was on regulation event extraction from full text, the F-scores drop to 0.38–0.40, again showing the complexity of the relationship extraction task.

In its first years, the BioNLP shared tasks focused on protein–protein relationships but, in 2013, a task for chemical and protein relationships, the pathway curation task, was provided. This task specifically addressed metabolic reactions and chemical compound relationships. The event representation was sufficiently expressive to be able to extract the pathway information necessary for systems biology representation languages such as SBML or BioPax. Although more than 30 different teams participated in the GE task, only two teams developed solutions for the new and more complex task. Nevertheless, for both these teams, the overall results were similar to the results for the GE tasks with overall F-scores above 0.50 and F-scores for regulation events below 0.40. However, until now, none of these methods is ready to deliver fully automatically extracted high-quality networks. It is our hope that more systems that can support and accelerate the generation of structured information from literature will become available. In other text mining areas, experience has shown that the enhancement of training data can also lead very quickly to improvements in the methods.

In this section, the many different tools that have been developed for different extraction tasks have been described. For the more complex tasks, workflows with many components have to be deployed. In the next section, a framework for the assembly of complex systems is described.

4.4 The UIMA Text Mining Framework

UIMA is one of the most commonly used frameworks for the integration of information extraction software. UIMA was developed (and is still used) by IBM (e.g., in the Watson project), and the Apache Software Foundation currently maintains it as an open-source project. The Apache license model (http://www.apache.org/licenses/) allows many commercial applications to be developed on top of the provided framework. UIMA is designed for the analysis of large volumes of unstructured information.

The simplest concept of a UIMA component is the *Analysis Engine* (AE), which is a program that can analyze so-called *artifacts* (e.g., documents) and infer information from them, thereby creating additional data (metadata) about the artifact. It is a goal of UIMA that AEs need not be concerned with anything other than their analysis logic, for example, the details of their deployment or their interaction with other annotators. AEs produce the analysis results in the form of typed feature structures, which are data structures that have a type and a set of attribute-value pairs. An annotation is a particular type of feature structure that is attached to the region of the artifact that is being analyzed (e.g., a span of text in a document). For instance, a complete UIMA application might ingest plain text and identify entities, such as genes, diseases, cell lines, or relationships such as "binds-to" or "induce-expression." These entities can be adapted and extended easily to other entities and relationships. UIMA supports the configuration and running of pipelines of annotator components that do the actual work of analyzing the unstructured information. Users can write their own annotators, or configure and use pre-existing annotators. Some annotators are available as part of this project and others are contained in various online repositories.

In contrast to other text mining architectures, UIMA implements the first official OASIS standard (www.oasis-open.org/standards) for unstructured information management architectures. Compared with other frameworks, UIMA offers features including *annotation-type inheritance* (important for ontologies), is parallelizable, is distributable, and supports workflow management [64]. Additionally, UIMA provides capabilities to wrap components as network services, and can scale to very large volumes by replicating processing pipelines over a cluster of networked nodes. The UIMA-AS Scaleout Framework is an add-on to the base Java framework, supporting a very flexible scale-out capability based on JMS (Java Messaging Services) and ActiveMQ.

4.5 Chemical Structure Recognition Tools

As described in Sect. 2.1, the two-dimensional depiction of molecular structures, which is based on the molecular graph, is a universally recognized representation of molecules. Such images can be found in nearly all electronic sources of chemical information (e.g., journals, patents, and web interfaces of chemical databases). Currently, these images are generated with special drawing programs, either automatically from computer-readable file formats (i.e., collections of molecules) or manually through a graphical user interface. Although these programs can produce and store the information in a computer-readable format, the information is published as bitmap images (e.g., GIF for web interfaces or BMP for text documents) and, as a result, the structure information can no longer be used as input to chemical analysis software packages. The extracted images usually have to be converted manually by redrawing every structure, which is a time-consuming and error-prone process. To make the chemical information contained in the drawings of chemical structures accessible for computer programs, projects were started to convert bitmap images of molecules back into machine-readable form. This conversion process is called chemical structure reconstruction. In general, two different strategies (not restricted to the chemical domain) have been used for the recognition of patterns in images: statistical pattern recognition and structural pattern recognition [65]. Some of the available tools are listed here:

Kekulé [66] uses the vectorization of TIFF images. Optical character recognition (OCR) techniques and neural networks are used to identify special symbols such as chiral bonds and text representations, and a connection table is computed for the atoms and bonds. The last step comprises a post-processing phase that normalizes the reconstructed graph (e.g., bond lengths).

chemoCR [68] uses images of chemical molecules that can contain text annotations or images of reaction schemes as the input. The output is a standard chemical table file in SDF format. The algorithm is based on an image pre-processing phase comprising binarization, vectorization, and OCR followed by an assembly phase based on a chemical rule system.

CLiDE [67] uses segmentation algorithms for monochrome black-white bitmaps to identify connected components that are grouped into graph primitives. There are special primitives for the different chemical contexts (e.g., super atoms and bond types).

OSRA [69] reuses the existing software created by the open-source community to the fullest extent possible. OSRA is the first open-source program for optical structure recognition.

In principle, chemical structure recognition is a well-studied problem and over the last 25 years about ten different systems have been developed. However, the general problem of converting chemical structure images to totally correct connection tables is still not solved. All of the systems built follow the same design principles: the input image is binarized, line segments are vectorized, and potential characters are processed by OCR. Atoms and bonds are assembled from the resulting graphic primitives of lines and characters and converted into the connection table. Unfortunately, the pre-processing steps of OCR and vectorization produce a lot of errors and ambiguities. The correction of fused symbols (letters with other letters or bonds) or broken lines and characters is a huge challenge. Current methods for the conversion of graphic primitives are too inflexible to fix problematic decisions made by the system in an earlier step (e.g., sending a false connected component to the OCR or the vectorizer). Hence, recognition rates of all tools drop drastically in case of images with bad resolution, scan quality, or complicated chemical depictions like reaction schemes or Markush structures. One of the main reasons for the failure of the current systems is the lack of relevant benchmark sets and gold standards to develop better methods—the existing ones have been generated by images drawn automatically from uniform molecule collections.

5 Integration of Unstructured Data Resources

In this section, two strategies for semiautomatic information extraction are discussed. To generate networks for systems biology and toxicology we firstly present a workflow using the methods described above for NER and relationship extraction. As output, BEL as a promising language for toxicology network representation is generated. For the workflow, several constraints to meet the needs for toxicology network generation have to be imposed:

- Normalization of the recognized concepts, ideally to external resources such as chemical and gene protein databases or known terminology/ontology repositories; mere recognition of entity classes cannot be used for the generation of relationship networks.
- Flexible workflows that allow for different settings and integration of new modules.
- Context annotation of the toxicology information such as anatomy, organism, or cell system.
- End-user interfaces to start/manage extraction processes without the need for programming skills.
- Curation of results for easy readability and editing of formats.
- Framework that can be used for modeling

To meet all these conditions a UIMA workflow that allows for easy exchange and updating of methods was set up [70]. The currently developed workflow can use either abstracts or full text, includes state-of-the-art NER and relationship extraction tools, and can output BEL. We recommend UIMA since it allows unstructured information management, modular structure with clear interfaces, and a defined type system to be used for all modules and good scalability (cf. Chap. 4.4). But even the integration of already developed UIMA components from other pipelines is not always straightforward. The usage of different type systems is hindering such an easy integration. In our workflow we have developed a more general type system to allow an easier extension of new annotation types (such as new NER classes) without any change of the type system.

The normalization of entities is a fundamental problem that concerns both the automatic and manual generation of networks. In many cases, it is not possible to include only one resource as the reference for an entity class because no single comprehensive resource exists. We recommend to rely on public resources as far as possible. Publicly supported resources become more extensive over time and additionally, valuable information such as descriptions, synonyms, hierarchical information, and links to databases are often provided. In BEL, and similarly in the extraction workflow, different namespaces for one entity class may exist. Currently, for chemical compounds, references to CHEBI, ChEMBL, and a BEL-internal chemical dictionary (SCHEM) are given, and known equivalences have been established between the different entries. If different resources are used, equivalences are of high importance. This is not only true for the extraction of unstructured information but also for the semantic integration of structured resources. Information available from different resources addressing the same entity can only be merged if those equivalences are established. Sometimes, no public resource that can represent the entity is available, for example, if the entity is a new chemical compound that has not yet been registered in a public database. For those cases, there is the possibility of adding custom namespaces into a customized BEL framework.

Similarly, the automatic UIMA workflow uses different dictionaries to ensure higher recall. ProMiner was used for NER because it has good performances for established NER in the Life Sciences. Moreover, it can be easily adapted for the recognition of new entity classes and can work with large dictionaries. An overview of the entities currently supported in the described workflow by the automatic extraction system is given in Table 2. ProMiner is a commercial NER platform providing support for integration and regular updates for a number of common entity classes such as genes/proteins or diseases. Open-source NER developed in the area of Life Sciences can be used as well (c.f. Sect. 4.2). However, users should verify that external terminology resources used in

Table 2
Dictionaries integrated into the text mining workflow for the automatic extraction of BEL statements

Entity class	Resource	OpenBEL namespace
Human gene/protein	EntrezGene/Uniprot	HGNC
Mouse gene/protein	EntrezGene/Uniprot	MGI
Rat gene/protein	EntrezGene/Uniprot	RGD
Protein family name	OpenBel	PFH
Protein complex name	OpenBel	NCH
Protein complex name	Gene Ontology	GOCCTERM
Chemical name	OpenBel	SCHEM
Chemical name	ChEBI	CHEBI
Chemical name	ChEMBL	SCHEM
Disease name	MeSH	MESHD
Anatomy name	MeSH	MESHA

NER are updated and curated regularly. Otherwise, the NER tools get outdated over time.

For relationship extraction, two types of methods have been integrated in the first version: a binary classifier [71] and the BioNLP shared task trained tool TEES [72]. Experiences from assessments (described in Sect. 4.2) have shown that a combination of different methods in the majority of cases leads to a higher performance. The combination of the binary classifier and TEES results in a recall of 72 % for protein–protein relationships (under the condition that all proteins have been recognized before). The binary classifier is a simpler method that finds more relations (60 % recall in our experience) but needs more curation effort to generate BEL statements. TEES provides more granular annotations—but at the cost of recall (42 % recall). The output of both tools is used to generate BEL documents that contain BEL statements and context annotations together with the evidence sentences and citation information.

The described workflow is functional and already in production mode for several use cases. Nevertheless, in future developments relationship extraction has to be optimized in many ways. Experience from integrating available software for relationship extraction shows that this is not an easy task. Often, the tools contain complex closed workflows lacking clear interfaces for the exchange of NER methods. They are partly poorly documented and error handling is insufficient. Standardization efforts and further support for academic groups in future assessments are necessary to improve usability of the tools. The currently available relationship extraction modules focus on certain subgroups of relations. More efforts are necessary to generate appropriate training data and support further development of modules necessary for toxicology information extraction. The currently implemented extraction workflow creates valid BEL statements. However, in a number of cases the statements were incomplete or had to be interpreted to establish relevant relationships. Thus, the curation of the automatic generated statements is a necessary step in the extraction process to generate high-quality networks.

User interfaces are of utmost importance for the acceptance of information extraction in a user community. Instantly, they are needed for managing extraction processes and to view and edit the extraction results. Thus, for the workflow described above, a dashboard was developed to allow an easy upload of new text resources and starting the processing of documents. The processed documents can be viewed in a curation interface that presents the extracted BEL information to the user and enables easy editing of automatically extracted results. Visualization of all NER annotations in the relevant text parts supports the user further to validate and adapt the extracted statements. A first study showed that network construction can be supported substantially using a semiautomatic approach of this type [73]. Curation efficiency does depend not only on the relation extraction performance but also on how far and in which way all necessary information is available for the user. Moreover, the time needed to view and curate extracted information influences the curation performance. Experiences in first applications are currently used to optimize the interface. The text mining community is aware of the relevance of end-user interfaces and has already initiated tasks such as the BioCreative IV task described in Sect. 4.2 to support the CTD curation workflow.

The second example describes an information extraction workflow that supports the search for relevant information in patents. The development of new chemicals or pharmaceuticals is usually preceded by an in-depth analysis of the published patents in this field. For patent analysis the information retrieval step of finding relevant information within those partly very long documents (up to 600 pages) alone is costly and time consuming when done manually. Yet, it is mandatory for the potential success of an investment. The goal of the UIMA-HPC research project was to automate and hence speed up the process of patent knowledge mining. Multi-threaded analysis engines, developed according to UIMA standards, were used to process texts and images in thousands of documents in parallel. As training corpus for the development of the system a text corpus of 60 chemical synthesis patents from the European Patent Office was chosen based on relevance, diverse scan quality, and different notations of IUPAC terms. The following steps were performed to extract information from the corpus:

- Find all relevant chemical terms and depictions, annotate their position, and reconstruct the chemical structure.
- Highlight the annotations inside the original PDF document for visual inspection.
- Enrich with additional information (i.e., references to online databases).
- Add a chemical bookmark structure for easier accessibility of compounds and zones.
- Extract cross-references inside the document.

To give an impression of the extent of annotations found in the corpus: The system extracted an absolute number of 58,109 IUPAC terms and trivial names from the 60 patents that were enriched with structure information and of these, 9.45 % represented single-chemical elements (5491 occurrences of elements) and the remaining 52,618 identified chemical entities represented 10,523 unique chemical structures [74].

The goal of the automatic annotation strategy was to semantically enrich patents and to develop a suitable document viewer on top to assist the chemical expert. This patent viewer allows faster browsing and easier finding of relevant areas (such as synthesis information or relevant references within the patent text) in the semantically enriched patents. We tested the system on a real use case where our partner from chemical industry searched a certain synthetic procedure. This information was not available in any commercial database system such as REAXYS or Scifinder. The task was to find a feasible synthesis for a defined class of new image dyes and photographic developer compounds. As a starting point we retrieved 452 relevant patent documents by a key word-based patent search in Google patents. The manual analysis of the document corpus took a chemist about 5 working days. In the end a synthesis protocol could be identified in a German patent from 1984. In parallel, the documents were processed through the annotation workflow and results were stored into XCAS files containing the original full text, the extracted chemical depictions, and extracted textual concepts. When assisted with a graphical front end to browse and search the patents, the same chemist could easily narrow down the search to a smaller set of documents and was able to find the important passage within one afternoon. This is a significant reduction of tedious manual work. However, major issues making chemical searching in patents a particular challenge can be identified: hundreds of synonyms for many compounds, multimodal information such as chemical drawings, Markush formulas, or spelling errors in OCR.

Both examples show that text mining systems in current state of the art are ready to support experts in their work. Even if the systems are not able to recognize and normalize all entities and could not extract complex relations perfectly they can help to reduce extraction time. Furthermore, both examples emphasize the relevance of user interfaces for successful systems. Currently such systems are mostly used by small groups of specialized users such as database curators or professional search experts in industry. Close co-operations between end users and method developers as well as the preparation of more real scenario training data are necessary to enhance the current systems further.

6 Summary and Outlook

Over the last 15 years, industries have spent significant resources to integrate public data and information sources and to align this information with internal, proprietary data. At the same time, the academic research community has suffered from lack of access to large data sets, especially those that include curated bioactivity data. In contrast to bioinformatics data such as genomes, protein sequences, and protein structures that are available to everyone, chemoinformatics data have traditionally been closed and proprietary. The NIH roadmap drastically changed this setting with the creation of PubChem, a publicly available depository of screening data. Other databases such as DrugBank, ChemBank, and ChEMBL followed soon after, and today a series of databases that can be searched for compounds and associated biological data is available. Integration initiatives such as Open PHACTS are under way to simplify the access to chemical information dispersed in different resources. Data quality, standardization of terminology, as well as data interoperability are still major issues. Toxicological information is currently far more distributed in various resources and is harder to access. A high amount of information can only be extracted from unstructured text resources. The development of automatic methods for the extraction of the large volume of unstructured chemistry information has just begun. The outlined strategies for the integration of text mining workflows indicate ways to use the currently available text mining tools. They can speed up the process of information extraction from text resources. Further dedicated methods and tool development can be supported by the chemistry and toxicology community through the annotation of training data and the setting up or supporting of relevant tasks in public assessments.

References

1. Committee on toxicity testing and assessment of environmental agents et al (2007) Toxicity testing in the 21st century: a vision and a strategy. The National Academies Press, Washington, DC, http://www.nap.edu/openbook.php?record_id=11970. Accessed 16 Aug 2014
2. Kavlock R et al (2012) Update on EPA's ToxCast program: providing high throughput decision support tools for chemical risk management. Chem Res Toxicol 25:1287–1302
3. Sipes NS et al (2013) Profiling 976 ToxCast chemicals across 331 enzymatic and receptor signaling assays. Chem Res Toxicol 26:878–895
4. Uehara T et al (2010) The Japanese toxicogenomics project: application of toxicogenomics. Mol Nutr Food Res 54:218–227
5. McNaught A, Wilkinson A (1997) Compendium of chemical terminolgy - The gold book. Blackwell Science, Oxford, http://old.iupac.org/publications/books/author/mcnaught.html. Accessed 16 Aug 2014
6. Eller GA (2006) Improving the quality of published chemical names with nomenclature software. Molecules 11:915–928
7. Bauerschmidt S, Gasteiger J (1997) Overcoming the limitations of a connection table description: a universal representation of chemical species. J Chem Inf Model 37: 705–714
8. Li Z et al (2004) Personal experience with four kinds of chemical structure drawing software: review on ChemDraw, ChemWindow, ISIS/Draw, and ChemSketch. J Chem Inf Comput Sci 44:1886–1890
9. Dalby A et al (1992) Description of several chemical structure file formats used by computer programs developed at Molecular Design Limited. J Chem Inf Model 32(3):244–255
10. Heller S et al (2013) InChI – the worldwide chemical structure identifier standard. J Cheminform 5:7, http://www.jcheminf.com/content/5/1/7
11. Weininger D (1988) SMILES, a chemical language and information system. 1. Introduction to methodology and encoding rules. J Chem Inf Comput Sci 28:31–36
12. Siani MA, Weininger D, Blaney JM (1994) CHUCKLES: A method for representing and searching peptide and peptoid sequences on both monomer and atomic levels. J Chem Inf Model 34:588–593
13. Siani MA et al (1995) CHORTLES: A method for representing oligomeric and template-based mixtures. J Chem Inf Model 35: 1026–1033
14. Zhang T et al (2012) HELM: a hierarchical notation language for complex biomolecule structure representation. J Chem Inf Model 52:2796–2806

15. Chen WL et al (2011) Self-contained sequence representation: bridging the gap between bioinformatics and cheminformatics. J Chem Inf Model 51:2186–2208
16. Wang Y et al (2014) PubChem BioAssay: 2014 update. Nucleic Acids Res 42(Database issue):D1075–D1082
17. Bento AP et al (2014) The ChEMBL bioactivity database: an update. Nucleic Acids Res 42(Database issue):D1083–D1090
18. Hastings J et al (2013) The ChEBI reference database and ontology for biologically relevant chemistry: enhancements for 2013. Nucleic Acids Res 41(Database issue):D456–D463
19. Knox C et al (2011) DrugBank 3.0: a comprehensive resource for "omics" research on drugs. Nucleic Acids Res 39(Database issue): D1035–D1041
20. Kanehisa M et al (2014) Data, information, knowledge and principle: back to metabolism in KEGG. Nucleic Acids Res 42(Database issue):D199–D205
21. Qin C et al (2014) Therapeutic target database update 2014: a resource for targeted therapeutics. Nucleic Acids Res 42(Database issue): D1118–D1123
22. Wishart DS et al (2013) HMDB 3.0–The Human Metabolome Database in 2013. Nucleic Acids Res 41(Database issue):D801–D807
23. Williams AJ (2008) Internet-based tools for communication and collaboration in chemistry. Drug Discov Today 13:502–506
24. Davis AP, Murphy CG et al (2013) The Comparative Toxicogenomics Database: update 2013. Nucleic Acids Res 41(Database issue):D1104–D1114
25. Davis AP, Wiegers TC, Roberts PM et al (2013) A CTD-Pfizer collaboration: manual curation of 88,000 scientific articles text mined for drug-disease and drug-phenotype interactions. Database (Oxford) 2013:bat080. doi: 10.1093/database/bat080
26. Kohonen P et al (2013) The ToxBank Data Warehouse: supporting the replacement of in vivo repeated dose systemic toxicity testing. Mol Inform 32:47–63
27. Nyström-Persson J et al (2013) Toxygates: interactive toxicity analysis on a hybrid microarray and linked data platform. Bioinformatics 29:3080–3086
28. Waters M et al (2008) CEBS–Chemical Effects in Biological Systems: a public data repository integrating study design and toxicity data with microarray and proteomics data. Nucleic Acids Res 36(Database issue):D892–D900
29. Kuhn M et al (2014) STITCH 4: integration of protein-chemical interactions with user data. Nucleic Acids Res 42:D401–D407
30. Nelson, Stuart J. (2009) Medical terminologies that work: the example of MeSH. In: Proceedings of the 10th international symposium on pervasive systems, algorithms, and networks (ISPAN 2009), Kaohsiung, Taiwan. p 380–384
31. Harland L (2012) Open PHACTS: a semantic knowledge infrastructure for public and commercial drug discovery research. In: Teije A et al. (eds) Knowledge engineering and knowledge management. Lecture Notes in Computer Science, vol 7603. Springer, Heidelberg, pp 1–7
32. Sushko I et al (2011) Online chemical modeling environment (OCHEM): web platform for data storage, model development and publishing of chemical information. J Comput Aided Mol Des 25:533–554
33. Demir E et al (2010) The BioPAX community standard for pathway data sharing. Nat Biotechnol 28:935–942
34. Hucka M et al (2003) The systems biology markup language (SBML): a medium for representation and exchange of biochemical network models. Bioinformatics 19(4):524–531
35. Chaouiya C, Bérenguier D, Keating SM et al (2013) SBML qualitative models: a model representation format and infrastructure to foster interactions between qualitative modelling formalisms and tools. BMC Syst Biol 7:135
36. Slater T, Song, D (2012) Saved by the BEL: ringing in a common language for the life sciences. http://www.ddw-online.com/informatics/p191019-saved-by-the-bel:ringing-in-a-common-language-for-the-life-sciences-fall-12.html
37. Slater T (2014) Recent advances in modeling languages for pathway maps and computable biological networks. Drug Discov Today 19:193–8, http://www.ncbi.nlm.nih.gov/pubmed/24444544. Accessed May 16, 2014
38. Schlage WK et al (2011) A computable cellular stress network model for non-diseased pulmonary and cardiovascular tissue. BMC Syst Biol 5:168
39. Gebel S et al (2013) Construction of a computable network model for DNA damage, autophagy, cell death, and senescence. Bioinform Biol Insights 7:97–117
40. Chindelevitch L et al (2012) Causal reasoning on biological networks: interpreting transcriptional changes. Bioinformatics 28:1114–1121
41. Huang C-L et al (2012) Correlation set analysis: detecting active regulators in disease populations using prior causal knowledge. BMC Bioinformatics 13:46
42. Manning CD, Raghavan P, Schuetze H (2008) Introduction to information retrieval. Cambridge University Press, ISBN 9780521865715. http://nlp.stanford.edu/IR-book/. Accessed 16 Aug 2014

43. Lupu M et al. (2011) Overview of the TREC 2011 Chemical IR Track. In: The twentieth Text REtrieval Conference (TREC 2011) proceedings. pp 1–11 A
44. Spaerck-Jones K (1972) A statistical interpretation of term specificity and its application in retrieval. J Doc 28:11–22
45. Gurulingappa H et al. (2010) Prior art search in chemistry patents based on semantic concepts and co-citation analysis. In: Nineteenth Text REtrieval Conference, TREC 2010 proceedings
46. Hirschman L et al. (2012) Text mining for the biocuration workflow. Database (Oxford) 2012: bas020. doi: 10.1093/database/bas020
47. Smith L et al. (2008) Overview of BioCreative II gene mention recognition. Genome Biol 9 (Suppl 2): S2
48. Morgan AA et al. (2008) Overview of BioCreative II gene normalization. Genome Biol 9 (Suppl 2): S3
49. Hakenberg J et al (2011) The GNAT library for local and remote gene mention normalization. Bioinformatics 27:2769–2771
50. Schuemie M, Jelier R, Kors J (2007) Peregrine: lightweight gene name normalization by dictionary lookup. In: Hirschmann L, Krallinger M, Valencia A (eds) Proceedings of the second BioCreative challenge evaluation workshop
51. Fluck J et al. (2007) ProMiner: Recognition of human gene and protein names using regularly updated dictionaries. In: Hirschmann L, Krallinger M, Valencia A (eds) Proceedings of the second BioCreative challenge evaluation workshop. pp 149–151 A
52. Dowell KG et al. (2009) Integrating text mining into the MGI biocuration workflow. Database (Oxford) 2009: bap019
53. Jessop DM et al (2011) OSCAR4: a flexible architecture for chemical text-mining. J Cheminform 3(1):41
54. Rocktäschel T, Weidlich M, Leser U (2012) ChemSpot: a hybrid system for chemical named entity recognition. Bioinformatics 28: 1633–1640
55. Klinger R et al (2008) Detection of IUPAC and IUPAC-like chemical names. Bioinformatics 24:i268–i276
56. Davis AP, Wiegers TC, Johnson RJ et al (2013) Text mining effectively scores and ranks the literature for improving chemical-gene-disease curation at the comparative toxicogenomics database. PloS One 8:e58201, P J Janssen
57. Krallinger M et al. (2013) Overview of the chemical compound and drug name recognition (CHEMDNER) task. In: Krallinger M et al. (eds) Proceedings of the fourth BioCreative challenge evaluation workshop. 2: 1–33
58. Leaman R, Wei C-H, Lu Z (2013) NCBI at the BioCreative IV CHEMDNER task: recognizing chemical names in PubMed articles with tmChem. In: Krallinger M et al. (eds) Proceedings of the Fourth BioCreative Challenge Evaluation Workshop, vol 2. pp 34–41
59. Lowe D, Sayle R (2013) LeadMine: a grammar and dictionary driven approach to chemical entity recognition. In: Krallinger M et al. (eds) Proceedings of the fourth BioCreative challenge evaluation workshop, vol 2. pp 47–54
60. Pyysalo S et al. (2008) Comparative analysis of five protein-protein interaction corpora. BMC Bioinformatics 9 (Suppl 3): S6
61. Tikk D et al (2010) A comprehensive benchmark of kernel methods to extract protein-protein interactions from literature. PLoS Comput Biol 6:e1000837
62. Segura-Bedmar I, Martinez P, Herrero-Zazo M (2013) SemEval-2013 Task 9: extraction of drug-drug interactions from biomedical texts (DDIExtraction 2013) – S13-2056.pdf. In: Proceedings of the seventh international workshop on semantic evaluation (SemEval 2013)
63. Kim J-D, Wang Y, Yasunori Y (2013) The Genia Event extraction shared task, 2013 Edition – Overview. In: Nédellec C et al. (eds) Proceedings of the BioNLP Shared Task 2013. Sofia, Bulgaria: Association for Computational Linguistics (ACL), W13–2002
64. Bank M, Schierle M (2012) A survey of text mining architectures and the UIMA standard. In: Proceedings of the eight international conference on language resources and evaluation (LREC'12), Istanbul, Turkey, pp 23–25
65. Bunke H (ed) (1993) Advances in structural and syntactic pattern recognition. Proceedings of the international workshop. Bern, Switzerland, 26–28 Aug 1992, pp 163–209
66. McDaniel J, Balmuth J (1996) Automatic interpretation of chemical structure diagrams. In: Graphics recognition methods and applications. pp 148–158
67. Algorri M-E, Zimmermann M, Hofmann-Apitius M (2007) Automatic recognition of chemical images. In: eighth Mexican international conference on current trends in computer science (ENC 2007), IEEE, Morelia, Michoacan, Mexico, 24–28 Sept 2007, pp 41–46
68. Valko AT, Johnson AP (2009) CLiDE Pro: The latest generation of CLiDE, a tool for optical chemical structure recognition. J Chem Inf Model 49:780–787
69. Filippov IV, Nicklaus MC (2009) Optical structure recognition software to recover chemical information: OSRA, an open source solution. J Chem Inf Model 49:740–743

70. Fluck J et al. (2014) BELIEF – a semiautomatic workflow for OpenBEL network creation. In: Proceedings of the international symposium on semantic mining in medicine (SMBM2014), Aveiro, Portugal, 7–8 Oct 2014, pp 64–69
71. Bobic T et al (2012) Improving distantly supervised extraction of drug-drug and protein-protein interactions. In: Proceedings of ROBUS-UNSUP 2012: joint workshop on unsupervised and semi-supervised learning in NLP. Avignon, France, 23–27 Apr 2012, pp 35–43
72. Björne J, Salakoski T (2013) TEES 2.1: Automated annotation scheme learning in the BioNLP 2013 shared task. In: Proceedings of the BioNLP shared task 2013 workshop. Association for Computational Linguistics, Sofia, Bulgaria, 9 Aug 2013, pp 16–25
73. Ansari S et al. (2014) A semi-automated curation process for causal knowledge extraction. In: Abstracts of the 7th international biocuration conference, University of Toronto, Toronto, 6–9 Apr 2014, Poster abstract 31
74. Bergmann S, Romberg M, Zimmermann M et al (2012) Information extraction from chemical patents. Comp Sci 13:21–32

Chapter 6

Using a Crowdsourcing Approach for Network Verification

Natalia Boukharov and Jennifer S. Park

Abstract

Biological network models have many applications in systems toxicology and in biomarker and target identification for drug discovery. A critical challenge is to keep the models relevant and representative of current scientific knowledge. Harnessing community intelligence in knowledge curation holds great promise in dealing with the flood of biological information. Another important challenge is to be able to share and communicate networks with other scientists to aid in furthering similar scientific goals. To try to address both of these challenges, we designed a crowd verification platform to verify a comprehensive set of networks in a standardized biological expression language. This platform can increase the value of the network models, and promote their use in research applications such as drug discovery, personalized medicine, and toxicological risk assessment. In addition, we have provided practical recommendations for designing intuitive tools for network curation, identifying and recruiting the "target crowd," and using various approaches to engage and motivate participants.

Key words Crowdsourcing, Community curation, Crowd verification, Biological network models, Reputation system, Biological expression language, Knowledge curation, Molecular pathways

1 Biological Network Models

Biological network models provide a graphical representation of current scientific knowledge in the field while giving an intuitive view of the interactions underlying various biological processes. There are several public repositories of biological networks including the Kyoto Encyclopedia of Genes and Genomes (KEGG), Reactome, Pathway Commons, WikiPathways, and BioModels Database [1–5]. The qualitative biological network models described here are available at https://bionet.sbvimprover.com/. These models were constructed using the biological expression language (BEL), which allows the representation of causal relationships between biological entities and associated evidence including reference, tissue, species, disease context, and direct quotation supporting the edge from published scientific research. These models can be used for computational analysis of biological data for applications

Julia Hoeng and Manuel C. Peitsch (eds.), *Computational Systems Toxicology*, Methods in Pharmacology and Toxicology,
DOI 10.1007/978-1-4939-2778-4_6,

in systems toxicology, and in biomarker and target identification for drug discovery.

A major challenge in network construction is keeping the networks relevant and representative of current scientific knowledge. More biological data than ever is being generated; for example, the PubMed journal database contains more than 24 million citations. A huge volume of new scientific data, some of which can be used to support or reject entity relationships or to add new nodes and/or edges to the network models, is published every day. The sheer volume of the findings makes it impossible for individual scientists to review and identify relevant information. Harnessing community intelligence in knowledge curation holds great promise in dealing with the flood of biological knowledge.

2 Application of Crowdsourcing to Solving Biological Problems

Traditionally, biomedical science has been based on the achievements of individual research groups. Knowledge exchange is carried out by the publication of scientific articles and the presentation of research results at science conferences. Now, the increased speed of knowledge accumulation requires innovative approaches to collaborative research and information sharing. Connected by the Internet, scientists around the globe are increasingly exploring "crowdsourcing" approaches that allow large communities of researchers to directly address challenges of massive scale through collaboration. Crowdsourcing harnesses human collective intelligence and changes the way scientists use the Internet to collect ideas, solve complex problems, and build high-quality repositories of scientific information. One benefit of crowdsourcing is that information can be disseminated across the Internet very quickly, and people can work on projects in shared collaborative spaces across continents.

In 1714, the British Government offered £20,000 for people to invent a solution to what they called "The Longitude Problem," which made sailing difficult and perilous (killing thousands of sailors every year). This is possibly the first ever example of crowdsourcing. The problem, considered almost unsolvable, was solved by John Harrison, the son of a carpenter. Harrison invented the marine chronometer [6]. This example of crowdsourcing highlights one of the key principles of crowdsourcing; that is, innovation and creativity can come from anywhere.

The term "crowdsourcing" was coined by Jeff Howe in a 2006 issue of WIRED magazine in the article "The Rise of Crowdsourcing" [7]. In this article, crowdsourcing was defined as the outsourcing of a task to an undefined (and generally large) network of people in the form of an open call.

2.1 Basic Crowdsourcing Components

A basic crowdsourcing scheme has three primary components:

1. A customer who wants to solve a certain problem or gather information. This can be a university or a big company, a small research group, a public agency, or an individual science enthusiast.
2. A crowd—people who can help solve this problem if engaged and motivated to contribute information and/or solution to the cause.
3. A simple intuitive platform on which the customer can provide the crowd with tools to solve/submit solutions for the problem posted and/or to contribute valuable information to a repository.

There are numerous examples of completed and ongoing crowdsourcing biological projects that have benefited and are benefiting the scientific community at large.

2.2 Approaches to Motivate Participation

The approaches to motivate individuals to contribute voluntarily to crowdsourcing initiatives range from appealing to human altruism or ego to offering an educational opportunity or a monetary reward.

2.2.1 Gaming

One emerging trend has expanded the list of motivators to include the pursuit of fun and enjoyment through games [8]. Online games have emerged as an effective way to recruit, engage, and organize large numbers of volunteers to help address difficult biological challenges.

One such example is FoldIt, a visual puzzle game [9]. Players are presented with a primary protein sequence and challenged to find its lowest energy three-dimensional structure. Players can also communicate and share their approaches with each other. Collectively developed by FoldIt participants, the Blue Fuse protein strategy showed superior performance to an algorithm developed by professional structural biologists and was touted as a big success of crowdsourcing [10].

Another example of a crowdsourcing biology game that is generating valuable information is Dizeez, which is aimed at cataloging gene-disease associations documented in the literature but not yet reflected in structured annotation databases [11, 12]. Players are presented with a disease drawn from the Human Disease Ontology [13] and a multiple-choice selector with five genes, only one of which has prior evidence linking it to the disease. If a player correctly guesses the known disease from the list, they receive points. Regardless, all player answers are logged by the system. "Incorrect" gene selections by multiple players often represent new, unknown associations. Users can review the game log and submit new evidence for gene–disease links.

The gaming approach is also used in Phylo to improve large multiple sequence alignments [14], EyeWire to collect data about neurons in the retina [15], EteRNA to design RNAs that fold well [16], and The Cure, a card game to capture gene sets associated with breast cancer prognosis [17].

2.2.2 Altruism

Pure altruism is driving scientists to share their knowledge in BioGPS, a free extensible and customizable gene annotation portal. Users can register any Web-based biological resource as a "plug-in" in BioGPS, and then other users can use this resource and mix and match entries with any other desired resources to build their own customized gene annotation page. BioGPS receives millions of hits per year [18]. Gene Wiki is another such initiative. It is a collection of collaboratively written, community-verified, and continuously updated review articles for every human gene [19].

2.2.3 Education

A number of annotation projects have incorporated the annotation of new sequences directly into the curriculum of undergraduate courses employing the educational motivation approach. Students were excited to be able to contribute to the creation of knowledge. These initiatives have generated thousands of high-quality annotations [20, 21].

Universities and other entities have been launching crowdsourcing idea challenges, often with a cash prize attached. Harvard Catalyst, IBM (DREAM Challenge), the Pistoia Alliance [22–24], and others have engaged the scientific community in solving various biological problems by posting them online and organizing competitions.

2.2.4 Reputation System

Another form of motivation for crowdsourcing projects is a reputation system that allows users to gain a reputation based on their contributions measured by a reputation score and/or badges. Two examples of platforms with reputation systems are ResearchGate [25] for researchers with publications and StackOverflow [26] for programmers who answer questions. Both of these sites track a score associated with a participant that encourages them to contribute more to the Web site to obtain a higher score and look attractive among their peers. The Network Verification Challenge employs a reputation system to motivate participants to verify biological networks.

3 Network Verification Challenge

The Network Verification Challenge (NVC) is the third challenge in the sbv IMPROVER project (systems biology verification: Industrial Methodology for PROcess VErification in Research), a collaborative initiative between IBM Research and Philip Morris International R&D, which is designed to develop a robust, transparent, and practical process for assessing complex scientific data [27, 28]. NVC is a unique crowdsourcing challenge, implementing a reputation-based approach for motivating the scientific public to verify a comprehensive set of biological networks to review, edit, and download via an easy-to-use Web site.

Fifty networks were presented in the NVC for crowd verification. These networks were based on previously constructed non-diseased networks that described cell proliferation, cell stress, DNA damage, autophagy, cell death and senescence, pulmonary inflammation, tissue repair, and angiogenesis [29–33]. The networks used in the NVC were enhanced with lung and chronic obstructive pulmonary disease (COPD)-relevant mechanisms using literature and data approach [34]. The models were constructed based on existing literature and data-driven hypotheses, and encoded in BEL, a human-readable and machine-computable language that captures causal and correlative relationships between biological entities. Exact relationships are encoded at the highest level of granularity possible as edges between entities. Supporting literature evidence for each edge is also captured in BEL [35]. The structure of a BEL node consists of three parts, the function, namespace, and entity (Fig. 1). This structure allows for a well-defined, computable language that can be used across research groups, lending itself well to collaboration.

The goal of the NVC is to improve and expand these biological network models by challenging participants to review, approve, or reject existing evidence, and to add new evidence, nodes, or edges. Participants compete for a high score in a reputation system, collecting points and badges for the chance to win a prize [28]. The reputation system is designed to be self-monitoring, with participants reviewing each other's contributions by approving, rejecting, and commenting. The Web site provides a forum with a variety of actions that facilitates scientist communication and collaboration on a scale that is not achieved with traditional collaboration approaches.

function(namespace:entity)

Example: p(HGNC:AKT1)

Description: The protein AKT1,
whose symbol is defined in the
HGNC database

Fig. 1 Structure of a BEL node: The function describes an abundance or process related to the entity, the namespace represents a public database containing the entity, and the entity represents a biological concept (gene, protein, chemical, biological process) being described by the network. HGNC = Human Genome Nomenclature Committee

4 Bionet

The 50 biological networks were posted on Bionet [36], a platform that provides access to an interactive graphical representation of each network (Fig. 2). This platform allows participants to quickly search a network of interest and navigate through the network by selecting nodes or edges from the graphical interface or from a drop-down menu.

Bionet supports the full set of actions allowed in curating. Participants can use a list of nodes and edges to view and sort network elements and evidence associated with each edge, review and vote on each piece of evidence, add new evidence, and add comments to edges and nodes. When an edge is selected in the network, the associated evidence can be viewed, new evidence can be added, or a comment can be added (Fig. 3). Clicking on a piece of evidence associated with an edge allows participants to view that evidence's particular context and vote and comment on that particular evidence (Fig. 4). Clicking on New Evidence opens a window that allows to input a new literature reference supporting that edge, documenting the reference and context that will be associated with the BEL edge (Fig. 5).

For a selected node, participants can add new edges by formulating new BEL statements based on relationships they have identified in published scientific literature (Fig. 6). The BEL Web-based statement editor provides guidance in entering correct protein names, chemical compound names, Gene Ontology terms, and other biological entities (Fig. 7). A drop-down menu allows selection of edge type. Alternatively, a Convert Text to

Fig. 2 Bionet homepage: 50 networks are available for crowd verification

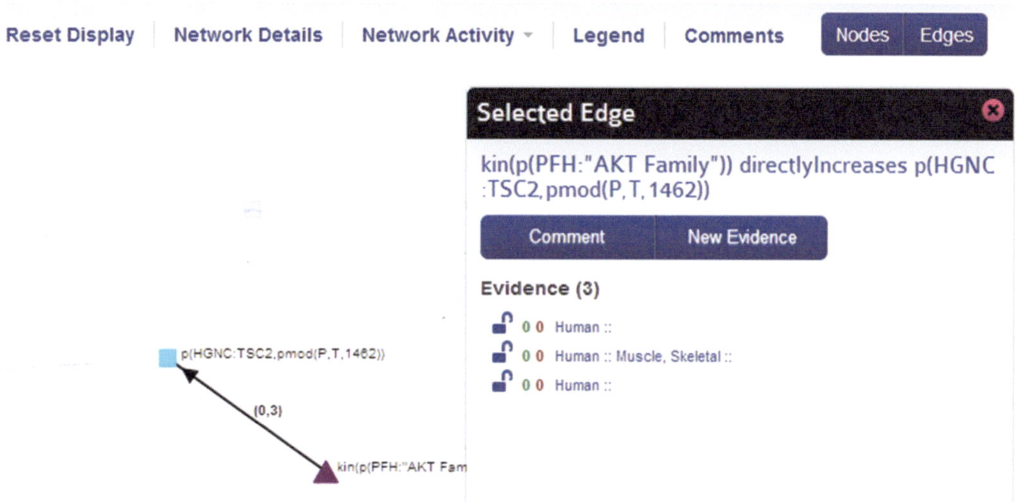

Fig. 3 Selected edge: Clicking on an edge in a network allows you to drill down to the evidence level. Three pieces of evidence support the selected edge

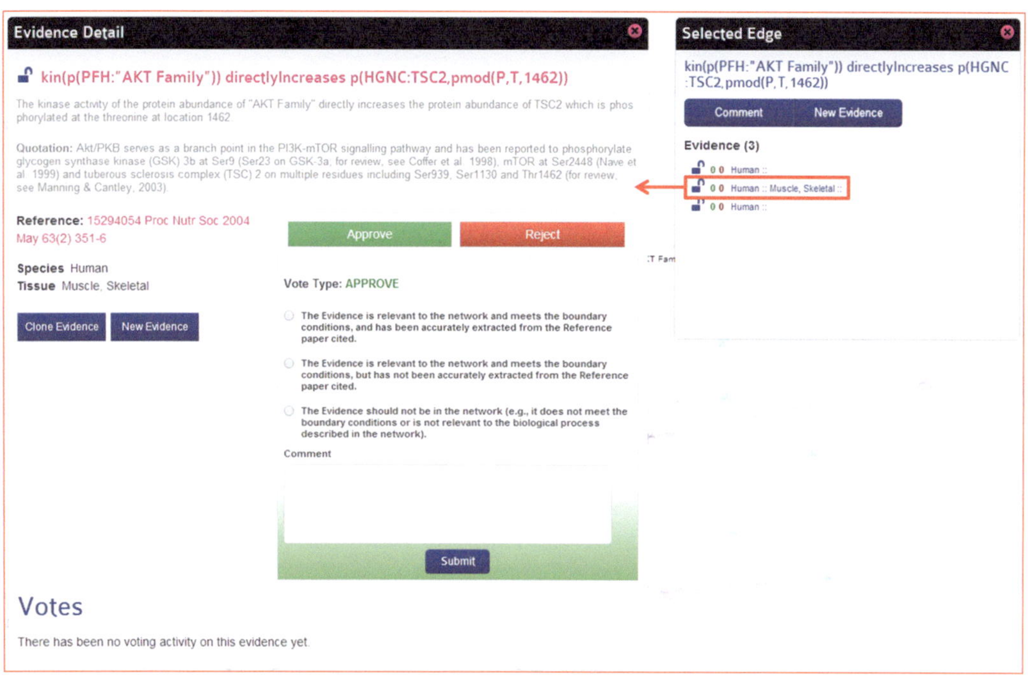

Fig. 4 Evidence detail: Clicking on one of the evidences for a selected edge opens an Evidence Detail window which lists information about the evidence including the reference, quotation, species, and tissue. In this window you can vote on the evidence or select the New Evidence button to add a new evidence

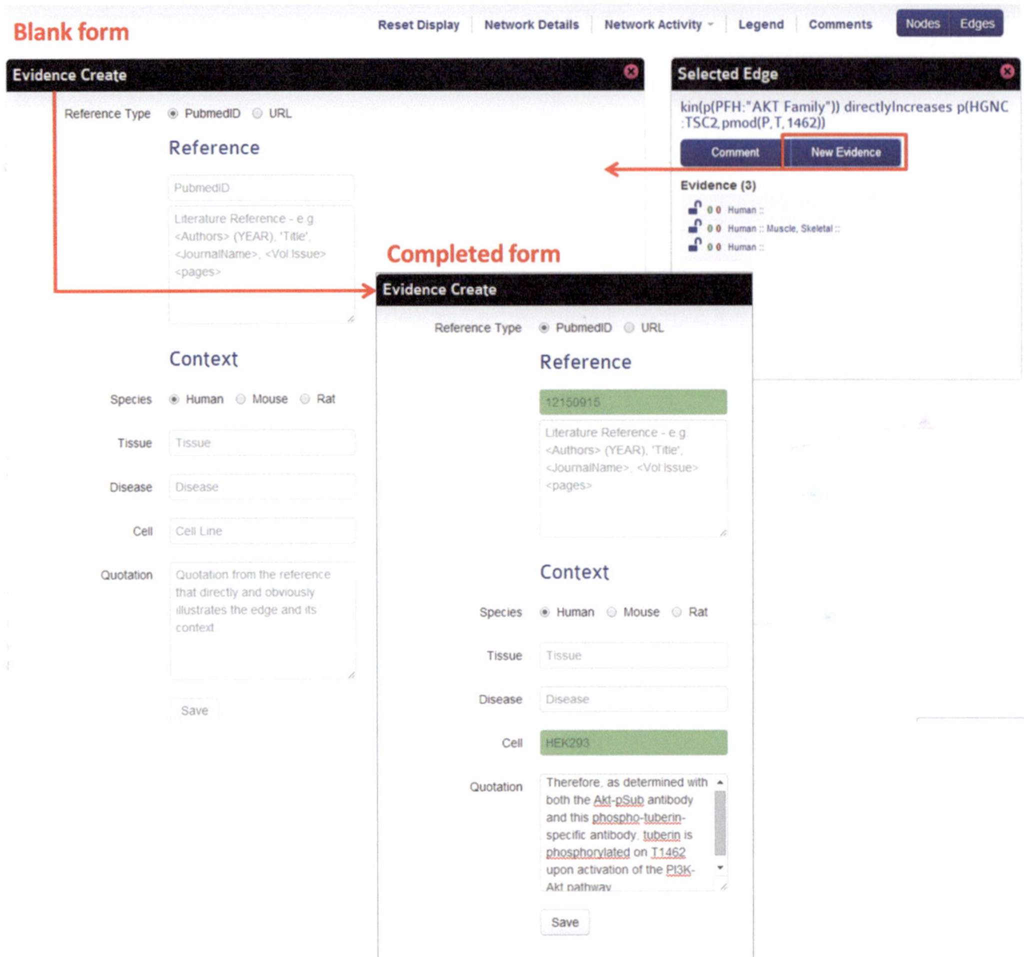

Fig. 5 New evidence: In the Evidence Create window you can add a new evidence from a particular literature reference that you find that supports the selected edge, adding contextual information from the literature reference including tissue, disease, or cell type and the quotation from the reference that supports the edge

BEL tool (currently in beta version) allows participants to simply paste a quotation from the published literature into the tool, which then translates the text to a valid BEL edge.

An important feature of BEL is its ability to save contextual information for each piece of evidence added to the network. The next step after creating the edge is adding evidence, including the reference, tissue, species, and the quotation from the published text describing the relationship (Fig. 8). Here, it is important to remind participants of the boundary conditions of the networks; in this case, lung and COPD-relevant tissues were requested.

All contributions are labeled as suggested changes until verified and approved by other participants through voting. All actions by participants are tied directly to a reputation system [28].

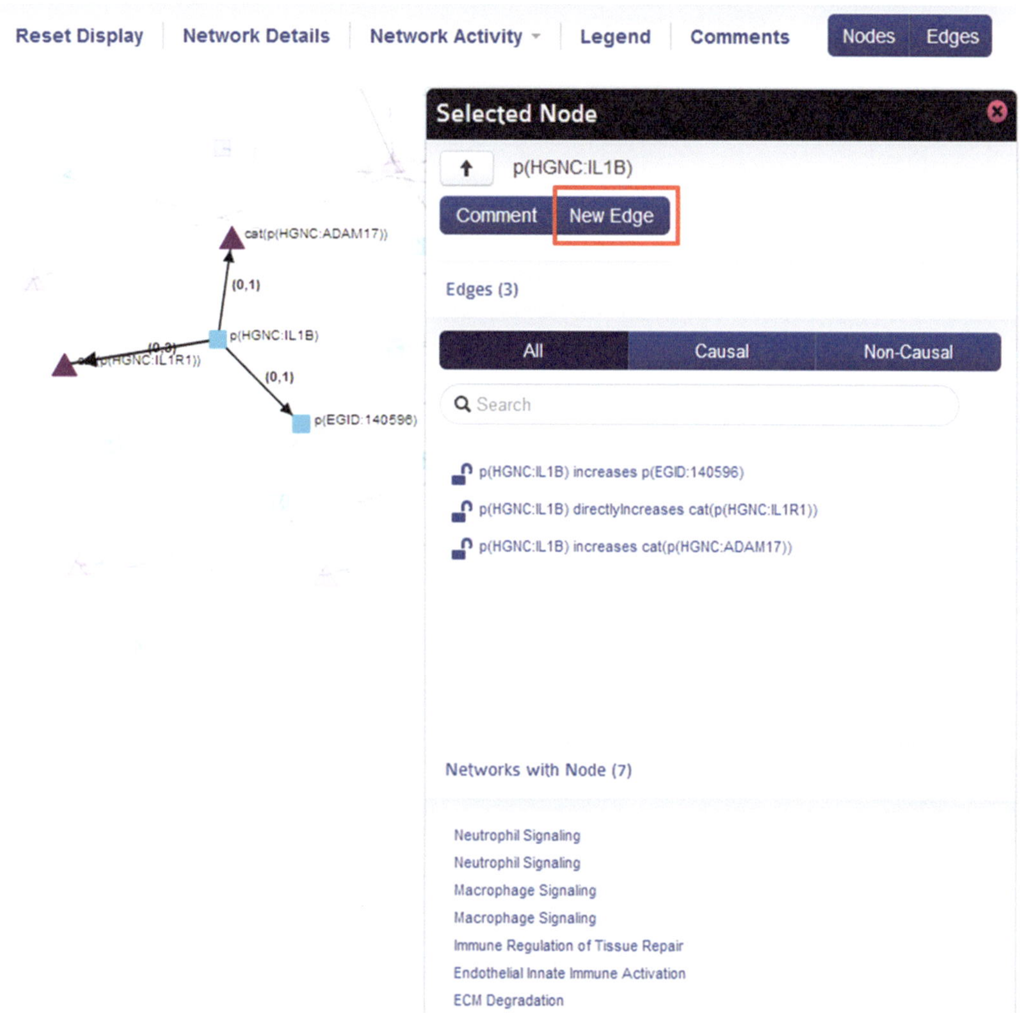

Fig 6 Selected node: For a selected node in the network, all of the associated edges containing that node are listed. This window also lists other networks in which the selected node resides. There is an option to create a New Edge for that node representing a relationship between the selected node and another node. Use this option to add a relationship that is relevant to the network but not yet represented by existing edges in the network

A participant gains reputation points by performing different tasks: approving/rejecting evidence, and/or adding new evidence and edges. Additional reputation points are gained when the participant's created edges and/or new evidences are approved by other participants. The progress of each participant can be viewed on a Community Leaderboard (Fig. 9) and participants can use this to gauge their ranking in relation to their peers by reputation points, and /or by number of added edges, pieces of evidence, and votes. A team ranking is also displayed and participants can create or join teams when they register for the NVC [28].

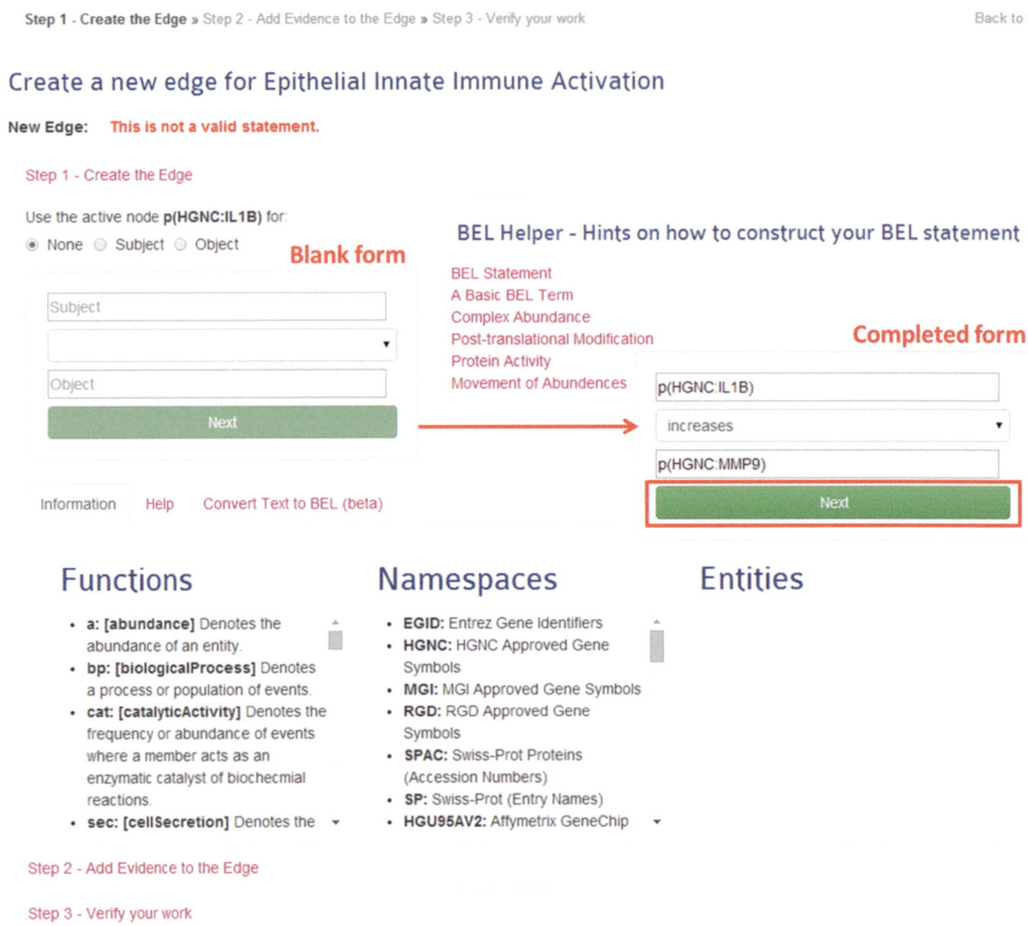

Fig 7 Create a new edge. This window provides guidance on how to create a new edge in the BEL format by entering the subject, relationship, and object. Functions, namespaces, and entities are suggested as you type in the subject or object fields. The relationship is selected in a drop-down menu. The Convert Text to BEL tool can also be used to generate the correct BEL term automatically

To encourage self-management of the platform, a Community Activity page offers an efficient way to find the most recent activity performed by participants. Because there are many networks in the NVC, it was important to have a quick way of finding the most recent actions. The Community Activity page allows for filtering based on network name and action type, which helps participants monitor the networks of interest to them (Fig. 10). On this page, participants can vote directly on actions, facilitating Web site activity and peer review.

Participants consisted of scientists from many different countries who shared an interest for signaling pathways. Many different approaches were used to identify and recruit the crowd.

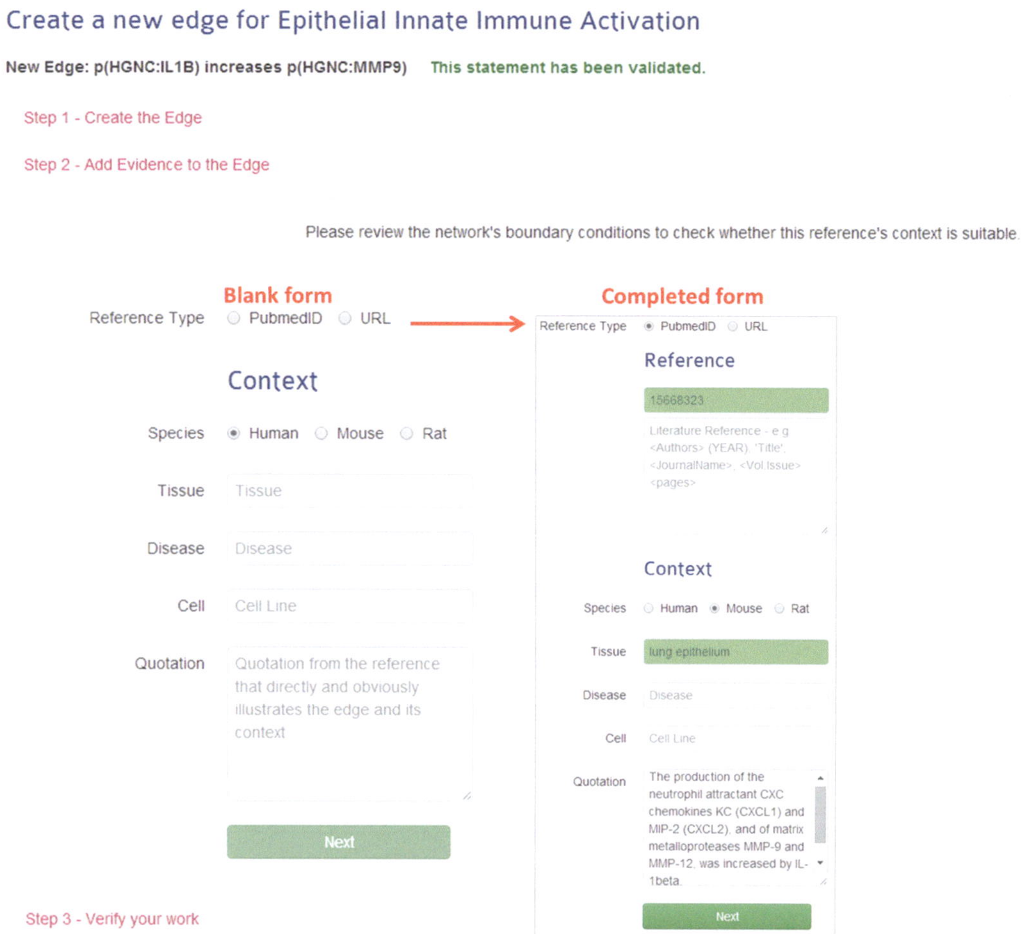

Fig 8 Add evidence to the new edge. After creating the BEL edge, add the reference and context associated with the evidence including the species, tissue, disease, or cell type and quotation from the literature reference supporting the edge

5 The Crowd

5.1 Recruitment

Several approaches were used to identify and recruit the "target crowd" and encourage participation. The target crowd consisted of scientists and students with knowledge in the field represented by the networks. Information about NVC and a call to participate was disseminated through e-mails to participants of past sbvIMPROVER challenges and to other scientists based on their research record, presentations at relevant scientific conferences, and publications. Several NVC seminars were conducted at selected network biology-focused laboratories. An open call for participation was also posted on several biological science information Web sites including the American Laboratory, Force11, Biospectrum, the International Society of Service Innovation

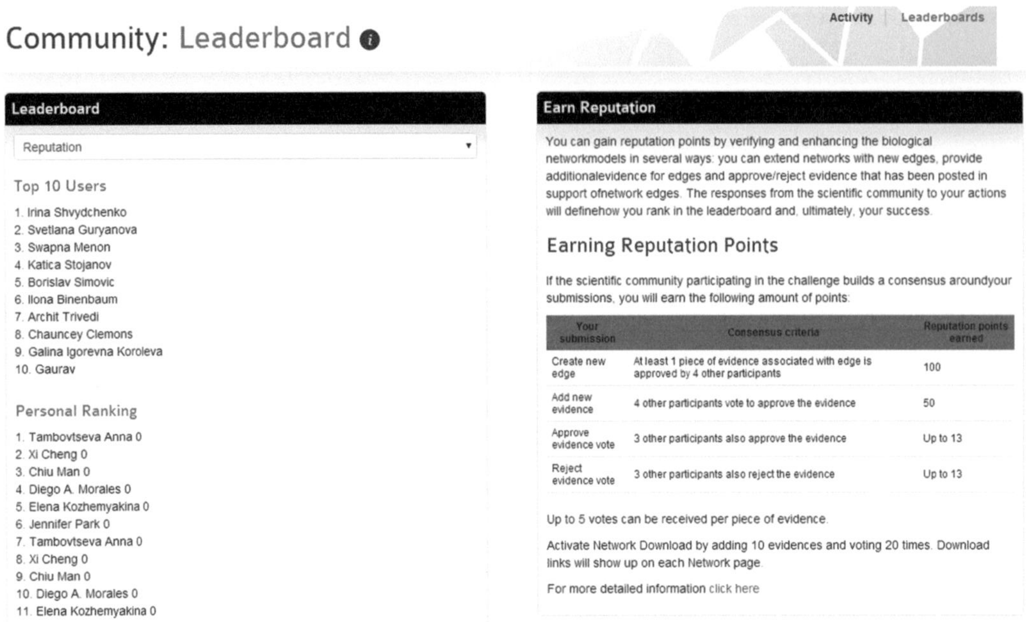

Fig. 9 Leaderboard tracks participant rank on the Bionet Web site. Users earn points for all Web site activity and compete for ranking on the Leaderboard

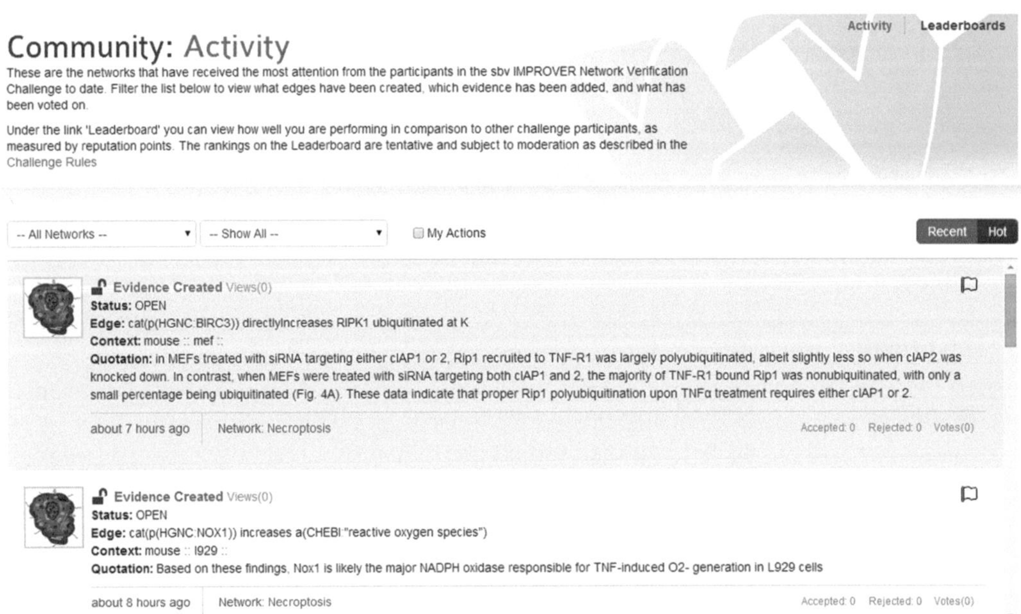

Fig. 10 Community Activity page on the Bionet Web site summarizes the most recent actions and allows users to vote from this page

Professionals, and EMBnet. A paper presenting the NVC and the Bionet Web site was published in Bioinformatics and Biology Insights [27]. Extensive training material was made available on the NVC Web site, including short videos, and regular training Webinars were offered by the NVC team.

The most effective approach in recruiting participants proved to be outreach through NVC Ambassadors, i.e., scientists who were familiar with the networks and who promoted the NVC through seminars and one-on-one training. In this more personal approach, Ambassadors contacted potential participants with personalized messages that emphasized the benefits of participating in the challenge and were tailored to each individual's scientific and/ or career interests. Ambassadors recruited, trained, and provided technical support to the participants, using their own scientific networks to promote the NVC.

We estimated that the advertising campaign during the first 5-month pilot phase of the project resulted in 1,000,000 NVC Bionet Web site views. From this, there were 1,298 unique visitors to the collaborative Web site, which resulted in 132 Bionet registrations. Finally, 26 highly active participants were selected as Best Performers in the NVC. About 85 % of all registered participants and 75 % of the Best Performers were recruited and/or trained and supported by NVC Ambassadors.

5.2 Motivation

NVC participants were motivated by several different factors. The most active contributors were invited to a 3-day meeting in Montreux, Switzerland, to discuss the final changes and additions to the next version of models and to network with experts in the field who were also invited. This was a powerful motivator for some of the participants. Other participants were excited by the opportunity to learn BEL for knowledge representation and biological network model building. BEL is gaining acceptance in industry and academia and knowledge of this language can be advantageous to new graduates aiming to enter the field of network biology or to seasoned scientists looking for new approaches to compile existing knowledge and better represent signaling pathways [30, 31]. The NVC Help material and Webinars, as well as the one-on-one training that was offered by the Ambassadors, were effective tools in the educational motivation approach.

Educational motivation was also used to engage Toledo University (Toledo, OH, USA) students in the Biosciences Graduate Program in Translational Research, and NVC has been incorporated into the curriculum at this university. After initial tutorials led by an NVC Ambassador, students were asked to complete assignments to review evidence autonomously and at their own pace. The NVC Ambassador was available for online consultations. Each student submitted a report justifying their "approve" or "reject" votes. For each report, the NVC Ambassador provided feedback that was sent to the course instructor and the student.

Participants in the NVC were allowed to download models after they had contributed a preset required number of new evidence pieces. This was a motivating factor for several scientists working on signaling represented by the network models.

Several scientists reported becoming "addicted" to the evidence review process, perhaps because gamification principles, which allowed participants to gain points and compete with their peers, were incorporated into the NVC. Both the competitive challenge and the pursuit of fun and enjoyment while playing a science game served as powerful motivators.

However, the top motivator for NVC participants, according to survey responses, was the possibility of co-authoring an academic publication. Indeed, the most active participants together with the NVC organizers have prepared and submitted a manuscript on the biological networks and improvements made during the first NVC.

6 Challenges and Future Prospects

One of the biggest challenges of any crowdsourcing project is recruiting a critical mass to participate in the activity in a consistent manner. We quickly learned that our target crowd was biologists with knowledge of biological processes covered in the networks who were motivated to add biology to the networks, as opposed to computational biologists who may have found the technology behind the NVC interesting, but who were not the right crowd to contribute and vet biology within the networks. To participate, biologists had to learn BEL, a computable language that might have discouraged people who did not want to commit time to learn a new language. NVC Webinars and Ambassadors offered BEL training. Nevertheless, many participants' entries required correction by the NVC team because of the incorrect use of BEL. This issue can be addressed in future challenges by enhancing the Bionet BEL statement entry page with controlled vocabulary that will help prevent incorrect terms being entered and make it a seamless task for the participants. Improvements to and additional advertising of the automated Convert Text to BEL tool will also encourage participants to contribute and avoid the extra step of the NVC team having to correct BEL statements.

Overall, many valuable biological contributions were made to the networks by the crowd. However, there were cases where the added evidence had missing information or included contexts outside the boundary conditions. To minimize these quality issues, additional Web site improvements will help guide users by specifically stating the boundary conditions at the point of evidence creation and requiring information such as tissue context before a form can be submitted.

The long-term goal is to have participants self-manage the Web site and flag quality issues (similarly to Wikipedia). As a reputation-based system, the Bionet Web site is designed to motivate participants to contribute biology to improve their scores and reputation among their peers. By becoming a standard in the network biology community, Bionet can draw on a crowd of scientists to review biological networks comprehensively and become the primary place where the most up-to-date networks can be sourced, thereby stimulating collaboration between scientific groups.

References

1. KEGG: Kyoto Encyclopedia of Genes and Genomes (2014) http://www.genome.jp/kegg/. Accessed 7 Aug 2014
2. Reactome: a curated pathway database (2014) http://www.reactome.org/. Accessed 7 Aug 2014
3. Pathway Commons (2014) http://www.pathwaycommons.org/about/. Accessed 7 Aug 2014
4. WikiPathways^{BETA} (2014) http://wikipathways.org/index.php/WikiPathways. Accessed 7 Aug 2014
5. BioModels Database (2014) http://www.ebi.ac.uk/biomodels-main/. Accessed 7 Aug 2014
6. Ahsan S (2014) UK public will choose which scientific puzzle to investigate for 2014 Longitude Prize. BMJ 348:g3417. doi:10.1136/bmj.g3417
7. Howe J (2006) The rise of crowdsourcing. WIRED Magazine 14(6):1–4
8. Good BM, Su AI (2013) Crowdsourcing for bioinformatics. Bioinformatics 29(16):1925–1933. doi:10.1093/bioinformatics/btt333
9. Parslow GR (2013) Commentary: crowdsourcing, Foldit, and scientific discovery games. Biochem Mol Biol Educ 41(2):116–117. doi:10.1002/bmb.20686
10. Good BM, Su AI (2011) Games with a scientific purpose. Genome Biol 12(12):135. doi:10.1186/gb-2011-12-12-135
11. Loguercio SI, Good BM, Su AI (2013) Dizeez: an online game for human gene-disease annotation. PLoS One 8(8), e71171. doi:10.1371/journal.pone.0071171
12. Dizeez: a multichoice quiz (2014) http://www.sciencegamecenter.org/games/dizeez. Accessed 7 Aug 2014
13. http://bioportal.bioontology.org/ontologies/DOID
14. Kawrykow A, Roumanis G, Kam A et al (2012) Phylo: a citizen science approach for improving multiple sequence alignment. PLoS One 7, e31362
15. Eyewire: a game to map the brain (2014) https://eyewire.org/. Accessed 7 Aug 2014
16. Rowles TA (2013) Power to the people: does Eterna signal the arrival of a new wave of crowd-sourced projects? BMC Biochem 14:26. doi:10.1186/1471-2091-14-26
17. The Cure: play games, defeat cancer (2014) http://genegames.org/cure/. Accessed 7 Aug 2014
18. Wu C, Macleod I, Su AI (2013) BioGPS and MyGene.info: organizing online, gene-centric information. Nucleic Acids Res 41(Database issue):D561–565. doi:10.1093/nar/gks1114
19. Good BM, Clarke EL, de Alfaro L, Su AI (2012) The Gene Wiki in 2011: community intelligence applied to human gene annotation. Nucleic Acids Res 40(Database issue):D1255–D1261. doi:10.1093/nar/gkr925
20. Brister JR, Le Mercier P, Hu JC (2012) Microbial virus genome annotation-mustering the troops to fight the sequence onslaught. Virology 434:175–180
21. Hingamp P, Brochier C, Talla E et al (2008) Metagenome annotation using a distributed grid of undergraduate students. PLoS Biol 6, e296
22. Harvard Catalyst (2014) http://catalyst.harvard.edu/. Accessed 7 Aug 2014
23. Wan Q, Pal R (2014) An Ensemble based top performing approach for NCI-DREAM drug sensitivity prediction challenge. PLoS One 9(6), e101183. doi:10.1371/journal.pone.0101183
24. Holland RC, Lynch N (2013) Sequence squeeze: an open contest for sequence compression. Gigascience 2(1):5. doi:10.1186/2047-217X-2-5

25. ResearchGate (2014) http://researchgate.net, Accessed 7 Aug 2014
26. StackOverflow (2014) http://stackoverflow.com. Accessed 7 Aug 2014
27. sbv IMPROVER project team (in alphabetical order), Ansari S, Binder J, Boue S et al (2013) On crowd-verification of biological networks. Bioinform Biol Insights 7:307–325. doi:10.4137/BBI.S12932
28. svb IMPROVER (2014) https://sbvimprover.com. Accessed 7 Aug 2014
29. Gebel S, Lichtner RB, Frushour B et al (2013) Construction of a computable network model for DNA damage, autophagy, cell death, and senescence. Bioinform Biol Insights 7: 97–117
30. Schlage WK, Westra JW, Gebel S et al (2011) A computable cellular stress network model for non-diseased pulmonary and cardiovascular tissue. BMC Syst Biol 5:168
31. Westra JW, Schlage WK, Frushour BP et al (2011) Construction of a computable cell proliferation network focused on non-diseased lung cells. BMC Syst Biol 5:105
32. Westra JW, Schlage WK, Hengstermann A et al (2013) A modular cell-type focused inflammatory process network model for non-diseased pulmonary tissue. Bioinform Biol Insights 7:167–192. http://www.ncbi.nlm.nih.gov/pubmed/23843693
33. Park JS, Schlage WK, Frushour BP et al (2012) Construction of a computable network model of tissue repair and angiogenesis in the lung. J Clinic Toxicol S12:002. doi:10.4172/2161-0495.S12-002
34. Catlett NL, Bargnesi AJ, Ungerer S et al (2013) Reverse causal reasoning: applying qualitative causal knowledge to the interpretation of high-throughput data. BMC Bioinformatics 14:340
35. The BEL Language (2014) http://www.openbel.org/. Accessed 7 Aug 2014
36. sbv IMPROVER Network Verification Challenge (2014). https://bionet.sbvimprover.com/. Accessed 18 Aug 2014

Chapter 7

Toxicological Assessment Via Gene Network Analysis

Simona Iacobacci, Francesco Gregoretti, Gennaro Oliva, Roman Polishchuk, and Vincenzo Belcastro

Abstract

Network inference and the analysis of a sub-network involved in the toxicological response are described in this chapter. The chapter introduces the project and the scientific context in which it was conceived and reviews the materials and methods used in the experiment. The methods section is organized into three parts that describe the data gathering process and inference of a gene network, the use of two network visualization tools (Cytoscape and NetView), and the analysis of a gene whose dysfunction implies a toxic response. The data gathering process consists of the selection of genome-wide gene expression profiles from public repositories and a subsequent data-formatting step. This part includes all necessary instructions indicating how to install and run the inference algorithm to generate a gene network from the data retrieved in the previous step. The second part of the section overviews the two network visualization tools used in the experiment, Cytoscape and NetView. Cytoscape was loaded with a mock gene network whose source files for its inference are distributed with the chapter. Alternately, NetView is pre-loaded, as it serves as a web interface to access mammalian gene networks. Lastly, this section describes how information of interest is extracted from the human NetView gene network. The chapter presents the findings of a study case, during which the gene of interest, the copper (Cu)-transporting ATPase gene *ATP7A*, is investigated. Mutations in *ATP7A* lead to the toxic accumulation of Cu in the intestinal epithelium and deficit of the metal in other tissues, thus causing Menkes disease. By identifying a network of genes that co-express with *ATP7A*, we identified targets that may disrupt trafficking of the protein product of the gene, which thus influence the toxicological outcomes of the disease. The last part of the chapter assesses a biological validation of NetView predictions, and provides wet lab protocols that may be used to identify the molecular players involved in ATP7A trafficking.

Key words Gene network, Reverse-engineering, Mutual information, Copper accumulation, Menkes disease

1 Introduction

Two decades have passed since high-throughput technologies were first applied to biological research. The advent of genomics has accelerated the characterization of organisms and allowed for comprehensive modeling of entire biological systems. Pioneering microarray technologies have allowed scientists to measure simultaneously the mRNA expression levels for thousands of genes.

Julia Hoeng and Manuel C. Peitsch (eds.), *Computational Systems Toxicology*, Methods in Pharmacology and Toxicology,
DOI 10.1007/978-1-4939-2778-4_7, © Springer Science+Business Media New York 2015

The availability of gene expression profiles at the genomic level has drastically changed the way researchers address biological investigations because a genome-wide expression profile is considered to be the cell's fingerprint; that is, it expresses the status of a cell, at the transcriptional level, under specific experimental conditions. At the beginning of an investigation, expression profiles can be used to narrow down the research space to the "active" parts of the cell. Currently, hundreds of thousands of genome-wide gene expression profiles are stored in public repositories such as the NCBI Gene Expression Omnibus (GEO) [1] and the EBI ArrayExpress [2].

In the first section of this chapter we discuss how gene expression data can be gathered from the ArrayExpress repository and organized into the format required for the inference algorithm. In the second section, we describe an inference algorithm based on mutual information (MI), which is a way to discriminate between pairs of genes whose expression profiles change coherently over a set of experiments. An example of this technique is delineated in Fig. 1, in which the expressions of three genes, *X*, *Y*, and *Z*, are plotted over different experiments. The example shows that the expressions of *X* and *Y* are directly related, undergoing downregulation and upregulating in a coherent manner. Genes *Z* and *Y* have unrelated expressions (the example of the expression of *X* versus that of *Z* is not reported). In this scenario the MI between *X* and *Y* is significant, indicated by an arrow that connects the two, while *Z*'s expression is independent from the others and is thus not connected with the other two genes.

MI is a probabilistic measure of relatedness [3] and measures the statistical dependencies between two variables (*see* **Note 1**).

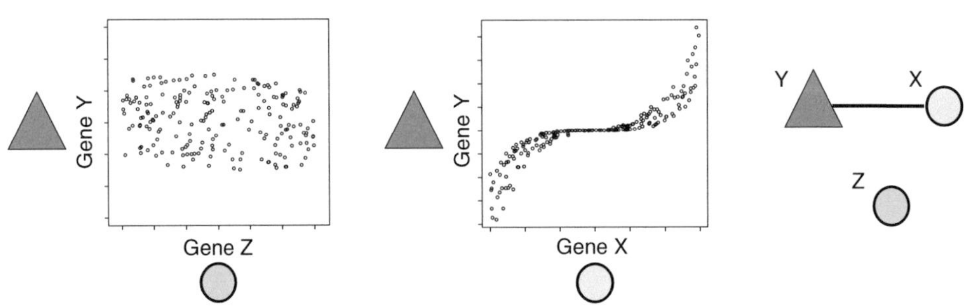

Fig. 1 Independent genes versus co-expressed genes: This example shows two gene pairs and their corresponding expression values across multiple experiments. Each *circle* within the plotting areas indicates the expression values of genes *Z* and *Y*, and genes *X* and *Y*, as indicated. Hence, each *circle* is associated with the outcome of an experiment/sample (axis ticks are deliberately left blank because they will have no effect on correlation or MI). Genes *Z* and *Y* are independent because the circles are distributed randomly within the plotting area. Thus, we can state that genes *Z* and *Y* have expression profiles for which the correlations and MIs will be close to zero. On the other hand, genes *X* and *Y* have dependent expression profiles because the expressions of both genes change coherently across the set of samples. Genes *X* and *Y* are thus co-expressed and show a positive correlation and MI. Therefore, when constructing the gene network, genes *X* and *Y* are connected with an edge to denote the fact that their profiles are not independent

In the context of gene network inference, the MI measures whether two genes tend to be co-expressed over a set of experiments. A higher tendency implies a higher MI value, and vice versa.

The algorithm and the manner in which it is applied to the project are described in Belcastro et al. [4]. The algorithm was first implemented to infer mammalian gene networks from a massive set of experimental data derived from different studies [5], and hence data were not directly comparable (as for data within the same experiment). Current methodologies such as ARACNe [6] and BANJO [7] cannot be applied since they require normalized or comparable expression values. However another highly used algorithm, NIR [8], required the gene perturbed in each experiment as an additional input value/variable. This information is usually only known for specific experimental designs. The algorithm overcomes the limitations of performing a preliminary normalization procedure that allows experimental data to be compared across independent experiments [5]. In the Methods section, we describe how to install and run the algorithm and how to collect the output. The output consists of a list of gene pairs (gene *X* and gene *Y*) and the corresponding MI value (*see* **Note 1**).

In the second section, we describe software that can be used to visualize a representative gene network and a platform to extract information from an existing gene network. To this aim, we considered two systems, Cytoscape [9], which is a visualization tool, and NetView [5], which is a system used to interrogate an existing gene network. Some researchers, us included, have already described how to query NetView to extract informative content, i.e., the identification of protein interactions [10], and the discovery of associations between de novo disease genes and pathways [5]. Although here we only describe how NetView was queried to evaluate toxicological responses, it is broadly applicable for other objectives. For example, NetView has been queried in the context of genetic diseases to identify interactors of the dystrophy-causing gene DYSF [10] and to investigate cilia dysfunctions [11].

In the third section we show how to query NetView to find potential modulators of Cu-transporting ATPase gene (*ATP7A*) function. *ATP7A* encodes a transmembrane pump that plays a major role in regulating the homeostasis of Cu, an essential cofactor for a substantial number of vitally important enzymes [12, 13]. Mutations in the *ATP7A* gene frequently affect the protein product's ability to traffic from the Golgi to the basolateral surface of enterocytes, where it participates in the excretion of excess Cu into the blood circulation. Consequently, ingested alimentary Cu that is absorbed in the intestine accumulates in the enterocytes and induces serious toxicity, while the rest of the body suffers a deficit of this essential metal [12, 13]. NetView provided a network of genes that were co-expressed with *ATP7A*, which can be targeted to modulate ATP7A function and ATP7A-associated toxicity in Menkes disease [14–16].

2 Materials

2.1 Cell Culture

HeLa cells were grown in Dulbecco's modified Eagle medium supplemented with 10 % fetal calf serum, 1 mM glutamine, and 1 mM penicillin and streptomycin (Invitrogen Life Technologies, Grand Island, USA). To analyze ATP7A trafficking, the cells were treated overnight with 500 μM copper-chelating agent bathocuproine disulfonic acid (BCS) (Sigma-Aldrich, St. Louis, MO, USA) and 200 μM $CuSO_4$ (Sigma-Aldrich) in culture medium at 37 °C for 2 h.

2.2 Antibodies and Small Interfering RNA

For immunofluorescence experiments the following antibodies were used: rabbit anti-human ATP7A (Hycult Biotech, Pennsylvania, USA), mouse anti-human golgin 97, secondary Alexa Fluor 488 goat anti-rabbit, and secondary Alexa Fluor 568 goat anti-mouse IgG antibodies (all antibodies were from Invitrogen Life Technologies).

Small interfering RNA (siRNA) oligonucleotides targeting the coiled-coil domain containing gene 2 (*GCC2*) were purchased from Sigma-Aldrich (a combination of two different human siRNA duplexes) (*see* **Note 2**). The following siRNAs were used: siGCC2-1, 5′-GGAGUUGGAACAAUCACAU-3′; and siGCC2-2, 5′-CUCGUAUUGAGACAUUACA-3′.

HeLa cells were transfected using oligofectamine (Invitrogen Life Technologies) according to the manufacturer's instructions. Transfection of siCONTROL non-targeting duplexes (Sigma-Aldrich) was used as a negative control. The cells were incubated with siRNAs for at least 72 h before further treatment.

2.3 Quantitative Real-Time PCR

Total RNA from the cells was purified using QIAshredder (Qiagen spa, Milan, Italy) and extracted with RNeasy Protect Mini Kit (Qiagen) under standard conditions. Total RNA (1 μg) was reverse transcribed using the QuantiTect Reverse Transcription kit (Qiagen) according to the manufacturer's instructions. Quantitative real-time PCR (qPCR) experiments were performed using Light Cycler 480 Syber Green MasterMix (Roche spa, Monza, Italy) for cDNA amplification and in LightCycler 480 II (Roche) for signal detection. The qPCR results were analyzed using the comparative Ct method normalized against the housekeeping gene, β-actin.

The specific primer pairs were b-ACTIN forward (5′-AAGA GCTACGAGCTGCCTGA-3′); b-ACTIN reverse (5′-GACTCC ATGCCCAGGAAGG-3′); GCC2 forward (5′- GCGTACGCT AAGTGCATACCA-3′); and GCC2 reverse (5′-TCGGACTTT GTAGCTCTCGAA-3′).

3 Methods

3.1 Gene Network Inference

The process of gathering gene expression data from ArrayExpress and how to install and run it to implement the inference algorithm are described here.

3.1.1 Gathering Gene Expression Data

Gene expression profiles can be retrieved from the EBI ArrayExpress repository [2]. Here we describe two different procedures; in one, we download a single experiment with multiple samples, and in the other, we fix a microarray chip and retrieve all the experiments in which that chip was used.

Procedure 1

(a) Connect to http://www.ebi.ac.uk/arrayexpress/experiments/browse.html.

(b) Select an organism of interest (e.g., "*Homo sapiens*").

(c) Select "Array assay" as the technology.

(d) Sort the results in descending order by "Assays."

(e) Select the dataset of interest for which a processed data file is available, and download it by clicking on the corresponding icon in the "Processed" column (e.g., "E-GEOD-28878," *see* **Note 3**).

(f) Format the data to reflect the input format required by the algorithm (*see* Sect. 3.1.2 below).

Procedure 2

(a) Connect to http://www.ebi.ac.uk/arrayexpress/arrays/browse.html.

(b) Search for a microarray chip by name (e.g., "HG-U133A").

(c) Click the link in the "Accession" column (it will be "A-AFFY-33," if "HG-U133A" was selected in the previous step).

(d) Click the hyperlink in the "Links" row.

(e) Download the processed data of interest and combine them to obtain a file that reflects the input format required by the algorithm (*see* Sect. 3.1.2 below).

3.1.2 Parallel Mutual Information Tool

The parallel mutual information (pmutualinformation) tool computes the pairwise MI for all pairs of genes from a massive and heterogeneous dataset. This dataset contains gene expression profiles that may have been obtained from different experiments collected from public repositories. Input data must be formatted as explained below. The pmutualinformation tool runs an MIMD system, and has several CPU that work independently, which gives it the capacity to execute different instructions on different data. A typical example of an MIMD system is a cluster of multi-core computers, and this has been implemented in C using the Message Passing Interface (MPI) standard (*see* **Note 4**).

Installation: The installation process is achieved easily by following these steps:

(a) Install the prerequisite software:
   - The gnu scientific library (gsl)
   - Any MPI library that is compatible with version 1.0 of the MPI standard (the software has been tested with mpich and openmpi).

(b) Download the latest release of the .tar.gz file from http://sourceforge.net/projects/pmutualinformation/

(c) Unpack the source code with the command

```
gunzip -c pmutualinformation-VERSION.tgz | tar xf -
```

(d) Change the working directory to the newly created directory with the command

```
cd pmutualinformation-VERSION
```

(e) Configure the installation with the command

```
./configure
```

This will check the system and automatically configure pmutualinformation. Run it with the --help option to obtain a list of command-line arguments to modify the installation.

(f) Compile pmutualinformation with the command

```
make
```

(g) Install the binary with the command (optional)

```
make install
```

or copy the pmutualinformation executable that can be found under the src subdirectory into a suitable location for execution with the MPI library (typically under the shared home directory).

Execution of the pmutualinformation tool (*see* also **Note 5**): To run the pmutualinformation tool users need to provide two input files, which are described graphically in Fig. 2:

- The gene expression file (called *data_matrix.txt* in the tarball archive) is a space-delimited text file that contains the gene expression values for all selected experiments arranged in a matrix format. The matrix dimension is given by the number of probe sets (rows of the matrix), and by the sum of the overall number of gene expression profiles (columns of the matrix).
- The input dataset info file (called *data_vector.txt* in the tarball archive) is a space-delimited text vector file that contains the number of samples for each experiment selected.

Expressions file

```
____________1st experiment__________  ___2nd experiment__     ______m-th experiment_______
|                                    |                   |   |                           |
0.85712 0.17775 0.59025 0.73301 0.9532 0.94919   ... 0.9862 0.45947 0.89961
0.18961 0.71907 0.22892 0.80476 0.86682 0.012479 ... 0.19807 0.24791 0.98917
.
.
.
0.10711 0.59011 0.02697 0.72015 0.67066 0.90053  ... 0.27624 0.18676 0.62662
```

Dataset info file

```
4 2 ... 3
```

Fig. 2 Representative samples of pmutualinformation input files: The expression file contains the gene expression values; each row is a gene and each column is a sample, separated by a space. Expression values from multiple experiments are bound column-wise. In the input matrix, "m" experiments are concatenated, each with a certain number of samples. In the dataset info file, the numbers of samples, one for each experiment, are arranged in a single row (*data_vector.txt*); the first experiment has four samples, the second experiment has two samples, and the m-th experiment has three samples

To test the pmutualinformation distribution a sample run should be performed. The two input files needed are included in the distribution of the code. The expression matrix that we have included was generated from a synthetic gene network of 100 genes, and used in the context of the DREAM5 challenge ([17], http://www.the-dream-project.org/).

The pmutualinformation tool is executed via a command line with the mpirun command:

mpirun [mpirun options] pmutualinformation -n 100 -d *data_vector.txt* -e *data_matrix.txt* -c 3 -o *output.txt*

The mpirun options are specific to the MPI library implementation (typically the number of parallel processes to be executed with the -np option can be specified).

Users can then specify a set of options to the pmutualinformation tool:

- n, set the number of probes.
- d, set the dataset info file name.
- e, set the expressions file name.
- c, set the discretization intervals (default 3, *see* **Note 6**).
- o, set the output file name.

The output file *output.txt* is a space-delimited text file arranged in a tabular format that contains only significant MI values (*see* **Note 3**). Each row of the file contains a pair of genes (geneX and geneY) and the corresponding MI value.

3.1.3 Comparison with Other Inference Algorithms

The MI algorithm that we have previously described overcomes some of the shortcomings of other gene network inference tools. The closest tool in terms of the adopted statistical model is ARACNe [6]. ARACNe computes the MIs between pairs of genes by fitting a probability distribution over gene expression values as the first step. Since such is the case, it requires a comparable set of expression values; all samples have to be generated within the context of the same experiments, or data from different experiments have to be normalized all together. Both cases limit the ARACNe's applicability to hundreds of samples. Hence, we could not directly compare ARACNe and pmutualinformation performances. However, as outlined in an extensive comparative study [18, 19] from the DREAM challenge organizers [17] among gene network inference algorithms, tools for relevance network inference, such as MI, show comparable inference performance.

BANJO [7] is a gene network inference algorithm that implements Bayesian networks A Bayesian network is a graphical model for probabilistic relationships among a set of random variables (genes in the context of gene network inference). Given a set of connections, the relationships between the nodes of the network are described by joint and conditional dependences whose distributions are fitted from the expression data.

As for ARACNe, its applicability is limited only to comparable expression data. BANJO performances were tested by Bansal et al. (2007) [20]. Results have shown that BANJO has a high precision rate but very low recall; in other words, BANJO recovers very few connections, although most of them are exact. In the same study, the authors were unable to run the java implementation of BANJO (the only one available) to recover networks of thousands of genes, given that BANJO required too much RAM to populate all data structures.

In the next section we introduce two gene network visualization tools, Cytoscape and NetView. Both tools are only adopted for visualization purposes and cannot be used as inference algorithms.

3.2 Network Visualization

In the following two subsections we describe Cytoscape and NetView. Cytoscape is a powerful network visualization and analysis tool, and NetView is an online platform that browses two mammalian co-expression networks.

3.2.1 Cytoscape

Cytoscape [9] is a Java application used to analyze biological data. Among the different functionalities and plug-in tools Cytoscape offers the possibility to load and visualize the users' own gene networks. Here we describe the steps to load a network (*see* **Note 7**) stored as a list of edges (the output of the inference tool from Sect. 3.1.2).

The network we visualize in the following steps is stored in the file *output.txt*, and is a mock gene network, generated from the pmutualinformation tool in Sect. 3.1.2. This mock network has been inferred from a set of computer-simulated gene expression profiles using the same simulator adopted to generate datasets for the DREAM challenge initiative [21].

(*We assume that Cytoscape version 3.x is used, and the output file name is output.txt.*)

(a) Download and open the Cytoscape application from http://www.cytoscape.org/.

(b) Click on "From Network File..." and locate the file "output.txt."

(c) Click "Show Text File Import Options."

(d) Select the correct delimiter (*output.txt* is a space-delimited file).

(e) Under the "Interaction Definition" options, select "Column 1" as "Source Interaction" and "Column 3" as "Target Interaction," and leave "Interaction Type" as it is. Then in the "Preview" click on "Column 2."

(f) Select menu option "Layout→Files Layout→Organic."

If the process is correct a network similar to the network in Fig. 3 should be obtained. The network in Fig. 3 has more information with respect to that of the network obtained following the previous steps. This network was obtained by merging information from both *output.txt* (predicted network) with the original in silico network, which was used to simulate gene expression for the inference in Sect. 3.1.2. The 100 nodes of the network represent the 100 mock genes, and edges represent significant MI between genes. Edges in red are also present in the original network and hence represent correct predictions. The software distribution we provide also includes a Cytoscape session file *cyto_net.cys*, along with the original file used to load the network, *net.txt* (*see* **Note 8**).

3.2.2 NetView

NetView [5] is a web platform that allows registered users (free for noncommercial use) to query and browse gene networks and perform gene ontology and enrichment analyses. The tool is available at http://netview.tigem.it. A step-by-step description of how to build your own NetView tool and how to use the existing tool to perform the different types of analyses is available in [22]. The gene identifier *ATP7A* was used to query the gene network. Co-expression between some genes and *ATP7A* (Fig. 4) was confirmed experimentally by their physical interactions, as reported in Sect. 3.3.

To reproduce the results in Fig. 4, perform the following steps:

(a) Connect to http://netview.tigem.it.

(b) Click the NETVIEW menu entry. (Only registered users can perform this analysis. Registration is free for noncommercial use.)

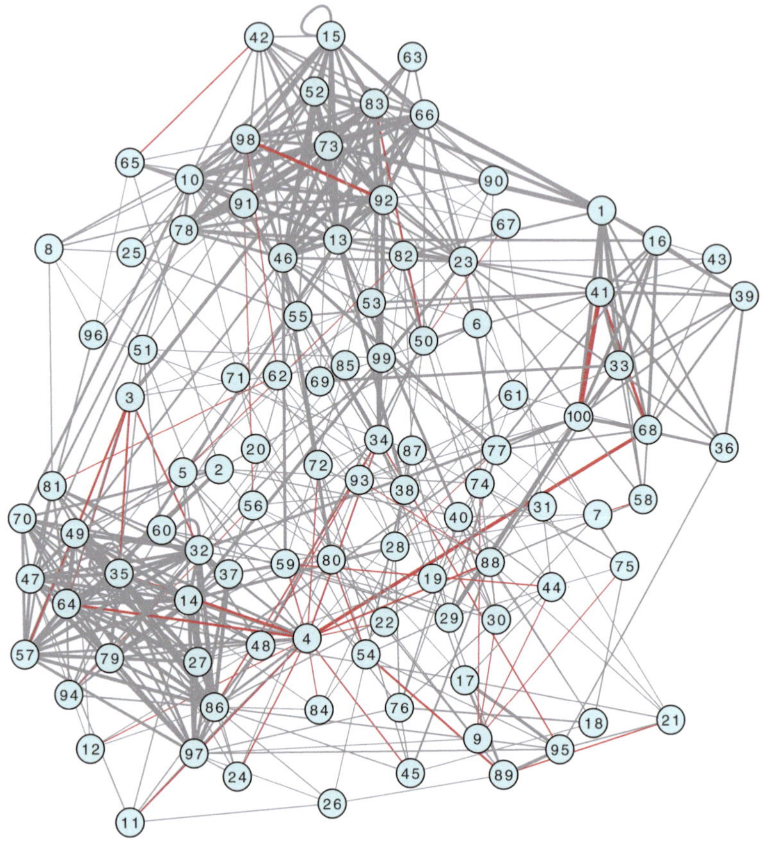

Fig. 3 Gene network visualization using the Cytoscape software: The network contains 100 nodes and 528 edges. Edges predicted by pmutualinformation are shown in *grey*. The edges shown in *red* were present in the original in silico network and also predicted by pmutualinformation. The nodes were arranged automatically by selecting the layout function "yFiles Layout→Organic"

(c) Type "atp7a" in the box "Identifier."

(d) Select the second "Probeset ID" (205198_s_at; the results for the other probeset ID are a subset of this).

(e) Select "5" as "Neighbors," and "1" as Depth.

(f) Click the "Show Predictions" button.

The result of the query is shown in Fig. 4 (*see* **Note 9**).

3.3 Experimental Validation

To investigate the molecular players involved in ATP7A trafficking we used the NetView tool (http://netview.tigem.it) to find the co-expressors that had a documented role in protein trafficking processes.

In the NetView query form we selected "human" as species, "Gene Symbol" as identifier type, "ATP7A" as identifier, "50" as number of neighbors, and "1" as depth. We performed two different queries using two possible probes (205197_s_at and

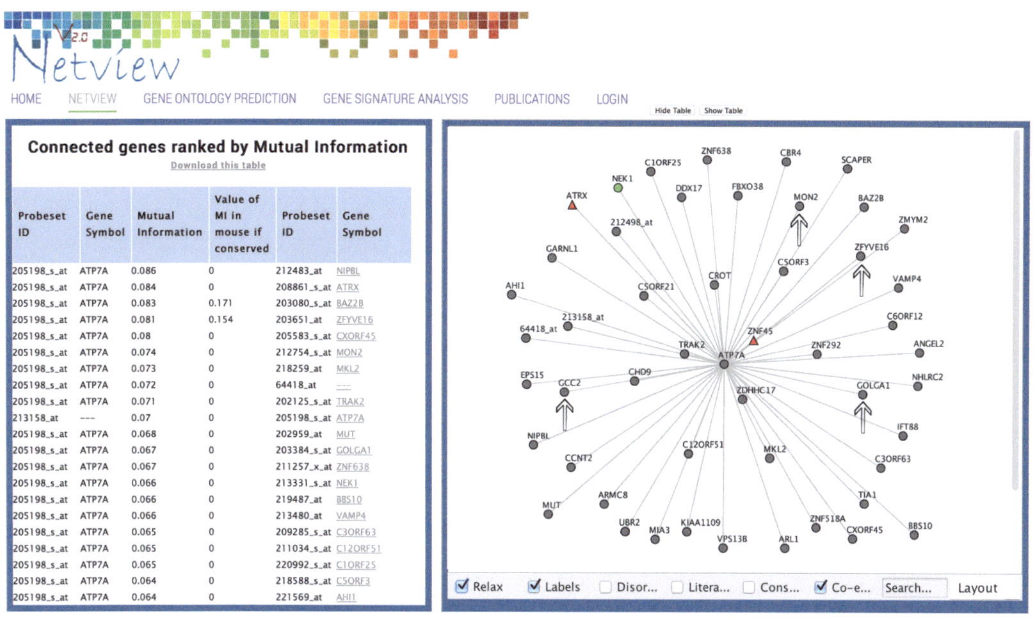

Probeset ID	Gene Symbol	Mutual Information	Value of MI in mouse if conserved	Probeset ID	Gene Symbol
205198_s_at	ATP7A	0.086	0	212483_at	NIPBL
205198_s_at	ATP7A	0.084	0	208861_s_at	ATRX
205198_s_at	ATP7A	0.083	0.171	203080_s_at	BAZ2B
205198_s_at	ATP7A	0.081	0.154	203651_at	ZFYVE16
205198_s_at	ATP7A	0.08	0	205583_s_at	CXORF45
205198_s_at	ATP7A	0.074	0	212754_s_at	MON2
205198_s_at	ATP7A	0.073	0	218259_at	MKL2
205198_s_at	ATP7A	0.072	0	64418_at	---
205198_s_at	ATP7A	0.071	0	202125_s_at	TRAK2
213158_at	---	0.07	0	205198_s_at	ATP7A
205198_s_at	ATP7A	0.068	0	202959_at	MUT
205198_s_at	ATP7A	0.067	0	203384_s_at	GOLGA1
205198_s_at	ATP7A	0.067	0	211257_x_at	ZNF638
205198_s_at	ATP7A	0.066	0	213331_s_at	NEK1
205198_s_at	ATP7A	0.066	0	219487_at	BBS10
205198_s_at	ATP7A	0.066	0	213480_at	VAMP4
205198_s_at	ATP7A	0.065	0	209285_s_at	C3ORF63
205198_s_at	ATP7A	0.065	0	211034_s_at	C12ORF51
205198_s_at	ATP7A	0.065	0	220992_s_at	C1ORF25
205198_s_at	ATP7A	0.064	0	218588_s_at	C5ORF3
205198_s_at	ATP7A	0.064	0	221569_at	AHI1

Fig. 4 NetView output predictions for the *ATP7A* gene. The *table on the left* contains the list of genes predicted to be co-expressed with the *ATP7A* gene. The table contains the MI values for the human samples and the conserved MI values for the mouse samples, in the third and fourth columns, respectively. On the *left*, the predictions are displayed using a Java applet. The nodes, indicated by *arrows*, have been shown experimentally to interact physically with *ATP7A*

205198_s_at), and found that several genes related to membrane trafficking (*GCC2*, *MON2*, *GOLGA1*, *ARL1*, *see* Fig. 4) were co-expressed with *ATP7A*. How these genes were relevant to ATP7A localization and trafficking was determined through their silencing roles in RNA interference (RNAi, see below).

3.3.1 Cell Culture

In the silencing experiment HeLa cells were detached with trypsin 1× and plated onto a glass cover slip in a 24-well plate with 20 % confluence (around 15,000 cells/well) for immunofluorescence experiments, and in a 12-well plate (around 30,000 cells/well) for qPCR experiments. The cells were grown without antibiotics to avoid inhibition of RNAi.

The next day, genes of interest were silenced in the HeLa cells, and then the cells were left for 3 days with the siRNA. For each sample, three different conditions were analyzed: steady state, low Cu, and high Cu. Under steady-state conditions the cells were kept in the culture medium without addition of any drug. Under low Cu conditions the cells were treated with 500 μM Cu-chelating agent BCS overnight. In high Cu condition the cells were treated with 500 μM Cu-chelating agent BCS overnight, and the day after they were incubated with 200 μM $CuSO_4$ for 2 h.

The cells from each of the three conditions underwent immunofluorescence or qPCR experiments.

3.3.2 RNA Interference

For the silencing experiments HeLa cells were transfected with oligofectamine using the following protocol. Cells were plated 1 day before transfection to ensure 30–50 % confluence at the time of transfection. For each transfection sample, complexes were prepared as follows:

(a) Dilute 1 μL of a 20 μM stock of siRNA in 16 μL of Opti-MEM I Reduced Serum Medium (Invitrogen Life Technologies) to a final volume of 17 μL. Mix gently.

(b) Mix oligofectamine reagent gently before use, and then dilute 0.4–0.8 μL in Opti-MEM® I Medium to a final volume of 3 μL. Mix gently and incubate for 5–10 min at room temperature.

(c) Combine the diluted oligonucleotide with diluted oligofectamine reagent (total volume = 20 μL). Mix gently and incubate for 15–20 min at room temperature (the solution may appear cloudy).

(d) While complexes are forming, remove the growth medium from the cells and wash once with medium without serum. Add 80 μL of medium without serum to each well containing cells.

(e) Mix the 20 μL of complexes (from step c of this procedure) gently and add to the cells.

(f) Incubate the cells at 37 °C in a CO_2 incubator.

3.3.3 RNA Preparation and qPCR

RNA extraction took place 3 days after the silenced cells were rinsed twice with phosphate-buffered saline (PBS) containing 0.1 mM $CaCl_2$ and 1 mM $MgCl_2$.

The cells were then lysed at room temperature with 350 μL of lysis buffer (RLT buffer, Qiagen), which was added directly to the cell culture dish. Then 3.5 μL of 14.3 M β-mercaptoethanol was added to the lysis buffer. Cells were disrupted with a cell scraper. The lysate was pipetted and collected in a QIAshredder spin column (Qiagen). The QIAshredder column was spun at 10,000 × *g* for 2 min. Then, one volume of 70 % ethanol (Sigma-Aldrich) was added and mixed with the homogenized lysate.

The sample was transferred to an RNeasy spin column placed in a 2 mL collection tube. The column was spun at 8000 × *g* for 30 s. The flow-through was discarded. Next, 700 μL of supplied buffer RW1 was added to the RNeasy spin column and it was spun at 8000 × *g* for 30 s to cleanse the column membrane. Then, 500 μL of wash buffer RPE was added twice to the RNeasy spin column, which was spun at 8000 × *g* for 30 s. The RNeasy spin column was placed in a new 2 mL collection tube and spun at 10,000 × *g* for 1 min to eliminate any possible carryover of ethanol contained in buffer RPE, because ethanol may interfere with downstream reactions of the RNA. The RNeasy spin column was placed in a new sterile 1.5 mL Eppendorf, and 30 μL of RNase-free water was added to the spin column membrane, which was spun at 8000×*g* for 1 min to elute the RNA. Purified RNA was stored at −80 °C in RNase-free water.

The concentration of RNA was determined by measuring the absorbance at 260 nm (A_{260}) in a NanoDrop 2000c Spectrophotometer (Thermo Scientific, Massachusetts, USA). We quantified 1 μL of each sample of RNA. As a sample for NanoDrop calibration, 1 μL of RNase-free water was used. We estimated the ratio of the readings at 260 and 280 nm (A_{260}/A_{280}) to assess the purity of RNA with respect to that of contaminants such as proteins that absorb in the UV spectrum. Pure RNA has an A_{260}/A_{280} ratio of about 2.0.

For the qPCR experiments, cDNA was synthesized from an RNA template using the QuantiTect Reverse Transcription kit (Qiagen). This kit integrates quick cDNA synthesis with rapid removal of genomic DNA contamination. From each sample, 1 μg of RNA was used for reverse transcription.

The genomic DNA elimination reaction was prepared as follows:

- 2 μL of genomic DNA Wipeout Buffer 7× (Qiagen).
- 1 μg of template RNA.
- RNase-free water (Qiagen) to final volume of 14 μL.

The reaction was incubated for 2 min at 42 °C in the PCR machine (Applied Biosystems, Life Technologies) and then it was placed immediately on ice.

The reverse transcription reaction mixture was prepared as follows:

- 1 μL of reverse transcriptase.
- 4 μL of reverse transcriptase buffer.
- 1 μL of reverse transcriptase primer mix.

This mixture was added to the RNA template obtained from the genomic DNA elimination reaction (14 μL). The samples were incubated for 30 min at 42 °C and then for 3 min at 95 °C in the PCR machine. The reverse transcription reactions were placed on ice, followed by qPCR.

For the qPCR, primer pairs were designed using Primer 3 [23, 24] using the following settings:

- Primer length: Primers should have a length of 18–22 nucleotides.
- Primer T_m: Primers should have a melting temperature of 60–63 °C. The T_m of both primers in a primer pair should be within 2 °C.
- Primer G+C content: Primers should have a G+C content of 40–60 %.

- Secondary structure: Primers should lack any inverted repeat sequences, which can form a stable hairpin and result in inefficient primer binding.
- Genomic DNA avoidance: In real time PCR assay primers should be designed to flank an intron or to span an exon-exon junction.
- Product size ranges: The amplicon consists of 100–120 base pairs.

The qPCR reaction was prepared for each sample as follows:

- 1 μL of primer forward 10 μM.
- 1 μL of primer reverse 10 μM.
- 10 μL of LightCycler 480 Syber Green MasterMix.
- 6 μL of sterile nuclease-free water.

The reaction was assembled in a sterile 96-well plate (Roche), and then 2 μL of cDNA (threefold diluted) was added to each well. β-Actin was used as the endogenous reference gene to normalize the qPCR experiments. Each qPCR assay sample was tested and analyzed in triplicate to increase the reproducibility and to minimize the variability.

The qPCR gave us as result the threshold cycle (C_T) of each sample. The C_T value corresponds to the initial concentration of the target nucleic acid, which can be expressed as the cycle number required to achieve a preset threshold of amplification. To convert the C_T value to a meaningful number, the data are analyzed using relative quantification. In this analysis, the C_T value of the target nucleic acid in the test sample is compared with that of a control sample (generally called the calibrator). The results are expressed as a ratio (or fold difference) of the amount of a target nucleic acid in the test sample relative to that in the reference sample. This method is used to determine the difference in target levels between two different samples. For example, it indicates how much the expression level of a gene changes with a particular treatment. In relative quantification, it is important to ensure that the target levels in the test sample and calibrator are compared from equivalent amounts of starting material. The most common normalization method consists of normalizing the expression of the target gene to that of an endogenous reference gene (in this case β-actin) for which the expression is constant in all samples tested. First, the target gene in the test sample and calibrator are normalized to the endogenous reference gene. Then, the normalized values are compared to obtain a fold difference. Typically, the expression of the normalized target gene in the calibrator is set to "1," and the expression of the normalized target gene in the test sample is expressed as an *n*-fold increase or decrease relative to the calibrator. This technique is preferred over absolute quantification because using an internal standard (the endogenous reference gene) minimizes potential variations in sample preparation and handling and circumvents the need for accurate quantification and loading of the starting material. The results of the qPCR experiment were analyzed using the C_T of each sample and expressed as a ratio (or fold change) of the amount of a target nucleic acid in the test sample relative to the reference. The qPCR results are shown in Fig. 5a. The cells transfected with siRNA for GCC2 showed decreased mRNA levels compared with those in the control cells, indicating that a fair amount of silencing had occurred.

3.3.4 Immunofluorescence Analysis

Cells were fixed for 10 min with 4 % paraformaldehyde (PFA) (Sigma-Aldrich) in 0.2 M Hepes (Sigma-Aldrich), followed by incubation with a blocking/permeabilizing solution of 0.5 % bovine serum albumin (BSA) (Sigma-Aldrich), 0.1 % saponin (Sigma-Aldrich), 50 mM NH_4Cl (Sigma-Aldrich), and 0.01 % sodium azide (Sigma-Aldrich) in PBS for 20–30 min. Primary antibodies were diluted in blocking/permeabilizing solution and added to the cells for 1 h at room temperature. Alternatively, primary antibodies can be added to cells overnight at 4 °C. Then, primary antibodies were removed, and the fixed cells were rinsed six times with PBS. Secondary antibodies were diluted in blocking/permeabilizing solution and added to the cells for 45 min. Afterwards, secondary antibody was removed, and the cells were rinsed six times with PBS. Finally, the cover slips were mounted onto glass slide with Mowiol (Merck Millipore, Massachusetts, USA) (in PBS).

Samples were examined with a Zeiss LSM 700 confocal microscope equipped with a 63× 1.4 NA oil objective. The colocalization module of the Zeiss Zen 2008 software was used to measure the colocalization of ATP7A with different intracellular markers.

3.3.5 Phenotype Evaluation

NetView provided a network of genes (Fig. 4) that co-express with *ATP7A* and cooperate with ATP7A within the same pathway/function. Several co-expressors belong to the membrane trafficking machinery, and can be targeted to modulate ATP7A intracellular distribution and, thus, function to overcome Cu-associated toxicity in Menkes disease. We paid particular attention to siRNAs, which retain ATP7A at the Cu excretion sites (mainly cell surface) and thus help prevent toxic accumulation of intracellular Cu via its efflux from the cytosol to extracellular space across the plasma membrane. To test whether any co-expressors can impact ATP7A trafficking we used an RNAi approach. We began by analyzing the roles of glutamate receptor-interacting protein (GRIP) and GCC2, which is a member of the golgin family of putative tethering proteins that maintain Golgi structure. Additionally, GCC2 is a RAB9 effector required for mannose-6-phosphate receptor (MPR) recycling from endosomes to the trans-Golgi network in living cells and in vitro [25, 26]. The expression of GCC2 was inhibited using

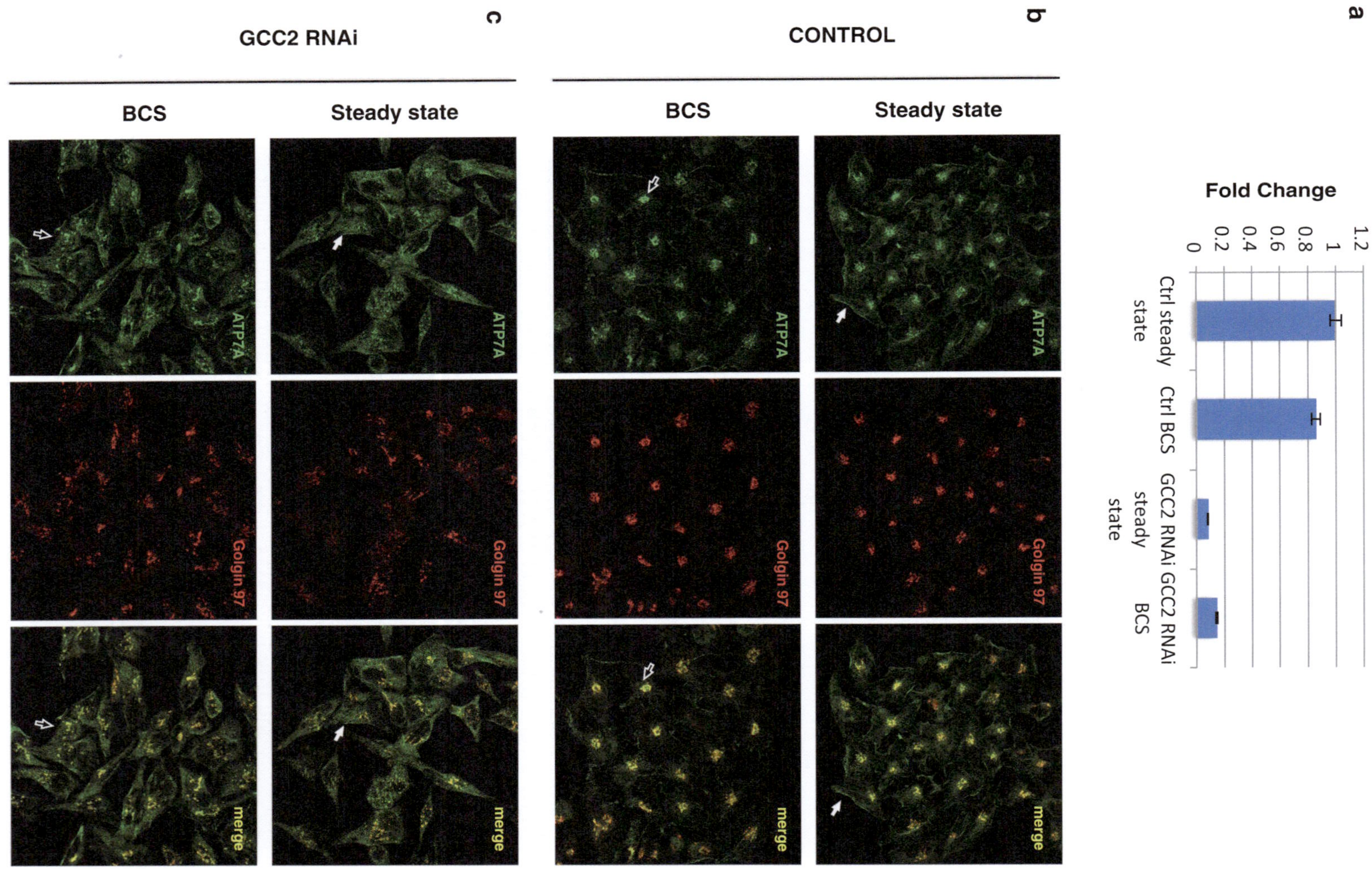

Fig. 5 Silencing co-expressors affect the localization of ATP7A. (**a**) Quantitative real-time PCR indicates a reduction of GCC2 mRNA levels in GCC2-silenced cells with respect to GCC2 mRNA levels in the control cells. (**b**, **c**). (**b**) Control cells exhibit some amount of ATP7A at the cell surface under steady-state conditions (*filled arrows*), while BCS-treated cells show complete retention of ATP7A within the Golgi (*empty arrows*). (**c**) GCC silencing results in a stronger ATP7A signal at the peripheral plasma membrane areas (*arrows in the top row*) compared with the control cells. This ATP7A pool is present in GCC-silenced cells even with BCS treatment (*arrows in the bottom row*), indicating that GCC depletion affects trafficking of ATP7A from the cell surface to the Golgi. The control (**b**) and GCC2-

RNAi, and how GCC2 silencing affected ATP7A localization with different Cu loads was investigated. In contrast to the control HeLa cells (Fig. 5b) that exhibited ATP7A at the plasma membrane and in the Golgi complex, the GCC2-silenced HeLa cells contained lower amounts of ATP7A in the Golgi area. When Cu was chelated by BCS, almost the entire pool of ATP7A moved to the Golgi complex, while in the GCC2-silenced cells, ATP7A remained in the peripheral cell surface areas (Fig. 5c). Although these experiments did not demonstrate physical interaction between co-expressors and ATP7A, this result showed that NetView-predicted GCC2 is required for ATP7A trafficking and can be targeted to correct localization of ATP7A mutants in Menkes disease models. We also noted that silencing of other genes that co-express with ATP7A (MON2 and GOLGA1) provided similar impact. Suppression of the other co-expressor ARL1 did not result in alteration of ATP7A trafficking despite the annotated role of ARL1 in exo- and endocytic events. Also silencing of co-expressor with no clear role in membrane trafficking (NEK1) did not produce any aberration in ATP7A localization. Thus we conclude that although NetView provides a substantial number of candidate genes that could be involved in ATP7A function only part of them is really involved in the protein activity.

4 Notes

1. A significant MI between two genes is represented as an edge in the network, and implies a coordinated response between the expressions of the two connected genes across a set of experiments; it does not necessarily imply causality.

   For each pair of genes the *pmutualinformation* discretizes expression values (*see* **Note 5**) in the number of *states* that the user selects (value of 3 = default), and then the tool computes the following formula:

$$\mathrm{MI}(g1, g2) = \sum_{i \text{ in states}} \sum_{j \text{ in states}} \frac{n_{ij}}{n} \log \frac{n_{ij}}{n_{i+} n_{+j}},$$

   where g1 and g2 are two genes; n_{ij} is the number of experiments where g1 is in state i and g2 is in state j; n_{i+} is the number of times g1 is in state i; n_{+j} is the number of times g2 is in state j; and n is the number of experiments.

   There are different ways to estimate MI. In general, the discrete or continuous MI formulae may be applied. MI computes the dependence/independence of two random variables. For discrete data, the computation includes an estimation of the marginal and joint probability distributions from the frequencies of outcome occurrences; in other words, the more data, the better. Hence, when selecting the dataset from ArrayExpress, a set containing many samples (≥80) is ideal.

Consider the following reviews [20, 27] for an extensive discussion on gene network inference algorithms, and dependency versus causality discussions.

2. Small interfering RNA (siRNA) duplexes are double-stranded RNA molecules, 20–25 base pairs in length, with 3′ overhangs (2 nucleotides) at each end. In order to get an efficient down-regulation of gene expression a combination of two different siRNA duplexes was used that target different regions of mRNA.
3. The inputs for the inference algorithm are genome-wide gene expression profiles. The algorithm has the capacity to infer networks with tens of thousands of genes. Since the algorithm considers all pairs of genes, it ends up computing hundreds of millions of MIs (there are ~100 millions of pairs potentially containing 10,000 genes). Since the majority of MIs are not statistically significant, the algorithm only considers the overall top 5 % MIs and the top 5 % MIs for each gene. This strategy drastically reduces the output length, all the while keeping information for all the genes within the input set. The rationale behind selecting the top 5 % connections is explicable by the following: in Belcastro et al. [4] a Gamma distribution was first fitted over all computed MIs, as theoretic studies suggest [28], and then a 0.05 FDR cutoff was selected. This resulted in the selection of roughly the most significant 5 % of all MIs.
4. MPI is a programming library specification used to develop parallel programs using the message passing model. The method requires that data between concurrent processes take place via sent and received messages. A broad community of vendors, developers, and users proposes that MPI be selected as the standard protocol for developing such programs. Several open-source implementations of MPI exist, such as OpenMPI and MPICH [29, 30].
5. Each process gets N/p genes where p is the number of processes available and N is the total number of genes. At the beginning of the computation each process computes the MI for each pair of its N/p genes. Then computation continues as an iterative process until MI values for all pairs of genes are computed: at each successive iteration step different pairs of processes exchange their own gene expression profiles and each process computes a subset of the all-pairs of MI values.
6. Gene expression profiles from ArrayExpress are normalized within experiments, but not across experiments. Therefore, expression values are discretized into a predetermined number

of bins (-c option) within each experiment. Then each expression value is replaced by the integer corresponding to the bin it falls into (see [5] for further details).

7. From our experience, Cytoscape crashes when attempting to load very large networks (tens of thousands of nodes with hundreds of thousands of edges). This is attributed to the amount of RAM required by the application. For this reason, we have provided instructions to load and visualize the top thousands of scored MIs.
8. The network and the expression matrix that we include can be downloaded from http://resources.bioinformatica.crs4.it/sysgensim_dataset/StatSeq_Datasets_Size100.zip. The *data_matrix.txt* file corresponds to the *Dataset_1_Network_100-1_Configuration_1_gene_expression_matrix.txt* file, while the *in silico* network from which the expression is simulated is included in the file *Dataset_1_Network_100-1_Configuration_1_edge_list.txt*.
9. Because of a recent security enhancement, uncertified Java applets are blocked by default. To display the network you have to add the URL (http://netview.tigem.it) to the exception site list following the steps provided at https://www.java.com/en/download/help/java_blocked.xml.

References

1. Barrett T, Troup DB, Wilhite SE et al (2011) NCBI GEO: archive for functional genomics data sets–10 years on. Nucleic Acids Res 39(Database issue):D1005–D1010. doi:10.1093/nar/gkq1184
2. Rustici G, Kolesnikov N, Brandizi M et al (2013) ArrayExpress update–trends in database growth and links to data analysis tools. Nucleic Acids Res 41(Database issue):D987–D990. doi:10.1093/nar/gks1174
3. Butte AJ, Kohane IS (2000) Mutual information relevance networks: functional genomic clustering using pairwise entropy measurements. Pacific Symposium on Biocomputing. pp 418–429
4. Belcastro V, Gregoretti F, Siciliano V et al (2012) Reverse engineering and analysis of genome-wide gene regulatory networks from gene expression profiles using high-performance computing. IEEE/ACM Trans Comput Biol Bioinform 9(3):668–678. doi:10.1109/TCBB.2011.60
5. Belcastro V, Siciliano V, Gregoretti F et al (2011) Transcriptional gene network inference from a massive dataset elucidates transcriptome organization and gene function. Nucleic Acids Res 39(20):8677–8688. doi:10.1093/nar/gkr593
6. Margolin AA, Nemenman I, Basso K et al. (2006) ARACNE: an algorithm for the reconstruction of gene regulatory networks in a mammalian cellular context. BMC bioinformatics 7 (Suppl 1): S7. doi:10.1186/1471-2105-7-S1-S7
7. Yu J, Smith VA, Wang PP et al (2004) Advances to Bayesian network inference for generating causal networks from observational biological data. Bioinformatics 20(18):3594–3603. doi:10.1093/bioinformatics/bth448
8. Gardner TS, di Bernardo D, Lorenz D et al (2003) Inferring genetic networks and identifying compound mode of action via expression profiling. Science 301(5629):102–105. doi:10.1126/science.1081900
9. Shannon P, Markiel A, Ozier O et al (2003) Cytoscape: a software environment for integrated models of biomolecular interaction networks. Genome Res 13(11):2498–2504. doi:10.1101/gr.1239303

10. Cacciottolo M, Belcastro V, Laval S et al (2011) Reverse engineering gene network identifies new dysferlin-interacting proteins. J Biol Chem 286(7):5404–5413. doi:10.1074/jbc.M110.173559
11. Liu YP, Tsai IC, Morleo M et al (2014) Ciliopathy proteins regulate paracrine signaling by modulating proteasomal degradation of mediators. J Clin Invest 124(5):2059–2070. doi:10.1172/JCI71898
12. Lutsenko S, Barnes NL, Bartee MY et al (2007) Function and regulation of human copper-transporting ATPases. Physiol Rev 87(3):1011–1046. doi:10.1152/physrev.00004.2006
13. Gupta A, Lutsenko S (2009) Human copper transporters: mechanism, role in human diseases and therapeutic potential. Future Med Chem 1(6):1125–1142. doi:10.4155/fmc.09.84
14. de Bie P, Muller P, Wijmenga C et al (2007) Molecular pathogenesis of Wilson and Menkes disease: correlation of mutations with molecular defects and disease phenotypes. J Med Genet 44(11):673–688. doi:10.1136/jmg.2007.052746
15. Tumer Z, Moller LB (2010) Menkes disease. Eur J Hum Genet 18(5):511–518. doi:10.1038/ejhg.2009.187
16. Kaler SG (2011) ATP7A-related copper transport diseases-emerging concepts and future trends. Nat Rev Neurol 7(1):15–29. doi:10.1038/nrneurol.2010.180
17. Stolovitzky G, Monroe D, Califano A (2007) Dialogue on reverse-engineering assessment and methods: the DREAM of high-throughput pathway inference. Ann N Y Acad Sci 1115:1–22. doi:10.1196/annals.1407.021
18. Prill RJ, Saez-Rodriguez J, Alexopoulos LG et al. (2011) Crowdsourcing network inference: the DREAM predictive signaling network challenge. Sci Signal 4(189):mr7. doi: 10.1126/scisignal.2002212
19. Marbach D, Costello JC, Kuffner R et al (2012) Wisdom of crowds for robust gene network inference. Nat Methods 9(8):796–804. doi:10.1038/nmeth.2016
20. Bansal M, Belcastro V, Ambesi-Impiombato A et al (2007) How to infer gene networks from expression profiles. Mol Syst Biol 3:78. doi:10.1038/msb4100120
21. Pinna A, Soranzo N, Hoeschele I et al (2011) Simulating systems genetics data with SysGenSIM. Bioinformatics 27(17):2459–2462. doi:10.1093/bioinformatics/btr407
22. Belcastro V, di Bernardo D (2014) Reverse engineering transcriptional gene networks. Methods Mol Biol 1101:179–196. doi:10.1007/978-1-62703-721-1_10
23. Koressaar T, Remm M (2007) Enhancements and modifications of primer design program Primer3. Bioinformatics 23(10):1289–1291. doi:10.1093/bioinformatics/btm091
24. Untergasser A, Cutcutache I, Koressaar T et al (2012) Primer3–new capabilities and interfaces. Nucleic Acids Res 40(15), e115. doi:10.1093/nar/gks596
25. Brown FC, Schindelhaim CH, Pfeffer SR (2011) GCC185 plays independent roles in Golgi structure maintenance and AP-1-mediated vesicle tethering. J Cell Biol 194(5):779–787. doi:10.1083/jcb.201104019
26. Reddy JV, Burguete AS, Sridevi K et al (2006) A functional role for the GCC185 golgin in mannose 6-phosphate receptor recycling. Mol Biol Cell 17(10):4353–4363. doi:10.1091/mbc.E06-02-0153
27. Penfold CA, Wild DL (2011) How to infer gene networks from expression profiles, revisited. Interface Focus 1(6):857–870. doi:10.1098/rsfs.2011.0053
28. Goebel B, Dawy Z, Hagenauer J et al. (2005) An approximation to the distribution of finite sample size mutual information estimates. In: Communications, 2005. ICC 2005. 2005 IEEE international conference, 16–20 May 2005, vol 1102, pp 1102–1106. doi: 10.1109/ICC.2005.1494518
29. MPI. http://www.sciencedirect.com/science/article/pii/0167819196000245
30. MPI springer chapter. http://link.springer.com/chapter/10.1007/978-3-540-30218-6_19

Chapter 8

Weaving Knowledge into Biological Pathways in a Collaborative Manner

Yukiko Matsuoka, Kazuhiro Fujita, Samik Ghosh, and Hiroaki Kitano

Abstract

Toxicity pathway modeling is an effective approach to understanding how biological systems function under chemical perturbations. Many efforts have been made to construct pathways by data-driven or literature-based approaches to elucidate the mechanisms of action of toxicity. In this chapter, we explain how to build a literature-based pathway map in a collaborative manner using in silico platforms such as CellDesigner to draw pathways and networks, Payao as the curation platform, iPathways+ as the publishing platform, and Garuda to integrate curated pathways while adopting model-descriptive standards such as Systems Biology Markup Language as a file format and Systems Biology Graphical Notation as the graphical representation.

Key words CellDesigner, Collaboration, Garuda platform, Pathway curation, SBGN (systems biology graphical notation), SBML (systems biology markup language)

1 Introduction

Knowledge accumulation is somewhat like weaving, i.e., selecting the materials, preparing threads, and interlacing them together to form the fabric. The work must be done in sequence and be repeated continuously and in collaboration to produce the final cloth. Information and/or knowledge can be stored in text and/or graphical formats. Text formats include printed books, online databases, and repositories, while graphical formats include maps, diagrams, charts, and pictures. Illustrations are often complementary to textual information and graphic adjuncts can increase comprehension of texts if the graphics assist cognitive activity during the encoding (and decoding) processes.

In biology, molecular interaction pathways represent knowledge in a graphical form. Interactions between molecules can be depicted, which allows the possible mechanisms of actions to be presented in a diagram. The potential role of pathway and network representations as means of tracking mechanisms of biological

Julia Hoeng and Manuel C. Peitsch (eds.), *Computational Systems Toxicology*, Methods in Pharmacology and Toxicology, DOI 10.1007/978-1-4939-2778-4_8,

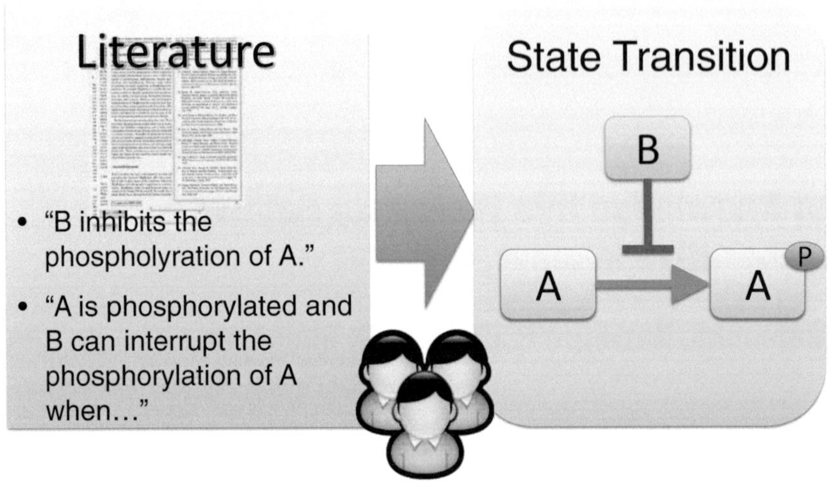

Fig. 1 Collaborative knowledge conversion from text to graphics

phenomena and the effects of chemical compounds has been widely recognized [1, 2]. The growing need to build pathways for toxicology-related applications has contributed to the push to develop a knowledge base of pathway formats. To this end, publications that contain relevant biological information should be collected and mined. To achieve this goal, textual information has to be converted into graphical knowledge, and tools that can help combine individual pieces of information into a bigger picture need to be available for use by pathway builders and curators. Knowledge-driven pathway construction of this kind entails the development of a detailed pathway knowledge base for a particular field of interest, in this case, toxicology.

At the same time, increasing diversity and volume of biological knowledge demand more collaborative efforts in knowledge aggregation. While data curation has been adopted community curation to accommodate with ever-expanding data size [3] and pathway/network curation such as sbv IMPROVER [4] have also approached the communities for quality verification.

Here we describe a collaborative approach to knowledge integration in the systems biology/toxicology domain, which aims to build pathways/networks purely from literature and literature-based curation (Fig. 1).

2 From Text to Pathways

Here we illustrate first the key points of literature-based pathway curation.

2.1 Visualize Knowledge: Biological Pathways

The pathway/network format is one approach used to depict biological knowledge. This format has the following advantages:

- *Simplification*: Text is summarized into a graphical form, which helps to simplify the text and disambiguate the relationships.
- *Navigability*: Pathways and networks are similar to geographical maps; that is, they help us find our way around a biological system.
- *Shareability*: Visualizing knowledge in a standard format can help integrate knowledge, navigate through a knowledge-scape, and can be analyzed to reveal the basic structure of the knowledge.

2.2 Standard Graphical Notations

A standard language within a community significantly contributes to the sharing of information and knowledge. For example, electrical engineers have set a standard notation for electric circuits, where all the symbols in electric circuits are specially designed and well defined so that anyone can use a particular circuit diagram to build the same device. Similar standardized graphical representations are required for biological pathway/network visualization.

Multiple styles have been used to depict pathway/network form of biological knowledge. For example, state transition notation can capture mechanisms of action. It can depict the phosphorylation event of protein A and protein B having an inhibitory effect on the event (Fig. 2a). Activity flow notation, on the other hand, can show that protein B has an inhibitory effect on protein A without mentioning particular event (Fig. 2c). Interrelationships between entities such as protein A, B, and C can be visualized by interaction notation (Fig. 2b).

The Systems Biology Graphical Notation (SBGN) [5] was a community effort to compile the standard graphical notation used to represent biological networks. In SBGN, three types of

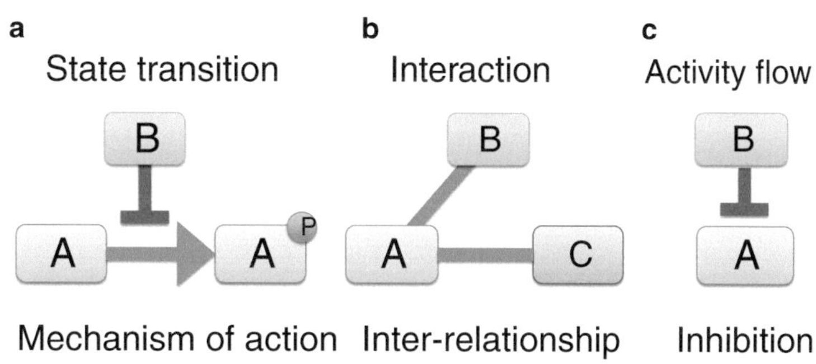

Fig. 2 Graphical notations for depicting representative biological processes: (**a**) State transition notation showing that protein B inhibits the phosphorylation of protein A. (**b**) Interaction notation showing that protein A interacts with proteins B and C. (**c**) Activity flow notation showing that protein B inhibits the activity of protein A

diagrams are defined: (1) activity flow diagram (AF), (2) process description diagram (PD), and (3) entity relationship diagram (ER). PD and AF are presented in Fig. 2. *SBGN* has been adopted by several major pathway databases including Panther Pathway [6, 7] and Reactome [8].

2.3 Standard File Formats

While SBGN can visualize information for humans, other languages can transform information into a form that computers can understand, thereby allowing the sharing of pathway/network model data. Several standard formats have been proposed for exchanging pathway information including SBML (Systems Biology Mark-up Language) [9], CellML [10], and BioPAX (Biological Pathway Exchange) [11]. Other formats to store molecular interactions have been developed (e.g., PSI-MI [Proteomics Standards Initiative Molecular Interaction]) [12] and formats to capture causal relationships are available (e.g., BEL [Biological Expression Language]; http://www.openbel.org [4, 13]).

SBML [9] was in particular designed to describe the systems of biological processes. SBML can store reactions, compartments, events, parameters, rules, and constraints information and is useful to interchange the models of metabolism and cell signaling and to run simulation. Over 250 tools and systems now support SBML.

2.4 Knowledge Collection

For knowledge-driven modeling and curation, the primary source of information is journal publications or books. The number of published articles increases daily, and it can be a challenging task to choose the most relevant papers on a specific domain among the collection. PubMed and other public resources that cover biomedical and life science publications in academic journals provide resources that can be searched for relevant articles. Various text mining technologies have been deployed to support the efficient and reliable collection of published information.

In addition to publications, existing pathway databases are valuable sources of knowledge that can be used to build, for example, a disease-specific pathway/network model. Individual groups (e.g., Kitano [14–19], Curie [20]), institutes (e.g., the Kyoto Encyclopedia of Genes and Genomes [KEGG] [21], Panther Pathway [22]), larger consortia (e.g., Reactome Project [23], Recon2 [24]), open platform projects (e.g., WikiPathways [25]), and commercial entities (e.g., Ingenuity [www.qiagen.com/ingenuit]) all provide pathways and databases. For the pathway databases, cross-reference resources such as PathGuide (http://www.pathquide.org) [26, 27] can also be useful.

While most of the databases such as Reactome and Panther provide their own curation protocols and platforms, there are independent curation platforms such as BioPP [28], Payao [29], and NaviCell [30] also available.

2.5 Interpretation: Deep Curation

Pieces of information need to be processed to obtain knowledge. This process can be regarded as "interpretation" of the data. When interpretation is carried out in a collaborative way, misinterpretation of the data can mostly be avoided.

The purpose of the visualization process is to extract relevant text from a published article, interpret it, and construct a graphical representation of the information. Often, the information in the publication is not presented in a form that can be easily represented graphically. It is therefore essential to read the entire contents of a published article to grasp the main points. Then, for cross-validation other literature relevant to the topics should be found. Here, we call this process of reading through the literature and graphically representing the contents "deep curation" of pathway information.

An advantage of the deep curation approach is that it can capture hypothesized parts of a pathway to generate an overall picture. In a previous study, an epidermal growth factor receptor (EGFR) map was compared with the information in the Reactome pathway database [31] (Fig. 3). It was concluded that the "*deeply curated*" pathway models in the EGFR pathway could represent a gold standard to evaluate this pathway in public databases.

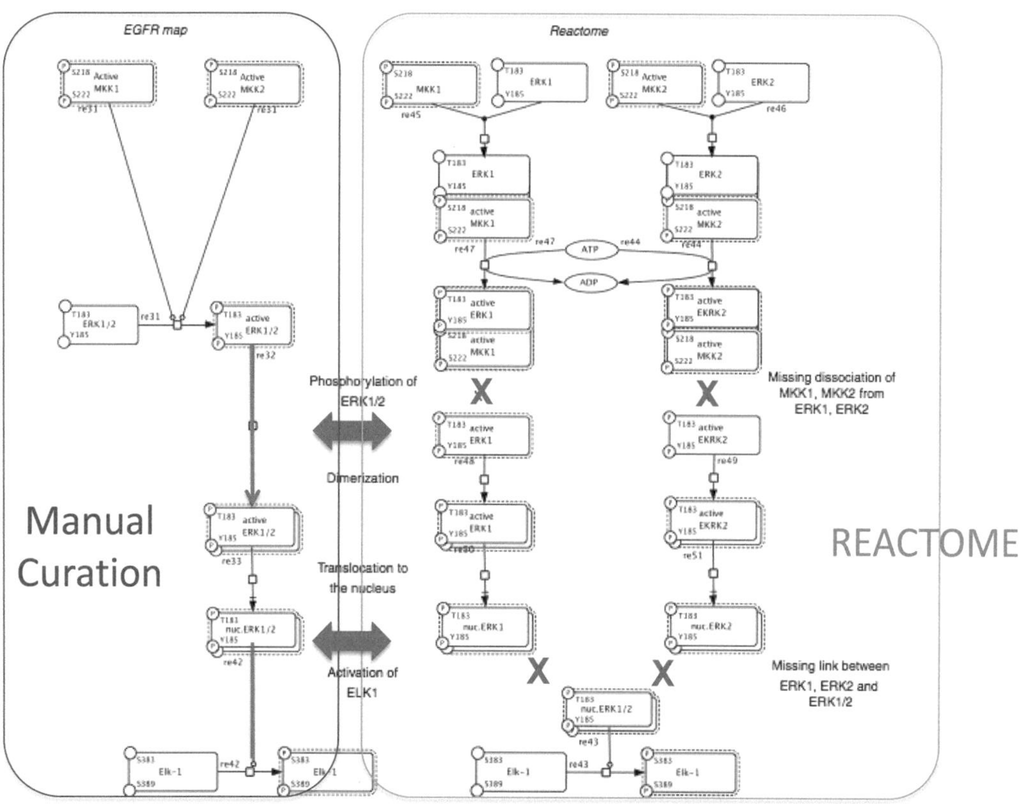

Fig. 3 Comparison of manually curated pathways. *Left*: The epidermal growth factor receptor (EGFR) pathway map reported by Oda et al. [13]. *Right*: Reactome pathway (as of 2009). Adopted from Bauer-Mehren et al. [23]

In deep curation processes, all possible pathways (including conflicting pathways) integrated within a network can be reexamined when new evidence is discovered. Thus, deep curation allows mechanisms of action of entire systems to be captured for specific domains of interest.

2.6 Significance for Collaboration

Collaboration has two major aims: one, to construct large-scale networks, and two, to accommodate differences in interpretations.

Large-scale modeling.

Depending on the size of the network map, the structure of the project team needs to be adjusted (Fig. 4). For example, the EGFR pathway map [18] published in 2005, which contained 322 nodes and 211 edges, was built mainly by a single person, while AlzPathway [32] with 1347 molecules and 1070 reactions was created by a group of geographically distributed teams with Alzheimer's disease experts as consultants. The latest Parkinson's disease map [33], published online in 2013, was constructed by an even larger number of members distributed in Tokyo and Luxembourg. It is advisable to have a concrete structure for the project team, e.g., a chief editor to edit the map so that all the curators' inputs are efficiently and reliably incorporated.

Consolidating interpretations.

Collective knowledge is sourced during discussions when new concepts or hypothesis are formed, consensus is reached, and uninvestigated areas of research are identified. When the pathway or network to be constructed is specialized, discussion with experts

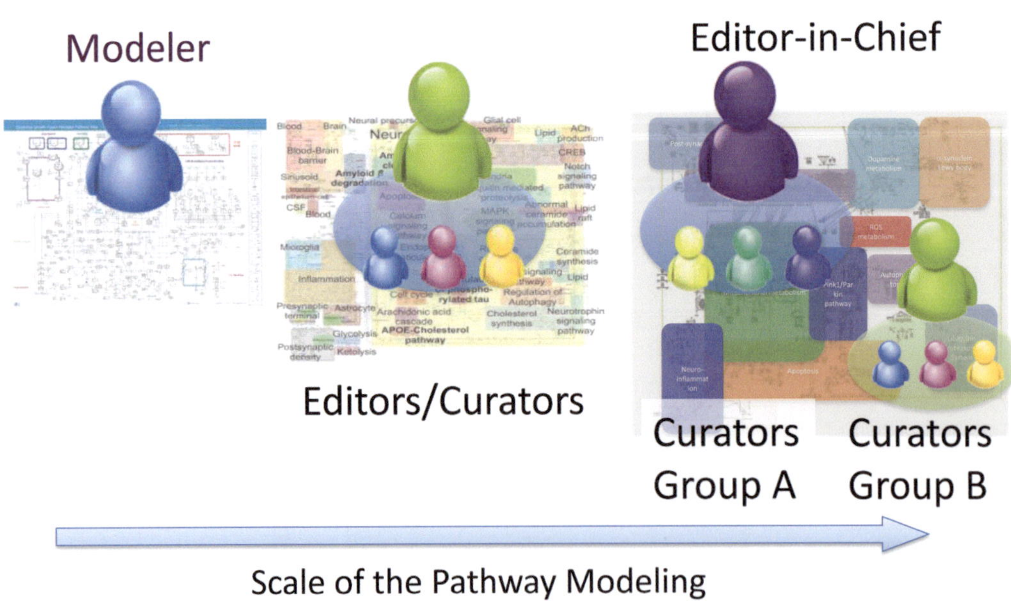

Fig. 4 Deep curation and project team formation: Members of the team are assigned based on the scale of the project and the size of the pathway to be built

who have domain knowledge is essential. Members of curation teams have unique roles, i.e., domain experts, curators who verify the pieces of evidence, editors who are familiar with graphical notations and editing tools, and moderators who drive the discussions. Curation guideline or triage scheme shared among the members assist to establish a common ground for discussions. It is also important to remember when obtaining consensus that participants in the pathway curation may come from different cultural backgrounds. Thus, social engineering [34] aspects must be considered in all pathway curation endeavors.

2.7 Community Pathway Curation Process

The basic workflow required to build deep-curated pathway maps including modeling, sources, and tools is shown in Fig. 5. First, relevant information is collected, and then the textual information is interpreted and represented in a graphical form. A basic map is drawn using the available software; annotations are assigned, and integrated into a model file, which is then open for review by other members of the project team. Team members search for and

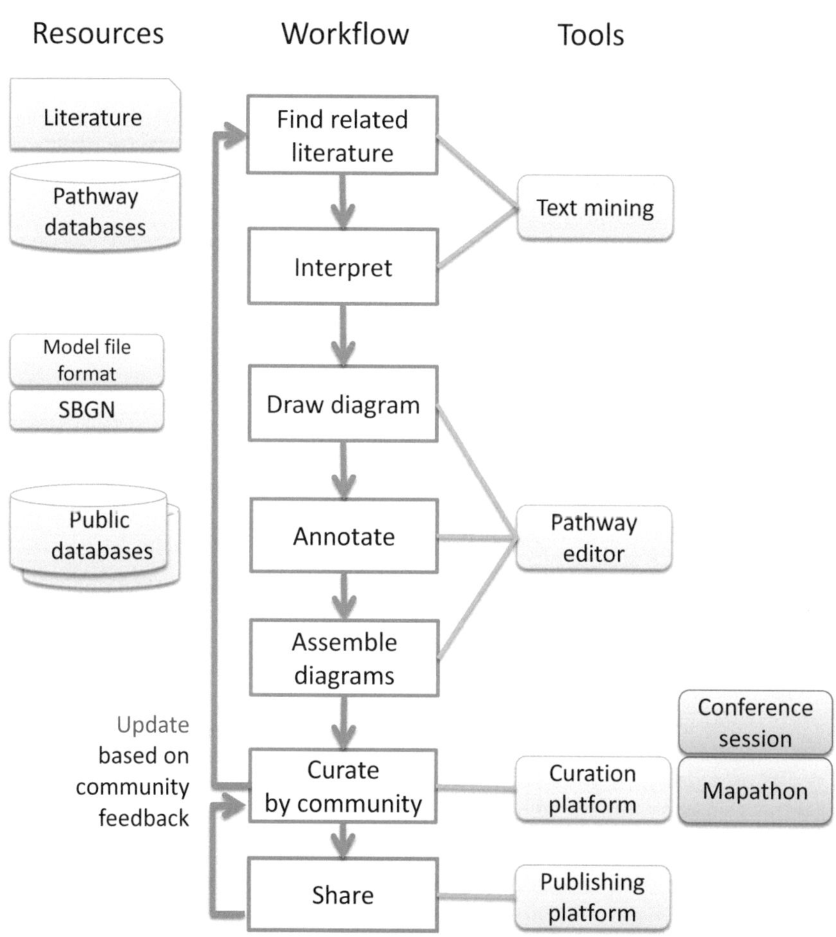

Fig. 5 Schematic flow chart of a generic distributed knowledge-driven pathway curation process

examine new knowledge and the latest information, and continue to update the map whenever necessary.

In the following sections, we illustrate the detailed procedures of community-driven pathway curation. We discuss how to build pathway maps using the CellDesigner editor [35] with SBGN as the common graphical notation, taking real examples of how to collate the discussion within the curation community.

3 Materials

3.1 Pathway Curation: Tools and Technologies

The pathway curation tools and resources that we used in the examples of knowledge-driven pathway curation processes discussed in this chapter are listed below:

1. Data that served as a basis for building the pathway were obtained from both review papers and primary research articles (see Table 1).
2. Public pathway databases such as KEGG, Reactome, and Panther were sourced for relevant information (see Table 1).
3. The availability of identifiers was checked in public reference databases such as UniProt [36] and Entrez Gene [37].
4. Facilities for modeling and curation platforms were prepared in advance. In this exercise, PC (Windows, Mac, or Linux) machine and the Internet access from these machines were required.

Table 1
Checklist of pathway curation tools and resources

Resources	Relevant literature	Full text if available Record doi/PubMed IDs • Review papers in the domain of interest • Research articles (e.g., pathway specific, disease specific)
	Databases	• Pathways • Identifiers: UniProt [36], Entrez Gene [37], Gene Symbol, MetaCyc [51]
	Pathway files	Manually curated custom-made pathway model files if any
Human resources	Curators Model editors	
Tools	Pathway editor	For example, CellDesigner (http://www.celldesigner.org) [35]
	Curation platform	For example, Payao (http://www.payaologue.org) [29]
	Publishing platform	For example, iPathways+ (http://www.ipathways.org/plus)
	Text mining tool (optional)	For example, iHOP [48] (http://www.ihop-net.org/UniPub/iHOP/)
Standards	SBML	http://sbml.org [9]
	SBGN	http://sbgn.org [5]

iHOP Information Hyperlinked over Proteins, *SBGN* Systems Biology Graphical Notation, *SBML* Systems Biology Markup Language

3.2 CellDesigner™: Pathway Editing Tool

CellDesigner™ [35] is a software tool for constructing and simulating models of gene regulatory and biochemical reactions. Using this tool, models can be drawn on a canvas in an intuitive manner. CellDesigner adopts a graphical notation of the process diagram compliant to the guidelines of SBGN (http://sbgn.org) [5]. The models are stored in SBML (http://sbml.org) format [9]. When the kinetic laws in a model are provided, CellDesigner can simulate the model by computing time courses using built-in simulation engines.

The major features of CellDesigner for pathway curation are the following:

- Intuitive graphical user interface for biochemical and gene regulatory network modeling.
- Consistent and easy-to-understand graphical notation of representing biochemical semantics (SBGN compatible) adopted the proposals by Kitano [38, 39] (**Note 1**).
- Export SBGN-ML [40], an exchange file format defined for SBGN graphics.
- Extensive descriptions of compartments, species, reactions, and proteins.
- Models in machine-readable SBML format, which can represent biochemical signaling, metabolic, and regulatory networks with rich enhanced function stored as SBML extended annotation [CD-SBML].
- Built-in simulators, direct integration with SBML ODE Solver [41], COPASI [42], and Systems Biology Simulation Core Library [43]. CellDesigner's built-in simulator UI "ControlPanel" allows users to specify parameters, change amounts, conduct parameter searches, and perform interactive simulations in an intuitive manner.
- CellDesigner supports connections to pathway databases such as BioModels.net [44], JWS Online [45], and Panther Pathway [6, 7, 22]. CellDesigner also supports connections to various reference databases, such as DBGET (a database retrieval system for a diverse range of molecular biology databases) [46], SGD (Saccharomyces Genome Database) [47], iHOP (Information Hyperlinked over Proteins) [48], the Genome Network Platform (http://genomenetwork.nig.ac.jp/), Entrez Gene [37], Gene Wiki [49], UniProt [36], ChEBI [50], MetaCyc [51], and the Panther website [6, 7, 22]. CellDesigner also connects to PubMed based on annotated PubMed IDs.
- Exports images in PNG, JPEG, SVG, or PDF formats.
- Supports a plug-in development framework; for example, BioPAX support is made available through a plug-in [52].

In summary, CellDesigner is designed to bridge machine-readable and human-readable formats, and to be a user-friendly mathematical modeling editor with a standard graphical representation.

The latest version (as of 2014) of CellDesigner is version 4.4. It is freely downloadable from http://www.celldesigner.org. CellDesigner can run on Windows, Mac, and Linux operating systems.

3.3 Curation Platform: Payao

Payao [29] is an online curation platform specially designed for pathway models generated by CellDesigner. Payao (http://www.payaologue.org) is a collaborative Web-based platform for the curation of gene-regulatory and biochemical pathway models. The system allows online pathway visualization and allows biological models to be annotated and curated by a community. Payao takes and displays models with the CellDesigner SBML format which is compliant notations to SBGNs. It also provides a user interface for model enrichment (adding tags and comments) to access controlled community members.

3.4 Publishing Platform: iPathways+

iPathways+ (iPathways Plus; http://www.ipathways.org/plus) is also a Web-based service that can be used to browse, publish, and share biochemical pathway models. iPathways+ can register pathway models built by CellDesigner (stored in extended SBML file format [CD-SBML]). It provides functionalities for users to share their customized pathway models with appropriate access control mechanisms. It also provides features to embed models in a research website, or to share links to the models via e-mail and social networking services to accelerate community discussion over the models.

3.5 Garuda Platform

It is essential to work efficiently with multiple data sources and software to build a comprehensive understanding of biological systems. A recent initiative for software integration is the Garuda Alliance (http://www.garuda-alliacne.org) [53], which was formed with the aim of developing a platform and guidelines to achieve a highly productive and flexible software and data processing environment that would serve as a one-stop service for systems biology and bioinformatics. The intention was to host increasing numbers of software and data or knowledge resources for building automated workflows and analysis pipelines.

The Garuda platform can be used to develop complex on-the-go workflows for integration of curated pathway models obtained from public domain databases (Panther, Reactome, KEGG). It provides data conversion tools and tools for visualizing the models in CellDesigner [54]. Garuda can also help users merge multiple models to build comprehensive pathway maps. Garuda Dashboard, the user interface, can be installed on Windows/Mac/Linux machines.

3.6 Text Mining

Using a text mining system attached to a curation platform can help find the appropriate publications [55]. In the case studies presented in this chapter, we explored a prototype text mining system along with the curation platform.

With the implementation of such a system, researchers can browse through the relevant papers to the nodes or edges in a pathway (nodes are molecular entities, edges represent reactions). Further, up-to-date publications can be found on specific components of the pathway after the original map is built and registered to the text mining system. Automating this kind of knowledge aggregation is in high demand, and various efforts to develop such systems are ongoing.

3.7 Sample Files

To check the CellDesigner graphical notation, install CellDesigner downloaded from http://www.celldesigner.org, and then check the sample files (e.g., Components44.xml and geneRNA.xml) available under the installed CellDesigner folder named "samples."

3.8 Curation Sheet and Face-to-Face Mapathon (Mapping Marathon)

While online curation platforms such as Payao and iPathways+ are extremely useful, face-to-face discussion is critical for resolving any issues around the interpretation of textual information and its conversion to a graphical format. The preparation of a curation sheet that contains, for example, details of the publication, referred sentences or paragraphs, keywords, and graphic diagrams is helpful in summarizing the information (Fig. 6).

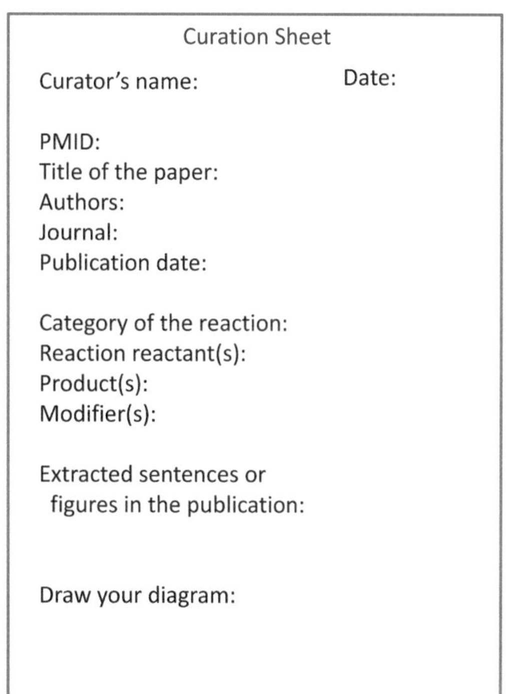
Curation Sheet

Curator's name: Date:

PMID:
Title of the paper:
Authors:
Journal:
Publication date:

Category of the reaction:
Reaction reactant(s):
Product(s):
Modifier(s):

Extracted sentences or figures in the publication:

Draw your diagram:

Fig. 6 Example of a curation sheet

4 Methods

4.1 Project Design

1. Identify the scope of the pathway modeling/curation.
   Set boundary conditions such as
   - Target disease, target tissue, target cell type, target biological processes
   - Experimental conditions (in vivo, in vitro)
2. Define the size of the project.
3. Identify the resources of information and collect the relevant literature/database resources.
4. Identify the members in the project team; assign specific tasks to each member, e.g., pathway editor and curator.
5. Recruit and assign the curators (e.g., paid staff, specialists, students, part-time staff).
6. Check where the team members are located, e.g., in the same physical location or distributed locations. It is important to select the tools/means for modeling and curation and to determine the communication method.
7. Identify the tools to be used to build and curate the pathway.
8. Train team members to use the tools before starting the project.
9. Define the annotation scheme to be used (e.g., MIRIAM [http://www.ebi.ac.uk/miriam/] [56]).
10. Decide the model file format to be used (here we employ CellDesigner-SBML).
11. Delineate the overall schedule.

4.2 Collect Information

Identify the sources of information and decide the means of collection.

1. Find the publications on the selected topic, and store them for future reference. If it is available, keep the full paper contents. It is useful to begin with review papers, which give an overview of the subject of interest.
2. Search the pathway databases for relevant pieces of information.
   Panther Pathway (http://www.pantherdb.org/pathway/), Reactome (http://www.reactome.org), and KEGG Pathway (http://www.genome.jp/kegg/pathway.html) are among the most popular resources for building pathways with CellDesigner. PathGuide (http://www.pathguide.org) provides a comprehensive list of all pathway resources.
3. For detailed information on each factor in the pathway, search reference databases: genes (e.g., Entrez Gene), proteins (e.g., UniProt), or chemical compounds (PubChem).

4. Use text mining tools such as iHOP (http://www.ihop-net.org/UniPub/iHOP/) and Stitch EMBL (http://stich.embl.de), which are useful especially for extending the network.

4.3 Read, Interpret, and Visualize

1. Read the paper, summarize the text, interpret it in a graphical notation, and write in the curation sheet (Fig. 7) (**Note 2**).
2. Check a relevant paper that refers to a similar event (e.g., state changes of an entity, or relationship between two entities). Then check the information against the previously drawn diagram.

Example 1: Text visualization.

Textual information can be in various forms; for example, "B inhibits the phosphorylation of A" or "A is phosphorylated. B can interrupt the phosphorylation of A when…." In this example, the SBGN diagram describing the process would be visualized as shown in Fig. 7.

Example 2: Interpretation.

The extraction of information contained in the NFkB pathway has been reported previously [57].

Three sentences describe the NFkB state transitions (Fig. 8):

(a) NFkB is activated.
(b) NFkB disassociates from IkBa in the cytosol, and then translocates to the nucleus.
(c) IkBa bound to NFkB is phosphorylated, ubiquitinated, and then degraded.

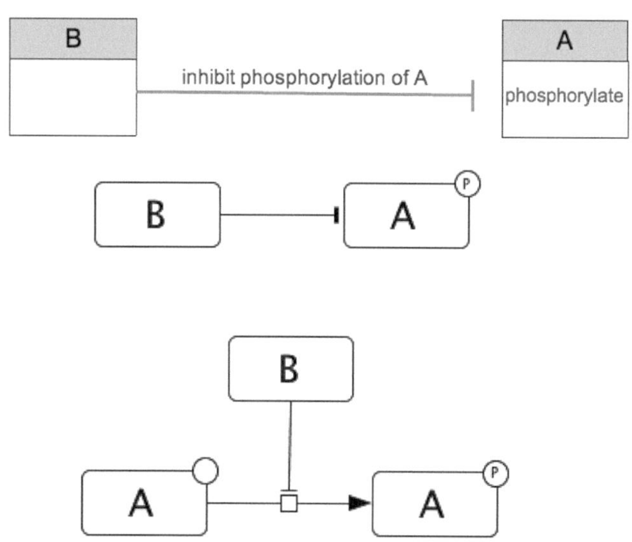

Fig. 7 Graphical notation represented the textual information that B inhibits the phosphorylation of A

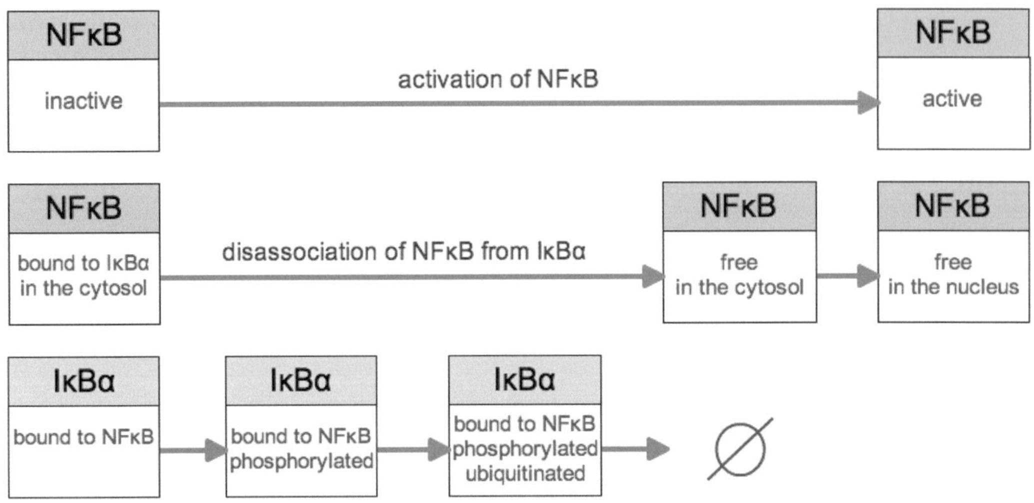

Fig. 8 Representation of the NFkB state transitions (sentences a, b, and c) in a graphical notation

4.4 Draw a Diagram with CellDesigner

1. Use CellDesigner to redraw the diagrams from the previous step and combine them into a model file (Fig. 9).
2. Draw each part of the diagram separately (Fig. 10).

4.5 Assemble Pieces of Knowledge

1. Integrate all the interpreted information into a combined diagram. Identify duplicated items among the pieces of information, and redraw the items accordingly (Fig. 11).

4.6 Annotate the Model: Mine Data

After drawing the combined diagram, the next step is to link the glyph to the database information and the original literature using CellDesigner. Annotations can be added in free text format or formatted MIRIAM table style (Fig. 12).

1. Annotate to species using, e.g., GeneID, UniProtID, and PubChemID.
2. Annotate to interaction edges (Reaction in SBML) using, e.g., PubMed ID and Reactome ID.

Note that in CellDesigner, annotations can be added in the notes or in MIRIAM format, as well as on the layer. Annotations can be stored in a separate wiki or website. Some annotation examples can be found in the BioModels Database at BioModels.net (http://www.ebi.ac.uk/biomodels-main/).

4.7 Merge and Extend the Model

1. After some sections of the diagrams have been built, they can be merged into a single model or extended to larger scale (Fig. 13).

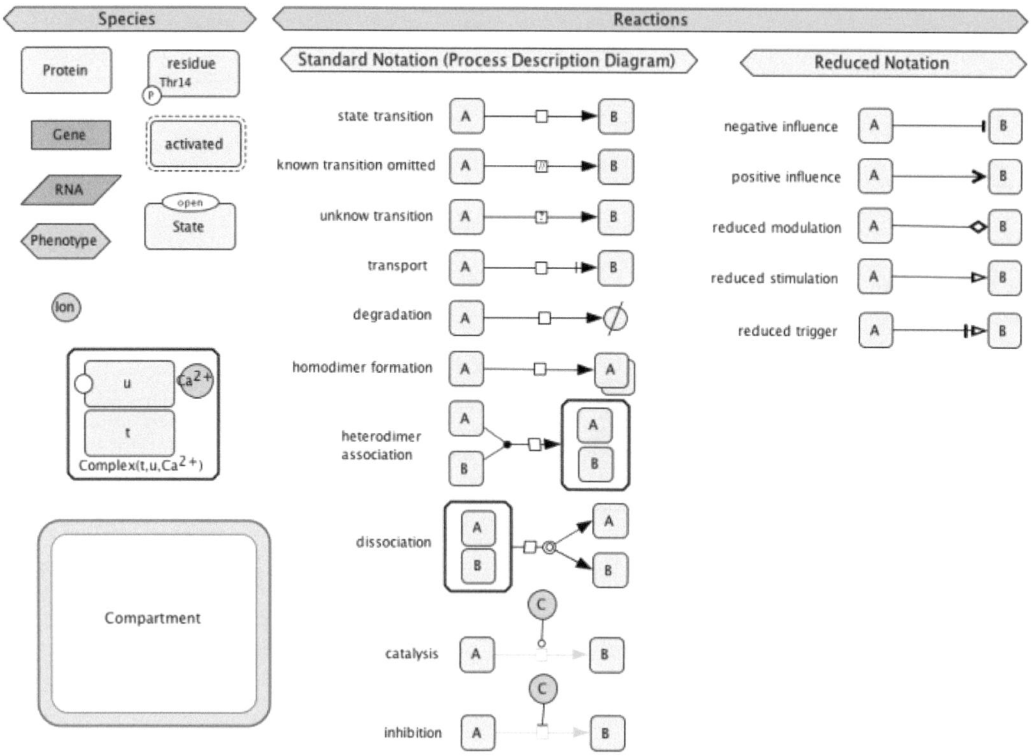

Fig. 9 Graphical notation of CellDesigner

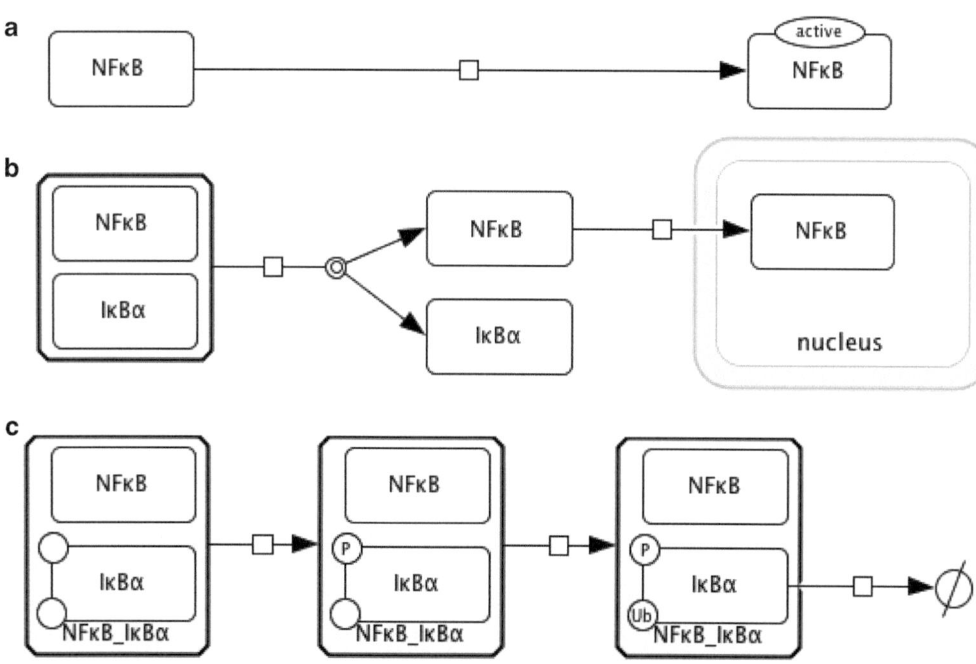

Fig. 10 Representation of the NFkB state transitions (from Fig. 9) in CellDesigner notation

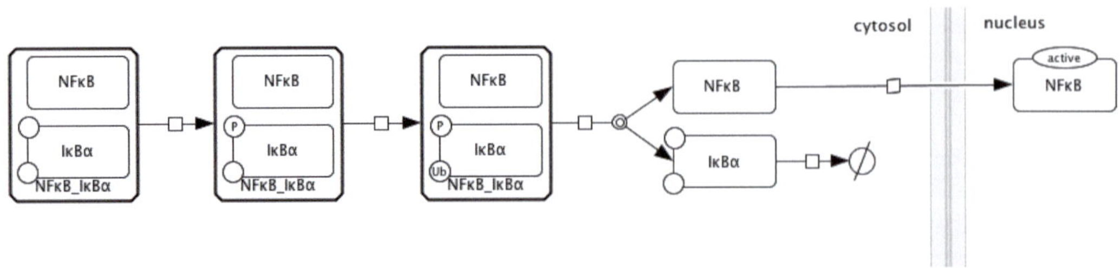

Fig. 11 Representative interpretation of the integrated NFkB activation pathway

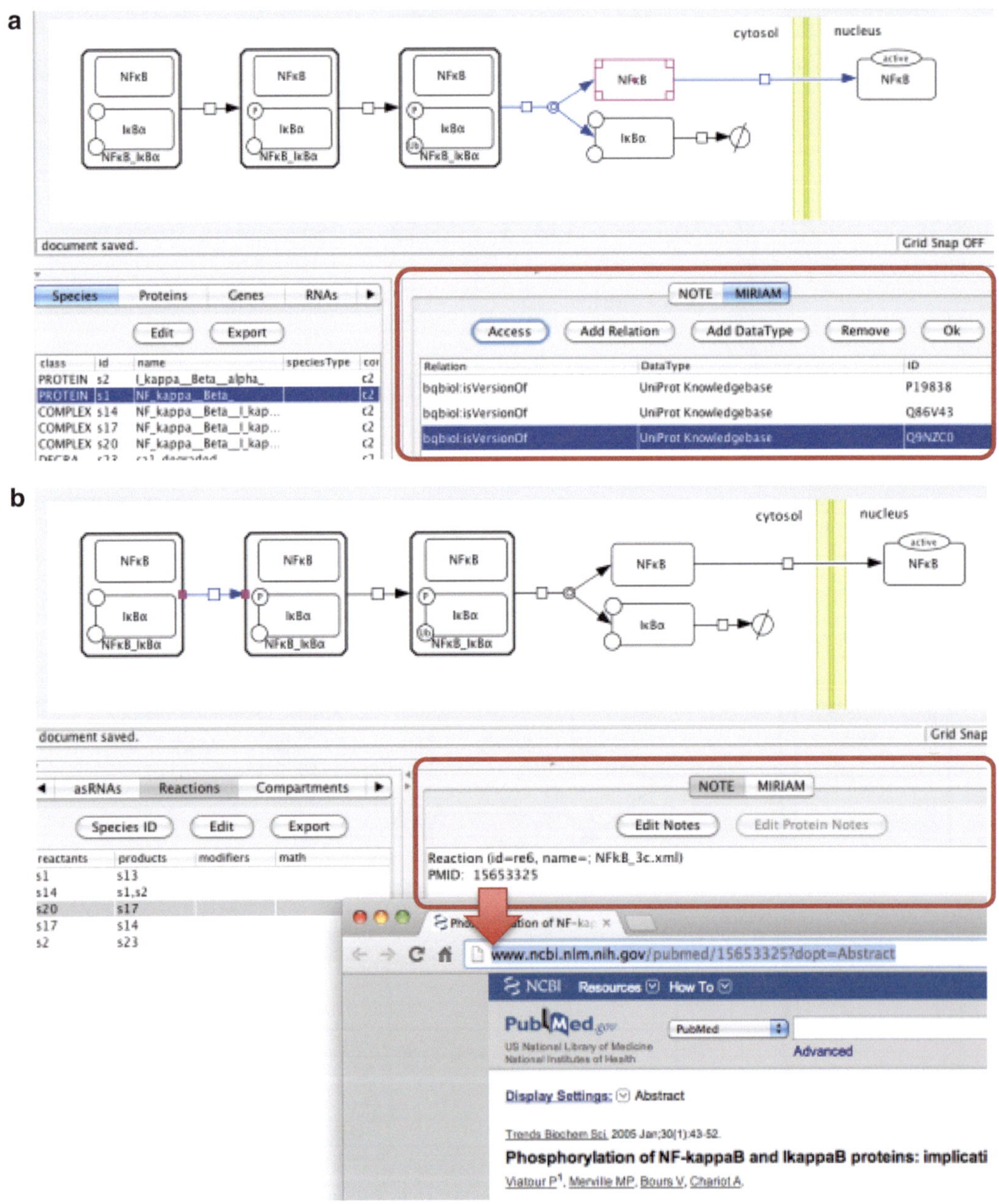

Fig. 12 Examples of annotations: (**a**) MIRIAM tab. (**b**) PubMed ID in the Notes tab of CellDesigner

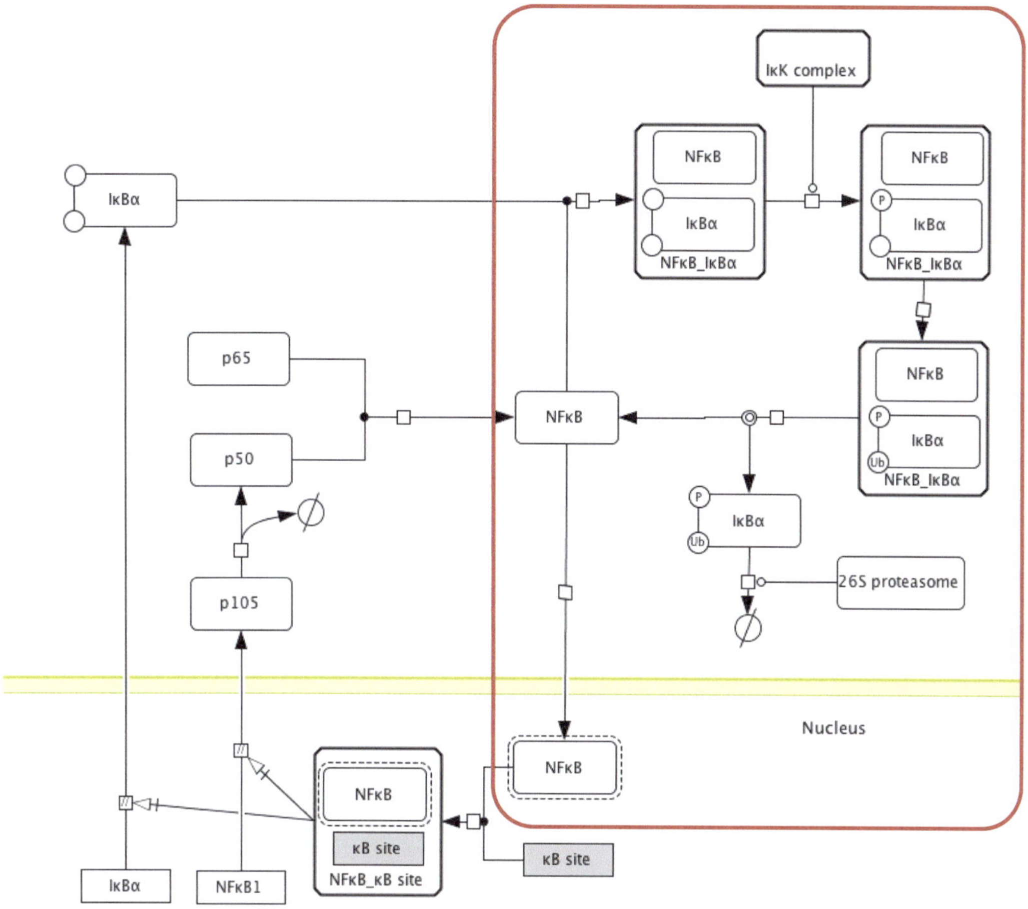

Fig. 13 Example of the extended NFkB diagram. The area enclosed by the *red line* indicates the diagram that was drawn in the previous step. Adapted from Oda et al. [48]

4.8 Lay the Model Out

1. The layout of the model provides another way of representing the interpretation of the knowledge. Sections can be consolidated to make the structure of the system easier to understand. An example of the relayouted model is shown in Fig. 14.
2. To merge model files, the "Merge Models" CellDesigner plug-in or the Garuda gadget can be used. It automatically identifies the duplicated components and interactions.

4.9 Add Drugs/ Chemicals to the Pathway

1. In systems toxicology, it is often important to identify where in the system a particular drug intervenes. When it is known, the information can be added to the pathway model (Fig. 15).

4.10 Curate Online

After the initial version of the map has been built, it can be subjected to further scrutiny with the aim of information refinement. Curation work of this type should not be restricted to primary inputs from one person, but should also allow inputs from a number of people as confirmation and consolidation of the information

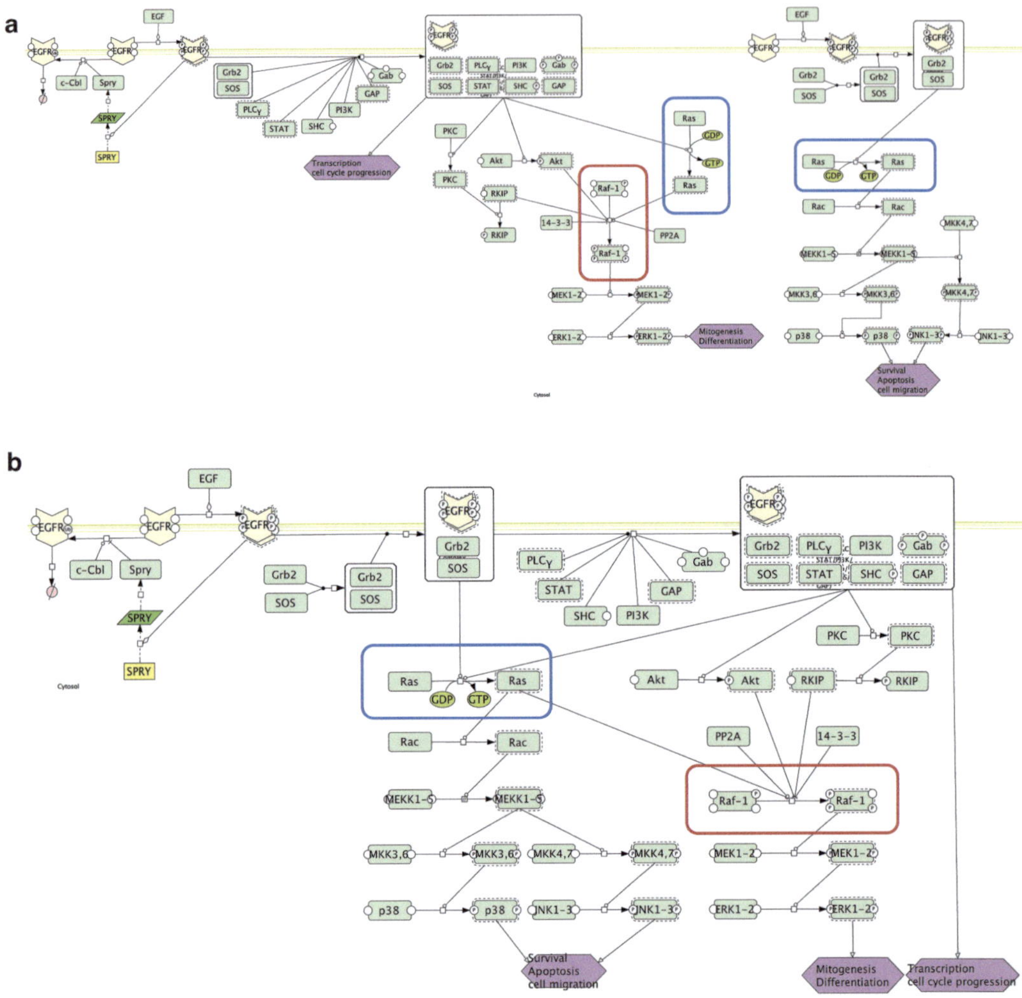

Fig. 14 Relayout process: The parts circled in *red* and *blue* were identified as duplications in the model representation. (**a**) The original pathway from the Panther Pathway database. (**b**) The identified duplicated reaction (Ras activation) was merged and the positions of the components were relayouted for highlighting the mechanisms of action

in the map. When curators are physically located in different places, the necessary tools for them to work online should be provided.

Note that any tools/systems that are used in the curation process should be familiar to the curators. When curators are not comfortable with the available online tools, it may be advisable to use widely employed methods such as e-mail, online discussion exchanges, and/or hand-written notes (**Note 3**).

4.10.1 Using Payao as the Online Curation System

1. Upload the model created by CellDesigner to the Payao system (http://www.payaologue.org).
2. Set the project team members and assign their roles.

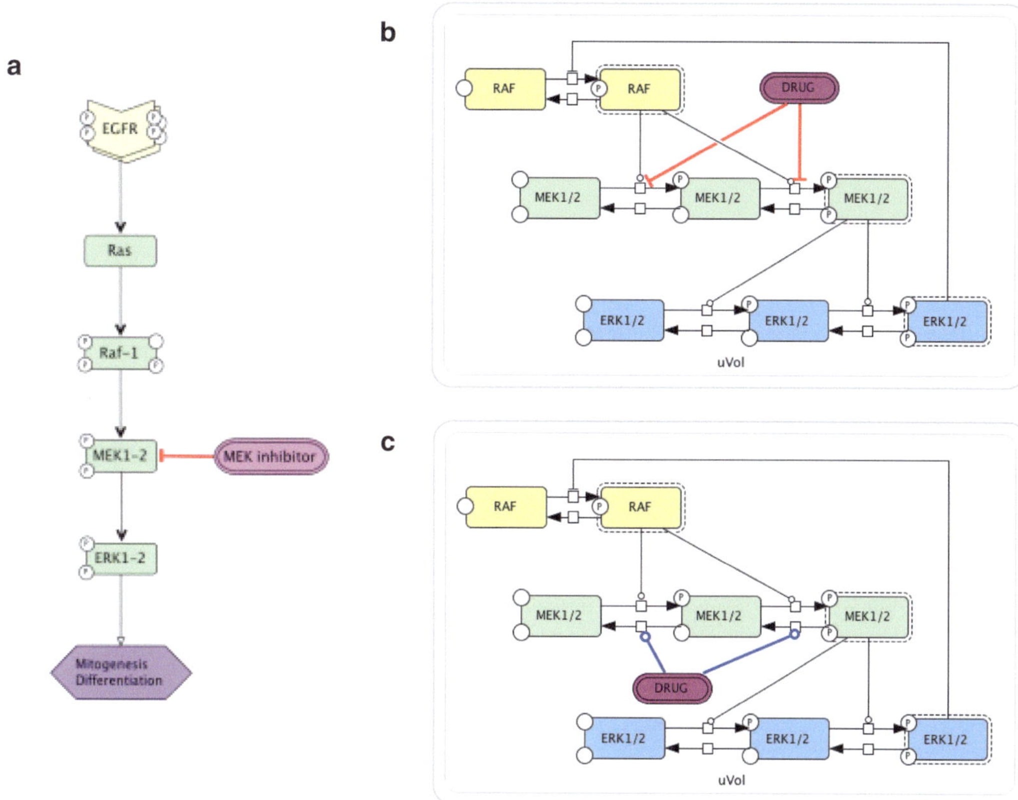

Fig. 15 Examples of how drug information can be added to biological pathways: (**a**) MEK inhibitor in an activity flow diagram and (**b** and **c**) in the process description diagram with inhibitory effects

3. Ask the curators to add tags and comments as shown in Fig. 16.
4. The model owner consolidates all the tags and comments.

4.11 Curate Through Face-to-Face Discussion

Online curation platforms assist in accumulating the comments from curators and other domain experts. However, it is always practical and beneficial to have a face-to-face discussion. Such events are called "Mapathons" or curation/verification workshops. It is advisable to invite domain experts as external reviewers. It is not easy to find the external reviewers who have expertise. Scheduling the Mapathon as a satellite workshop at the domain-related conferences is one of the practical approaches for recruiting the experts.

1. Make a checklist for discussion with the curators and domain experts.
2. Walk through the pathway model and check if there is any concern over particular parts of the model or annotation.

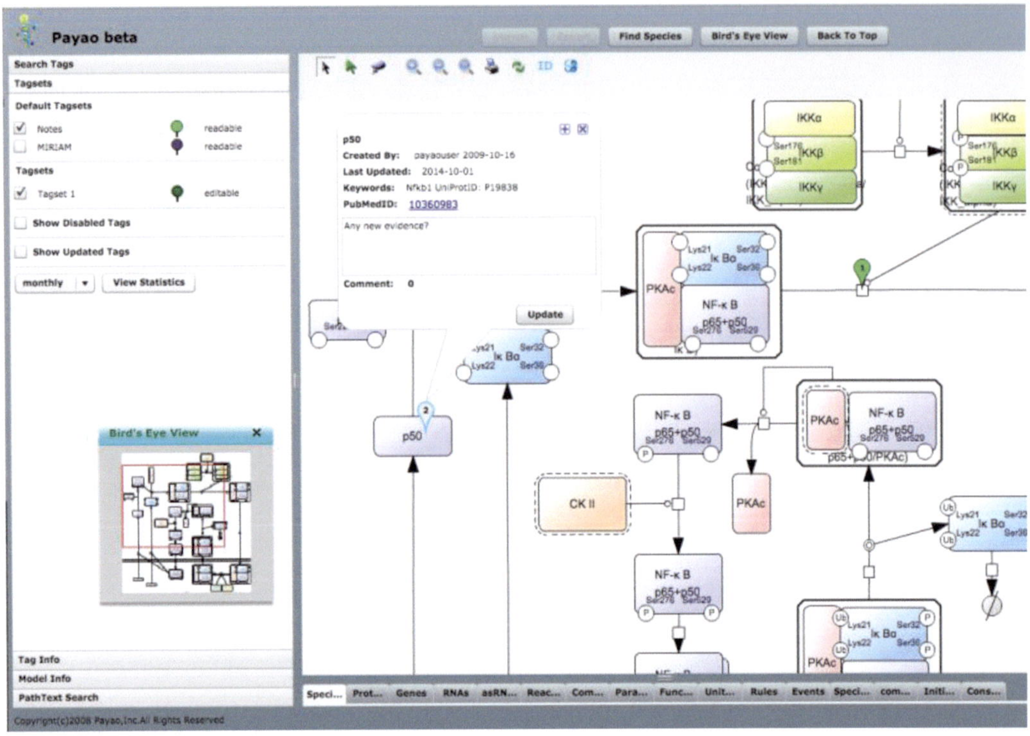

Fig. 16 Payao online pathway curation platform

3. Check the annotations and the associated pieces of evidence. Check that the graphical representation meets all the boundary conditions.
4. Discuss the possibility that there may be missing parts in the entire model.
5. When missing parts are identified, look for further pieces of evidence and publications that may help fill the information gaps.
6. Prepare curation sheets (Fig. 6), fill in the new information, and draw a diagram based on the newly found pieces of evidence and publications.
7. Discuss any issues of the interpretations among the experts to resolve during the curation/consolidation phase.
8. Consolidate the points and integrated into the base pathway diagram.

4.11.1 Interpretation Issues

An example of interpretation issues that are typically discussed during the curation phase is described here. The example depicted in Fig. 17 shows the potential interpretations of the role of eIF2.

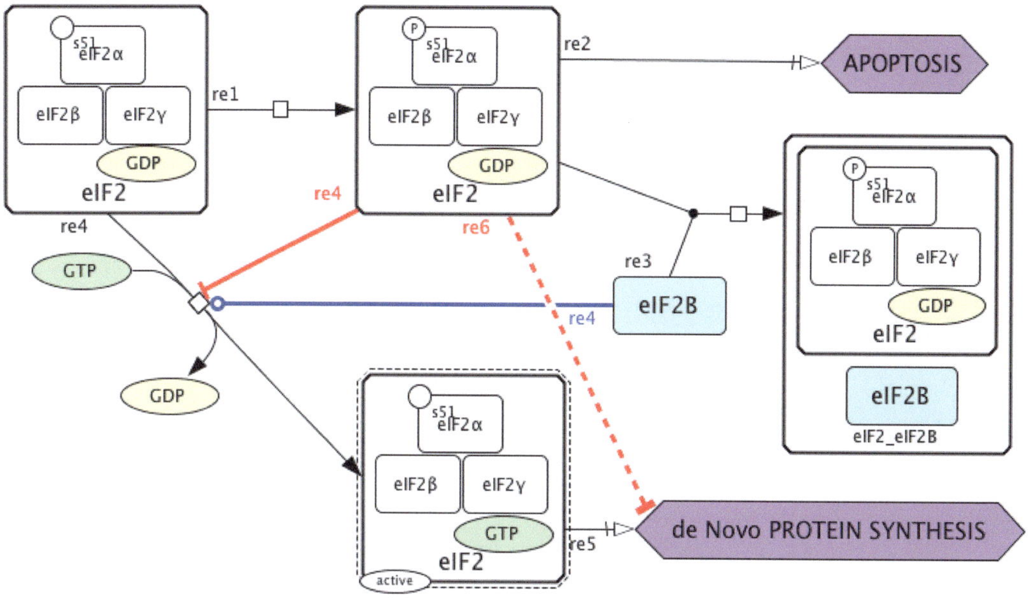

Fig. 17 The interpretation of the role of eIF2/eIF2B. The translation initiation factor eIF-2B interacts with the eIF2 alpha, beta, and gamma complex and binds to it. The exchange of GDP to GTP in the eIF2 complex is a function of eIF-2B, which leads to the activation of eIF2a. The inhibition of GDP to GTP exchange by eIF2 and the inhibition of de novo protein synthesis were deduced from the eIF2B activation of eIF2. Where the inhibitory reactions are placed in the diagram will depend on the interpretation of the text by the participants and what aspects of the pathway diagram are to be emphasized. *Blue line* indicates that the exchange of GDP to GTP in the eIF2 complex is a function of eIF-2B; *red solid line* indicates inhibition of the GDP to GTP exchange by eIF2; *red dashed line* indicates inhibition of de novo protein synthesis deduced from the eIF2B activation of eIF2

4.12 Share the Map

In addition to (or instead of) a curation platform that can be used by team members to comment and tag on the pathways, a publishing platform can be used to share the map with the wider community. A map published by the owner could be displayed on the platform as well as embedded on his or her website. Discussion and feedback over social network services (SNSs) can be encouraged. For example, while Payao is an online curation system, iPathways+ can be used for browsing and sharing with SNSs (e.g., Facebook and Twitter).

4.12.1 iPathways+ for Publishing Pathway

1. Upload the model created by CellDesigner to the iPathways+ system (https://ipathwaysplus.unit.oist.jp/).
2. Set the sharing setting. The model can be made open to the public by setting the public user sharing option to "can view."
3. Export the link and share it via mail/SNSs, or embed the pathway in a public Web page (Fig. 18).
4. Discuss and curate the contents of the model via e-mail/SNSs.

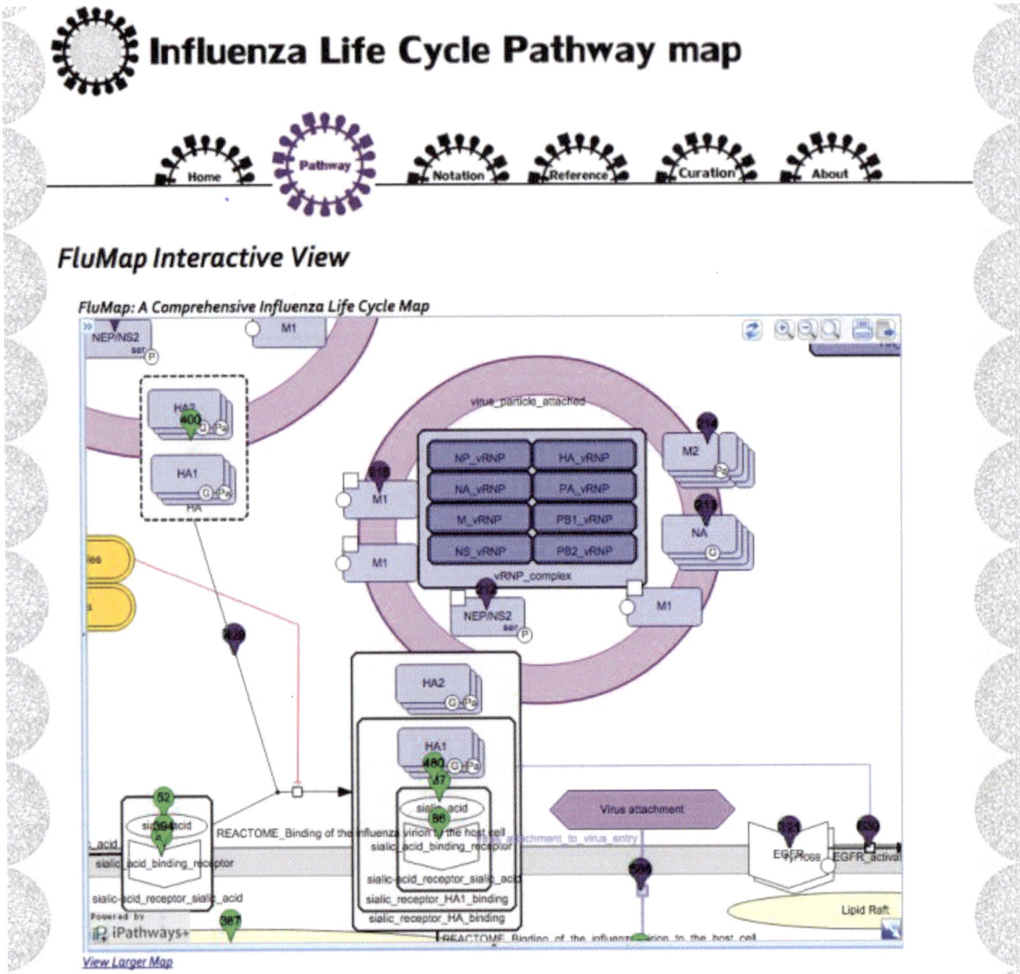

Fig. 18 Pathway embedded in a website using the pathway sharing platform iPathways+ (http://www.influenza-x.org)

5 Case Studies

5.1 Case 1: Students Jamboree for Curation

In 2010, the Open Source Drug Discovery project (OSDD; http://www.osdd.net/) was initiated in India with funding from the Council of Scientific and Industrial Research, India. The aim of the project is to accelerate the drug discovery process in an open-source manner. India suffers from neglected diseases, such as tuberculosis, for which major pharmaceutical companies seem reluctant to develop new drugs. The OSDD project aimed to start its own drug discovery project. A community website was set up, and multiple projects were undertaken by students who were recruited from across the country in a collaborative attempt to develop new drugs.

One of the sub-projects was a *Mycobacterium tuberculosis* metabolome challenge in which more than 400 students were recruited from across India to mine the literature as their 4-month summer project. The metabolite interaction data were collected in a table and converted to an SBML-based network using CellDesigner [58]. The students were highly motivated, and most of them knew at least one tuberculosis patient personally.

The OSDD project also has an educational aspect. Each student was given a laptop and assigned a set of publications to mine as input to their curated data. Because network connection speeds can be slow in India, the community website was not as widely used as expected.

A number of lessons were learnt from this challenge: (1) commitment and a sense of fulfillment from the participants are essential for large-scale curation projects, and (2) quality control is extremely important because student inputs necessarily require extra verification [26]. In this respect, we would recommend the trainings for the participants in advance for the project and establish the common grounds for discussion. It may be helpful to assign the role and build up the smaller team to monitor day-to-day activities and timely consultation.

5.2 Case 2: Curation by Geographically Spread Teams

In this case, a pathway curation project was initiated to build a map of the influenza virus life cycle to elucidate the host-pathogen interaction at the cellular level.

Existing pathway databases were first checked to identify any relevant pathways. Two major components of the influenza-related pathway were found in the Reactome pathway database: one for viral replication and the other for its host responses. About 400 factors and 53 interactions were covered in these two pathways. Other pathway databases, including KEGG and Panther Pathways, were searched for related modules such as toll-like receptors and immune response signaling.

The identified pathways were used as a base map, which was then enhanced. Based on information sourced for the published literature, the influenza virus life cycle map was enhanced and integrated with the host responses map into a single pathway. The pathway map now covers 964 factors and 448 reactions, citing about 500 papers [14].

The project was started in the autumn 2010 with one postdoc and a few students who build the base map. A Mapathon curation event was held during which influenza experts were invited and assigned a collection of selected papers from which they extracted information and constructed diagrams to represent the source data. To make their work easier, curation sheets were provided for them to write down the necessary information. The chief curator collected the curation sheets and compiled the information into the base map. After the Mapathon sessions, all the participants got

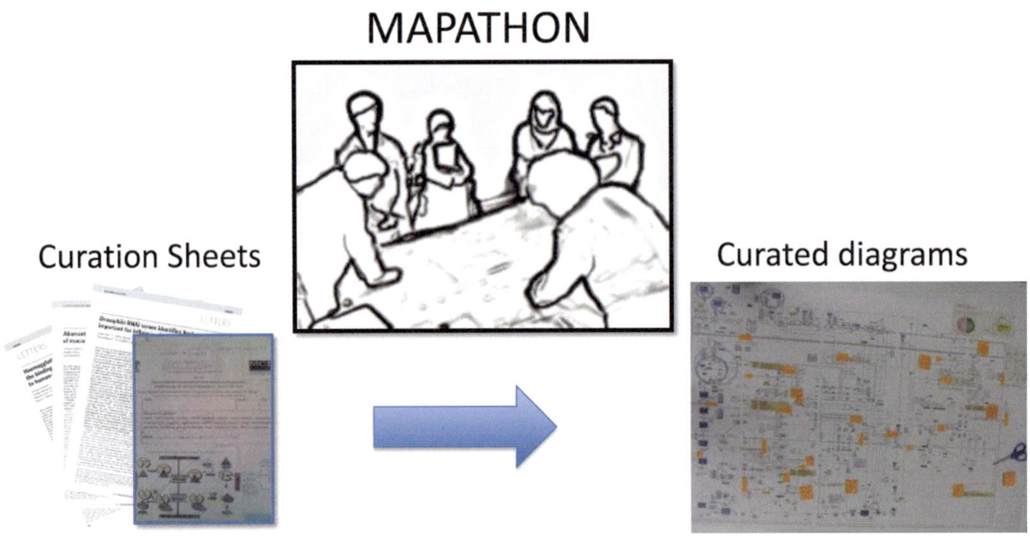

Fig. 19 Face-to-face Mapathon workflow

together to verify the data that had been newly added to the pathway map (Fig. 19).

This exercise also ensured that all the team members knew how to curate the map, and understood all the graphical notations used in the diagrams. The participants also reported that they had experienced difficulties in interpretation of the text into the graphics.

Although the face-to-face Mapathon was conducted in Tokyo, all the materials were shared with a team in the USA to whom both the PDF file of the map and the online version on the Payao curation platform were sent. In the USA, the network was fast enough, so it was expected that participants would validate the map online. However, their preference was to print it out and post it on a wall to browse and comment on. Feedback was returned by e-mail with photographs of printouts with the written comments.

An important lesson was learnt from this project, namely that the communication channels should be simple and comfortable for the curators to use so as to maximize the participation and knowledge collection.

6 Conclusions

Here we illustrated a collaborative approach to pathway modeling and curation purely from literature and literature-based curation, focusing on conversion of the textual information into graphical knowledge representation adopting standard languages. We took the example of building the customized pathways for systems biology and toxicology domain. Data-centric pathway verification was not in the scope of this chapter; yet data-driven verification approach would be essential for constructing the toxicological pathways. In the future, how to integrate knowledge obtained from the data-driven modeling with the literature-driven pathway construction need to be investigated.

For the future collaborative knowledge curation projects, primary challenges would lie in coping the scale while maintaining the quality.

With the exponential growth of data and publications in the life sciences, it is becoming increasingly difficult for individual laboratories to conduct "large-scale" research. Community-driven collaborative projects spanning multiple research groups across geographies are gaining prevalence in life sciences. Thus, social aspects of collaboration between scientific partners participating in a research project become increasingly important issues in governing the success of such initiatives [59–62]. While the life sciences has historically seen success in large-scale collaborative projects through the Human Genome Sequencing projects of the 1990s, the rise of data-sharing platforms [63], high-throughput technologies, and social media avenues [64] open up new challenges and opportunities, as outlined in [59].

New mechanisms for rewards [65], engagement [61], community-wide participation from industry and academia [66] and verification [4, 60], as well as attribution and funding [67] will need to be explored as life sciences research and roadmap for community systems biology efforts need to be developed [62].

Aspects from social networks [63] in managing and coordinating such open collaborative efforts, together with the computational platforms which enable big data sharing and analytics, will drive the success of future large-scale collaborative knowledge curation projects [34], in particular, toxicity pathway curation projects [1].

7 Notes

1. The differences between the CellDesigner graphical notation and SBGN can be checked in CellDesigner by selecting *View→ Show SBGN compliant*, and the graphics can be converted automatically by selecting *View→ Convert to SBGN PD (i.e., process description) view*.
2. When interpreting the text to graphics, it is not always the case that the interaction is described in a single sentence in the text. The biological events are often described in multiple sentences, not summarized in a single sentence. Description may be scattered in multiple paragraphs. This makes it difficult to deploy a text mining tool in a fully automated way to generate different pieces of the pathway.

3. It is important to use tools and/or methodologies that the domain experts are familiar with so that curation can continue to the next step of knowledge accumulation.

Acknowledgements

This work was supported, in part, by funding from the Genome Network Project of the Ministry of Education, Culture, Sports, Science and Technology, the New Energy and Industrial Technology Development Organization, the International Strategic Collaborative Research Program of the Japan Science and Technology Agency (JST), the Exploratory Research for Advanced Technology program of JST [to the Systems Biology Institute (SBI)], and a strategic cooperation partnership between the Luxembourg Centre for Systems Biomedicine and the SBI. Inspired by the sbv IMPROVER workshops.

References

1. Krewski D, Acosta D Jr, Andersen M et al (2010) Toxicity testing in the 21st century: a vision and a strategy. J Toxicol Environ Health B Crit Rev 13:51–138. doi:10.1080/10937404.2010.483176
2. Kleensang A, Maertens A, Rosenberg M et al (2014) t4 workshop report: Pathways of toxicity. ALTEX 31:53–61. doi:10.14573/altex.1309261
3. Howe D, Costanzo M, Fey P et al (2008) Big data: the future of biocuration. Nature 455:47–50. doi:10.1038/455047a
4. sbv IMPROVER project team, Ansari S, Binder J et al (2013) On crowd-verification of biological networks. Bioinform Biol Insights 7:307–325. doi:10.4137/BBI.S12932
5. Le Novère N, Hucka M, Mi H et al (2009) The systems biology graphical notation. Nat Biotechnol 27:735–741. doi:10.1038/nbt1558
6. Mi H, Lazareva-Ulitsky B, Loo R et al (2005) The PANTHER database of protein families, subfamilies, functions and pathways. Nucleic Acids Res 33:D284–D288. doi:10.1093/nar/gki078
7. Mi H, Guo N, Kejariwal A et al (2007) PANTHER version 6: protein sequence and function evolution data with expanded representation of biological pathways. Nucleic Acids Res 35:D247–D252. doi:10.1093/nar/gkl869
8. Croft D, Mundo AF, Haw R et al (2014) The Reactome pathway knowledgebase. Nucleic Acids Res 42:D472–D477. doi:10.1093/nar/gkt1102
9. Hucka M, Finney A, Sauro HM et al (2003) The systems biology markup language (SBML): a medium for representation and exchange of biochemical network models. Bioinformatics 19:524–531. doi:10.1093/bioinformatics/btg015
10. Lloyd CM, Halstead MD, Nielsen PF (2004) CellML: its future, present and past. Prog Biophys Mol Biol 85:433–450. doi:10.1016/j.pbiomolbio.2004.01.004
11. Demir E, Cary MP, Paley S et al (2010) The BioPAX community standard for pathway data sharing. Nat Biotechnol 28:935–942. doi:10.1038/nbt.1666
12. Stromback L, Lambrix P (2005) Representations of molecular pathways: an evaluation of SBML, PSI MI and BioPAX. Bioinformatics 21:4401–4407. doi:10.1093/bioinformatics/bti718
13. Slater T (2014) Recent advances in modeling languages for pathway maps and computable biological networks. Drug Discov Today 19:193–198. doi:10.1016/j.drudis.2013.12.011
14. Matsuoka Y, Matsumae H, Katoh M et al (2013) A comprehensive map of the influenza A virus replication cycle. BMC Syst Biol 7:97. doi:10.1186/1752-0509-7-97
15. Kaizu K, Ghosh S, Matsuoka Y et al (2010) A comprehensive molecular interaction map of the budding yeast cell cycle. Mol Syst Biol 6:415. doi:10.1038/msb.2010.73
16. Caron E, Ghosh S, Matsuoka Y et al (2010) A comprehensive map of the mTOR signaling

network. Mol Syst Biol 6:453. doi:10.1038/msb.2010.108

17. Oda K, Kitano H (2006) A comprehensive map of the toll-like receptor signaling network. Mol Syst Biol 2. doi:10.1038/msb4100057
18. Oda K, Matsuoka Y, Funahashi A et al (2005) A comprehensive pathway map of epidermal growth factor receptor signaling. Mol Syst Biol 1:2005.0010. doi:10.1038/msb4100014
19. Kitano H, Oda K, Kimura T et al (2004) Metabolic syndrome and robustness tradeoffs. Diabetes 53(Suppl 3):S6–S15
20. Calzone L, Gelay A, Zinovyev A et al (2008) A comprehensive modular map of molecular interactions in RB/E2F pathway. Mol Syst Biol 4:173. doi:10.1038/msb.2008.7
21. Kanehisa M, Goto S (2000) KEGG: Kyoto encyclopedia of genes and genomes. Nucleic Acids Res 28:27–30
22. Mi H, Thomas P (2009) PANTHER pathway: an ontology-based pathway database coupled with data analysis tools. Methods Mol Biol 563: 123–140. doi:10.1007/978-1-60761-175-2_7
23. Joshi-Tope G, Gillespie M, Vastrik I et al (2005) Reactome: a knowledgebase of biological pathways. Nucleic Acids Res 33:D428–D432. doi:10.1093/nar/gki072
24. Thiele I, Swainston N, Fleming RM et al (2013) A community-driven global reconstruction of human metabolism. Nat Biotechnol 31: 419–425. doi:10.1038/nbt.2488
25. Pico AR, Kelder T, van Iersel MP et al (2008) WikiPathways: pathway editing for the people. PLoS Biol 6, e184. doi:10.1371/journal.pbio.0060184
26. Bader GD, Cary MP, Sander C (2006) Pathguide: a pathway resource list. Nucleic Acids Res 34:D504–D506. doi:10.1093/nar/gkj126
27. Bockmann B, Heiden K (2013) PathGuide—model-based generation of guideline-compliant pathways for the use in different hospital information systems. Stud Health Technol Inform 192:1089
28. Viswanathan GA, Nudelman G, Patil S et al (2007) BioPP: a tool for web-publication of biological networks. BMC Bioinformatics 8:168. doi:10.1186/1471-2105-8-168
29. Matsuoka Y, Ghosh S, Kikuchi N et al (2010) Payao: a community platform for SBML pathway model curation. Bioinformatics (Oxford, England) 26:1381–1383. doi:10.1093/bioinformatics/btq143
30. Kuperstein I, Cohen DP, Pook S et al (2013) NaviCell: a web-based environment for navigation, curation and maintenance of large molecular interaction maps. BMC Syst Biol 7:100. doi:10.1186/1752-0509-7-100
31. Bauer-Mehren A, Furlong LI, Sanz F (2009) Pathway databases and tools for their exploitation: benefits, current limitations and challenges. Mol Syst Biol 5:290. doi:10.1038/msb.2009.47
32. Mizuno S, Iijima R, Ogishima S et al (2012) AlzPathway: a comprehensive map of signaling pathways of Alzheimer's disease. BMC Syst Biol 6:52. doi:10.1186/1752-0509-6-52
33. Fujita KA, Ostaszewski M, Matsuoka Y et al (2014) Integrating pathways of Parkinson's disease in a molecular interaction map. Mol Neurobiol 49:88–102. doi:10.1007/s12035-013-8489-4
34. Kitano H, Ghosh S, Matsuoka Y (2011) Social engineering for virtual 'big science' in systems biology. Nat Chem Biol 7:323–326. doi:10.1038/nchembio.574
35. Funahashi A, Matsuoka Y, Jouraku A et al (2008) CellDesigner 3.5: a versatile modeling tool for biochemical networks. Proceedings of the IEEE 96, pp 1254–1265
36. Apweiler R, Bairoch A, Wu CH et al (2004) UniProt: the Universal Protein knowledgebase. Nucleic Acids Res 32:D115–D119. doi:10.1093/nar/gkh131
37. Maglott D, Ostell J, Pruitt KD et al (2005) Entrez Gene: gene-centered information at NCBI. Nucleic Acids Res 33:D54–D58. doi:10.1093/nar/gki031
38. Kitano H (2003) A graphical notation for biochemical networks. BIOSILICO 1:169–176. doi:http://dx.doi.org/10.1016/S1478-5382(03)02380-1
39. Kitano H, Funahashi A, Matsuoka Y et al (2005) Using process diagrams for the graphical representation of biological networks. Nat Biotechnol 23:961–966. doi:10.1038/nbt1111
40. van Iersel MP, Villeger AC, Czauderna T et al (2012) Software support for SBGN maps: SBGN-ML and LibSBGN. Bioinformatics 28:2016–2021. doi:10.1093/bioinformatics/bts270
41. Machne R, Finney A, Muller S et al (2006) The SBML ODE Solver Library: a native API for symbolic and fast numerical analysis of reaction networks. Bioinformatics 22:1406–1407. doi:10.1093/bioinformatics/btl086
42. Hoops S, Sahle S, Gauges R et al (2006) COPASI—a COmplex PAthway SImulator. Bioinformatics 22:3067–3074. doi:10.1093/bioinformatics/btl485
43. Keller R, Dorr A, Tabira A et al (2013) The systems biology simulation core algorithm. BMC Syst Biol 7:55. doi:10.1186/1752-0509-7-55

44. Le Novere N, Bornstein B, Broicher A et al (2006) BioModels Database: a free, centralized database of curated, published, quantitative kinetic models of biochemical and cellular systems. Nucleic Acids Res 34:D689–D691. doi:10.1093/nar/gkj092
45. Olivier BG, Snoep JL (2004) Web-based kinetic modelling using JWS Online. Bioinformatics 20:2143–2144. doi:10.1093/bioinformatics/bth200
46. Fujibuchi W, Goto S, Migimatsu H et al (1998) DBGET/LinkDB: an integrated database retrieval system. Pac Symp Biocomput 683–694. http://www.ncbi.nlm.nih.gov/pubmed/9697222
47. Cherry JM, Adler C, Ball C et al (1998) SGD: Saccharomyces Genome Database. Nucleic Acids Res 26:73–79
48. Fernandez JM, Hoffmann R, Valencia A (2007) iHOP web services. Nucleic Acids Res 35:W21–W26. doi:10.1093/nar/gkm298
49. Huss JW III, Lindenbaum P, Martone M et al (2010) The Gene Wiki: community intelligence applied to human gene annotation. Nucleic Acids Res 38:D633–D639. doi:10.1093/nar/gkp760
50. Degtyarenko K, de Matos P, Ennis M et al (2008) ChEBI: a database and ontology for chemical entities of biological interest. Nucleic Acids Res 36:D344–D350. doi:10.1093/nar/gkm791
51. Caspi R, Altman T, Dreher K et al (2012) The MetaCyc database of metabolic pathways and enzymes and the BioCyc collection of pathway/genome databases. Nucleic Acids Res 40:D742–D753. doi:10.1093/nar/gkr1014
52. Mi H, Muruganujan A, Demir E et al (2011) BioPAX support in Cell Designer. Bioinformatics 27:3437–3438. doi:10.1093/bioinformatics/btr586
53. Ghosh S, Matsuoka Y, Asai Y et al (2013) Toward an integrated software platform for systems pharmacology. Biopharm Drug Dispos 34:508–526. doi:10.1002/bdd.1875
54. Matsuoka Y, Funahashi A, Ghosh S et al (2014) Modeling and simulation using Cell Designer. Methods Mol Biol 1164:121–145. doi:10.1007/978-1-4939-0805-9_11
55. Hirschman L, Burns GA, Krallinger M et al (2012) Text mining for the biocuration workflow. Database 2012:bas020. doi:10.1093/database/bas020
56. Le Novère N, Finney A, Hucka M et al (2005) Minimum information requested in the annotation of biochemical models (MIRIAM). Nat Biotechnol 23:1509–1515. doi:10.1038/nbt1156
57. Oda K, Kim JD, Ohta T et al (2008) New challenges for text mining: mapping between text and manually curated pathways. BMC Bioinformatics 9(Suppl 3):S5. doi:10.1186/1471-2105-9-S3-S5
58. Ghosh S, Matsuoka Y, Asai Y et al (2013) Software platform for metabolic network reconstruction of Mycobacterium tuberculosis. In: Beste DJV, Kierzek AM, McFadden J (eds) Systems biology of tuberculosis. Springer, New York, pp 21–35. doi:10.1007/978-1-4614-4966-9_2
59. Jasny B (2013) Realities of data sharing using the genome wars as case study—an historical perspective and commentary. EPJ Data Sci 2:1
60. Meyer P, Alexopoulos LG, Bonk T et al (2011) Verification of systems biology research in the age of collaborative competition. Nat Biotechnol 29:811–815. doi:10.1038/nbt.1968
61. Friend SH, Norman TC (2013) Metcalfe's law and the biology information commons. Nat Biotechnol 31:297–303. doi:10.1038/nbt.2555
62. Zengler K, Palsson BO (2012) A road map for the development of community systems (CoSy) biology. Nat Rev Microbiol 10:366–372. doi:10.1038/nrmicro2763
63. Aral S, Walker D (2012) Identifying influential and susceptible members of social networks. Science 337:337–341. doi:10.1126/science.1215842
64. Bohon W, Robinson S, Arrowsmith R et al (2013) Building an effective social media strategy for science programs. Eos 94:237–238. doi:10.1002/2013EO270001
65. Lakhani KR, Boudreau KJ, Loh PR et al (2013) Prize-based contests can provide solutions to computational biology problems. Nat Biotechnol 31:108–111. doi:10.1038/nbt.2495
66. Sheridan C (2011) Industry continues dabbling with open innovation models. Nat Biotechnol 29:1063–1065. doi:10.1038/nbt1211-1063a
67. Open to interpretation (2013). Nat Biotechnol 31: 661. doi:10.1038/nbt.2665

Chapter 9

Pathway Analysis and Machine Learning as Tools in Systems Toxicology

Adi Laurentiu Tarca

Abstract

An important aspect of toxicity testing is the identification of signaling pathways perturbed by biologically active substances or their metabolites that can cause adverse health effects in humans. This chapter describes analysis methods that can use (1) human transcriptomics data to identify perturbed signaling pathways and (2) animal model data to infer the pathway activity in human for the same stimuli. For the former goal, we describe Pathway Analysis with Down-weighting of Overlapping Genes (PADOG) and Pathway Level Analysis of Gene Expression (PLAGE), which have been shown to perform well for pathway prioritization. For the second goal, we describe two strategies that rely on transcriptomic changes induced by a set of training stimuli in both human and animal models to learn how to predict pathway activity in human for new stimuli from similar data in an animal model. One strategy is based on machine learning that maps pathway activity results from animal models to human, and the other strategy works by deriving functional homology between the animal model and human genes so that a gene set analysis method (e.g., PADOG and PLAGE) can be applied to the animal data to identify significantly perturbed pathways in human. In addition to describing the methods and their assumptions and limitations, we illustrate how to apply these methods using the R statistical environment and open-source Bioconductor packages.

Key words Gene set analysis, Pathway analysis, Machine learning, Data-driven functional homology

1 Introduction

Identification of functionally related gene sets (e.g., biological processes and pathways) that are perturbed by biologically active substances or their metabolites is a common and important aim in systems toxicology. For example, human hepatoma (HepG2) cells were exposed to sertraline and then subjected to whole-genome gene expression microarray analysis. Pathway analysis of the gene expression data revealed that endoplasmic reticulum stress gene set was perturbed by this treatment [1]. The assessment of the toxicity potential of new molecules usually cannot be performed directly in humans; hence, use of surrogate models and "omics" technologies is required [2]. The direct translation of findings from animal models to humans is often unreliable [3], prompting the need for systematic studies to assess the

Julia Hoeng and Manuel C. Peitsch (eds.), *Computational Systems Toxicology*, Methods in Pharmacology and Toxicology, DOI 10.1007/978-1-4939-2778-4_9, © Springer Science+Business Media New York 2015

extent to which pathway activation in animal models is informative for pathway activation in humans and to find the best approaches to make the pathway activity inference between organisms [4].

The *objective* of this chapter is to describe approaches that can predict human pathway under *two scenarios*: (I) when human expression data for the stimulus of interest and corresponding control samples are available, and (II) when human expression data for the treatment of interest are not available but data for other stimuli and corresponding controls are available in both human and an animal model. The use of these methods with R (www.r-project.org) software packages and functions is also described. The R statistical language and environment are freely available on most operating systems including Windows, Linux, and Mac.

1.1 Scenario I

Two methods that can be applied on gene expression data measured in control and treated cells to identify pathways perturbed by the treatment are first described (see Fig. 1). Gene sets that are to be tested for association with a phenotype can be biological pathways (e.g., the Kyoto Encyclopedia of Genes and Genomes (KEGG) [5] and Reactome [6] databases), or genes grouped by gene ontology [7], or simply custom gene sets such as some sets that are included in the Molecular Signatures Database (MSigDB) [8]. There are at least three classes of methods that can be used to perform gene set analysis, and these have been described and compared elsewhere [9]. The two methods described here are (1) Pathway Analysis with Down-weighting of Overlapping Genes (PADOG)

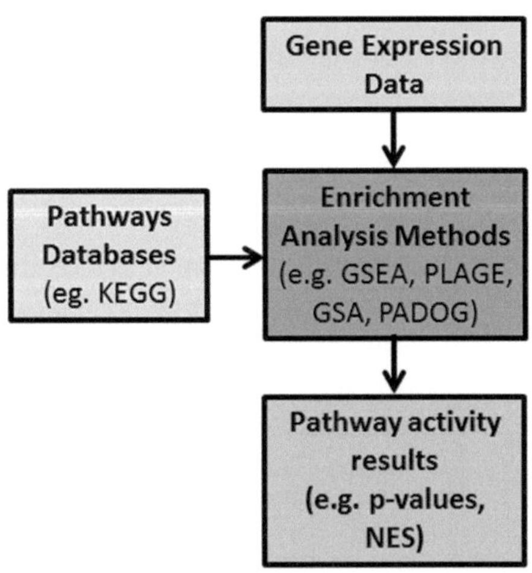

Fig. 1 Identifying perturbed pathways from gene expression data. Gene expression levels in treated and control cells as well as the phenotype of the samples are used as input into an enrichment method, together with a collection of predefined gene sets (such as biological pathways). The result is an ordered list of pathways and their significance *p*-values, and eventually enrichment scores. *KEGG* Kyoto encyclopedia of genes and genomes, *GSEA* gene set enrichment analysis, *PLAGE* Pathway Level Analysis of Gene Expression, *GSA* gene set analysis, *PADOG* Pathway Analysis with Down-weighting of Overlapping Genes, *NES* normalized enrichment score

[10], which combines differential expression evidence from all genes in a gene set and then establishes the significance of the pathway level score by permutation of the samples, and (2) Pathway Level Analysis of Gene Expression (PLAGE) [11], which computes a gene set score (metagene) for each sample, and then the scores are compared between groups using appropriate models (e.g., a *t*-test). These methods provide a significance *p*-value, and eventually pathway enrichment scores, for the association between phenotype and predefined sets of genes available in public databases. Although both methods (described in more detail in Sect. 2) sometimes rely on different null hypotheses and statistics, they basically assume that when a gene set is indeed relevant to a given treatment, a sizeable proportion of the genes will show some amount of differential expression between groups in either one or both directions, depending on the method.

1.2 Scenario II

When gene expression data are not available for the treatment of interest in humans, we describe two methods identified in the IMPROVER Species Translation Challenge [4] that can use animal model data to make pathway activity predictions in human. The first approach (Fig. 2) relies on a set of training stimuli for

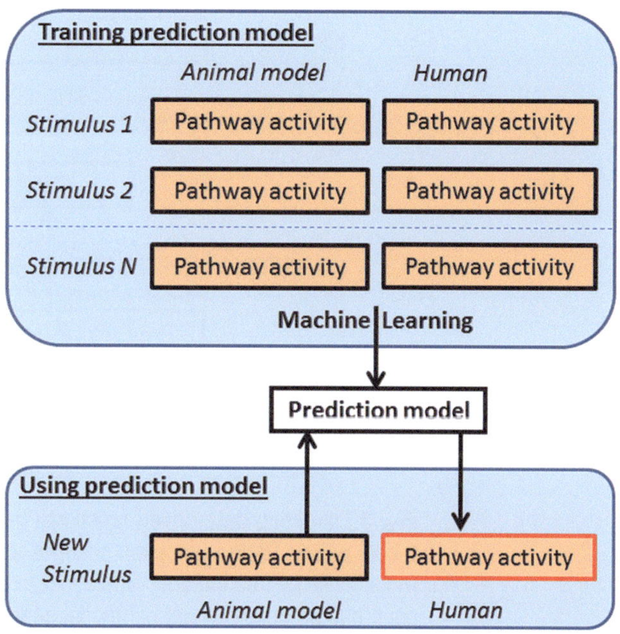

Fig. 2 Machine learning approach to predict human pathway activity from animal model data. Pathway activity data are obtained for human and animal model cells by treating both types of cells with a set of training stimuli. Then a machine learning approach is used to obtain a prediction model for each human pathway based on the activity of one or more pathways in the animal model (predictors/features). The trained model is then applied on pathway activity data obtained for a new stimulus available only for the animal model to predict the human pathways that will be perturbed by that stimulus

which pathway activity data are obtained by applying a gene set analysis method to expression data from treated and untreated cells in both human and an animal model. This approach generates a prediction model for each pathway of interest, which then can be used to make predictions for a new stimulus for which only the animal model pathway activity data are available.

The second approach uses gene expression changes between treated and untreated cells in both organisms to derive data-driven functional homology that maps each human gene to one animal model gene. Based on this homology, the animal model pathway activity data for a new stimulus are then used in conjunction with a gene set analysis method to identify perturbed gene sets and pathways in human (Fig. 3).

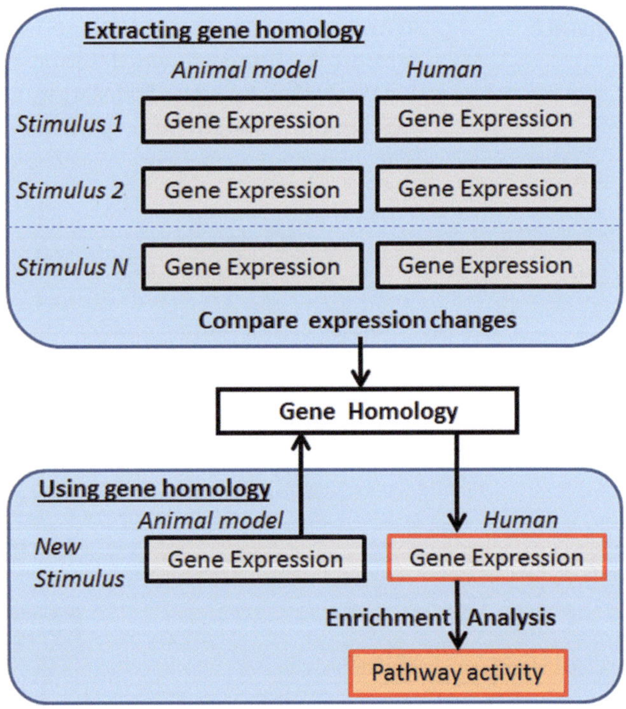

Fig. 3 Extracting data-driven functional homology between human and animal model genes. Gene expression changes are determined for human and animal model cells by treating both types of cells with a set of training stimuli. Then for each human gene one animal model gene is selected that best mimics the differential expression of the human gene across all training stimuli. For a new stimulus, the animal model data are used to replace the human data based on the one-to-one homology already established. Finally, a gene set analysis method is applied on the animal model expression data to determine the pathway activity status in human

2 Methods

2.1 Gene Set Analysis Methods

The PADOG and PLAGE approaches for gene set analysis are described in this section. The choice between these methods depends on the available sample size and experimental design, as described in **Note 1**. The performance of these two methods as well as of 14 others for pathway was compared using 42 real datasets [9]. The phenotype in each of the 42 datasets corresponded to an existing disease pathway/gene set in KEGG or Metacore (Thomson Reuters, NY) databases, e.g., *Colorectal cancer* (http://www.genome.jp/kegg-bin/show_pathway?hsa05210) and *Alzheimer's disease* (http://www.genome.jp/kegg-bin/show_pathway?hsa05010). Analysis methods were compared in terms of their ability to generate a small *p*-value (*sensitivity*) and rank close to top based on the *p*-values (*prioritization*) the disease-specific pathway that corresponded to the phenotype while not producing more false positives than expected when the same analyses were performed with randomly assigned phenotypes. PLAGE was ranked first based on combined sensitivity and prioritization criteria, while PADOG was ranked first based on prioritization alone. The ranking of the 16 methods in terms of sensitivity and prioritization is shown in Fig. 4. Of note, 4 of the methods included in this comparison (MRGSE, GSEAP, GAGE, and SIGPATH.Q1) use genes as sampling unit, unlike the remaining 12 that use subjects as sampling unit, and hence they produce more false positives than expected when phenotypes are randomly assigned to samples, as described in [9]. This benchmark of methods using real datasets and disease pathways has been adopted also by others [23] to assess novel pathway analysis methods.

2.1.1 Pathway Analysis with Down-Weighting of Overlapping Genes

Let GS_i with $i = 1 \ldots N_{GS}$ be the collection of gene sets to be analyzed, each containing $N(GS_i)$ genes, and G be the set of all the genes measured on the array that can be mapped to at least one of the gene sets to be analyzed. Then let T_g be the value of a moderated *t*-score [24] of the gene g between the two conditions of interest. A gene set score can be determined as follows:

$$S_0\left(GS_i\right) = \frac{1}{N\left(GS_i\right)} \sum_{g \in GS_i} \left|T(g)\right| \cdot w(g), \tag{1}$$

where gene weight $w(g)$ is computed as

$$w(g) = 1 + \sqrt{\frac{\max(f) - f(g)}{\max(f) - \min(f)}}, \tag{2}$$

where $f(g)$ is the frequency of gene g across all gene sets to be analyzed. The gene weights are 1.0 for the most popular gene (highest frequency across all gene sets) and 2.0 for a gene that appears in only one given gene set, and hence is specific to that gene set.

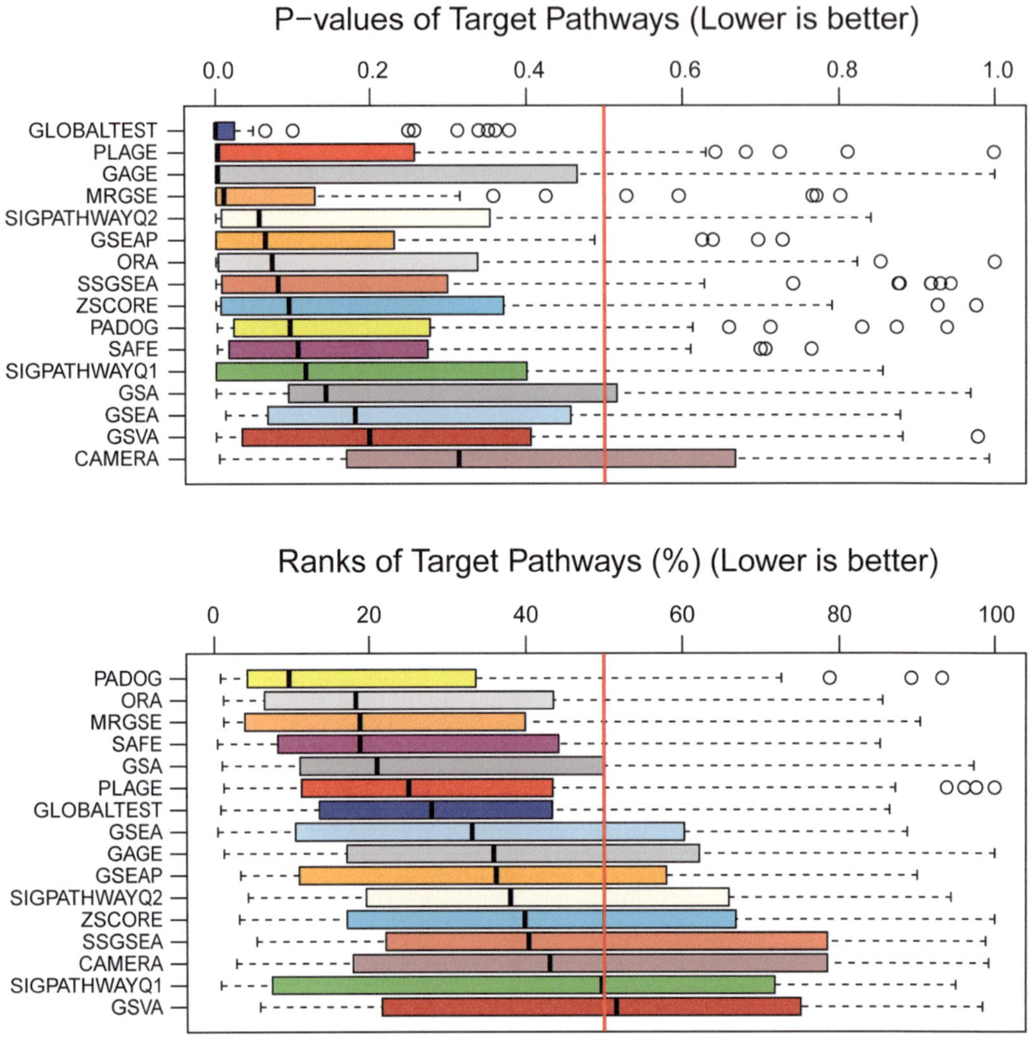

Fig. 4 A comparison of sensitivity and prioritization ability of 16 gene set analysis methods adapted from [9]. Each box contains 42 data points representing the *p*-value (*top*) and the rank (%) (*bottom*) that the target pathway received from a given method when using as input an independent dataset and a collection of gene sets (either KEGG or Metacore). Since the target pathways were designed by KEGG and Metacore for those diseases we expected that, in average, they will be found relevant by the different methods. Methods are ranked from best to worst according to the median *p*-value (*top*) and median rank (*bottom*). The 16 methods are Over-Representation Analysis (ORA) [12], Gene Set Enrichment Analysis (GSEA) and GSEA Pre-Ranked (GSEAP) [8], GLOBALTEST [13], SIGPATHWAY with two null hypotheses (Q1 and Q2) [14], Gene Set Analysis (GSA) [15], Generally Applicable Gene set Enrichment (GAGE) [16], Significance Analysis of Functional categories in Gene Expression studies (SAFE) [17], Mean-Rank Gene Set Enrichment tests (MRGSE) [18], Pathway Analysis with Down-weighting of Overlapping Genes (PADOG) [10], Correlation Adjusted Mean Rank gene set test (CAMERA) [19], Pathway Level Analysis of Gene Expression (PLAGE) [11], *Z*-score (ZSCORE) [20], Single Sample GSEA (SSGSEA) [21], and Gene Set Variation Analysis (GSVA) [22]

The gene set scores obtained with the formula above are first standardized using a row randomization approach [16] to obtain $S_0'(GS_i)$. A second standardization is applied by subtracting the mean and dividing to the standard deviation of $S_0'(GS_i)$ scores across all *NGS* gene sets to obtain the observed standardized scores $S_0^*(GS_i)$. The probability $P_{\text{PADOG}}(GSi)$ of observing such a large or larger standardized score is determined by permuting the array/sample labels a large number of times (e.g., $N_{\text{ite}} = 1000$) as

$$P_{\text{PADOG}}\left(GS_i\right) = \frac{\sum_{\text{ite}} I\left(S_{\text{ite}}^*\left(GS_i\right) \geq S_0^*\left(GS_i\right)\right)}{N_{\text{ite}}}, \qquad (3)$$

where I is a function that returns 1 when the argument is true and 0 otherwise, and $S_{\text{ite}}^*(GS_i)$ represents the standardized score obtained with the ite-th permutation of the samples for gene set *GSi*.

Method usage in R: Assuming that the PADOG package is installed and an R session is open, the following R commands can be used to run PADOG on KEGG signaling and metabolic pathways provided: a normalized and log-transformed gene expression matrix *X*, a vector of phenotypes of the samples (*g*), and assuming a non-paired two group design. Note that the gene expression data matrix *X* needs to be suitably preprocessed as described in **Note 2**.

```
outpadog=padog(esetm=X, group=g, paired=FALSE, block=NULL,
annotation="hgu133plus2.db", gslist="KEGG.db",
organism="hsa", Nmin=3,NI=2000)
```

In this function call, we assumed that the gene expression data matrix, *X*, was obtained on the Affymetrix HG-U133 Plus 2.0 microarray, information that is used to map the row names identifiers of the expression data *X* (Affymetrix probe set IDs) to ENTREZ gene identifiers that are mapped to *homo sapiens* (organism="hsa") KEGG pathways using the KEGG.db package of Bioconductor. We also specified in the function call that only pathways with Nmin=3 or more genes measured on this array be used in the analysis. The number of permutations of the samples to compute *p*-values for each pathway was *NI*=2000. Alternatively, to use PADOG with a custom list of gene sets (e.g., gslist= mygslist) each element of mygslist needs to be a vector with ENTREZ gene identifiers, case in which a) either annotation=NULL and the row names of X are ENTREZ IDs or b) annotation is set to a valid installed annotation package that can be used to map the row names of *X* to ENTREZ IDs.

2.1.2 Pathway Level Analysis of Gene Expression

Unlike PADOG, which computes a pathway score based on differential expression statistics for each gene in the gene set, PLAGE computes a pathway activity score in each sample and then compares the scores between sample groups. PLAGE uses singular value decomposition (SVD) on the expression data matrix *X* that includes all samples (rows) and only the genes in a given pathway (columns) to obtain eigenvectors or metagenes of *X*. Among all possible metagenes, the first metagene is the one that corresponds to the largest eigenvalue. The square of the eigenvalues gives the amount of variation accounted for by the corresponding metagene. Before applying the SVD transformation, the log-transformed gene expression data are standardized so that the level of each gene across all samples has mean 0 and unit standard deviation.

Method usage in R: The gene set variation analysis (GSVA) package [22] also implements the PLAGE method.

```
library(GSVA)
activ =gsva(X, gset.idx.list= mygslist, min.sz=3,
max.sz=1000, method="plage")
```

The function call above requires a gene expression matrix *X* with rows that correspond to genes and columns that correspond to samples. The *gset.idx.list* specifies the collection of gene sets as a list. The arguments *min.sz* and *max.sz* set the minimum and maximum gene set size to avoid analyzing gene sets that are too small or too large.

The resulting *activ* matrix contains the activity score of each pathway (rows) for each sample (columns). These can be compared between groups by applying a moderated *t*-test using the *limma* package as follows:

```
library(limma)
design <- model.matrix(~ factor(g))
fit <- lmFit(activ, design)
fit <- eBayes(fit)
outplage <- topTable(fit, coef=2, number=Inf)
```

where *g* is the vector indicating the phenotype of the samples and *outplage* contains the test results from comparing the pathway activity scores between sample groups. The function *model.matrix* creates a matrix whose rows correspond to samples and columns correspond to linear predictors derived from the phenotype variable *g*. The second column of this matrix corresponds to the contrast between the two phenotype groups. The function *lmFit* fits a linear model for each gene set (row of matrix *activ*) and *eBayes* performs the moderation of t-scores corresponding to the differences between the gene set score means between groups. In-depth details for using the *limma* package for more complex designs can be found elsewhere [24].

2.2 Using Animal Model Data to Predict Pathway Activity in Human

Most often than not, transcription data measured in human tissues or cells, obtained after treatment with given stimuli, are not available and hence data collected from animal models need to be used instead. A large-scale initiative to study the translatability of pathway activation between rat and human models was described recently [4].

In this study, gene expression data were measured in rat and human lung epithelial cells after treatment with 52 stimuli. Perturbed pathways were identified using pre-ranked gene set enrichment analysis (GSEAP) [8] for each stimulus and organism. The question asked was whether data from 26 stimuli in rat and human can be used to learn how to predict human pathway perturbation for the remaining 26 stimuli based on the corresponding data in rat. Below we describe two alternative approaches named *Team49_alt1* and *Team49*, which we proposed in the *Inter-species Pathway Perturbation Prediction* sub-challenge [4]. The performance of submission *Team49_alt1* was 2 % points lower on one of the three metrics than was needed for being tied for first place, while submission *Team49* was ranked second in this challenge. The first submission was based on a machine learning approach while the second involved the establishment of a data-driven functional homology between rat and human genes. Merits and disadvantages of both types of approaches are described elsewhere [25] while limitations of these approaches are provided in **Note 3**. Making pathway activity inferences between species requires that transcription data are measured in a similar experiment design to the one used in the *Inter-species Pathway Perturbation Prediction* sub-challenge. The methods below assume that gene expression profiles in both organisms, for control and stimuli-treated tissues or cells, are available.

2.2.1 Machine Learning Method to Infer Pathway Activity

The input data required for the machine learning approach are a pathway activity matrix for the model organism and a similar one for human. The rows in these matrices correspond to chemical stimuli that were applied to normal tissues/cells while the columns correspond to the different gene sets or pathways. The values in the matrices are some measure of the evidence that the corresponding pathway was activated when a given stimuli was applied, as determined by a gene set analysis method. For instance, if GSEAP is used to determine pathway activity, then normalized enrichment scores could be used as pathway activation data. Alternatively, if PLAGE is used instead, the t-scores for comparing PLAGE scores for a given pathway between groups can be used as evidence for pathway activation data.

Let A_m and A_h denote the pathway activation matrix for the model organism and human, respectively. The machine learning approach uses some or all of the columns in A_m as predictors for each column in A_h. If instead of a continuous pathway activity score in A_h binary data are used (1 if the pathway was significantly activated and 0 if not), then this prediction problem is akin to a supervised learning approach. In this approach [26, 27], the pathway activity for one or a few animal model pathways is used as a predictor in a linear discriminant analysis (LDA) model that is trained to predict the activation status of one human pathway at a time. If the given pathway is not activated by two or more stimuli in the training set, such a prediction model cannot be trained. To predict human pathway activation status for a new stimulus, the trained model is applied to the pathway activation data obtained in the animal model for that stimulus.

To overcome the applicability limitation of the machine learning methods (to pathways that were activated by at least a few training stimuli), an alternative approach, described in Sect. 2.2.2, can be used to identify a human pathway as activated even when that pathway was never activated in the animal model.

2.2.2 Approach Based on Functional Homology and Gene Set Analysis

Alternative approaches to the machine learning technique can be based on the idea of finding an "orthologous" animal model gene for each human gene by learning from the gene expression changes induced by the stimuli in the training set [25]. Then, for new stimuli, the animal model gene expression data can be used to perform gene set analysis on gene sets constructed from homologous animal model genes and hence infer activity of human pathways. Gene set analysis can be conducted with any general gene set analysis method such as GSEA, PADOG, or PLAGE.

The procedure can be detailed as a series of steps:

1. Consider a collection of gene sets and pathways from a public repository, such as KEGG or MSigDB. Discard those entries with fewer than three genes present on the platform used for gene expression profiling.
2. For each stimulus in the training set,[1] compute the rank of each gene based on the moderated t-scores comparing gene expression levels between stimuli-treated and control samples. This analysis is done separately for the human and animal model data. Let $Rank_{h,i}$ and $Rank_{m,i}$ denote the ranks that a given human gene h and a given animal model gene m receive for stimulus i. The $Rank$ values are normalized so that 0 corresponds to the most downregulated gene, while 1.0 corresponds to the most up-regulated gene. For each human gene h:

   (a) Compute distance

$$D(h,m) = \sum_{i=1}^{26} \left| Rank_{h,i} - Rank_{m,i} \right| w_i \tag{1}$$

   between the human gene h and all animal model genes. The weight w_i is defined as

$$w_i = \left| Rank_{h,i} - 0.5 \right|^2 / \sum_{i=1}^{26} \left| Rank_{h,i} - 0.5 \right|^2 . \tag{2}$$

[1] We assume here that there are 26 training stimuli as in the IMPROVER Species Translation challenge [4].

(b) Choose as ortholog for the human gene *h* the animal gene that minimizes the *D(h,m)* distance defined above.

The weight *w* gives more importance to the rank differences when the human gene is differentially expressed (rank far from 0.5) as opposed to when the human gene is not differentially expressed (rank closer to 0.5). This choice of weighting function *w* captures our belief that genes that are not differentially expressed are more likely to be affected by noise than the genes ranked close to the top or bottom of the list which are changing between treatment groups.

3. For a new stimulus for which gene expression data are available only in the animal model, use animal model gene expression data in conjunction with a gene set analysis method (e.g., PADOG, PLAGE, GSEA) by substituting the unknown human gene expression profiles with the available animal model profiles using the homology determined in step 2 above.

Method usage in R: An implementation of the data-driven functional homology described above was implemented in R and is available from http://bioinformaticsprb.med.wayne.edu/?page_id=50. The implementation uses human gene set and pathway definitions in the *gmt* format of MSigDB, and uses GSEAP as the gene set analysis method. There are several prerequisites to run the analysis:

(a) Download the SC3 folder from http://bioinformaticsprb.med.wayne.edu/?page_id=50 and unzip it.

(b) Obtain GSEAP from the Broad Institute (http://www.broadinstitute.org/gsea/index.jsp), as a java jar file (e.g., *gsea2-2.09.jar*). If a different version is used, then modify the provided R script file *gseapF.r* to replace *gsea2-2.09.jar* with the version you are using. Place the jar file in the root of the SC3 folder.

(c) R statistical environment v 3.0.0 or higher is needed on an operating system that supports the *parallel* and *limma* packages that have to be installed first.

The entry point to run the analysis is the R script file *sc3prb.r* that is available in the SC3 folder. The code reads the animal model (in this example rat) and human gene expression files for stimuli in the training set ("GEx_rat_train.txt," "GEx_human_train.txt") as well as the rat test dataset "GEx_rat_test.txt" and finds for each human gene the rat gene that best mimics its behavior. This homology mapping is saved as a named vector *respf* in an *RData* file named *respf.RData*. The vector names are the human genes, while vector values are the corresponding rat genes. For each stimulus in the test set, a moderated t-score is computed from animal model data to rank genes from the most downregulated to the most up-regulated, and then GSEAP is called to obtain adjusted *p*-values (false discovery rate) for each pathway. These false discovery rate values are transformed to confidence levels that the pathway is activated, as described elsewhere [25], and written into the file "PRB_SC3_predictions.txt" that follows the template provided by the organizers of the Species Translation Challenge ("template_prediction_SC3_gene_sets_human.txt").

3 Notes

Note 1: Choosing a gene set/pathway analysis method

There is a plethora of gene set analysis methods, 16 of which are compared in terms of sensitivity, prioritization, and specificity elsewhere [9]. When the sample size is rather small (i.e., 2–5 samples per group) none of the methods are expected to perform very well, while some (e.g., PADOG) are not feasible since there are not enough unique permutations of the sample labels to accurately estimate *p*-values, and hence PLAGE can be applied instead. Alternatives to PLAGE for small sample sizes are the MRGSE method [18] and GSEAP. When there are six or more samples per group, PADOG is a better choice because it has a better ranking capability; that is, PADOG ranks meaningful pathways closer to the top more often than other methods do.

Note 2: Data preparation

Typically, high-throughput gene expression data are obtained using microarrays and more recently by RNA sequencing. Microarray data obtained using the Affymetrix GeneChip® need to be background corrected and normalized using, for instance, the robust multichip average (RMA) [28] which also log transforms the data to make it suitable for differential expression analyses and gene set analysis. For Illumina Bead Arrays (Illumina Inc., San Diego, CA), typically log transformation followed by quantile normalization [29] is used. Sequencing count data summarized at gene level by counting the number of reads that align to the genome [30] can be further transformed [31] to make them suitable for differential expression and pathway analysis methods initially developed for normally distributed data, like microarrays.

Note 3: Limitations in using animal model data to predict pathway activity in human

The use of animal model gene expression data to predict pathway perturbation in human for a given stimuli was demonstrated elsewhere [4]. The rather limited success in inferring pathway activation from animal models to human (about 56 % average sensitivity and specificity for the best method according to these criteria) means that most pathways that will be activated by a stimulus of interest in human may not be identified by using animal model data. This is because all approaches used to make the inference (including the two described in this chapter) require that multiple of the training stimuli activates each given pathway or at least results in sensible differential expression for several genes on these pathways so that either (1) one model can be fit for each human pathway or (2) reliable gene ortholog pairs be identified between organisms, respectively. Such requirements are hard to meet in practice especially for a completely new type of stimuli.

References

1. Chen S, Xuan J, Couch L et al (2014) Sertraline induces endoplasmic reticulum stress in hepatic cells. Toxicology 322C:78–88
2. Ahuja V, Sharma S (2014) Drug safety testing paradigm, current progress and future challenges: an overview. J Appl Toxicol 34:576–594
3. Seok J, Warren HS, Cuenca AG et al (2013) Genomic responses in mouse models poorly mimic human inflammatory diseases. Proc Natl Acad Sci U S A 110:3507–3512
4. Rhrissorrakrai K, Belcastro V, Bilal E et al (2014) Understanding the limits of animal models as predictors of human biology: lessons learned from the sbv IMPROVER Species Translation Challenge. Bioinformatics pii:btu611
5. Ogata H, Goto S, Sato K et al (1999) KEGG: Kyoto encyclopedia of genes and genomes. Nucleic Acids Res 27:29–34
6. Joshi-Tope G, Gillespie M, Vastrik I et al (2005) Reactome: a knowledgebase of biological pathways. Nucleic Acids Res 33:D428–D432
7. Ashburner M, Ball CA, Blake JA et al (2000) Gene ontology: tool for the unification of biology. The gene ontology consortium. Nat Genet 25:25–29
8. Subramanian A, Tamayo P, Mootha VK et al (2005) Gene set enrichment analysis: a knowledge-based approach for interpreting genome-wide expression profiles. Proc Natl Acad Sci U S A 102:15545–15550
9. Tarca AL, Bhatti G, Romero R (2013) A comparison of gene set analysis methods in terms of sensitivity, prioritization and specificity. PLoS One 8, e79217
10. Tarca AL, Draghici S, Bhatti G, Romero R (2012) Down-weighting overlapping genes improves gene set analysis. BMC Bioinformatics 13:136
11. Tomfohr J, Lu J, Kepler TB (2005) Pathway level analysis of gene expression using singular value decomposition. BMC Bioinformatics 6:225
12. Tavazoie S, Hughes JD, Campbell MJ et al (1999) Systematic determination of genetic network architecture. Nat Genet 22:281–285
13. Goeman JJ, van de Geer SA, de Kort F et al (2004) A global test for groups of genes: testing association with a clinical outcome. Bioinformatics 20:93–99
14. Tian L, Greenberg SA, Kong SW et al (2005) Discovering statistically significant pathways in expression profiling studies. Proc Natl Acad Sci U S A 102:13544–13549
15. Efron B, Tibshirani R (2007) On testing the significance of sets of genes. Ann Appl Stat 1:107–129
16. Luo W, Friedman MS, Shedden K et al (2009) GAGE: generally applicable gene set enrichment for pathway analysis. BMC Bioinformatics 10:161
17. Barry WT, Nobel AB, Wright FA (2005) Significance analysis of functional categories in gene expression studies: a structured permutation approach. Bioinformatics 21:1943–1949
18. Michaud J, Simpson KM, Escher R et al (2008) Integrative analysis of RUNX1 downstream pathways and target genes. BMC Genomics 9:363
19. Wu D, Smyth GK (2012) Camera: a competitive gene set test accounting for inter-gene correlation. Nucleic Acids Res 40, e133
20. Lee E, Chuang HY, Kim JW et al (2008) Inferring pathway activity toward precise disease classification. PLoS Comput Biol 4, e1000217
21. Barbie DA, Tamayo P, Boehm JS et al (2009) Systematic RNA interference reveals that oncogenic KRAS-driven cancers require TBK1. Nature 462:108–112
22. Hanzelmann S, Castelo R, Guinney J (2013) GSVA: gene set variation analysis for microarray and RNA-Seq data. BMC Bioinformatics 14:7

23. Goel G, Conway KL, Jaeger M et al (2014) Multivariate inference of pathway activity in host immunity and response to therapeutics. Nucleic Acids Res 42(16):10288–10306
24. Smyth GK (2004) Linear models and empirical Bayes methods for assessing differential expression in microarray experiments. Stat Appl Genet Mol Biol 3:Article 3
25. Hafemeister C, Romero R, Bilal E et al (2015) Inter-species pathway perturbation prediction via data driven detection of functional homology. Bioinformatics 31:501–508. doi:10.1093/bioinformatics/btu570
26. Tarca AL, Than NG, Romero R (2013) Methodological approach from the best overall team in the sbv IMPROVER Diagnostic Signature Challenge. Syst Biomed 1, e25980
27. Dayarian A, Romero R, Wang Z et al (2014) Predicting protein phosphorylation from gene expression: top methods from the IMPROVER Species Translation Challenge. Bioinformatics pii:btu490
28. Irizarry RA, Hobbs B, Collin F et al (2003) Exploration, normalization, and summaries of high density oligonucleotide array probe level data. Biostatistics 4:249–264
29. Bolstad BM, Irizarry RA, Astrand M et al (2003) A comparison of normalization methods for high density oligonucleotide array data based on variance and bias. Bioinformatics 19:185–193
30. Anders S, McCarthy DJ, Chen Y et al (2013) Count-based differential expression analysis of RNA sequencing data using R and Bioconductor. Nat Protoc 8:1765–1786
31. Law CW, Chen Y, Shi W et al (2014) Voom: precision weights unlock linear model analysis tools for RNA-seq read counts. Genome Biol 15:R29

Chapter 10

Quantifying the Biological Impact of Active Substances Using Causal Network Models

Alain Sewer, Florian Martin, Walter K. Schlage, Julia Hoeng, and Manuel C. Peitsch

Abstract

In this chapter a five-step strategy is described that provides comparative evaluations of the effects of biologically active substances. These evaluations constitute an integral part of the determination of the risks for the human population to exposure to these substances. The strategy is based on the concept of biological impact quantification for which novel computational methodologies have been developed in the past few years; these methodologies are reviewed in this chapter. The effects of the active substances are then described in terms of networks containing the biological mechanisms involved in the response to the exposure. As a consequence, the biological impact assessment represents a systems-wide metric of network-based perturbed biological mechanisms. The implementation of the strategy involves the generation of transcriptomics data following the exposure experiment and their evaluation in the context of causal network models. After the five-step strategy for biological impact quantification has been described in some detail, its application in a concrete case of a mouse smoking-cessation experiment is presented. The results show how mechanistic insights into the potential toxic effects of exposure to active substances can benefit the safety assessment.

Key words Biological impact of exposure, Causal biological network models, Network perturbations, Mechanistic fingerprinting

1 Introduction

In this chapter, we present a quantitative methodology to evaluate high-throughput omics data, which represents an innovative implementation of systems biology and the principles of "Toxicity Testing in the 21st Century" promoted by the US Environmental Protection Agency (EPA). This methodology paves the way for the development of a computational systems toxicology discipline that produces reproducible, scientifically sound, and exploitable results. A key element is the concept of biological impact assessment, which consists of quantifying the effects of the exposure of a biological system to an exogenous factor in terms of the perturbations

Julia Hoeng and Manuel C. Peitsch (eds.), *Computational Systems Toxicology*, Methods in Pharmacology and Toxicology, DOI 10.1007/978-1-4939-2778-4_10,

of a priori selected biological mechanisms. The mechanisms are represented by networks of interacting biological entities, which have emerged during the past 10 years as the most adequate way of describing the responses of cells and tissues. Even though, in a toxicology context, the term "active substances" traditionally refers to chemicals that induce a response in an exposed system, the methodology described here can be applied to other types of exposures for which a biological impact assessment is of interest, from drug development to consumer product testing and environmental impact analysis.

In developing our novel methodology we were driven by the constantly increasing interest of the public and regulatory authorities in the evaluation of the health impacts of drugs, consumer products, and environmental factors. For instance, the "Critical Path Initiative" of the US Food and Drug Administration (FDA) fostered the modernization of the scientific and technical tools used to evaluate medical products, particularly their safety [1]. Similarly, the "Toxicity Testing in the 21st Century" vision recommended the evaluation of the changes in biological processes using the most recent technologies as the main focus [2]. The resulting data are expected to provide more robust evidence and thereby help researchers and organizations make better decisions about the potential risks to humans posed by exposure to the tested products or substances [3]. In fact, the process of assessing health risks faces two main challenges. First, the conclusions of laboratory (or even clinical) investigations need to be translated to the level of the entire human population. Second, the observations made after the tested short-term exposures must be indicative of the symptoms of the actual diseases, which are likely to appear only after realistic longer term exposures that cannot be reproduced in the laboratory. These two challenges are addressed currently by interpreting the experimental observations using prior biological knowledge, with the aim of understanding the mechanisms associated with the response to the exposure [4]. The overall strategy underlying the quantitative methodology presented in this chapter follows this line through the use of biological networks as the fundamental substrate for reaching a mechanistic understanding [5].

Networks provide a relevant description of cellular processes because the key to the mechanistic understanding of these processes is not the exhaustive enumeration of the involved genes but rather the identification of their mutual interactions [6]. These conclusions supported the applicability of systems biology concepts to the understanding of cellular processes [7]. Biological networks have not only been used in explaining "normal state" properties, but have also been applied to decipher the mechanisms that lead to "perturbed states" such as diseases, which were viewed subsequently as network phenomena (top left panel of Fig. 1) [8, 9]. Further, the application of biological networks to

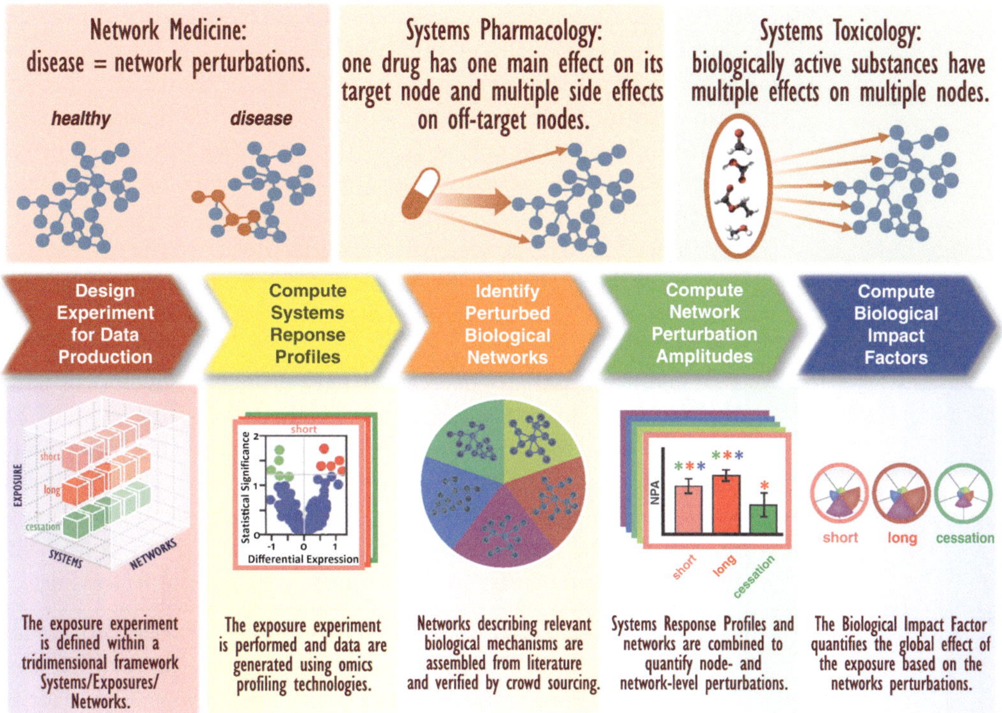

Fig. 1 Systems toxicology and the network-based five-step strategy for assessing biological impact. The *upper panels* show how systems toxicology relates to network medicine and systems pharmacology. The *lower panel* displays our five-step strategy for comparative evaluations of the biological effects of active substances using causal network models. Each step is presented in more detail in the next Figs. 2–6. It is important to keep in mind that these figures contain plots and simplified mathematic formulae meant for illustrative purposes. The real data of the presented experiment are shown in Fig. 7 and the accurate mathematical definitions are given in Box 2

pharmacology led to an integrated view of the main and side effects of a drug (top middle panel of Fig. 1) [10]. As a corollary of "Systems Pharmacology," our "Systems Toxicology" view can be conceptualized simply and involves the multiple molecules that constitute an exposure matrix, which induce multiple "side effects" leading to a collective systems-level response (top right panel of Fig. 1).

In our quantitative "Systems Toxicology" methodology, we used exclusively "causal network models," which constitute a type of biological networks that is becoming increasingly popular for the interpretation of transcriptomics data [11–13]. The reasons for selecting this particular type of biological networks were twofold. First, their content is of high confidence. As indicated by their name, the edges in these networks contain mostly "causal" relationships, as opposed to "associative" or "correlative" relationships, which describe weaker degrees of connections between the two end nodes. The "causal" relationships are supported by one or more explicit experimental statements extracted from the specialized literature, which provide high confidence in the relationship captured by an edge. Second, their content is genuinely mechanistic in that the nodes correspond to biological functions or activities at various levels of detail, such as "kinase activity of CDK2" or "cell cycle." Causal networks are not restricted to transcript levels or protein abundances like other network types used in systems biology. This feature satisfied the essential need of the mechanistic understanding expected from our "Systems Toxicology" strategy mentioned previously [4]. Details about the specific format of the biological network models are given in Sect. 2.

Having chosen to use causal networks, we developed a five-step strategy to implement the "Toxicity Testing in the 21st Century" principles in a systems biology context (bottom panel of Fig. 1) [5, 14, 15]. The main purpose of our strategy is to support health risks assessment by providing a systematic framework for integrating the experimental evidence obtained on tested substances and systems. Its main characteristics are the following:

- *A truly quantitative nature.* The strategy is based on the concept of biological impact assessment that enables explicit comparisons between the different responses of the systems exposed to the active substances.
- *Objectivity.* Untargeted systems-wide collection of experimental data using transcriptomics technologies ensures that the exposed systems response is captured in an unbiased way. In parallel, assembly of the causal network models describing the a priori relevant biological mechanisms is performed to best represent the current consensus of the scientific community.
- *Specificity.* The strategy clearly favors specificity over sensitivity. Statistical tests are performed, which indicate whether the results of the biological impact assessment specifically reflect the content of the network models and do not incidentally capture the background noise. This aspect ensures reliable and exploitable results that can be investigated with standard "phenotypic" toxicological endpoints. Possible sensitivity issues are identified by running a "standard" explorative analysis on the transcriptomics data.
- *Reproducibility.* This crucial aspect is facilitated by the choice of standard experimental and computational tools, as well as by the publication in open-access journals and the sharing of the network models and methodologies [16]. This transparency is also important for the defensibility of the results.

This chapter describes our five-step strategy for assessing biological impact and is organized as follows. Section 2 describes the elements needed for the concrete implementation of the strategy. Additional aspects of the causal biological network (CBN) models are discussed in Chap. 3. Section 3 explains each of the five steps in more detail and gives a breakdown of their structures. Technical boxes are included to display the computational protocols. In the "Notes" (Sect. 4) we included a concrete application of our methodology to an inhalation study in which the effects of smoking cessation vs. continued smoking were investigated in a C57 mouse model for respiratory diseases such as chronic obstructive pulmonary disease (COPD).

2 Materials

2.1 Causal Biological Network Models

To represent the biological processes relevant for our purpose and to assemble the causal network models describing them, we adopted the two following formats or conventions:

- *Biological Expression Language (BEL)*. BEL is a semantic programming language that allows for a flexible representation of biological processes in a computable format [17].
- *Causal Networks*. The nodes in the networks are biological entities such as protein abundances, mRNA expressions, and protein activities. Network nodes can also represent biological processes (e.g., apoptosis), as well as chemicals or small molecules whose transcriptional signatures represent signaling that is similar to the signaling induced by exposure to a compound of interest. Edges are the relationships between the nodes, and are categorized as either causal or semantic. Causal edges are directional cause-effect relationships between nodes (e.g., catalytic activity of MYD88 directly increases kinase activity of IRAK4), whereas semantic edges connect different forms of a biological entity, such as the abundance of a particular protein to the abundance of its phosphorylated form (e.g., TP53 protein abundance to TP53 phosphorylated at serine 15). It is useful to note that "causal" does not mean "direct" in the sense of a physicochemical interaction, as "indirect" but causal relationships supported by a literature statement are included in our biological network models.
- *The Selventa Knowledgebase*. The "raw material" for assembling the networks comes mainly from the Selventa Knowledgebase, which is a comprehensive repository containing over 1.5 million nodes (biological concepts and entities) and over 7.5 million edges (connections between nodes). The Knowledgebase is derived from peer-reviewed scientific literature as well as other public and proprietary databases. Small networks that causally link an upstream controller node that represents a biological mechanism to a set of downstream transcriptomics nodes are termed "HYPs" [13, 18]. By design, a large fraction of the network model nodes in the Selventa

Knowledgebase are HYPs (see Step 3). The Selventa Knowledgebase is licensable, like similar sources of curated biological knowledge that required an extensive mining of the scientific literature (e.g., Biobase and Ingenuity Pathway Analysis [12]). However, the openBEL portal contains an increasing amount of publically available material extracted from the Selventa Knowledgebase such as the "BEL Large Corpus" [13, 17].

- *Backward Reasoning.* This modeling approach is based on the idea that the differential expressions measured in transcriptomics data are the consequences of changes in the biological activities of upstream controlling entities (such as, but not limited to, transcription factor proteins). Backward reasoning represents an alternative to the often implicit assumption that transcript differential expressions reflect changes in the activity of the corresponding proteins, which is used when integrating transcriptomics data and pathways and networks [19]. In our network models, HYP nodes allow the application of backward reasoning and thereby the integration of the transcriptomics data (see Step 4) [18, 20]. The acronym "HYP" reflects the fact that these small networks constitute mechanistic hypotheses to explain the transcript differential expressions.
- *Modeling Assumptions.* The biological networks do not constitute only a practical representation of prior biological knowledge; they also entail a precise modeling approach of the considered biological processes. Besides the backward reasoning, the networks also involve directed signed edges "→" and "⊣," which have a well-defined meaning. Such a highly informative content makes these networks extremely valuable; however, appropriate biological expertise is also required to meaningfully interpret results based on them (see Step 5).

Additional details about the causal biological network models are presented in Chap. 3.

2.2 Biological Expertise

Modeling molecular or cellular biological processes is a difficult task. As mentioned above, the use of BEL and causal network models implies several contextual modeling assumptions, which need to be taken into account when interpreting the results of our impact assessment methodology. For this reason, we have included "suitable biological expertise" as a fundamental requirement for the meaningful application of our five-step strategy.

2.3 Computational Environment

We implemented the computational methodology in the R statistical computing environment [21] because it constitutes a logical extension of the bioinformatics methods we used, all of which were taken from the Bioconductor software (see Step 2) [22]. Although Step 5 of our methodology can be computationally intensive, it can still be executed on a computer that can preprocess the transcriptomics data by running the standard Bioconductor pipeline in Step 2. This aspect is reexamined following a more detailed description of the computations required by the five-step strategy (Step 5 of Sect. 3). Using the open R language and its software packages is of clear benefit for the reproducibility of our results [23]. Currently, we are implementing advanced standardization steps to further reinforce the reproducibility of the methodology [24, 25].

3 Methods: A Five-Step Strategy for Assessing the Biological Impact of Active Substances

3.1 Step 1: Design Experiments for Data Production

The first step of our strategy is developing the experimental design that will produce data that can be used to assess the biological impact of the tested substances. This task requires the definition of a specific framework that can position the assessment experiments with respect to the ideal case of the actual human population (Fig. 2).

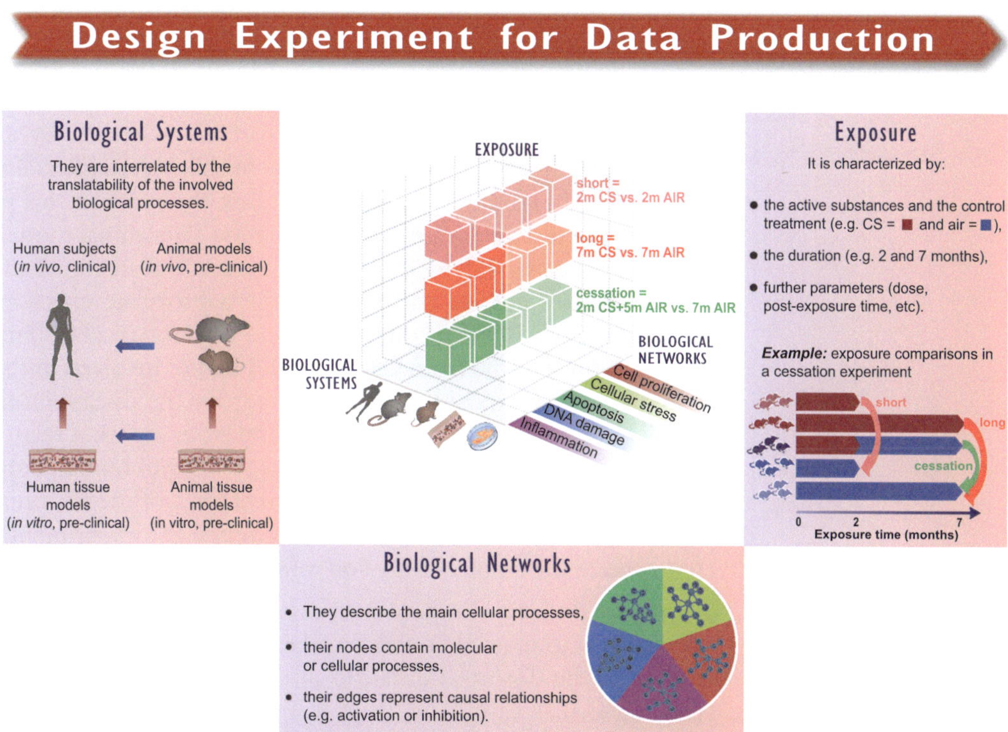

Fig. 2 Step 1: The framework for designing the experiments for data production. The "design cube" and its three dimensions "biological systems," "exposure to the biologically active substances," and "biological networks" are represented in the case of a smoking-cessation experiment performed on a particular mouse strain relevant for respiratory diseases (see also Fig. 7a)

This strategy supports systems toxicology in achieving its main goal of evaluating the risks associated with long-term exposure to biologically active substances [5]. It implies the ability to extrapolate short-term observations to long-term outcomes and to translate the potential risks identified from a combination of in vivo and in vitro experimental systems to human populations. Achieving these tasks first requires laboratory models for both exposure and biological systems that best mimic the real situation. Despite the shortcomings of models of this type, they allow biological mechanisms perturbed by the exposure to be identified. The candidate mechanisms can then be refined in dedicated experiments and gradually integrated together with other evidence to build the causal chain of events that lead to the onset of diseases in human subjects. These considerations require a systematic framework for designing the data-generating experiments from which the causal statements can be extracted. This framework is our "design cube" represented in Fig. 2. The design cube dimensions are (1) the biological systems, (2) the biological networks, and (3) the exposure to the biologically active substance. These three dimensions are discussed in more detail below.

The "Design Cube" Framework (Fig. 2)

1. *Biological Systems.* We usually rely on preclinical models to generate data, i.e., in vivo animals and conventional or organotypical in vitro cultures. These experimental systems allow us to gain insights into the biological perturbations caused by an active substance, identify mechanism-specific biomarkers for use in human studies, and eventually link the mechanisms to the onset of disease for impact assessments. The animal models that are used should reproduce at least some features of the human disease and be adequate for the appropriate exposure regimens [26]. Similarly, the cellular and organotypical systems should represent the main cell types and tissues involved in the response to the exposure and be relevant for the disease etiology. This leads to the "translational parallelogram" that relates both in vivo animal models and in vitro cultures to human biology, which also constitutes an implementation of the 3R principles: reduce the number of animals, refine the experiments, and replace the animals with non-animal systems [27–30].
2. *Exposure.* Because the goal is to recreate in vivo and in vitro exposure regimens that most realistically mimic the human situation, the dose and duration of exposure should be tuned with particular care. This usually implies exploratory "dose-range finding" experiments that will identify the optimal exposure response of the system and lead to the definition of standard exposure regimens that will contribute to the reproducibility of the impact assessment experiments [31].

3. *Biological Networks.* Although the biological network dimension is not an experimental factor, it nevertheless represents a fundamental component of our experimental design "cube." Indeed, it contains the "mechanistic dimension" of the experiment that enables the specific aspects of the systems responses to the exposures that are described by the networks to be focused on separately. This approach is of great help in the translational context associated with the "biological systems" dimension because it provides "mechanistic bridges" between the different systems. Biological networks allow the explicit investigation of how the human biology is recapitulated in the preclinical experiments and provide a solid support for the obtained results in case of confirmed translatability. The modeling approach underlying the networks used in our approach has already been described in Sect. 2 and the principles guiding the determination of their biological content will be explained in Step 3.

Table 1 contains a list of the transcriptomics datasets accompanying the exposure experiments we have performed during the past few years and that were deposited in public databases [32, 33]. It reveals the different biological systems and exposures we used and shows how the "design cube" framework enables the mechanistic translatability into the response of human subjects. In "Notes" (Sect. 4), our approach is illustrated with concrete data taken from a smoking-cessation experiment performed on a particular mouse strain relevant for respiratory diseases such as COPD [15]. The execution of this experiment and details of the application of the chosen exposure on the specific biological system will not be discussed in detail in this chapter, because these factors often depend strongly on the particular toxicological context and are therefore less generic than the broadly applicable network-based strategy described here. However, it does not mean that they are not challenging, as mentioned in "Notes" (Sect. 4) below.

In summary, Step 1 of our systems toxicology strategy uses the "design cube" that provides the framework for experimental data production. Besides the "Exposures" and "Biological Systems" factors, the design cube specifically includes a "mechanistic" dimension constituted by the biological network models that is the key for mechanistic inter-systems translatability.

3.2 Step 2: Compute Systems Response Profiles

The second step consists of applying state-of-the-art high-throughput technologies for molecular profiling to the biological material obtained from the exposure experiment conceived in Step 1 and using these data to reliably compute the systems-wide impact of exposure at the basic level of biological organization (Fig. 3).

Besides the biological material extracted from the appropriately executed exposure experiment (in vivo tissues or body fluids;

Table 1
List of the transcriptomics datasets that have been deposited in public databases

ArrayExpress/GEO identifier	Biological system	Exposure
E-MTAB-874 [71]	Organotypic culture of human primary bronchial epithelial cells	Cigarette smoke (CS)
E-MTAB-1027 [20]	Normal human bronchial epithelial (NHBE) cells	Tumor necrosis factor α (TNFα)
E-MTAB-1173 [72]	Human umbilical vein endothelial cells (HUVEC), human coronary arterial endothelial cells (HCAEC), monocytic Mono Mac 6 (MM6) cells	Aqueous CS extract
E-MTAB-1272 [58]	NHBE cells	Inhibitors for cyclin-dependent kinases 4 and 6
E-MTAB-1312 [60]	NHBE cells	TNFα
E-MTAB-1313 [60]	Lung from apolipoprotein E-deficient (ApoE) mice	CS/CS cessation
E-MTAB-1390 [73]	Pulmonary and nasal tissues from ApoE mice	CS/CS cessation
E-MTAB-1426 [74]	Lungs from A/J mice	CS
E-MTAB-1721 [69]	Organotypic cultures of human bronchial and nasal epithelial cells	CS
E-MTAB-1842, GSE50254 [68]	Pulmonary and nasal tissues from Sprague-Dawley rats	CS, prototypic modified risk tobacco product (pMRTP)
E-MTAB-1871 [75]	Lungs from A/J mice	CS
E-MTAB-2080 [64]	NHBE cells	Acrolein, catechol, formaldehyde
E-MTAB-2091 [76]	NHBE and rat bronchial epithelial cells	>50 different compounds
E-MTAB-2250 [77]	Organotypic cultures of the human bronchial and nasal epithelial cells	CS
E-MTAB-2251 [78]	Organotypic cultures of the human buccal and gingival epithelial cells	CS

They illustrate the "biological system" and "exposure" dimensions of the "design cube" from Step 1 and results from the concrete implementation of Step 2

in vitro tissues or cultured cells), several specific capabilities are necessary to generate the systems response profiles (SRPs). These capabilities cover both the molecular biology and the bioinformatics aspects of data analysis and are often implemented as relatively standardized service centers in both academic and industrial organizations (or outsourced to biotechnology companies that specialize in providing such services) [34]. Therefore, it is

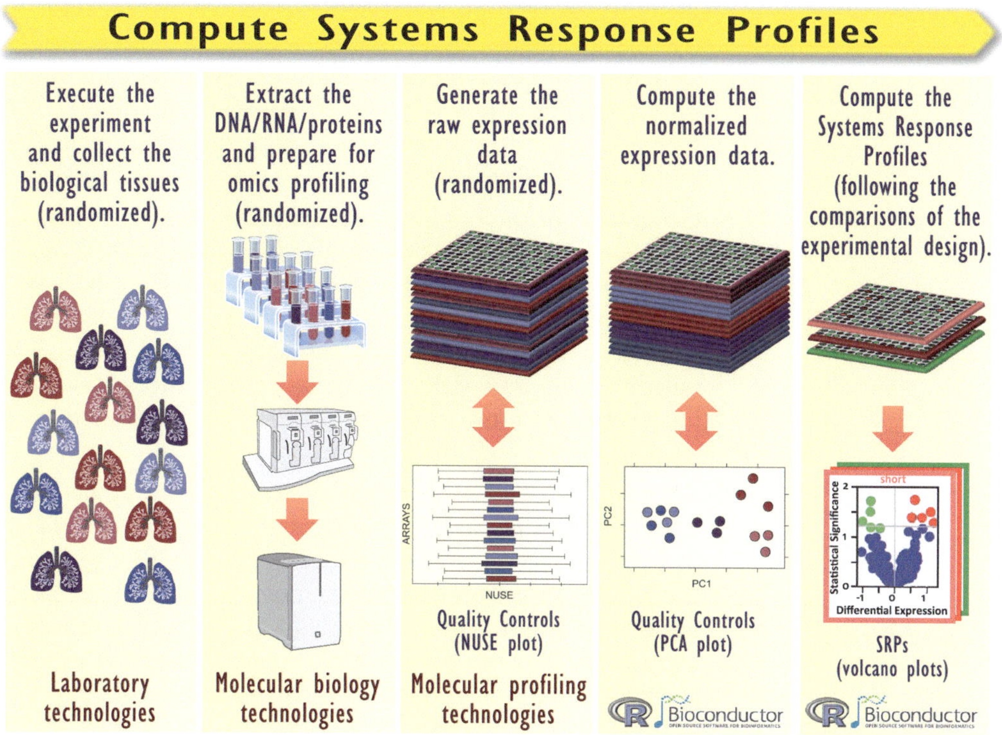

Fig. 3 Step 2: Computing systems response profiles (SRPs). The breakdown of this step involves both experimental and computational activities matching the current technological standards. Our implementation adds the fine-tuning of several specific aspects such as sample randomization or computational reproducibility, which are essential in the perspective of health risk assessments (see also Box 1)

reasonable to follow and possibly fine-tune the "community standards" in executing this second step, because this will greatly help in ensuring the quality and reproducibility of the results [35].

The Workflow (Fig. 3)

1. The biological material extracted from an exposure experiment is subject to several constraints. First, the material needs to be sampled in a way that ensures the statistical power of the experiment, which fixes the number of biological replicates in the various treatment groups (balanced against the financial resources) [36]. Second, the material must be available in an amount sufficient to provide high-quality inputs to the molecular profiling machines. Finally, the different samples must be randomized not only to avoid confounding effects with the experimental design, but also to distribute the technical variability introduced by the profiling system as uniformly as possible across the treatment groups [37].
2. The appropriately generated biological material is then used for molecular profiling. Because of its lower cost, transcriptomics (mRNA profiling by microarray or RNA sequencing) is

the most widely used approach [38, 39]. We have mainly used microarray technology consisting of sequential steps of DNA extraction and amplification, followed by microarray hybridization and scanning [4, 40]. Untargeted quantitative proteomics is another source of high-quality systems-wide molecular profiles [41].

3. Generation of the raw sequence data constitutes the interface between the "analogic" molecular biology and the "digital" bioinformatics. For microarray data, this interface involves the transformation of the color intensities on the hybridized microarray images into raw transcript levels. Because the numeric transcript data can be processed more easily than image data, systematic quality controls are performed at this stage to ensure high-quality data.
4. Because raw data are generally generated on a single-sample basis, the data can contain associated biases. It is important, therefore, to readjust all the intensity values before generating the global expression matrix. The raw data normalization step performs this task quite efficiently [42]. The global expression matrix can then be used for multivariate principal component analysis, which should reflect the experimental grouping of the samples and allow the detection of outliers that should be removed from the data [43].
5. Calculation of the SRPs follows and essentially consists of univariate calculations of comparisons between the sample groups contained in the design of the experiment. For each mRNA or protein (both generically referred to as "gene"), a differential expression is derived, and classified by its statistical significance, which is obtained based on a statistical model that calculates the null distribution of differential expressions against which the actual values are compared to derive a *p*-value [44]. Because the number of detected genes is "large' (typically 20,000), a multiple testing correction needs to be applied on these *p*-values [45].

As already mentioned, this workflow has now become standard in biomedical research and we implemented it with widely used and recurrently tested components (Box 1) [46]. We used the Affymetrix GeneChip™ technology for the transcriptomics aspect and set up the bioinformatics processing pipeline in the R statistical computing environment using Bioconductor software packages [21, 22, 47]. These tools ensure the quality that is necessary to quantify biological impact and they are continuously reassessed using the various QC metrics included in the processing pipeline. Another advantage of using the open-access, versioned, and archived Bioconductor software is that it contributes significantly to the reproducibility of the SRP results, because it allows the user

Box 1 Workflow for calculating SRPs using R and oconductor software. Each operation is given a graphical label that indicates its type: ↘ = input data, ↗ = output results, ↓ = execute instructions, ↻ = repeat the execution until the following test is successfully completed, and ▲ = complete the test.

Prerequisites

↘ 1. Install R and Bioconductor from www.r-project.org and www.bioconductor.org.

↘ 2. Install the following packages: *affy, affyplm, limma.*

Compute SRPs (Step 2)

↘ 1. The raw data are delivered (CEL files for Affymetrix transcriptomics platform).

↘ 2. The metadata table describing the experimental samples is delivered.

▲ 3. Check for consistency between raw data and metadata.

↓ 4. Read the CEL files and create an *affybatch* object.

↻ 5. Normalize the raw data using *gcrma* and compute QC metrics (NUSE, RLE, RNA degradation, pseudo-images, images, correlations, etc.).

▲ 6. Discard bad quality arrays and re-execute step 5 until all arrays pass the QC.

↗ 7. Save the matrix of log2 normalized expression data.

↓ 8. Calculate the probeset-level treatment/exposure effects β using linear models and the *limma* package.

↓ 9. Convert probesets to Gene Symbols using *biomaRt* or *Confero.*

↗ 10. Save the vector of differential expressions β and their variances var(β) in idMAPS format.

to control his or her software version and thereby record all the parameters of a given calculation [23]. Table 1 contains a list of the datasets that have been generated following Step 2 and the cited open-access publications provide more detailed descriptions of the experimental settings. To efficiently manage all the resulting SRPs, we also developed a software platform for storing such data in a simple and standard format called "idMAPS" that is used as input for the downstream analyses (see Boxes 1 and 2) [48].

In summary, our workflow for generating SRPs follows and adjusts current research standards with special care given to sample randomization and quality controls at each intermediate level.

3.3 Step 3: Identify Perturbed Biological Networks

The third step consists of assembling the biological networks that most objectively describe the a priori relevant mechanisms involved in the response of an exposed system. We used causal network models encoded in the Biological Expression Language (BEL), two modeling approaches that have been presented in Sect. 2. The design and assembly of causal network models is an iterative, multi-step process that was complemented recently by a crowdsourcing initiative (Fig. 4).

The process of assembling our network models can be characterized by both "container" and "content" features, which is similar to the participative "Web 2.0" context. The "container" aspects are determined by the modeling conventions constituted by the BEL semantics and the requirements of literature-supported causal

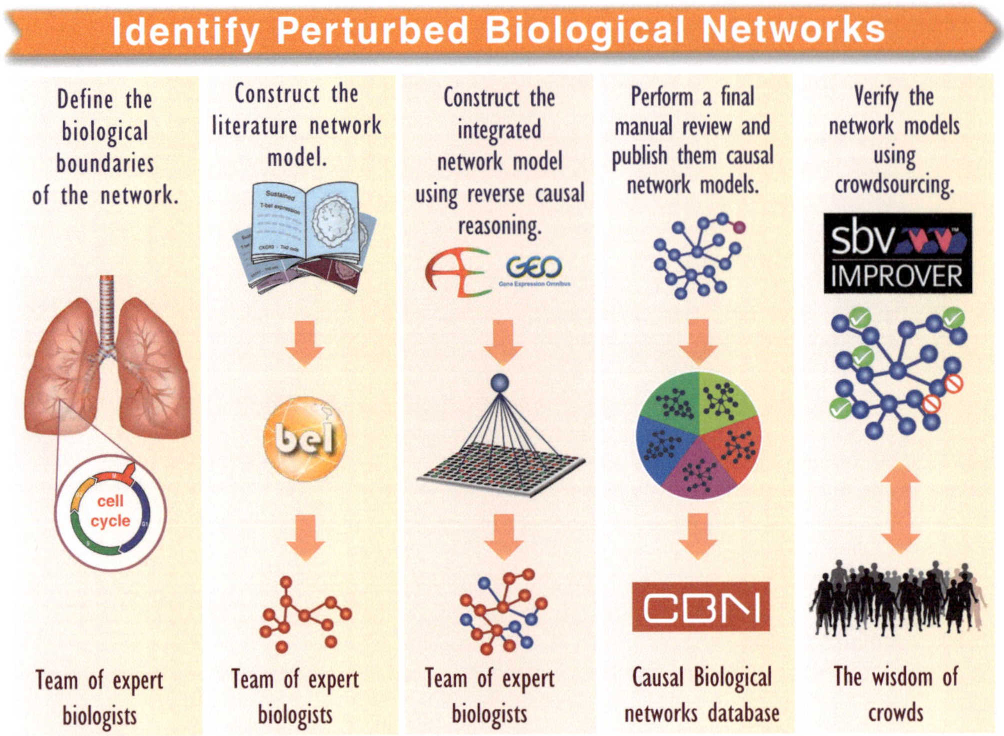

Fig. 4 Step 3: Identifying a priori relevant biological mechanisms and assembling causal network models. This step relies strongly on the biological expertise mentioned in Sect. 2 to collect and organize prior biological knowledge into causal network models. In order to enforce their objectivity and thereby their acceptance, an innovative crowdsourcing initiative is used to update them based on the recommendations of an extended community of biological experts

relationships for each edge of the networks (see Sect. 2). The "content" covers the actual biological processes described by the network. The validity of the content is first ensured by subject matter experts (as in Wikipedia), and then an innovative crowdsourcing approach that provides an even broader acceptance of the network content taking advantage of the "wisdom of crowds" (as in Google's PageRank) has been tested [49, 50]. Our biological impact assessment strategy also imposes the rule that each biological process corresponds to a well-defined network, even if nodes common to several networks can lead to nonzero overlaps [51]. As a consequence, the representation of the biological mechanisms by separated network models can induce minor artifacts that must be taken into account when interpreting the results of perturbation quantifications resulting from the Steps 4 and 5 of our general approach.

The Workflow (Fig. 4)

1. The biological boundaries of the network are first defined by a team of discipline-specific field experts guided by a literature survey of the relevant signaling pathways related to the process of interest (e.g., cell proliferation in lung). As a general rule,

the tissue context is explicitly taken into account and the exposure response is described for the healthy tissue (healthy lung in our case). This process led to four groups of networks: Cell Proliferation [51], Cellular Stress [52], Cell Fate (DNA damage, autophagy, apoptosis, necroptosis, and senescence—DACS) [53], and Pulmonary Inflammation [54]. For the conditions of chronic disease, separate network models have been built or are under construction [55]. Within the defined boundaries, literature-based network models were built initially by subject matter experts from the published literature based on known, often canonical, signaling pathways relevant to the kind of exposure (inhaled cigarette smoke) and to the expected exposure-related pathobiological changes (lung inflammation and emphysema) that were relevant to the investigation.

2. BEL was used to capture the biological entities and their causal relationships underlying the nodes and edges in the networks. All the relationships are stored in the Selventa Knowledgebase (see Sect. 2). BEL-encoded causal relationships describing the pathways covered by the biological boundaries of the network are extracted from the Selventa Knowledgebase. These relationships are assembled to nucleate the "literature network model" with causal relationships within the boundaries of tissue context and biological mechanisms. If necessary, a modular design is used, where the content of each module is constrained to discrete areas of signaling. For example, in the network describing cell proliferation, a module depicting cell cycle regulation was combined with modules describing the pathways known to influence cell cycle progression in the lung (e.g., Wnt, Hedgehog, and prostaglandin E2 (PGE2) signaling).
3. The content of the literature network model is augmented by performing reverse causal reasoning (RCR) on relevant context-specific datasets available from public transcriptomics data repositories such as the GEO or ArrayExpress databases [32, 33]. RCR evaluates HYPs from the Selventa Knowledgebase as hypotheses to explain the observed differential expressions (see Sect. 2) [13]. HYPs predicted by RCR, which were not included in the literature network model, are inserted into an augmented version, provided that there is strong mechanistic connection to the biological processes underlying the network. Using this approach, including iterative steps of expert curation, a more comprehensive "integrated network model" is generated from the combined use of molecular profiling data and prior biological knowledge of causal relationships.
4. Discipline-specific scientific experts conduct a terminal round of manual review and refine the content and connectivity of the integrated network model. The resulting network model is then submitted to a peer-reviewed open-access publication

where reviewers' comments can help to further improve the quality of its content. To better disseminate the results of network model assembly efforts and to get more numerous feedback, the CBN repository has been set up to enable external users to use and comment on all the published network models (see Chap. 3) [56].

5. As mentioned above, a subset of the published network models have undergone a further step of curation in the framework of the Network Verification Challenge of the Systems Biology Verification project (SBVImprover) [50]. This challenge was based on an innovative crowdsourcing approach where a large community (>50) of contributors (mainly biologists) were given tools that allowed them to vote on various edges and nodes of the networks via a dedicated web interface. A "moderator" supervised the votes for each models and made decisions to include or exclude node and edges based on the community "choices." After the completion of the challenge, the crowdsourcing infrastructure evolved into a reputation-based platform providing self-moderation and enabling the continued development and verification of the network models [57].

In the case of the Cell Proliferation network model, the above workflow was supplemented by a dedicated experiment in which normal human bronchial epithelial (NHBE) cells were released into the cell cycle in a synchronized manner [58]. RCR analysis was performed on the resulting transcriptomics data that allowed the validity of our network model for this specific process to be confirmed. This experience showed that "validating" a network model is not a simple task that can be achieved in a single experiment testing one specific process, even if it is executed in a controlled manner. Mining the literature and its "assimilation" by the scientific community remains the main source of high-quality content for the network models, which can foster their acceptance. We also note that because scientific knowledge increases constantly, the network models need to be updated with novel findings. Therefore careful versioning of their content is required for the already mentioned reproducibility of the results obtained with the biological impact strategy described in this chapter.

In summary, the network assembly process of Step 3 produces causal network models that describe the main biological processes perturbed by exposure in a very reliable manner. These network models provide the input for the next two steps of the five-step strategy (bottom panel of Fig. 1).

3.4 Step 4: Compute Network Perturbation Amplitudes

The fourth step of our strategy for assessing biological impact of exposure consists of integrating the SRPs (obtained in Step 2) into the biological network models (assembled in Step 3) to quantify the amplitudes of the network-level effects or the perturbations that result from the exposure treatments applied to the biological system.

Unlike Step 2, which uses standard computational approaches, Step 4 requires the development of an ad hoc novel methodology. The idea of quantifying the relevance of prior biological knowledge structured in pathways or networks has been explored previously [19]. However, most of the available approaches were based on enrichment statistics that generated *p*-values, which are not adequate to quantify the *amplitude* of the treatment-induced perturbations in a given network. In addition, to match the transcriptomics SRPs with the network node entities, these previous approaches are all based on the approximate "forward assumption," which consists of equating changes in the protein activities to the differential expressions of the corresponding transcripts. This "forward assumption" is not used in the causal network models obtained in Step 3, because a more powerful "backward approach" can be developed by taking full advantage of the fact that most of the network nodes are HYPs (as mentioned in Step 3 and in Sect. 2). This feature permits a consistent integration of the transcriptomics SRPs into the causal network models. It also induces a "two-layered" structure for the prior biological knowledge, as illustrated in the first column of Fig. 5, where the "functional layer" contains the CBN models from Step 3, and the "transcript layer" contains the downstream measurable transcripts of the HYP nodes of the network models [20]. This situation constitutes an extension of previous approaches where changes in transcription factor activities were inferred from the differential expressions of the transcripts they regulate [59]. However, the specific problem of the "two-layered" prior knowledge was not addressed when we started developing the methodology behind Step 4 of our strategy. We proceeded in two phases by first introducing the concept of the "network perturbation amplitude" (NPA) in a proof-of-principle study [18], and then expanding it by fully taking into account the "two-layered" structure of the prior knowledge [20].

The Workflow (Fig. 5 and Box 2)

1. As discussed above, the starting point is the integration of the SRPs obtained in Step 2 and the causal biological network models assembled in Step 3. Integration is achieved by taking advantage of the downstream entities of the HYP nodes of the network models, which results in the two-layered structure shown in the first column of Fig. 5. In this initial situation, only the "transcript layer" can be assigned experimental values, which correspond to a large subset of the SRPs (often around 80 %).
2. The first operation consists of calculating the perturbation amplitudes of the nodes of the functional layer. This is achieved by a fitting procedure where the largest SRP-based

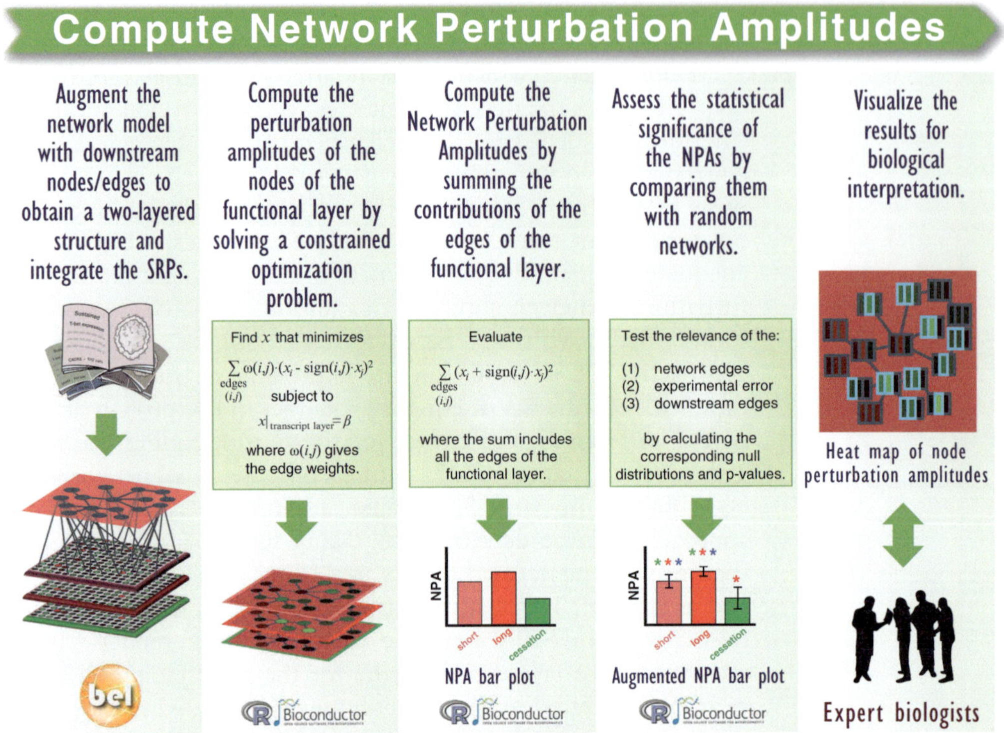

Fig. 5 Step 4: Computing the network perturbation amplitudes (NPAs). This step builds on network models augmented by their downstream transcriptional edges and nodes to compute node- and network-level perturbation amplitudes. These results are associated with three statistics $***$ testing their significance with respect to the network structure and the experimental uncertainties, which facilitates their mechanistic interpretation by biological experts

(multidimensional) "shape" that is compatible with the (multidimensional) "shape" of the network model is calculated. This procedure results in the constrained optimization problem indicated in the second column of Fig. 5.

3. The second operation consists of calculating the perturbation amplitudes of the full network for a given SRP. This is achieved by summing the contributions of each edge of the network model. The perturbation amplitude of an individual edge depends on the perturbation amplitudes of its start and end nodes. Our experience is that the squared average of the amplitudes is the most suitable definition of the corresponding edge perturbation amplitude.

4. To assess the statistical significance of the obtained network-level perturbation amplitudes, their null distributions are calculated by repeating the two previous operations 500 times for two distinct ways of rearranging the network models: either by permuting the edges between the functional and the transcript layers or by rearranging the edges of the functional layer.

Box 2 Workflow for calculating NPAs and biological impact factor (BIF) using R software. The same graphical conventions are used as in Box 1 (full mathematical details can be found in our publications [15, 20]).

Compute NPAs (Step 4)

For a given subnetwork and all the SRPs, apply the following steps.

1. Import into the R environment the two-layered structure of the subnetwork given by the blockwise adjacency matrix of the corresponding (undirected signed) graph: $A = \begin{pmatrix} A_{FF} & A_{TF}^t \\ A_{TF} & 0 \end{pmatrix}$ where *"F"* and *"T"* refer to the functional and transcript layers, respectively; *"t"* means transposed; $(A_{FF})_{ij} = (A_{FF})_{ji} \in \{-1,0,1\}$; $(A_{TF})_{ij} \in \left\{\frac{-1}{n_i}, 0, \frac{1}{n_i}\right\}$ for n_i=number of downstream transcripts of HYP node i.
2. Construct the corresponding signed Laplacian matrix $L = D - A \equiv \begin{pmatrix} L_{FF} & L_{TF}^t \\ L_{TF} & L_{TT} \end{pmatrix}$ where D is the diagonal degree matrix $D_{ii} = \sum_j |A_{ij}|$.
3. Import the SRPs in idMAPS format that contains the log2 differential gene expressions $\boldsymbol{\beta}$ and their variances $\mathrm{var}(\boldsymbol{\beta})$, among others.
4. Compute the node-level perturbations $\boldsymbol{\pi} = -L_{FF}^{-1} L_{TF}^t \boldsymbol{\beta}$ and their variance-covariance matrix $\mathrm{var}(\boldsymbol{\pi}) = L_{FF}^{-1} L_{TF}^t \,\mathrm{diag}(\mathrm{var}(\boldsymbol{\beta}))\, L_{TF} L_{FF}^{-1}$.
5. Compute the NPAs by summing the contributions of the N_{FF} edges (i,j) of the functional layer: $\mathrm{NPA} = N_{FF}^{-1} \sum_{\mathrm{edges}\ (i,j)} (\pi_i + (A_{FF})_{ij}\, \pi_j)^2 \equiv \boldsymbol{\pi}^t Q_{FF} \boldsymbol{\pi}$.
6. Compute the *-statistics by randomly permuting the transcript layer ($\boldsymbol{\beta} \rightarrow \boldsymbol{\beta}_{RAND}$):
    6.1 re-execute steps 4-5 500 times to get the null distribution of the NPAs;
    6.2 determine the p-value of the actual (non-permuted) NPA.
7. Compute the *-statistics by propagating the uncertainties of the SRPs:
    7.1 compute the variance of the NPAs as $\mathrm{var}(\mathrm{NPA}) = 2\,\mathrm{tr}(Q_{FF}\,\mathrm{var}(\boldsymbol{\pi})\, Q_{FF}\,\mathrm{var}(\boldsymbol{\pi})) + 4\,\boldsymbol{\pi}^t Q_{FF}\,\mathrm{var}(\boldsymbol{\pi})\, Q_{FF}\, \boldsymbol{\pi}$;
    7.2 test whether 0 belongs to the 95%-confidence interval of the NPA values.
8. Compute the *-statistics by randomly permuting the functional layer ($A_{FF} \rightarrow A_{FF,RAND}$):
    8.1 re-execute steps 2-4-5 500 times to get the null distribution of the NPAs;
    8.2 determine the p-value of the actual (non-permuted) NPA.
9. Determine the "leading nodes" contributing to 80% of the NPAs to guide the interpretation of the perturbation (indicated in cyan on the network figures).
10. Generate the various plots for visualization and interpretation (augmented bar plots, network node heat maps, etc.).

Compute BIFs (Step 5)

1. Compute the NPAs and their associated statistics *** for all subnetworks and SRPs.
2. For a given subnetwork n (part of the network n') and SRP c, define the filtering weights as $\omega(n,c) = 1$ if $\mathrm{NPA}(n,c)$ achieves 5% significance in all three NPA statistics ***, and $\omega(n,c) = 0$ else.
3. Compute the network-level BIFs by summing the weighted NPAs: $\mathrm{nBIF}(n',c) = \sum_{n \in n'} \frac{\omega(n,c)}{\sum_{m''} \sum_{m \in m'} \omega(m,REF) \cdot \mathrm{NPA}(m,REF)} \mathrm{NPA}(n,c)$ where REF designates the reference SRP, which is usually chosen as the strongest expected treatment.
4. Compute the BIFs by summing the network-level BIFs: $\mathrm{BIF}(c) = \sum_{n'} \mathrm{nBIF}(n',c)$. Normalizing by the reference SRP in step 3 ensured that $\mathrm{BIF}(REF) = 1$.
5. Generate the various plots for visualization and interpretation (bar, radar, star, and snowflake plots).

The results of these two statistical tests are indicated by green and blue stars * and * on the bar plots when they are significant (i.e., p-value < 0.05; Fig. 5, fourth column). These tests provide an informed way of measuring the relevance of the prior knowledge underlying the NPA results and consequently allow false positives to be discarded when the NPA values do not lie outside their null distributions sufficiently. An additional statistic is obtained by propagating the uncertainties of the SRPs that result from the combination of the biological variabilities within and between the experimental groups. As a consequence, the quality of the SRPs directly influences the quality of the NPA results, even if the univariate p-values/FDR do not enter explicitly the NPA calculations. The significance for the uncertainty statistics is reached when the 95 % confidence interval around the (positive) NPA values does not contain 0, as indicated by a red star * on the bar plots. We consider a network to be "perturbed" when all three statistics *** are significant (Fig. 5).

5. The final operation consists of presenting the NPA results in a way that facilitates their interpretation by the biological experts (see Sect. 2). The three statistics *** are also used to prioritize the investigations by indicating which treatments and network models contain significant perturbations, while examination of the "leading nodes" within a perturbed network provides insights into the biological mechanisms. It is important to make all results available because reliable negative results might be as important as positive ones in the context of the assessment of biological impact of exposure.

In several cases the qualitative behavior of the computed NPAs was confirmed by the measurements of specific phenotypic endpoints using standard experimental techniques [18, 20, 60]. They enabled explicit comparisons between the network-level perturbation amplitudes derived from untargeted systems-wide transcriptomics data and the targeted results obtained from the appropriate assays. These successful comparisons showed that the NPAs have the potential to complement the phenotypic endpoints measured by standard toxicological assays in providing insights into the underlying perturbed mechanisms using the causal network models assembled in Step 3.

Although the NPA methodology described above was primarily meant to be used in the context of assessment of biological impact, other applications are possible (Table 2) [20]. One such application was the explorative investigation of transcriptomics experiments, where the NPA methodology provided a useful tool to gain a first glimpse of the biological processes activated by the experimental treatments. However, much care is required in such a situation, because the methodology cannot indicate when a causal

Table 2
Descriptions of the methods can be found in the cited publications and detailed explanations of the chosen features are given in our latest NPA publication [20]

Reasoning	Category [65]	Method	Qualitative	Quantitative	Use topology	Interpretability	Diagnostic signatures	Threshold-free
Forward	ORA	Hypergeometric test [79]	✓	✗	✗	✗	✗	✗
	FCS	GSEA [80]	✓	(✓)	✗	✓	✗	✓
		GSA [81]	✓	(✓)	✗	✗	✗	✓
		PLAGE [82]	✓	(✓)	✗	✗	✗	✓
		CORG [83]	✗	✗	✗	✓	✓	✓
	PT	SPIA [84]	✓	✗	✓	✗	✗	✗
		NetGen [85]	✓	✗	✓	✗	✗	✓
		NetWalker [86]	✓	✗	✓	✓	✗	✓
		SVM based [87]	✗	✗	✓	✗	✓	✓
Backward	ORA	RCR [13]	✓	✗	✗	✗	✗	✗
	FCS	Modified GSA [88]	✓	✓	✓	✗	✗	✓
		MARINa [89]	✓	✓	✗	✗	✗	✓
	PT	NPA [18, 20]	✓	✓	✓	✓	✓	✓

Comparison of the NPA-based framework with existing analysis approaches that integrate a priori biological knowledge with transcriptomics data network model is a priori inapplicable in a particular biological context. Another application was to assess individual values (rather than group averages) and the use of such individual NPAs for patient stratification, diagnosis purposes, or biomarker identification [61]. We also note that our presentation was restricted to transcriptomics SRPs; however, proteomics and other omics data modalities can be in principle integrated into the network models as complementary or confirmatory data.

In summary, the developed NPA methodology provides a quantitative framework in which the perturbations of biological processes upon experimental treatments can be assessed and compared. The methodology is complemented by the mechanistic information extracted from the causal biological network models.

3.5 Step 5: Compute Biological Impact Factors

The last step of our strategy consists of calculating the BIF, a systems-wide and pan-mechanistic measure of the impact of the exposure treatment applied to the experimental biological system. This step also allows the detailed mechanistic fingerprint of the tested substances to be unraveled by exploring the "shape" of the underlying treatment-induced NPAs across the hierarchical structure of the network models for any given SRPs.

Similarly to Step 4 where the development of a novel methodology for computing NPAs was required, the calculation of BIFs required a genuinely innovative approach for which no equivalent published work existed. Therefore, we introduced early on a "prototype BIF" that demonstrated the feasibility and promise of our complete network-based strategy [5]. Although the prototype BIF was somewhat simplistic because it combined the NPAs of only two generic network models (cell proliferation and inflammation), it nevertheless captured relevant mechanistic information at the systems level and displayed two essential features of our global strategy. First, it appropriately quantified the impact of the applied experimental treatment (rats exposed during 3 months to formaldehyde), thereby enabling comparisons between the distinct quantitative doses. Second, it showed that long-term disease symptoms (tumorigenesis rate after 2 years) could be meaningfully correlated to the short-term exposure response captured by the BIF. This aspect constituted a concrete illustration of the usefulness of our strategy to assess disease risk upon long-term exposure, as discussed in Sect. 1 and in Step 1.

The extension of the prototype BIF to the final BIF that can cope with the full collection of tissue-specific network models required the ability to meaningfully combine the contributions of the individual networks. Additionally, similar to the way the NPAs of one network could be compared across several SRPs, the BIF and its individual network-level contributions should be comparable across several SRPs for the same network collection. Some of these requirements were satisfied in an earlier toxicogenomics study, where radar charts were used to display the different contributions of the individual pathways across several SRPs [62]. However, in that study, the authors did not combine all the pathway contributions into one single number, which would have been the equivalent of the BIF. The specific methodology that we developed showed that the BIF was not a trivial sum of NPAs but a more subtle combination thereof [15, 60, 63]. In these calculations, we needed to ensure the specificity of the BIF values by appropriately taking into account the contributions of the perturbed biological processes. In parallel, we introduced a weighting scheme that enabled the various network contributions to be aligned with another, taking into account the structure of the network models and of the treatment-induced perturbations for each SRP. The recently published BIF workflow concretely illustrates our network-based strategy from which we have borrowed the example discussed in the "Notes" (Sect. 4) [15].

The Workflow (Fig. 6 and Box 2).

1. The starting point for computing the BIFs is the SRP data obtained in Step 2 and a suitable collection of biological network models assembled in Step 3, as well as the ability to

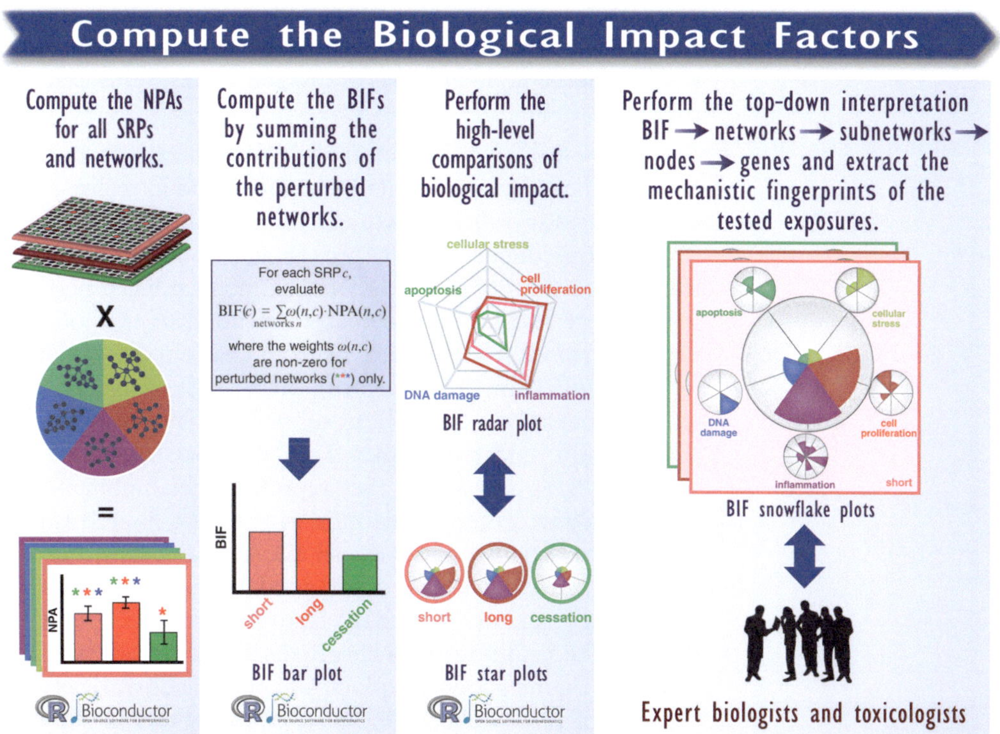

Fig. 6 Step 5: Computing the biological impact factors (BIFs) and constructing the mechanistic fingerprint of the tested exposures. In this final step the NPAs are computed for all the SRPs of the experiment and all the a priori relevant network models. These results are summarized into the high-level BIF values, which enable multidirectional comparative assessments of biological impacts: between the tested exposures, between the network models, and in top-down fashion within one network by following its hierarchical structure

compute NPAs as explained in Step 4. The subsequent calculations of the NPAs and their three associated statistics *** for all SRPs and all network models are a computationally intensive step because it involves multiplications of matrices of dimensions up to 10,000.

2. The next step consists of successively applying a specificity filter and a weighting scheme on the NPA values so that their combinations yield specific BIF values that can be compared across multiple SRPs. We used a stringent specificity filter by retaining only the contributions of the "perturbed" networks, i.e., the NPA values that display the *** statistical significance. The between-SRPs weighting scheme is based on the choice of a "reference treatment" (and thereby reference SRP) for which the BIF is set to 1, which helps to calibrate all the NPA value combinations.
3. Besides the magnitude of the BIF, the distribution of its network-level contributions is the other valuable biological information obtained from our calculations. Similar to the functional toxicogenomics study mentioned above, we used radar plots that show all treatments on the same plot [62]. We also introduced star plots that not only display the distributions of the network-level contributions to the BIF, but also indicate the actual BIF values for the multiple SRPs because the sum of the segment areas of one star corresponds exactly to the BIF.
4. For clarity, we have not yet explicitly taken into account the fact that the network models are modular and in reality consist in several distinct "subnetworks," as briefly mentioned in Step 3. As a consequence, our approach allows the description of the biological processes and their treatment-induced perturbations at five levels of detail that correspond to the following entities (from top to bottom, or from coarse to fine grained): network collection, BIF; network, network-level BIF; subnetwork, subnetwork-level NPA; subnetwork node, node-level perturbation; and gene, SRP. The "network-level BIFs" are similar to BIFs except that they represent the combinations of the suitably weighted contributions of all the filtered subnetworks of a particular network. To display graphically this additional level of detail, we augmented the BIF star plots by adding the "network-level BIF" star plots into their periphery. Because the resulting plots clearly display some self-similarity, we called them "snowflake plots." They judiciously display the "shape" of the treatment-induced perturbations in the three top levels of our hierarchy of biological description: the overall BIF as the sum of the segment areas of the central star, the network-level BIFs as the segment areas of the central star, and the NPAs of the "perturbed" subnetworks as the segment

areas of the peripheral stars. These plots, when completed with more detailed node-level and gene-level results from Steps 4 and 2, provide the material for the biological experts to construct the "mechanistic fingerprint" of the different exposures tested in the underlying experiment (Step 1) [64]. These mechanistic fingerprints can be compared between SRPs and provide not only a meaningful quantification of the global impact of the exposure with the BIF, but also give access to a multi-scale mechanistic characterization of the perturbed biological processes.

The computation of the BIF and the construction of the mechanistic fingerprint of the tested substances completed in this final step provided all the material necessary for the assessment of their biological impact. However, even if we processed systems-wide transcriptomics data in Step 2 and assembled detailed network models describing the a priori most relevant biological processes in Step 3, the completion of the five-step strategy does not mean that all the significant biological aspects of the considered experiment have been covered. Other systems-wide experimental data modalities can be measured and other biological processes could become a posteriori relevant in a particular experimental context. Quite generally, we also recommend that besides the assessment of biological impact a more "standard" functional interpretation of the omics data is to be conducted [65]. Such an interpretation can not only confirm the assessment results but also lead to the generation of novel biological hypotheses, which can be tested eventually in a dedicated experiment, following the discovery cycle of a system biology approach [7]. Ultimately, this augmentation of biological knowledge will benefit us directly by helping to further refine the network model collection and thereby improve the credibility of our results. To help make the considerations of our five-step assessment strategy more concrete, a concrete illustration is given in "Notes" (Sect. 4).

Regarding the computational infrastructure enabling the implementation of the NPA and BIF computations, we mentioned in Step 2 that we used the R statistical environment and the Bioconductor software [21, 22]. Our experience showed that a computer that can process the raw data of a medium-sized dataset (50–100 samples) following the standard R/Bioconductor pipeline presented in Box 1 suffices for running the corresponding NPA and BIF computations. The bottleneck of these computations is the matrix multiplications $L_{TF}t$ diag(var(β)) L_{TF} involved in the calculation of the $*$ statistics of the NPAs because β can contain up to 10,000 downstream transcript differential expressions for some networks (see Box 2). An efficient coding and a careful memory management kept the running time within acceptable limits: for ten SRPs, it does not exceed 30 min for the largest networks and 4 h for the whole collection of networks. Additional optimizations are possible, in particular a parallelization over the more than 50 subnetworks, which can further reduce the running time of the NPA and BIF computations.

In summary, our five-step impact assessment strategy is completed by computing BIFs building on the NPA methodology. The application of BIFs to the SRPs and network models obtained in the earlier steps enabled the construction of a "mechanistic fingerprint" to characterize quantitatively the treatment-induced perturbations at a multiple scale of description detail.

4 Notes: Application of the Five-Step Strategy to a 7-Month Cigarette Smoke Inhalation and Smoking Cessation Study on Mice

The aim of this study was to investigate the effects of smoking cessation vs. continued smoking in a mouse model that exhibits typical features of COPD, one of the major smoking-related diseases, which is known not to revert but to slow down its progression upon smoking cessation [15, 66].

4.1 Step 1: Design the Experiment for Data Production

- *Biological System.* Following the rationale to be as closely comparable to the human disease as possible, we chose a mouse strain (C57BL/6) that is susceptible to develop smoke-induced lung inflammation and emphysema, two important manifestations of COPD that are also observed in humans. To capture the long-term multiple cell type interactions between resident lung cells and recruited systemic inflammatory cells that lead to issue destruction, an in vivo mouse model was preferred over in vitro models that would only be able to reproduce isolated steps or interactions, even as complex organotypical cultures.
- *Exposures* (Fig. 7a). In the smoking arms of the study, mice were exposed in whole-body inhalation chambers for 2 or 7 months to diluted mainstream cigarette smoke (CS) according to the Health Canada intense smoking regimen [67]. This type of exposure is well tolerated by the mice, while the delivered dose is high enough to elicit measurable symptoms of smoking-related emphysema within 2 months. In the cessation arm, mice were exposed to fresh air for 5 months following the 2-month smoke exposure. Control air-exposed groups that received fresh air only for 2 or 7 months were included as age-matched, untreated controls.
- *Biological Networks.* A collection of transcriptomics-based networks was chosen to describe the biology related to the exposure-induced disease mechanisms. The causal networks were selected to cover the biological processes in the lung that are known to play major roles in smoking-related responses

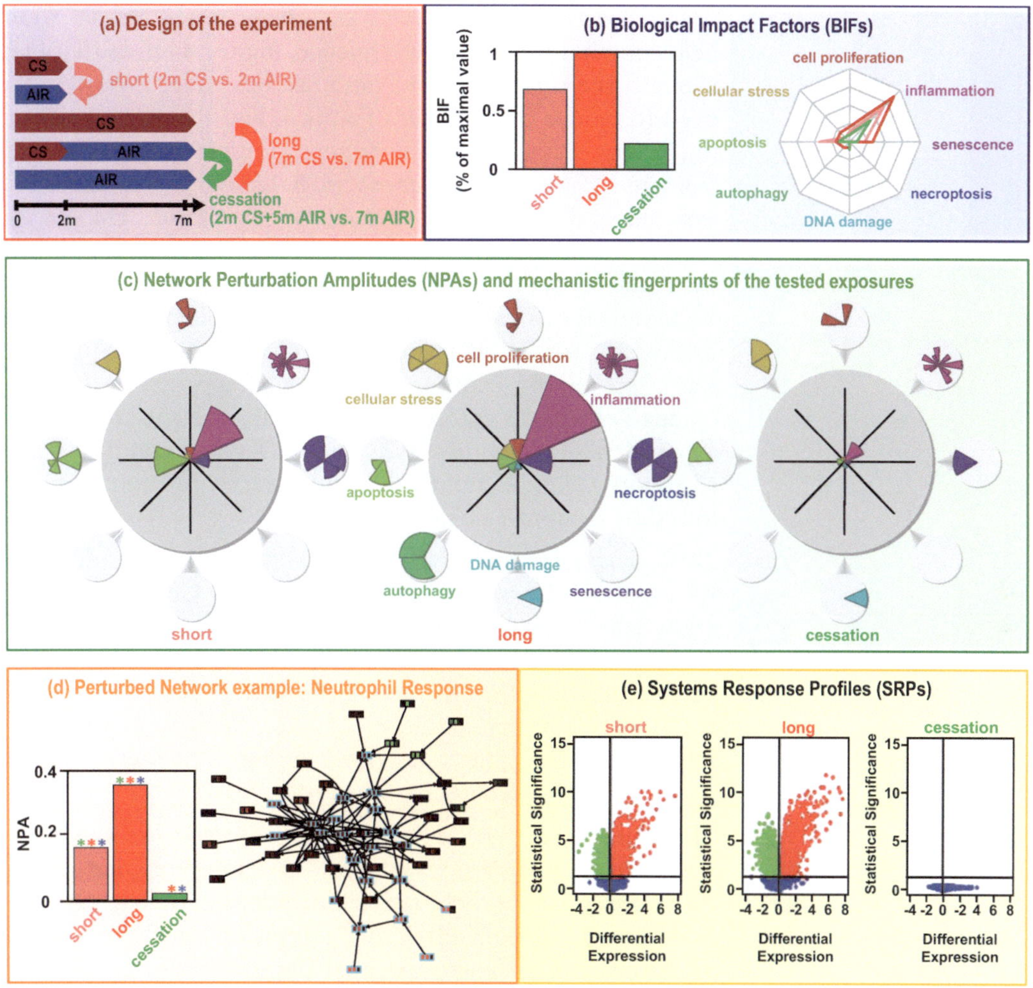

Fig. 7 (**a**) Application of the five-step strategy in an animal smoke-cessation experiment. *Panels* **b–e** display the results in increasing mechanistic details, starting with the high-level BIF values and their breakdown into their respective network- and subnetwork-level contributions (*panels* **b**, **c**). The perturbation amplitudes for subnetworks and its nodes provide more detailed insights into the perturbed mechanisms (*panel* **d**). Furthermore, the SRP results show that the statistical significances of the differential expressions of individual genes are not a good indicator of the biological impact (e.g., the "cessation" treatment results in *panels* **b**, **e**)

and COPD pathogenesis, such as pulmonary inflammation, DNA damage, apoptosis, cellular stress, and cell proliferation (see also Step 3 below) [51–54].

4.2 Step 2: Compute the Systems Response Profiles

- *Design of experiment considerations.* For the generation of the transcriptomics data, the design requirements concerned first the number (n) of biological replicates per experimental group that allowed statistically significant exposure-induced gene differential expressions to be obtained. While the interindividual differences and high-throughput technology are inescapable sources of variability, standardized dissection and RNA extraction protocols, blocking, and use of an inbred animal strain help containing it. As a consequence, higher *n* values that help increase the signal-to-noise ratio are desirable; however, the need to reduce the number of animals and the costs are limiting factors. Based on our experience in respiratory tract tissues from mice and rats, we chose $n \sim 8$ biological replicates per experimental group as a suitable compromise. The design requirements also concerned the execution of experiment from the tissue collection schedules to the RNA hybridization order. Randomizations taking into account the blocking were performed to help avoid confounding effects. Finally, to capture the effect of the exposures specifically, appropriate pairwise comparisons were performed along the relevant arms of the experiment (Fig. 7a): *short* (2-month CS vs. 2-month air), *long* (7-month CS vs. 7-month air), and *cessation* (2-month CS followed by 5-month air vs. 7-month air).

- *Data generation and SRP calculations.* For the hybridization, we used the Affymetrix GeneChip® Mouse Genome 430 2.0 array, which simultaneously measures the expression of over 39,000 transcripts. The data quality check and analysis were performed using the Bioconductor software suite available for the R statistical environment [21, 22]. SRPs were generated following the workflow described in Step 2 of Sect. 3 and its implementation details are displayed in Box 1.
- *SRP results* (Fig. 7b). The three pairwise comparisons *short*, *long*, and *cessation* are presented in volcano plots that display the statistical significance (which includes multiple testing corrections) vs. the log 2 differential expression for all the detected transcripts. The results show that the number of significantly differentially expressed genes increased with the continuation of the CS exposure from 2 to 7 months. In contrast, the remaining differentially expressed genes in the cessation group did not reach statistical significance, indicating that their gene expression was similar to that of the control mice that were exposed to fresh air during all 7 months.

4.3 Step 3: Identify Perturbed Biological Networks

The causal biological network models available in our collection have been designed to be tissue and context specific (see Step 3 in Sect. 3). Therefore, the well-defined biological boundaries of the network models must be compatible with the biological processes that can be potentially activated in our study in C57BL/6 mice. In the lung tissues shown in this example, these considerations resulted in an a priori selection of eight networks (Cell Proliferation, Cellular Stress, Apoptosis, DNA Damage, Necroptosis, Autophagy, Senescence, Inflammation), each of which contains several subnetworks, which add to a total of 64 (=13+6+7+8+2+3+6+19 when split into the respective network contributions). The application of the same network models to the respiratory tract tissues of rats exposed to cigarette smoke for a shorter period of time (28 days) has already been published [68].

4.4 Steps 4 and 5: Compute Network Perturbation Amplitudes and Biological Impact Factors

We follow a top-down examination of the NPA and BIF results following Steps 4 and 5 (described in Sect. 3) to show the advantages of our multi-level assessment of biological impact based on the hierarchically structured network models. These results agree with the phenotypic endpoints measured by standard toxicological assays [66], and therefore confirm the ability of the NPA and BIF methodologies to provide insights into the biological mechanisms that are relevant for the tested exposures (see Step 4 in Sect. 3).

- *Network collection level* (Fig. 7b). The BIF is our systems-wide and pan-mechanistic measure of biological impact that allows various exposures to be concisely compared. For the CS arms, the BIF of the *long* exposure has about twofold the value of the BIF of the *short* exposure; for *cessation*, the BIF dropped to less than 10 % of the *short* value. However, this residual effect is statistically significant and shows that the absence of a signal in the *cessation* SRP is not necessarily indicative of the actual biological impact.
- *Network level* (Fig. 7b, c). The star plots display the network-level BIF values that can be compared across network models and exposures. They indicate that the *short* exposure causes a major response in the lung inflammatory processes, followed by apoptosis, and weak perturbations in cellular stress, cell proliferation, and senescence. After the *long* exposure the relative contribution of the inflammatory processes was further increased compared with *short*, while the senescence, cellular stress, and cell proliferation contributions were more prominent, and the contribution of apoptosis was decreased; very weak autophagy and DNA damage signaling were also detected. In the *cessation* group, the 2-month CS effect on lung inflammatory processes, senescence, cellular stress, apoptosis, and cell proliferation was strongly attenuated after 5 months of fresh air exposure, but was still detectable (i.e., statistically significant). This finding further illustrates the increased sensitivity of our multivariate network-based approach compared to the univariate SRP calculation that did not reveal any significant gene-level responses for the *cessation* group.
- *Subnetwork level* (Fig. 7c). A further level of mechanistic detail is provided by the breakdown of network-level BIF values into the corresponding subnetwork-level NPAs that are displayed on the periphery of the snowflake plots. For the lung inflammatory processes, a relatively constant pattern of subnetwork-level NPA values is evident for all three experimental treatments,

while for, e.g., cell proliferation, the subnetwork-level NPAs are similar for both *short* and *long* exposure while two different subnetworks contribute to the minimal cell proliferation result in the *cessation* group.

- *Node level* (Fig. 7d). The subnetwork-level NPAs can be broken down further and node-level perturbation values can be calculated for each single node, as shown in the network diagram using the Neutrophil Response subnetwork part of the inflammatory process network as an example.

5 Conclusions and Future Prospects

We have described a five-step strategy for comparative evaluations of the effects of biologically active substances. Its purpose is to support health risk assessment by providing a systematic framework for integrating the experimental evidence obtained on tested substances and systems. Its central components are the causal network models describing a comprehensive set of a priori selected biological mechanisms on which the effects of the tested exposures can be assessed. This assessment of biological impact was made possible by the development of the NPA and BIF methodologies, which go beyond the standard correlative or associative approaches used in computational systems toxicology. We mentioned several experiments where the NPA and BIF results agreed with the available measurements of the corresponding phenotypic endpoints, which confirmed the ability of our strategy for biological impact assessment to provide mechanistic understanding that can complement the standard toxicological assays. A concrete example of a mouse smoking-cessation experiment also showed that our network-based methodology can be more sensitive than the gene-level differential expressions constituting the SRPs: while no statistically significant genes were observed in the *cessation* arm, the NPA and BIF results showed that several subnetworks were perturbed (Fig. 7). This sensitivity property is a direct consequence of a network-based approach that captures the concerted effects of multiple genes. It also enables the mechanistic inter-system translatability that underlies our "design cube" framework (Fig. 2), as recently illustrated in the case of the xenobiotic metabolism response between in vivo human smokers and in vitro smoke-exposed human bronchial epithelial cell cultures [69].

Although we have focused our publications mainly on CS exposure and respiratory biology/diseases (Table 1), the approach can be applied to all biologically active substances or treatments in the relevant experimental context. It is therefore one of our main goals for the next future to actively share it with the scientific community, so that it can be further verified and improved to become more widely accepted [70]. In order to do so, we recapitulate all the features intended to facilitate its partial or complete adoption by interested external scientists. The overall strategy itself can be applied in other contexts ranging from drug development to consumer product testing and environmental impact analysis, and therefore benefit to a larger community. Because the data generation in Steps 1 and 2 requires significant technical and financial resources, we made our transcriptomics datasets publicly available in the well-annotated MAGE-TAB format and published the accompanying article in open-access journal (Table 1). Interested computational biologists can therefore reprocess the raw data, compare the conclusions to published findings, as well as derive new insights. The causal network model suite can be downloaded from the CBN repository [56], while part of their downstream transcriptional entities are contained in the "BEL Large Corpus" freely available on the openBEL portal [13, 17], to which should be added several complete two-layered subnetworks included in our methodological publications: IKK/NF-κB Signaling [18], Cell Cycle [60], and Xenobiotic Metabolism Response [20]. The SBVImprover project, which uses crowdsourcing techniques to address fundamental challenges in systems biology, is continuing to develop a platform that will offer the research community a high-performance environment with the goal of providing a link between model generators and biological experts to verify and extend the network models [57]. Finally, the NPA and BIF methodologies explained in Box 2 and in the dedicated publications [18, 20, 60] have been implemented in the free R statistical environment using the data structures of the open-source Bioconductor software packages [21, 22], which are both standard in the bioinformatics community.

The hereby described methodology evaluates the biological impact of exposures in an objective, systematic, and quantifiable manner, enabling the computation of a systems-wide and pan-mechanistic biological impact measure for a given active substance or mixture. Our results suggest that various fields of human disease research, from drug development to consumer product testing and environmental impact analysis, could benefit from using this methodology.

References

1. U.S. Food and Drug Administration (2004) The critical path initiative. http://www.fda.gov/ScienceResearch/SpecialTopics/CriticalPathInitiative/
2. Committee on Toxicity Testing and Assessment of Environmental Agents, National Research Council (2007) Toxicity testing in the 21st century: a vision and a strategy. The National Academies Press, Washington, DC, http://www.nap.edu/openbook.php?record_id=11970
3. Krewski D, Westphal M, Andersen ME et al (2014) A framework for the next generation of risk science. Environ Health Perspect 122(8):796–805. doi:10.1289/ehp.1307260
4. Waters MD, Fostel JM (2004) Toxicogenomics and systems toxicology: aims and prospects. Nat Rev Genet 5(12):936–948

5. Hoeng J, Deehan R, Pratt D et al (2012) A network-based approach to quantifying the impact of biologically active substances. Drug Discov Today 17(9-10):413–418, doi:10.1016/j.drudis.2011.11.008 S1359-6446(11)00425-9 [pii]
6. Barabasi A-L, Oltvai ZN (2004) Network biology: understanding the cell's functional organization. Nat Rev Genet 5(2):101–113
7. Kitano H (2002) Systems biology: a brief overview. Science 295(5560):1662–1664
8. Del Sol A, Balling R, Hood L et al (2010) Diseases as network perturbations. Curr Opin Biotechnol 21(4):566–571
9. Barabási A-L, Gulbahce N, Loscalzo J (2011) Network medicine: a network-based approach to human disease. Nat Rev Genet 12(1): 56–68
10. Berger SI, Iyengar R (2009) Network analyses in systems pharmacology. Bioinformatics 25(19):2466–2472
11. Chindelevitch L, Ziemek D, Enayetallah A et al (2012) Causal reasoning on biological networks: interpreting transcriptional changes. Bioinformatics 28(8):1114–1121. doi:10.1093/bioinformatics/bts090
12. Kramer A, Green J, Pollard J Jr et al (2014) Causal analysis approaches in ingenuity pathway analysis. Bioinformatics 30(4):523–530. doi:10.1093/bioinformatics/btt703
13. Catlett NL, Bargnesi AJ, Ungerer S et al (2013) Reverse causal reasoning: applying qualitative causal knowledge to the interpretation of high-throughput data. BMC Bioinformatics 14(1):340
14. Sewer A, Hoeng J, Deehan R et al. Systems biology approaches for compound testing. Data Min Drug Discov 291–316
15. Hoeng J, Talikka M, Martin F et al (2014) Case study: the role of mechanistic network models in systems toxicology. Drug Discov Today 19(2):183–192. doi:10.1016/j.drudis.2013.07.023
16. Miller GW (2014) Improving reproducibility in toxicology. Toxicol Sci 139(1):001–003
17. Selventa The openBEL portal. http://www.openbel.org/
18. Martin F, Thomson TM, Sewer A et al (2012) Assessment of network perturbation amplitudes by applying high-throughput data to causal biological networks. BMC Syst Biol 6:54. doi:10.1186/1752-0509-6-54
19. Mitrea C, Taghavi Z, Bokanizad B et al (2013) Methods and approaches in the topology-based analysis of biological pathways. Frontiers Physiol 4
20. Martin F, Sewer A, Talikka M et al (2014) Quantification of biological network perturbations for mechanistic insight and diagnostics using two-layer causal models. BMC Bioinformatics 15(1):238
21. The Core R team (2012) R: a language and environment for statistical computing
22. Gentleman RC, Carey VJ, Bates DM et al (2004) Bioconductor: open software development for computational biology and bioinformatics. Genome Biol 5(10):R80
23. Gentleman R, Lang DT (2007) Statistical analyses and reproducible research. J Comput Graph Stat 16(1)
24. Xie Y (2014) Knitr: a comprehensive tool for reproducible research in R. Implement Reprod Res 1:20
25. Liu Z, Pounds S (2014) An R package that automatically collects and archives details for reproducible computing. BMC Bioinformatics 15(1):138
26. Rhrissorrakrai K, Belcastro V, Bilal E et al (2014) Understanding the limits of animal models as predictors of human biology: lessons learned from the sbv IMPROVER Species Translation Challenge. Bioinformatics 31(4):471–483, 10.1093/bioinformatics/btu611
27. Wikipedia. The three Rs (animals). http://en.wikipedia.org/wiki/The_Three_Rs_(animals)
28. Commission TE (2009) Alternative testing strategies – progress report 2009 – replacing, reducing and refining use of animals in research. Office for Official Publications of the European Communities, Luxembourg
29. Edwards SW, Preston RJ (2008) Systems biology and mode of action based risk assessment. Toxicol Sci 106(2):312–318. doi:10.1093/toxsci/kfn190
30. Russell WMS, Burch RL, Hume CW (1959) The principles of humane experimental technique
31. Pleil JD, Sheldon LS (2011) Adapting concepts from systems biology to develop systems exposure event networks for exposure science research. Biomarkers 16(2):99–105
32. Rustici G, Kolesnikov N, Brandizi M et al (2013) ArrayExpress update—trends in database growth and links to data analysis tools. Nucleic Acids Res 41(D1):D987–D990
33. Barrett T, Wilhite SE, Ledoux P et al (2013) NCBI GEO: archive for functional genomics data sets—update. Nucleic Acids Res 41(D1): D991–D995
34. Allison DB, Cui X, Page GP et al (2006) Microarray data analysis: from disarray to

consolidation and consensus. Nat Rev Genet 7(1):55–65
35. Shi L, Shi L, Reid LH et al (2006) The MicroArray Quality Control (MAQC) project shows inter-and intraplatform reproducibility of gene expression measurements. Nat Biotechnol 24(9):1151–1161
36. Yang YH, Speed T (2002) Design issues for cDNA microarray experiments. Nat Rev Genet 3(8):579–588
37. Churchill GA (2002) Fundamentals of experimental design for cDNA microarrays. Nat Genet 32:490–495
38. Lockhart DJ, Winzeler EA (2000) Genomics, gene expression and DNA arrays. Nature 405(6788):827–836
39. Ozsolak F, Milos PM (2010) RNA sequencing: advances, challenges and opportunities. Nat Rev Genet 12(2):87–98
40. Nuwaysir EF, Bittner M, Trent J et al (1999) Microarrays and toxicology: the advent of toxicogenomics. Mol Carcinog 24(3):153–159
41. Cox J, Mann M (2011) Quantitative, high-resolution proteomics for data-driven systems biology. Annu Rev Biochem 80:273–299
42. Irizarry RA, Hobbs B, Collin F et al (2003) Exploration, normalization, and summaries of high density oligonucleotide array probe level data. Biostatistics 4(2):249–264
43. Ringnér M (2008) What is principal component analysis? Nat Biotechnol 26(3):303–304
44. Smyth GK (2004) Linear models and empirical Bayes methods for assessing differential expression in microarray experiments. Stat Appl Genet Mol Biol 3(1)
45. Benjamini Y, Hochberg Y (1995) Controlling the false discovery rate: a practical and powerful approach to multiple testing. J R Stat Soc B 57(1):289–300
46. Suárez E, Burguete A, Mclachlan GJ (2009) Microarray data analysis for differential expression: a tutorial. P R Health Sci J 28(2)
47. Gershon D (2005) DNA microarrays: more than gene expression. Nature 437(7062): 1195–1198
48. Hermida L, Poussin C, Stadler MB et al (2013) Confero: an integrated contrast data and gene set platform for computational analysis and biological interpretation of omics data. BMC Genomics 14(1):514
49. Surowiecki J (2005) The wisdom of crowds. Random House LLC, London
50. Ansari S, Binder J, Boue S et al (2013) On crowd-verification of biological networks. Bioinform Biol Insights 7:307
51. Westra JW, Schlage WK, Frushour BP et al (2011) Construction of a computable cell proliferation network focused on non-diseased lung cells. BMC Syst Biol 5(1):105
52. Schlage WK, Westra JW, Gebel S et al (2011) A computable cellular stress network model for non-diseased pulmonary and cardiovascular tissue. BMC Syst Biol 5(1):168
53. Gebel S, Lichtner RB, Frushour B et al (2013) Construction of a computable network model for DNA damage, autophagy, cell death, and senescence. Bioinform Biol Insights 7:97
54. Westra JW, Schlage WK, Hengstermann A et al (2013) A modular cell-type focused inflammatory process network model for non-diseased pulmonary tissue. Bioinform Biol Insights 7:167
55. De León H, Boué S, Schlage WK et al (2014) A vascular biology network model focused on inflammatory processes to investigate atherogenesis and plaque instability. J Transl Med 12(1):185
56. Boué S, Talikka M, Westra JW et al (2014) Causal Biological Network (CBN) database: a comprehensive platform of causal biological network models focused on the pulmonary and vascular systems. Submitted
57. The SBVImprover project team. The SBVImprover Bionet platform. http://bionet.sbvimprover.com
58. Belcastro V, Poussin C, Gebel S et al (2013) Systematic verification of upstream regulators of a computable cellular proliferation network model on non-diseased lung cells using a dedicated dataset. Bioinform Biol Insights 7:217
59. Liao JC, Boscolo R, Yang Y-L et al (2003) Network component analysis: reconstruction of regulatory signals in biological systems. Proc Natl Acad Sci 100(26):15522–15527
60. Thomson TM, Sewer A, Martin F et al (2013) Quantitative assessment of biological impact using transcriptomic data and mechanistic network models. Toxicol Appl Pharmacol 272(3):863–878. doi:10.1016/j.taap.2013.07.007
61. Tarca AL, Lauria M, Unger M et al (2013) Strengths and limitations of microarray-based phenotype prediction: lessons learned from the improver diagnostic signature challenge. Bioinformatics 29(22):2892–2899
62. Kiyosawa N, Manabe S, Yamoto T et al (2010) Practical application of toxicogenomics for profiling toxicant-induced biological perturbations. Int J Mol Sci 11(9):3397–3412. doi:10.3390/ijms11093397
63. Vasilyev DM, Thomson TM, Frushour BP et al (2014) An algorithm for score aggregation over causal biological networks based on random walk sampling. BMC Res Notes 7(1):516

64. Gonzalez-Suarez I, Sewer A, Walker P et al (2014) Systems biology approach for evaluating the biological impact of environmental toxicants in vitro. Chem Res Toxicol. doi:10.1021/tx400405s
65. Khatri P, Sirota M, Butte AJ (2012) Ten years of pathway analysis: current approaches and outstanding challenges. PLoS Comput Biol 8(2), e1002375. doi:10.1371/journal.pcbi.1002375
66. Phillips B, Veljkovic E, Peck MJ et al (2014) A 7-month cigarette smoke inhalation study in C57BL/6 mice demonstrates reduced lung inflammation and emphysema following smoking cessation or aerosol exposure from a prototypic modified risk tobacco product
67. Canada H (1999) Determination of "Tar" and nicotine in sidestream tobacco smoke. Health Canada Test Method T-115
68. Kogel U, Schlage WK, Martin F et al (2014) A 28-day rat inhalation study with an integrated molecular toxicology endpoint demonstrates reduced exposure effects for a prototypic modified risk tobacco product compared with conventional cigarettes. Food Chem Toxicol 68:204–217
69. Iskandar AR, Martin F, Talikka M et al (2013) Systems approaches evaluating the perturbation of xenobiotic metabolism in response to cigarette smoke exposure in nasal and bronchial tissues. BioMed Res Int 2013:512086
70. Meyer P, Hoeng J, Rice JJ et al (2012) Industrial methodology for process verification in research (IMPROVER): toward systems biology verification. Bioinformatics 28(9):1193–1201
71. Mathis C, Poussin C, Weisensee D et al (2013) Human bronchial epithelial cells exposed in vitro to cigarette smoke at the air-liquid interface resemble bronchial epithelium from human smokers. Am J Physiol Lung Cell Mol Physiol 304(7):L489–L503
72. Poussin C, Gallitz I, Schlage WK et al (2014) Mechanism of an indirect effect of aqueous cigarette smoke extract on the adhesion of monocytic cells to endothelial cells in an in vitro assay revealed by transcriptomics analysis. Toxicol In Vitro 28(5):896–908
73. Boué S, De León H, Schlage WK et al (2013) Cigarette smoke induces molecular responses in respiratory tissues of ApoE$^{-/-}$ mice that are progressively deactivated upon cessation. Toxicology 314(1):112–124
74. Xiang Y, Kogel U, Gebel S et al (2014) Discovery of emphysema relevant molecular networks from an A/J mouse inhalation study using reverse engineering and forward simulation (REFS™). Gene Regul Syst Biol 8:45
75. Luettich K, Xiang Y, Iskandar A et al (2014) Systems toxicology approaches enable mechanistic comparison of spontaneous and cigarette smoke-related lung tumor development in the A/J mouse model. Interdiscip Toxicol 7(2):73–84
76. Poussin C, Mathis C, Alexopoulos LG et al (2014) The species translation challenge—a systems biology perspective on human and rat bronchial epithelial cells. Scientific Data 1
77. Talikka M, Kostadinova R, Xiang Y et al (2014) The response of human nasal and bronchial organotypic tissue cultures to repeated whole cigarette smoke exposure. Int J Toxicol. doi:10.1177/1091581814551647
78. Schlage WK, Iskandar AR, Kostadinova R et al (2014) In vitro systems toxicology approach to investigate the effects of repeated cigarette smoke exposure on human buccal and gingival organotypic epithelial tissue cultures. Toxicol Mech Methods 24(7):470–487
79. Castillo-Davis CI, Hartl DL (2003) GeneMerge—post-genomic analysis, data mining, and hypothesis testing. Bioinformatics 19(7):891–892
80. Subramanian A, Tamayo P, Mootha VK et al (2005) Gene set enrichment analysis: a knowledge-based approach for interpreting genome-wide expression profiles. Proc Natl Acad Sci U S A 102(43):15545–15550
81. Efron B, Tibshirani R (2007) On testing the significance of sets of genes. Ann Appl Stat 1(1):107–129
82. Tomfohr J, Lu J, Kepler TB (2005) Pathway level analysis of gene expression using singular value decomposition. BMC Bioinformatics 6(1):225
83. Lee E, Chuang H-Y, Kim J-W et al (2008) Inferring pathway activity toward precise disease classification. PLoS Comput Biol 4(11), e1000217
84. Tarca AL, Draghici S, Khatri P et al (2009) A novel signaling pathway impact analysis. Bioinformatics 25(1):75–82
85. Shojaie A, Michailidis G (2009) Analysis of gene sets based on the underlying regulatory network. J Comput Biol 16(3):407–426
86. Komurov K, Dursun S, Erdin S et al (2012) NetWalker: a contextual network analysis tool for functional genomics. BMC Genomics 13(1):282
87. Rapaport F, Zinovyev A, Dutreix M et al (2007) Classification of microarray data using gene networks. BMC Bioinformatics 8(1):35
88. Ackermann M, Strimmer K (2009) A general modular framework for gene set enrichment analysis. BMC Bioinformatics 10(1):47
89. Lefebvre C, Rajbhandari P, Alvarez MJ et al (2010) A human B-cell interactome identifies MYB and FOXM1 as master regulators of proliferation in germinal centers. Mol Syst Biol 6(1)

Chapter 11

Analysis of Proteomic Data for Toxicological Applications

Bjoern Titz, Thomas Schneider, Ashraf Elamin, Florian Martin, Sophie Dijon, Nikolai V. Ivanov, Julia Hoeng, and Manuel C. Peitsch

Abstract

Toward a comprehensive characterization of toxicant responses, systems toxicology requires the integration of different data modalities. Proteomics approaches measure changes in the levels of proteins and their posttranslational modifications, which can closely reflect the biological effects of a toxicant. With a focus on mass spectrometry-based proteomics, we describe the isobaric tag-based approach for quantitative proteomics (iTRAQ® and TMT™) and describe computational approaches to derive biological/mechanistic insights. Specifically, we describe the generation and quantification of mass-spectrometry data and the identification of affected proteins, functional modules, and protein subnetworks. We illustrate these approaches using results from a 90-day rat inhalation toxicology study. Overall, we provide the foundation to employ quantitative proteomics within systems toxicology assessment strategies.

Key words Computational analysis, iTRAQ®/TMT®, Systems toxicology, Quantitative proteomics

1 Proteomics Within the Systems Toxicology Framework

Conventional toxicological assessment of chemical substances relies heavily on in vitro assays and animal studies to test and identify exposure doses at which relevant apical endpoints (e.g., reduction of body weight) are adversely affected [1]. Although this conventional toxicological approach has clearly proven its value, more recent discussions on the future requirements for toxicological assessment have highlighted some of its shortcomings and emphasized the need to further evolve toxicology assessment with new tools and approaches (such as the Tox21 and EPA ToxCast™ initiatives) [2, 3]. For example, there is an urgent need for deeper insights into toxicological mechanisms as the basis for improved toxicity predictions for different human exposure scenarios [2]. Recently, an integrative systems toxicology framework has been proposed [4], which combines the systems biology approach with its large-scale

Julia Hoeng and Manuel C. Peitsch (eds.), *Computational Systems Toxicology*, Methods in Pharmacology and Toxicology, DOI 10.1007/978-1-4939-2778-4_11, © Springer Science+Business Media New York 2015

measurements and computational modeling approaches with the requirements of toxicological studies. Specifically, this integrative approach relies on extensive measurements of exposure effects at the molecular level (e.g., proteins and RNAs), at different levels of biological complexity (e.g., cells, tissues, animals), and across species (e.g., human, rat, mouse).

Proteomic measurements are an important component of this framework [1]. Although mRNA and protein expression are tightly linked through translation, their correlation is limited, and mRNA transcript levels explain only about 50 % of the variation of protein levels [5]. This is because of the additional levels of protein regulation, including their rate of translation and degradation. In addition, protein activity is often controlled further through posttranslational modifications such as phosphorylation. For example, several cellular stress responses (e.g., oxidative stress) are regulated at the level of protein translation [6] and are controlled through protein phosphorylation cascades [7–9].

In this chapter, we give a practical overview on the experimental and computational analysis of proteomic data and how it can be applied for systems toxicology applications. A recent dataset from a 90-day rat cigarette smoke inhalation study will serve as an example for processing of a quantitative proteomic dataset obtained from lung tissues (Box 1). First, we discuss the mass-spectrometric generation of proteomic data (focused on the iTRAQ® workflow). Second, we discuss peptide and protein identification and

Box 1 90-day rat inhalation study used to demonstrate the analysis of proteomic data sets

In this chapter we demonstrate the analysis of proteomic data sets using results from a 90-day rat cigarette smoke inhalation study. Titz et al. describe these proteomic results in more detail [1]. This lung proteome analysis was part of a much more comprehensive lung toxicology assessment strategy that, for example, also included transcriptomic analyses and more traditional toxicological parameters such as gross pathology and pulmonary histopathology as required by the OECD test guideline 413 (OECD TG 413) for a 90-day subchronic inhalation toxicity study. Briefly, Sprague-Dawley rats were exposed to fresh air or two concentrations of a reference cigarette (3R4F) aerosol [8 μg/L (low) and 23 μg/L (high) nicotine] for 90 days (5 days per week, 6 h per day) (Fig. 3a). This exposure period was followed by a 42-day recovery period with fresh air exposure. For the two time points and exposure conditions, quantitative alterations in the lung proteome were analyzed by a multiplexed iTRAQ® approach. Cigarette smoke exposure clearly induced dose-dependent alterations in the lung proteome after 90 days and these alterations reverted largely after the 42-day recovery period (Fig. 3b). Overall, the same response pattern was observed at the gene-set level (GSEA, Fig. 4). The biological functions affected by cigarette smoke exposure included "oxidative phosphorylation" and immune-related functions (e.g., "neutrophil granule constituents" and "myeloid cell development"). The functional network analysis shows the affected biological functions more concisely (Fig. 5), including an "inflammatory response," "unfolded protein response," and a "xenobiotic metabolism" and "oxidative stress" response. Overall, our analysis of the quantitative changes of the lung proteome reflected the broad biological impact of cigarette smoke exposure and constitutes an important component of our integrative assessment strategy.

quantification approaches. Third, we demonstrate the analysis of protein sets and protein modules. Fourth, we show how proteomic results can be analyzed in the context of biological networks. Finally, we highlight approaches for integrative analyses. Overall, this chapter is intended to give the reader the theoretical and practical foundation to start integrating proteomics into a systems toxicology assessment framework.

2 Liquid Chromatography-Mass Spectrometry (LC-MS/MS)

Mass spectrometry (MS) is widely considered the central technology platform for toxicoproteomics [10]. MS approaches have brought many advantages including unsurpassed sensitivity, improved speed, and the ability to produce high-throughput datasets. Because of the high accuracy of MS, peptides in the femtomolar (10^{-15}) to attomolar (10^{-18}) range can be detected in tissues and biological matrices [11]. This is greatly beneficial in comparative analysis where simultaneous comparisons between control and treated samples are crucial to increasing the understanding of how toxicants affect the proteome and to the subsequent identification of potential biomarkers [12].

Currently, the most common workflow for the quantitative analysis of proteomes starts with the digestion of protein samples by a specific protease (e.g., trypsin). Optionally, the derived peptide mixture is affinity enriched, e.g., to facilitate the detection of posttranslational modifications [13]. Liquid chromatography (LC) is commonly used to separate the peptides (e.g., by hydrophobicity), which reduces sample complexity for the mass-spectrometric analysis [12]. Subsequently, MS is used to identify and quantify the peptides. Confident peptide identifications are based on both the measured mass of the peptide and its fragmentation spectrum.

Two general quantification approaches exist: label-free and label-based quantitation. In the label-free approach, the peptide mixtures are analyzed directly and quantified by MS. Peptide abundance is estimated either based on the number of times a peptide was identified (spectral counting) or based on the integrated peptide signal over the LC-elution peak. The main advantages of this approach are the following: (1) comparison of multiple samples is possible (no restriction in sample number); (2) a broad dynamic range of concentrations is covered; and (3) no further sample treatment is required. In the label-based approach, samples are modified prior to analysis. One of the most common label-based techniques is the use of isobaric tags with the iTRAQ® or TMT® methods (Fig. 1) [14]. For example, for the iTRAQ® approach, isobaric tags with eight different reporter group masses are available. This allows for labeling and subsequent pooling of up to eight samples with different reporter groups. The main idea is that

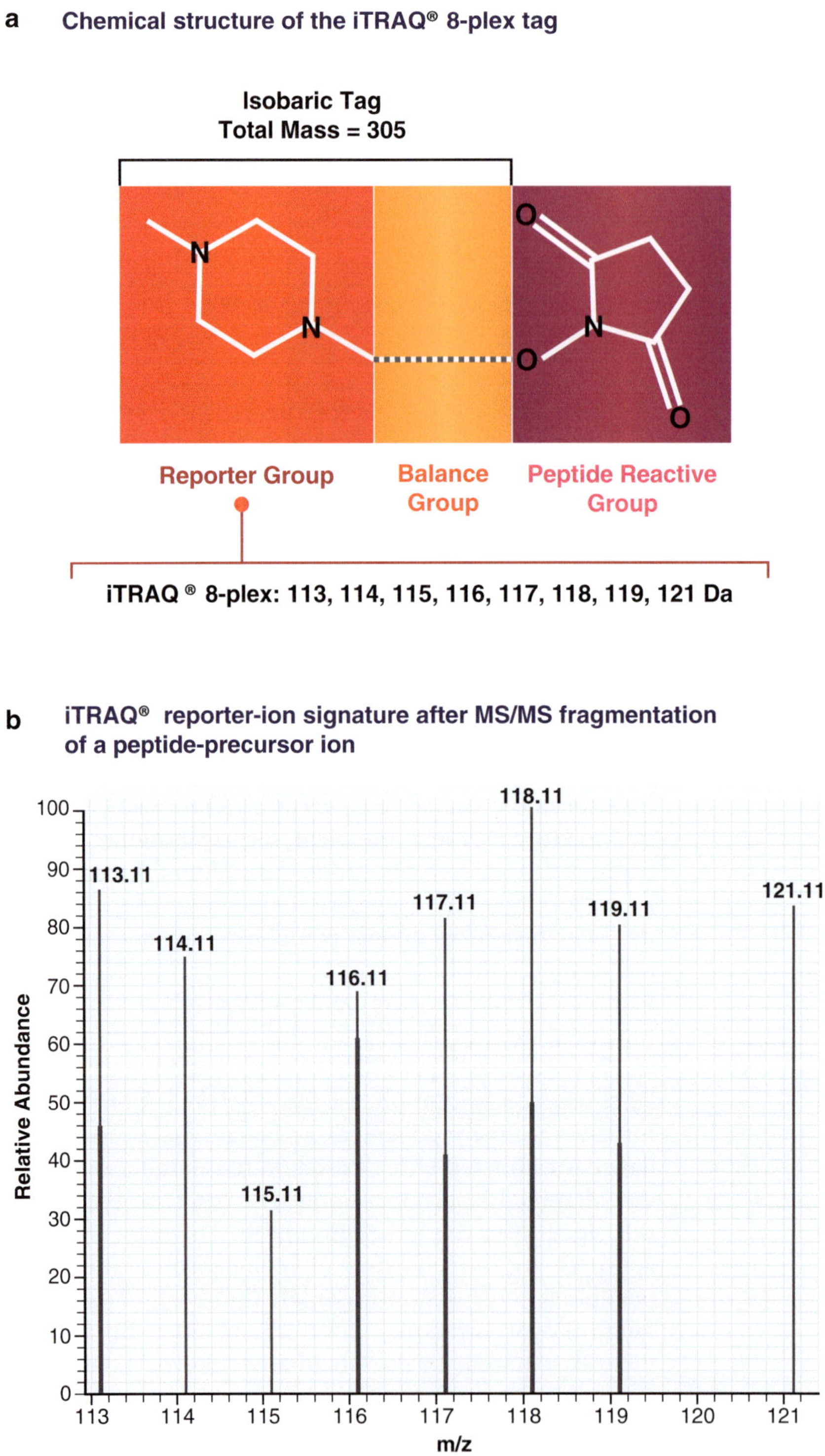

Fig. 1 Example of an iTRAQ® 8-plex experiment. iTRAQ® is an isobaric tag-based quantification method. (**a**) Chemical structure of a representative iTRAQ® 8-plex tag. Peptides are reacted with the isobaric label, which consists of a reporter and a balance group. Eight different reporter groups with specific masses (between 113 and 121 Da)

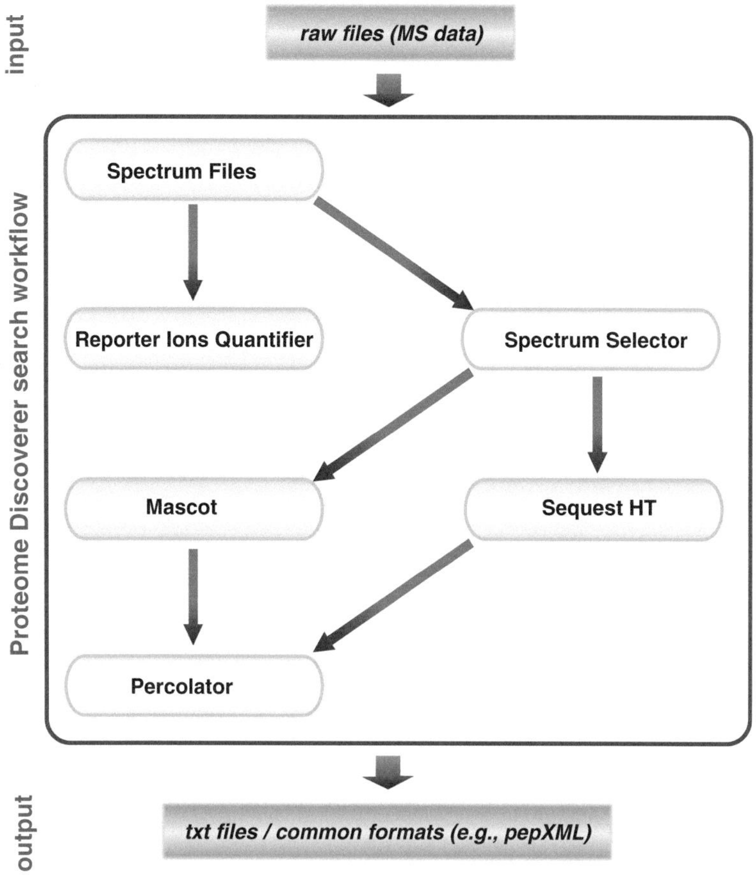

Fig. 2 Example of a Proteome Discoverer (PD) workflow: identification and quantification. The PD software allows for the graphical definition of processing workflows (*center box*): processing nodes are selected and connected according to the processing workflow. The workflow consists of the following nodes: spectrum file selection and merging, reporter ion quantification, spectrum selection, protein identification from spectrum data, and evaluation of identification probabilities. The settings in the respective nodes are as follows: spectrum selector (filter node): default; reporter ion quantifier node: iTRAQ® 8-plex; mass tolerance 20 ppm; and show raw-quan values option activated. Mascot and Sequest HT nodes (peptide search-engine nodes): enzyme used for protein digestion: trypsin; allowed missed cleavages: 2; parent mass tolerance: 10 ppm; fragment ion mass tolerance: 20 mmu; fixed modifications: methylthio, iTRAQ® 8-plex N-term, and lysine (K) residues; and variable modifications: methionine (M) oxidation and iTRAQ® modification of tyrosine (Y). Mascot Percolator node: high confidence 0.01 and medium confidence 0.05. MS spectra files (raw files) serve as the input; the analysis results are saved in a vendor-specific file format, and can be exported as text files or common file formats such as pepXML

Fig. 1 (continued) are available. These reporter groups are paired with corresponding mass balance groups so that all eight labels have the same (isobaric) total mass of 305 Da. Eight different samples can be reacted with the eight different labels. Subsequently, the labeled samples are pooled and analyzed in a multiplexed mass-spectrometry run. (**b**) Example of an iTRAQ® reporter-ion signature after MS/MS fragmentation of a peptide-precursor ion. The mass of each isobaric-labeled peptide is jointly detected in the mass spectrometer. The peptide is fragmented, which supports identification and simultaneously releases the reporter groups. As exemplified by the shown spectrum, the relative intensity differences of the eight reporter groups reflect the abundance differences of the selected peptide in the corresponding eight labeled samples. The data shown are an example from our own work

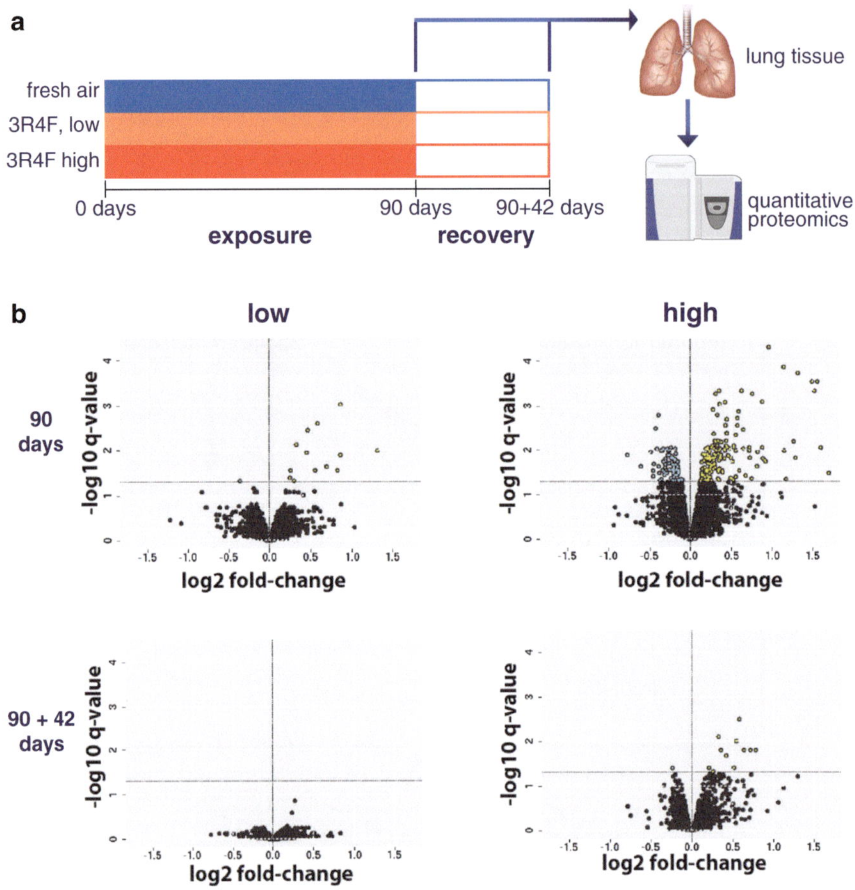

Fig. 3 Proteomic data analysis: impact of cigarette smoke exposure on the rat lung proteome. (**a**) Summary of rat inhalation study. Rats were exposed for 90 days to fresh air, or to a low (8 μg/L nicotine) or high (23 μg/L nicotine) concentration of reference cigarette smoke (3R4F) [89]. Following the exposure period, the rats were exposed to fresh air during a 42-day recovery period. Lung tissue samples were collected for the three exposure conditions at two time points in six replicates. Tissue lysates were processed for mass spectrometry analysis; derived peptides were labeled for isobaric quantification (iTRAQ®) and analyzed by high-resolution nanoLC-MS/MS. (**b**) Tobacco smoke exposure showed a strong overall impact on the lung proteome. The volcano plots show the log 2 fold change versus – log 10 of the FDR-adjusted *p*-values. Significantly up- or downregulated proteins are colored (adjusted p-value < 0.05). This figure is adapted from Titz et al. [1]

a balancer group in these labels adjusts them to the same total mass (irrespective of the reporter group mass). Thus, a peptide that has been labeled with different reporter groups shows the same mass (and retention time), and as a result pooled peptides from different samples are first selected and identified together in the MS instrument. In the peptide fragmentation step, the different reporter ions are then released and the relative quantification of the pooled samples is achieved by comparing the intensities of the reporter

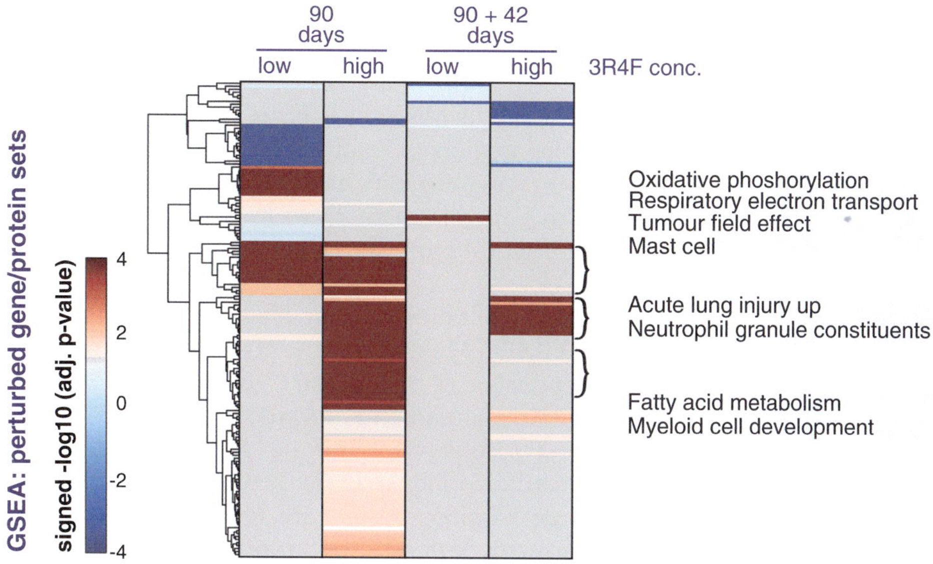

Fig. 4 Gene set enrichment analysis (GSEA) for a 90-day cigarette smoke inhalation study. The heatmap shows a concentration-dependent gene/protein set perturbation by cigarette smoke and a partial recovery after 42 days of fresh air exposure. The significance of association (−log 10 adjusted *p*-value) of up- (*red*) and down- (*blue*) regulated proteins with gene sets is shown. Gene sets enriched for proteins up-regulated by cigarette smoke exposure are highlighted for three different clusters. See Fig. 3a for the study design. This figure is adapted from Titz et al. [1]

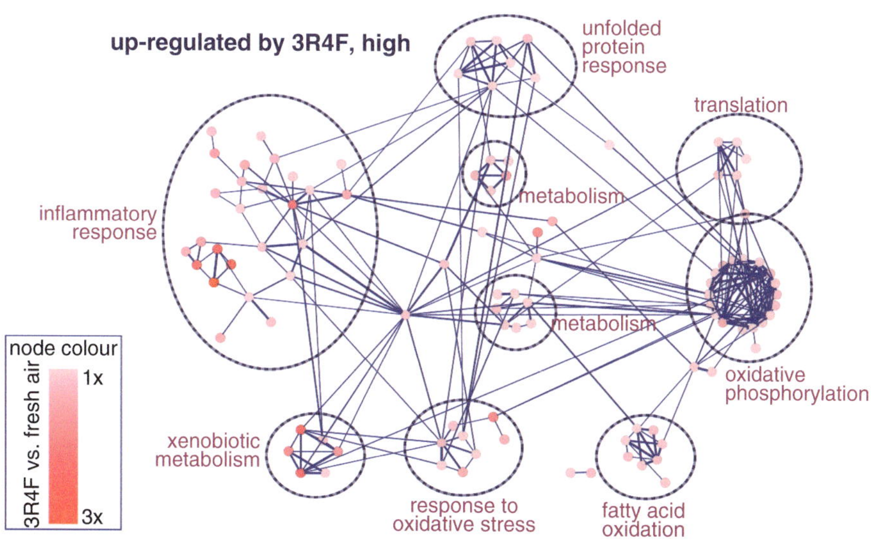

Fig. 5 Functional interaction network of significantly up-regulated proteins for a 90-day cigarette smoke exposure study. An example of a functional network analysis using the STRING database is shown. See Fig. 3a for the study design. The affected functional clusters include xenobiotic metabolism, response to oxidative stress, and inflammatory response. This figure is adapted from Titz et al. [1]

groups. The main advantages of isobaric tag-based quantification are the following: (1) simultaneous comparison of large numbers of samples (up to eight for iTRAQ®, up to ten for TMT®); (2) reduction of required MS runs (i.e., reduction of analysis time) because samples are pooled before MS analysis; and (3) low probability of introducing experimental errors during analysis because of pooling [15]. In this chapter, we focus on isobaric tag-based protein quantification using the iTRAQ® method.

2.1 Protocol for Quantitative MS (iTRAQ® 8-plex)

2.1.1 Materials

iTRAQ® 8-plex Multiplex reagent kit (AB Sciex, Framingham, MA, USA), iTRAQ® Multiplex Buffer kit (AB Sciex, content is sufficient for processing of two iTRAQ® 8-plex Multiplex Reagent kits), 1 M triethylammonium bicarbonate (TEAB), detergent removal spin columns with 100 μg peptide binding capacity (Thermo Scientific), and solid-phase extraction columns with 100 μg peptide binding capacity are required. Acetonitrile (MS grade), formic acid, SDS solution, trypsin (MS grade), urea, and MilliQ water are the reagents that are needed. For sample separation and analysis, a nanoflow LC system and a high-resolution/high mass-accuracy (HR/HM) mass spectrometer are required.

2.1.2 Description of the Process

The iTRAQ® protocol consists of three parts: (1) protein reduction, alkylation, and digestion; (2) peptide labeling and purification; and (3) sample analysis.

Protein Reduction, Cysteine Blocking, and Digestion

1. Transfer an aliquot of eight samples containing 50–100 μg total protein dissolved in resuspension buffer (0.5 M TEAB, 1 M urea, 0.1 % SDS) into eight 1.5 mL Eppendorf tubes. The protein amount and volume in all tubes have to be equal.
2. To each sample tube, add 2 μL reducing reagent provided in the iTRAQ® Multiplex Buffer kit.
3. Vortex to mix, and then spin.
4. Incubate the tubes at 37 °C for 1 h, and then spin.
5. To each tube, add 1 μL cysteine-blocking reagent provided in the iTRAQ® Multiplex Buffer kit.
6. Vortex to mix, and then spin.
7. Incubate the tubes at room temperature for 10 min.
8. Reconstitute a vial containing 100 μg of trypsin with 210 μL of resuspension buffer (final trypsin concentration: 0.5 μg/μL).
9. Vortex to mix, and then spin.
10. To each sample tube, add trypsin to a final amount of 1/10 of the total protein amount. Vortex to mix, and then spin.
11. Incubate the tubes at 37 °C overnight (12–16 h).
12. Spin to bring the sample digest to the bottom of the tube.

Peptide Labeling and Purification

1. Bring each vial of required iTRAQ® Reagent to room temperature (15 min at room temperature).
2. Add 50 μL of isopropanol provided in iTRAQ® Multiplex Buffer kit to each iTRAQ® Reagent vial.
3. Vortex each vial for 1 min to dissolve the iTRAQ® Reagent, and then spin.
4. Transfer the content of one iTRAQ® Reagent vial to the respective sample tube.
5. Add 25 μL of isopropanol to each tube.
6. Vortex each tube to mix, and then spin.
7. Incubate the tubes at room temperature for 2 h.
8. Combine the contents of all eight tubes belonging to each iTRAQ® set into a new 2 mL Eppendorf tube.
9. Vortex to mix, and then spin.
10. Dry samples in a SpeedVac (Thermo Scientific, Waltham, MA, USA) at 37 °C. Samples can be stored in the dry state at −20 °C.
11. Resuspend samples in 400 μL 0.5 M TEAB (final peptide concentration 1 μg/μL).
12. Remove storage liquid from detergent removal spin columns by centrifugation at 1,500 × *g* for 1 min.
13. Prepare detergent removal spin columns by adding 400 μL of 0.5 M TEAB. Centrifuge at 1,500 × *g* for 1 min. Repeat this step three times.
14. Load an aliquot of the resuspended sample containing 100 μg of peptides to a column. Incubate for 2 min. Centrifuge at 1,500 × *g* for 2 min. Collect flow through in a new Eppendorf tube.
15. Dry samples in a Speedvac.
16. Resuspend samples in 0.5 mL sample buffer (5 % acetonitrile, 0.1 % formic acid). Desalt samples via a solid-phase extraction cartridge that has been equilibrated to sample buffer prior to use according to the manufacturer's protocol.
17. Elute peptides by adding 2× elution buffer (60 % acetonitrile, 0.1 % formic acid). Collect eluate in a 1.5 mL Eppendorf tube.
18. Dry samples in a Speedvac.

Sample Analysis

1. Resuspend sample in nanoLC buffer A (5 % acetonitrile, 0.1 % formic acid in MilliQ) to a final peptide concentration of 0.5 μg/μL.
2. Analyze 1 μg of peptides using a nanoLC high-resolution/high mass accuracy mass spectrometer setup (LC-MS). We used the Easy nanoLC 1000 connected to a Q-Exactive mass spectrometer (Thermo Scientific). Samples were loaded to a

50-cm-long C_{18} reversed-phase nanoLC column (Thermo Scientific) and eluted by applying an acetonitrile gradient (ACN) from 5 % ACN, 0.1 % formic acid in MilliQ to 40 % ACN, and 0.1 % formic acid in MilliQ over 200 min. Each sample was analyzed in fast and sensitive instrument acquisition mode according to Kelstrup et al. [16].

Notes

1. Different solutions with different multiplexing capabilities are available. AB Sciex offers an iTRAQ® 4-plex and 8-plex kit (for the protocol provided here we used the 8-plex kit). Thermo Scientific offers Tandem Mass Tag (TMT®) 2-plex, 6-plex, and 10-plex kits. Depending on the experimental design, the appropriate kit can be chosen. The overall principle is the same for all solutions.
2. Before starting the procedure samples should be tested for protein integrity (e.g., by one-dimensional polyacrylamide electrophoresis).
3. Protein and peptide dissolution efficiency is increased when samples are treated for 10 min in an ultrasound bath at room temperature.
4. After each drying step, peptides can be stored at −20 °C until further processing.
5. Randomly check pH of the reaction mixture by pipetting 1 μL of the mix on top of pH paper (pH range has to cover 5–10). The pH has to be above 8.0. If the pH is too low, add 5 μL resuspension buffer to the mix. Check pH again.
6. To increase the number of protein identifications and to reduce an effect, described as precursor co-isolation, that influences the accuracy of quantification, samples can be fractionated before LC-MS analysis. Different matrices such as ion-exchange columns or reversed-phase material allow peptide fractionation according to peptide hydrophobicity, charge, or a combination of both.

3 Identification of Peptides and Proteins

3.1 Processing Workflow Packages

After the raw MS data are generated, the next steps are peptide identification and quantification. Several software packages that support these tasks are available, including several freely available software solutions (Table 1). While we exemplify the data analysis with the Proteome Discoverer software (Thermo Scientific) in this chapter, we would advise evaluating the different available software solutions for the specific needs of a study.

Table 1
Proteomic workflow software packages

Tool	License	Input format	Identification	Quantification	Comment	Reference/link
Trans-Proteomic Pipeline	Open source	.mzXML (others via conversion)	Sequest, Mascot, X!Tandem, SpectrumST, etc.	Isobaric tag, SILAC	Comprehensive suite of tools for MS-based proteomics	www.tools.proteomecenter.org [90]
CPAS	Open source	.mzXML	X! Tandem, Sequest, Mascot	Isobaric tag, spectral counting	MS raw data processing	www.labkey.org [91]
OpenMS	Open source	.mzML/.mzXML (others via conversion)	Sequest, Mascot, X!Tandem, SpectrumST, etc.	SILAC, isobaric tag, label-free	Flexible workflows for MS raw data processing	www.openms.de [17]
MaxQuant	Freeware, closed source	.RAW (Thermo)	Andromeda	SILAC, isobaric tag, label-free	Integrated package for quantitative proteomics analysis	www.maxquant.org [92]
Proteome Discoverer	Commercial	.RAW (Thermo)	Sequest, Mascot, X!Tandem, SpectrumST, etc.	SILAC, isobaric tag, (label-free)	Analysis software for Thermo MS instruments	www.thermoscientific.com

CPAS computational proteomics analysis system, *MS* mass spectrometry, *SILAC* stable isotope labeling by amino acids in cell culture

With the rapidly evolving MS technology, an important consideration is how adaptable the selected software solution is. For example, OpenMS provides a flexible, modular analysis framework that can be easily adapted for specific needs [17]. In OpenMS, specific processing modules can be generated using a scripting language interface and modules can be integrated in different workflows using the KNIME workflow manager [18, 19]. In addition, hybrid solutions are possible. For example, in our own workflow, we are using Proteome Discoverer for peptide/protein identification and basic reporter ion quantification. However, all subsequent processing steps (e.g., normalization, protein collapsing) are done within the open-source R statistical environment [20], which allows for more flexible extension of all the processing steps.

Another important consideration is how well the employed quantification approach is supported by the software. Nahnsen et al. provide a software survey for label-free quantification [21] and Leemer et al. provide both label-free and label-based quantification tools [22]. Again, custom scripts (e.g., in the R language) can compensate for limitations of the available tools (e.g., by providing additional normalization methods).

3.2 Spectra Matching Algorithms

The output from each MS run is a file with spectra of the peptide ions (MS1) and their selected fragments (MS2). The next step is the identification of peptides based on this spectral information. The most commonly used approach involves matching between the measured and theoretical peptide fragmentation spectra. For this, a relevant protein database is selected (e.g., a human proteome database), the proteins are digested in silico based on the cleavage specificity of the protease used during the actual sample digestion step (e.g., trypsin), and for each computationally derived peptide, a theoretic MS2 fragmentation spectrum is calculated. Taking the measured (MS1) precursor mass into account, each measured spectrum in the datasets is compared with the theoretical spectra of the proteome, and the best match is identified.

The most commonly employed spectra-matching algorithms are the commercial Sequest [23] and Mascot [24] algorithms and the open-source X!Tandem [25] and OMSSA [26] algorithms. Importantly, each derived spectrum to peptide match is associated with a score that reflects the match quality (e.g., a cross-correlation score [27]). Because these scores do not have a direct meaning, they need to be converted into probability *p*-values. After multiple testing correction, these probabilities can then be used to control for the false discovery rate (FDR) of the identifications (often at the 1 or 5 % level). For this, a commonly used approach is to compare the obtained identification scores for the actual analysis with results obtained for a randomized (decoy) protein database [28]. For example, this approach is used in the Percolator algorithm [29, 30], which is available through the Proteome Discoverer software. Another important consideration is how peptide identifications are converted into protein identifications. Proteome Discoverer takes a heuristic approach that assigns peptide identifications to protein groups based on their overlap.

4 Quantification and Identification of Differentially Abundant Peptides and Proteins

Following the identification process, quantitative values are derived for the identified peptides/proteins. As mentioned, several quantification methods are available, but here we focus on the isobaric tag-based quantification with iTRAQ® labels (Fig. 1). The Proteome Discoverer software that we use in our example workflow includes a dedicated module for isobaric tag-based peptide quantification (Fig. 2). Briefly, the reporter ion intensities are first extracted from the MS2 spectra (region between 113 and 121 Da). Peptide spectrum ratios are calculated (commonly, to a predefined reference sample), and then protein ratios are calculated as the median of their peptide spectrum ratios. Finally, the median of the protein ratios of each sample is normalized to 1 to correct for experimental bias.

Beyond this standard implementation, several adaptations are possible. In our own workflow, we use Proteome Discoverer to extract only the raw reporter ion intensities. We then conduct the further data processing in the R statistical environment. This allows us, for example, to use and evaluate the following more recently reported adaptations.

Different normalization methods can be used [31]. Because the optimal normalization method is dataset specific, comparisons of different normalization methods, as supported by the Normalizer tool, can be beneficial [32]. For example, we are using variance-stabilizing normalization [33, 34], which can compensate for the higher variance of lower intensity ions in iTRAQ® datasets.

Another question is how to cope with the ratio compression effect that has been observed for isobaric tag data. The ratio compression term refers to the observation that protein expression ratios measured by isobaric approaches are generally lower than expected. This effect has been explained by the co-isolation of other labeled peptide ions with similar parental mass. Because these co-isolated peptides tend not to be differentially regulated, they generate a common reporter ion background signal that decreases the ratios calculated for any pair of reporter ions. Approaches to cope with this phenomenon computationally include filtering out spectra with a high percentage of co-isolated peptides (e.g., above 30 %) [35]. Another approach attempts to directly correct for the measured co-isolation percentage [36]. In our own tests, this correction approach results in more accurate ratios (i.e., reduces the bias), but at the same time it increases the variance of the measurements. Thus, it will depend on the scientific objective whether this correction is beneficial.

Finally, the inclusion of a common reference sample is a standard procedure for isobaric tag quantification. The main idea is to express all measured values as ratios to the common reference sample so as to cancel out differences in ionization efficiencies and between sample runs. However, recently it has been demonstrated that this reliance on a single sample can increase the overall variance and that alternatively, it is beneficial to use the median of all measured reporter ions for spectrum normalization [37]. In support of this finding, in our studies we commonly detect a larger number of differentially expressed proteins without normalization to a common reference.

4.1 Peptide Identification and Quantification with Proteome Discoverer

4.1.1 Materials

Proteome Discoverer with Quantification option (Thermo Scientific), and a 64-bit workstation with minimum 16 GB RAM are used. Access to Mascot (Matrix Science, Boston, MA, USA) for protein database searches and organism-specific protein reference databases (e.g., reference proteomes available at www.uniprot.org) is required. A workflow containing the required tools to perform protein identification, evaluation of the identification results, and protein quantification is also needed (Fig. 2).

Protocol

1. Merge the spectra acquired in fast and sensitive analysis mode of each sample by loading the raw files into the Proteome Discoverer workflow (Fig. 2) using the "Spectrum Files" node.
2. Choose an appropriate protein reference database in the "Mascot" and "SequestHT" node (see Notes).
3. After the start of the analysis workflow, spectra will be searched automatically against the chosen protein reference database using Mascot and SequestHT (detailed search settings are provided in the legend of Fig. 2). Peptide identification probability is evaluated with the Percolator tool. Reporter ion intensities from each sample in the iTRAQ® 8-plex set are extracted from the "Reporter ions quantifier" node.
4. After finalization of the search, open the respective results file in Proteome Discoverer using medium peptide confidence settings and one peptide as the minimum protein-identification requirement in the results filter tab.
5. Export the data as tab-delimited text file without additional filtering. The resulting file will be used for further processing steps.

Notes

1. When more than one search engine is used for protein identification, search settings and the chosen protein reference database have to be identical in the settings of the different engines.

4.2 Identification of Differentially Expressed Proteins

The next common analysis step is the identification of differentially expressed proteins. Here, important considerations involve the selection of the protein-level statistics for differential abundance and how multiple hypothesis testing is taken into account. For example, Ting et al. tested a fold change approach, Student's *t*-test, and empirical Bayes moderated *t*-test as the protein-level statistics [38]. In addition, they used an approach that is commonly used in RNA microarray experiments to construct linear models that captured the relevant experimental factors (LIMMA model) [39]. They concluded that applying the empirical Bayes moderated *t*-test in the linear model framework resulted in a high-quality list of statistically significant differentially abundant proteins. Finally, the derived *p*-values need to be adjusted for multiple testing [40] and for this the Benjamini–Hochberg approach is most commonly used [41].

In principle, spreadsheet programs such as Excel can be used, but statistical software such as the R environment is highly beneficial [20]. Box 2 demonstrates how an R script can be used to conduct a basic LIMMA analysis for the identification of differentially abundant proteins in a simulated dataset. Figure 3 shows an example differential expression result from a 90-day rat inhalation toxicology study (see Box 1 for study background). A good description of the LIMMA package for more complex cases with multiple contrasts is given by Gillespie et al. [42].

5 Protein Set Enrichment Analysis

Beyond the simple manual interpretation of differentially abundant protein lists, it is often pertinent to ask whether any functional categories are significantly perturbed. The basis for this analysis is the modular organization of biological systems [43–45]. For example, under a certain cellular stress we can expect to observe the coordinated up-regulation of a specific stress response protein module or the activation of a particular signaling pathway [46].

Commonly, for protein set analysis the same algorithms as for gene sets are used. These algorithms consist of three main components: (1) a metric to score the level of perturbation for each protein (protein-level statistic); (2) a database of relevant protein modules/sets; and (3) an algorithm to score and evaluate the statistical significance of the module enrichment (module-level statistic).

5.1 Protein-Level Statistics

The protein-level statistic captures the level of deregulation of a given protein. Threshold-based approaches can, for example, be based on a multiple testing corrected *t*-test *p*-value (e.g., comparing a toxin-exposed group with a control group). Threshold-free approaches rely on continuous protein-level statistics such as fold

Box 2

Example R script for the identification of differentially expressed proteins (DEPs) with LIMMA models [39]. This script exemplifies the use of the R statistical environment for the identification of DEPs. Quantitative data are either read from a file or example data are generated as a mixture of background and differentially expressed proteins. The data are median centered ("normalized") and p-values from the empirical Bayes moderated t-test are calculated. Finally, the p-values are adjusted using the Benjamini–Hochberg approach

LimmaExample.R

```
####################################
## read data
####################################
data_DEP = read.csv("my_filename.csv") #comma-separated table with ratio expression data (sample groups A + B)
data_DEP = log2(data_DEP)
n_samplesPerGroup = 6 #number of samples per group
group_labels = as.factor(c(rep("A", n_samplesPerGroup), rep("B", n_samplesPerGroup))) #define sample groups: A and B
####################################
## Alternatively, generate example data
####################################
n_proteins = 1000 #number of proteins
n_samplesPerGroup = 6 #number of samples per group
n_deps = 20 #number of differentially expressed proteins
#generate normal distributed background data
data = matrix(rnorm(n_proteins*n_samplesPerGroup * 2, mean = 0, sd = 0.2), nrow = n_proteins, ncol = n_samplesPerGroup * 2)
rownames(data) = 1:n_proteins
#define sample groups: A and B
group_labels = as.factor(c(rep("A", n_samplesPerGroup), rep("B", n_samplesPerGroup)))
#simulate differentially expressed proteins (increased abundance in sample group B)
data_DEP = data
for (i in 1:n_deps) {
dep = t(as.matrix(c(rnorm(n_samplesPerGroup, mean = 0, sd = 0.2),rnorm(n_samplesPerGroup, mean = i/10, sd = 0.2))))
rownames(dep) = paste0("DEP",i)
data_DEP = rbind(data_DEP, dep)
}
####################################
## median sample normalization
####################################
medians = apply(data_DEP, 2, median)
data_DEP_norm = t(apply(data_DEP, 1, function(x) x - medians))
####################################
## LIMMA model
####################################
library(limma)
model.formula = " ~ 1 + group_labels"
lm.fit.res = lmFit(data_DEP_norm, model.matrix(as.formula(model.formula)))
lm.fit.res.eb = eBayes(lm.fit.res)
adj.p.value = p.adjust(lm.fit.res.eb$p.value[,2], method = "BH")
sort(adj.p.value, decreasing = FALSE) [1:20]
```

change, *t*-score, or signal-to-noise ratio. An important consideration for the selection of the protein-level statistics is whether up- and downregulation of module components are considered together or as distinct effects [47].

5.2 Protein Set Databases

Several gene set databases that can also be used for protein set analysis are available. The most prominent is the mSigDB database from the Broad Institute [48]. Here, the functional sets are grouped in different categories that range from canonical pathways (e.g., the Kyoto Encyclopedia of Genes and Genomes [KEGG] and Reactome databases) to gene ontologies [49–51]. Another popular database is GeneSigDB [52], which contains signatures that have been extracted manually from the published literature. In addition, gene ontology annotations are often used for set enrichment analyses [53].

5.3 Three Classes of Enrichment Algorithms

Finally, algorithms to evaluate the module enrichment are required. These algorithms can be grouped into three categories: (1) overrepresentation analysis (ORA) approaches; (2) functional class scoring (FCS) approaches; and (3) pathway topology (PT) approaches [54] (Table 2).

Briefly, ORA approaches depend on a threshold of the protein statistic to define a list of differentially expressed proteins. Subsequently, the overlap between the differentially expressed proteins and each set in the protein set database is statistically assessed (e.g., using Fisher's exact test and multiple hypothesis correction). The most prominent FCS approach is gene set enrichment analysis (GSEA) [55]. For this, the proteins are ranked based on a

Table 2
Protein set enrichment approaches

Approach	Principle	Advantages	Disadvantages	Examples
Overrepresentation analysis	Threshold-based statistical assessment of overlap between protein sets	Simplicity, quick run time, availability	Fixed threshold	DAVID Bioinformatics Resources [56], Enricher tool [93]
Functional class scoring	Statistical evaluation of protein set statistic (e.g., ranking statistic)	No fixed threshold, correlation structure can be taken into account	Topology not taken into account	GSEA [55]
Pathway topology	Scoring of consistence with pathway topology	Causal relationships taken into account	Rely on high-quality pathway models	SPIA [94], NPA [95]

GSEA gene set enrichment analysis, *NPA* network perturbation amplitude, *SPIA* signaling pathway impact analysis continuous protein-level statistic (such as fold change) and the enrichment of the sets in the database at the top or the bottom of the ranked list is statistically evaluated. The PT approaches take the actual topology of the evaluated biological pathways into account. In the following protocols, we demonstrate the use of the ORA (using David Bioinformatics [56]) and FCS (using GSEA [55]) approaches. An example GSEA result for a 90-day cigarette smoke exposure study is shown in Fig. 4.

5.4 Protocols for Protein Set Enrichment Analysis

The David Bioinformatics resources provide comprehensive support for ORA analysis through a convenient web interface. Detailed analysis protocols for gene expression data have been published by Huang et al. [56].

5.4.1 ORA with David Bioinformatics

1. Select a list of differentially regulated proteins. For example, this could be the list of all up-regulated proteins under a certain condition with an adjusted *t*-test *p*-value less than 0.05. Then, for example, a list of UniProt IDs or accession numbers is accepted. Official gene symbols are also accepted, but they require additional conversions and do not support the generation of long background lists.
2. Go to david.abcc.ncifcrf.gov/tools.jsp and select the "Upload" tab. Paste the protein list into the corresponding field. Select the protein identifier, e.g., "UNIPROT_ACCESSION." Select "gene list" as the "list type" and submit the query.
3. If not all identifiers are mapped, proceed by either submitting only the mapped identifiers or submit the partially mapped list to the conversion tool.
4. If requested, select the correct species.
5. Select the list of background proteins (see Notes). As noted above (step 1), gene symbols are not fully supported for the creation of long background lists. Select the "upload" tab and paste the identifiers of the background proteins. Select the identifier type and select "background" as the "list type" and submit the list. If not all identifiers are mapped, proceed by either submitting only the mapped identifiers or submit the partially mapped list to the conversion tool.
6. Select "Functional Annotation Tool" and proceed with "Functional Annotation Clustering."
7. The clustered enrichment result is shown. Overlapping sets are grouped to reduce redundancy. The clusters are ordered by the geometric mean (–log scale) of the *p*-values within a cluster, which is thought to rank their biological significance. The "Options" dialog allows the display of additional statistics such as the FDR and it is important to evaluate the statistical significance of the considered protein sets (e.g., based on multiple testing corrected *p*-values).

8. The results can be analyzed further by exploring the contributing proteins and their shared annotations.
9. "Download File" saves a copy of the results as a text file.

5.4.2 FCS with GSEA

1. Prepare input file with a pre-ranked protein list (rnk file):
   (a) First, the ranking statistic needs to be selected. For example, this can be the (log) fold change between the study groups or a *t*-test value. It is recommended to create a non-redundant list with a single entry for each protein. If duplicate entries for the same protein are maintained through the quantification procedure, they can, for example, be collapsed to the entry with the largest absolute value (e.g., the largest absolute fold change).
   (b) Save the list of the quantified proteins and their associated ranking statistics in the rnk file format. For example, Excel can be used for this purpose. Human gene symbols should be used as the protein identifiers. The rnk format has two columns that are separated by a tab character. The first column lists the (human) gene symbols, and the second column the corresponding (numeric) ranking statistic (http://www.broadinstitute.org/cancer/software/gsea/wiki/index.php/Data_formats#Ranked_Gene_Lists).
   (c) Save the file in text format and change the file ending to .rnk.
2. Go to www.broadinstitute.org/gsea/index.jsp and select "Download the GSEA software." Signup or log in. Launch the "javaGSEA Desktop Application" (up to 4 GB of memory can be selected for a 64-bit JAVA version).
3. Select "Load data" and drag the created rnk file onto the "Method 3" panel. Click "load these files."
4. Select "Tools/GseaPreranked" in the menu.
5. Select a gene set database with a human gene symbol annotation (e.g., "c2.all.v4.0.symbols.gmt").
6. Select the uploaded ranked list.
7. Collapse dataset to gene symbols: set to FALSE.
8. When in doubt whether the ranking statistics works with the weighted scoring scheme, set the *enrichment statistic* to *classic*.
9. Click "run" and wait for the analysis to finish.
10. Once the analysis is done, the status is changed to "success" in the "GSEA reports" section. The results can be accessed by clicking on the "success" status indicator and viewed in a web browser.

Notes

1. For ORA approaches, the selection of the protein universe/background is critical. The statistical enrichment is evaluated with respect to this background protein set. For proteomics experiments, which usually show a bias toward abundant proteins, it is advisable to use the set of detected and quantified proteins as the background.
2. The main challenge for the interpretation of results from these methods is often the high overlap (and partial redundancy) of the protein modules. Enrichment maps have been presented as a solution to better visualize enriched gene sets/modules and their relationships [57]. Here, the enrichment results are presented as a similarity network, where the nodes represent gene sets, the links represent the overlap between sets, and the node color encodes the enrichment strength.
3. It is beneficial to conduct the protein set enrichment analyses within the R statistical environment. For example, the piano package for R provides several FCS methods (including GSEA) and supports a basic ORA analysis [47]. For example, the topGO package supports gene ontology enrichment analysis [58].

6 Deriving Insights Through Network Analyses

Network biology approaches can highlight highly perturbed protein subnetworks that warrant further investigation [59]; they help to understand the modular organization of the cell [60], and can be applied for improved diagnostics and therapies [61, 62].

6.1 Biological Network Models

The STRING database is one of the most comprehensive, freely available databases of functional protein-protein links for a broad range of species [63]. It is based on a probabilistic model that scores each link based on its experimental or predicted support from diverse sources such as physical protein interaction databases, text mining, and genomic associations. The Reactome database is a manually curated database with a narrower scope of human canonical pathways [64]. Recently, Reactome has been supplemented with predicted functional protein associations from several sources including protein–protein interaction databases and co-expression data (Reactome Functional Interaction Network) [65]. In addition, several commercial curated pathway/network databases exist including KEGG, the Ingenuity® Knowledge Base, and MetaCore® [49, 66–68]. Moreover, for systems toxicology applications, we have developed and published a collection of specific mechanistic network models [69]. These models range from xenobiotic to oxidative stress, to inflammation-related, and to cell cycle models [70–73]. The networks are described in the Biological

Expression Language, which enables the development of computable network models based on cause-and-effect relationships [74].

6.2 The Cytoscape Platform

Although the existing approaches for network analyses can be overwhelming, they are helped by the availability of the common network analysis software platform Cytoscape [75]. At its core, Cytoscape allows for import, annotation, visualization, and basic analysis of molecular interaction networks. However, its functionality is significantly expanded by the many available plug-ins/apps (see Saito et al. for a "travel guide" [76]).

6.3 Functional Context Networks

To understand the biology altered under a specific condition, it is often helpful to visualize and analyze how the differentially expressed proteins are functionally connected and whether they form specific functional clusters. Several Cytoscape® apps support the integration of protein lists with different functional network resources. For example, the Reactome Functional Interaction (Reactome FI) plug-in allows construction of a subnetwork of the extensive Reactome FI Network for a given set of genes/proteins [65]. Similar functional network analyses are also a central component of the commercial Ingenuity and MetaCore analysis tools [66–68]. Here, we present a step-by-step protocol, in which we use the STRING database and Cytoscape for the discovery of functional network clusters.

6.4 Step-by-Step Procedure for a Functional Network Analysis with the STRING Database

This procedure supports the identification of functionally associated protein clusters, e.g., for a list of differentially expressed proteins. A similar procedure has been used by us to identify functional protein clusters that are affected by 90-day cigarette smoke exposure in rat lung tissue (Fig. 5). This analysis can yield a very compact representation of the affected biology under a certain condition.

1. Select the protein list of interest, e.g., the list of significantly up-regulated proteins. For this, a list of protein identifiers is needed and, generally, the official gene symbols work well.
2. Navigate to string.embl.de and open the "multiple names" tab. Paste the protein identifiers into the text field, select the organism, and click "go."
3. The result of the identifier mapping is displayed. Check the mappings and correct if necessary. Continue with the analysis.
4. The identified functional associated network is displayed. All mapped proteins are shown as the nodes of the network, and these are connected if (at least medium confidence) evidence for a functional link between two proteins is available.
5. At this point, the results can be further explored. For example, clicking on a functional link brings up a page with more detailed information on this link (e.g., the different supporting evidence sources). In addition, basic analysis options such

as clustering and enrichment analyses are accessible via the menu.

6. If the displayed network is rich enough for further analysis, proceed by exporting the network information. Go to the "Save/Other formats" menu option and save the tab-delimited "text summary." This file contains the information on all edges of the network.
7. Download and install the Cytoscape package from www.cytoscape.org. Run Cytoscape and install the "ClusterONE" [77] and "Bingo" [78] apps.
8. Import the file with the String DB network: select the menu item "File/Import/Network from Table," select the input file and the gene symbol columns (1 and 2), and import the network (skip the header).
9. Select a layout; for example, select the menu item "Layout/Cytoscape Layouts/Spring embedded."
10. Start the ClusterONE app: select menu item "Plugins/ClusterONE/Start." Keep the default parameters and start the cluster identification algorithm.
11. Identify functional clusters based on the clusters identified in the results panel.
12. Use the Bingo app or a Web-based enrichment tool such as David Bioinformatics (see above) to identify the core functionality of each defined cluster.

Notes

1. For a more automated analysis, the STRINGdb package for the R statistical environment is available [63]. This allows for the scripted identifier conversion and functional network extraction from the STRING database. In addition, this package supports identification of protein clusters and their functional annotation.

6.5 Activated Subnetworks

Whereas functional context networks are built from the bottom up, other approaches start with the entire biological network and aim to identify network regions (subgraphs) with a significant enrichment of differentially expressed proteins. The benefit of these methods is that they do not necessarily rely on strict cutoffs for the definition of protein lists; they can take more of the global topology of the network into account and be evaluated more globally for statistical significance. The first such algorithm, which is still in common use and available as a Cytoscape plug-in (jActiveModules), was published by Ideker et al. [59]. In a recent example, this algorithm has been applied to identify activated subnetworks upon early life exposure to mitochondrial genotoxicants [79].

Table 3
Algorithms for the detection of activated subnetworks

Algorithm	Comment	Availability	Reference
jActiveModules	Aggregate Z-scores to derive subnetwork score. Heuristic search algorithm to identify high-scoring subnetworks	Cytoscape app	[59]
Heinz/BioNet	Subnetwork scoring function based on a signal-noise mixture model. Linear programming for the identification of optimal subnetworks	R package	[96, 97]
GeNA	Approach optimized for large-scale weighted networks	Cytoscape app	[98]
NetWalk	Subnetwork identification based on data-biased random walks	Software (for noncommercial purposes)	[99, 100]
Walktrap-GM	Random-walk-based subnetwork identification with data-derived edge weights	R code	[101]
DEGAS	Identifies minimal subnetworks of proteins dysregulated in a selected number of cases	Software package	[102]
NetWeAvers	Significance scoring of predefined network clusters	R package	[103]
COSSY	Identification of subnetworks. Ranking by clustering approach. Optimized for RNA microarray results	R package	[104]
DIRAC	Quantitatively measures how network ordering differs within and between phenotypes	Matlab code	[105]

Several related methods have been developed more recently and are commonly available as Cytoscape apps, for the R statistical environment, or as stand-alone programs (Table 3).

7 Deriving Insights Through Data Integration

Integration of different datasets and data modalities (e.g., transcriptomics and proteomics data) can yield a more comprehensive picture and build up confidence in the obtained results. Different data modalities can be integrated at different biological levels ranging from individual molecules (proteins/transcripts) to functional modules, to biological networks. The most basic approach directly compares lists of differentially expressed gene/protein lists (e.g., [80, 81]). The advantage of directly integrating *p*-values is that consistently but weakly regulated molecules across datasets can be identified (e.g., using the Fisher [82] or Stouffer methods [83]). Integration at the functional module (pathway, gene set) level can reveal similarities that are not apparent at the level of individual biomolecules [84]. Finally, data can be integrated at the biological network level [170]. Moreover, different integration methods can be combined. For example, Balbin et al. generated transcriptomics, proteomics, and phospho-proteomics profiles for KRAS-mutated non-small-cell lung cancer cell lines [85]. These three omics data sources were first integrated using an approach based on the Stouffer method. Subsequently, a network-based approach was used to identify a novel targetable subnetwork in KRAS-dependent CRC cell lines centered on the LCK kinase.

8 Summary

MS-based proteomics methods are evolving rapidly toward higher sensitivity, higher throughput, higher coverage, and highly accurate quantification, and thus will constitute a central component of future integrative systems toxicology approaches [86]. Although there remain challenges to making full use of the richness of proteomic datasets [10, 87, 88], overall there is great opportunity for proteomics to contribute to an improved understanding of toxicant action, the linkages to accompanying dysfunction and pathology, and the development of predictive biomarkers and signatures of toxicity [1]. Here, we have provided a practical overview of the data generation and analysis steps needed for the quantitative analysis of toxico-proteomic data. However, this chapter can only serve as a starting point. Because of the richness of proteomic datasets, especially when combined with other data modalities, more extensive and specifically tailored data analysis approaches will be necessary to fully understand the biological impact of a toxicant.

Acknowledgment

The research described in this chapter of the book was funded by Philip Morris International.

References

1. Titz B, Elamin A, Martin F et al (2014) Proteomics for systems toxicology. CSBJ 11(18):73–90
2. Rowlands JC, Sander M, Bus JS (2014) FutureTox: building the road for 21st century toxicology and risk assessment practices. Toxicol Sci 137(2):269–277
3. National Research Council. Committee on Toxicity T (2007) Assessment of environmental A. Toxicity testing in the 21st century: a vision and a strategy. National Academies Press, Pittsburgh, PA
4. Sturla SJ, Boobis AR, FitzGerald RE et al (2014) Systems toxicology: from basic research to risk assessment. Chem Res Toxicol 27(3):314–329. doi:10.1021/tx400410s
5. Vogel C, Marcotte EM (2012) Insights into the regulation of protein abundance from proteomic and transcriptomic analyses. Nat Rev Genet 13(4):227–232. doi:10.1038/nrg3185

6. Spriggs KA, Bushell M, Willis AE (2010) Translational regulation of gene expression during conditions of cell stress. Mol Cell 40(2):228–237
7. Hardie DG, Ross FA, Hawley SA (2012) AMPK: a nutrient and energy sensor that maintains energy homeostasis. Nat Rev Mol Cell Biol 13(4):251–262
8. Shiloh Y, Ziv Y (2013) The ATM protein kinase: regulating the cellular response to genotoxic stress, and more. Nat Rev Mol Cell Biol 14(4):197–210
9. Runchel C, Matsuzawa A, Ichijo H (2011) Mitogen-activated protein kinases in mammalian oxidative stress responses. Antioxid Redox Signal 15(1):205–218
10. Merrick BA, Witzmann FA (2009) The role of toxicoproteomics in assessing organ specific toxicity. EXS 99:367–400
11. Mann M, Kelleher NL (2008) Precision proteomics: the case for high resolution and high mass accuracy. Proc Natl Acad Sci U S A 105(47):18132–18138
12. Mallick P, Kuster B (2010) Proteomics: a pragmatic perspective. Nat Biotechnol 28(7):695–709, doi:nbt.1658 [pii] 10.1038/nbt.1658
13. Yates JR III, Mohammed S, Heck AJ (2014) Phosphoproteomics. Anal Chem 86(3):1313
14. Wiese S, Reidegeld KA, Meyer HE et al (2007) Protein labeling by iTRAQ: a new tool for quantitative mass spectrometry in proteome research. Proteomics 7(3):340–350
15. Zieske LR (2006) A perspective on the use of iTRAQ reagent technology for protein complex and profiling studies. J Exp Bot 57(7):1501–1508, doi:erj168 [pii] 10.1093/jxb/erj168
16. Kelstrup CD, Young C, Lavallee R et al (2012) Optimized fast and sensitive acquisition methods for shotgun proteomics on a quadrupole orbitrap mass spectrometer. J Proteome Res 11(6):3487–3497
17. Sturm M, Bertsch A, Gröpl C et al (2008) OpenMS – an open-source software framework for mass spectrometry. BMC Bioinformatics 9(1):163
18. Röst HL, Schmitt U, Aebersold R et al (2014) pyOpenMS: a python-based interface to the OpenMS mass-spectrometry algorithm library. Proteomics 14(1):74–77
19. Berthold MR, Cebron N, Dill F et al (2008) KNIME: The Konstanz information miner. Springer, New York, NY
20. Team RC, Team RC (2005) R: a language and environment for statistical computing. R Foundation for Statistical Computing, Vienna
21. Nahnsen S, Bielow C, Reinert K et al (2013) Tools for label-free peptide quantification. Mol Cell Proteomics 12(3):549–556
22. Lemeer S, Hahne H, Pachl F et al (2012) Software tools for MS-based quantitative proteomics: a brief overview. Quantitative methods in proteomics. Springer, New York, NY, pp 489–499
23. Eng JK, McCormack AL, Yates JR (1994) An approach to correlate tandem mass spectral data of peptides with amino acid sequences in a protein database. J Am Soc Mass Spectrom 5(11):976–989, doi:10.1016/1044-0305(94)80016-2
24. Cottrell JS, London U (1999) Probability-based protein identification by searching sequence databases using mass spectrometry data. Electrophoresis 20(18):3551–3567
25. Craig R, Beavis RC (2004) TANDEM: matching proteins with tandem mass spectra. Bioinformatics 20(9):1466–1467
26. Geer LY, Markey SP, Kowalak JA et al (2004) Open mass spectrometry search algorithm. J Proteome Res 3(5):958–964
27. MacCoss MJ, Wu CC, Yates JR (2002) Probability-based validation of protein identifications using a modified SEQUEST algorithm. Anal Chem 74(21):5593–5599
28. Elias JE, Gygi SP (2007) Target-decoy search strategy for increased confidence in large-scale protein identifications by mass spectrometry. Nat Methods 4(3):207–214
29. Käll L, Storey JD, MacCoss MJ et al (2007) Assigning significance to peptides identified by tandem mass spectrometry using decoy databases. J Proteome Res 7(01):29–34
30. Spivak M, Weston J, Bottou L et al (2009) Improvements to the percolator algorithm for peptide identification from shotgun proteomics data sets. J Proteome Res 8(7):3737–3745
31. Cappadona S, Baker PR, Cutillas PR et al (2012) Current challenges in software solutions for mass spectrometry-based quantitative proteomics. Amino Acids 43(3):1087–1108
32. Chawade A, Alexandersson E, Levander F (2014) Normalyzer: a tool for rapid evaluation of normalization methods for omics datasets. J Proteome Res 13(6):3114–3120
33. Huber W, Von Heydebreck A, Sültmann H et al (2002) Variance stabilization applied to microarray data calibration and to the quantification of differential expression. Bioinformatics 18(Suppl 1):S96–S104
34. Karp NA, Huber W, Sadowski PG et al (2010) Addressing accuracy and precision issues in iTRAQ quantitation. Mol Cell Proteomics 9(9):1885–1897

35. Sandberg A, Branca RM, Lehtio J et al (2013) Quantitative accuracy in mass spectrometry based proteomics of complex samples: the impact of labeling and precursor interference. J Proteomics 96C:133–144. doi:10.1016/j.jprot.2013.10.035
36. Savitski MM, Mathieson T, Zinn N et al (2013) Measuring and managing ratio compression for accurate iTRAQ/TMT quantification. J Proteome Res 12(8):3586–3598
37. Herbrich SM, Cole RN, West KP Jr et al (2013) Statistical inference from multiple iTRAQ experiments without using common reference standards. J Proteome Res 12(2): 594–604
38. Ting L, Cowley MJ, Hoon SL et al (2009) Normalization and statistical analysis of quantitative proteomics data generated by metabolic labeling. Mol Cell Proteomics 8(10): 2227–2242
39. Smyth GK (2004) Linear models and empirical Bayes methods for assessing differential expression in microarray experiments. Stat Appl Genet Mol Biol 3(1):Article 3
40. Diz AP, Carvajal-Rodríguez A, Skibinski DO (2011) Multiple hypothesis testing in proteomics: a strategy for experimental work. Mol Cell Proteomics 10(3):M110.004374
41. Benjamini Y, Hochberg Y (1995) Controlling the false discovery rate: a practical and powerful approach to multiple testing. J R Stat Soc Ser B Methodol 57(1):289–300
42. Gillespie CS, Lei G, Boys RJ et al (2010) Analysing time course microarray data using Bioconductor: a case study using yeast2 Affymetrix arrays. BMC Res Notes 3(1):81
43. Ravasz E, Somera AL, Mongru DA et al (2002) Hierarchical organization of modularity in metabolic networks. Science 297(5586):1551–1555
44. Gavin A-C, Aloy P, Grandi P et al (2006) Proteome survey reveals modularity of the yeast cell machinery. Nature 440(7084):631–636
45. Titz B, Low T, Komisopoulou E et al (2010) The proximal signaling network of the BCR-ABL1 oncogene shows a modular organization. Oncogene 29(44):5895–5910
46. Martindale JL, Holbrook NJ (2002) Cellular response to oxidative stress: signaling for suicide and survival. J Cell Physiol 192(1):1–15. doi:10.1002/jcp.10119
47. Väremo L, Nielsen J, Nookaew I (2013) Enriching the gene set analysis of genome-wide data by incorporating directionality of gene expression and combining statistical hypotheses and methods. Nucleic Acids Res 41(8):4378–4391
48. Liberzon A, Subramanian A, Pinchback R et al (2011) Molecular signatures database (MSigDB) 3.0. Bioinformatics 27(12): 1739–1740
49. Kanehisa M, Goto S, Sato Y et al (2014) Data, information, knowledge and principle: back to metabolism in KEGG. Nucleic Acids Res 42(D1):D199–D205. doi:10.1093/nar/gkt1076
50. Croft D, Mundo AF, Haw R et al (2014) The Reactome pathway knowledgebase. Nucleic Acids Res 42(D1):D472–D477
51. Consortium GO (2013) Gene ontology annotations and resources. Nucleic Acids Res 41(D1):D530–D535
52. Culhane AC, Schröder MS, Sultana R et al (2012) GeneSigDB: a manually curated database and resource for analysis of gene expression signatures. Nucleic Acids Res 40(D1): D1060–D1066
53. Ashburner M, Ball CA, Blake JA et al (2000) Gene ontology: tool for the unification of biology. Nat Genet 25(1):25–29
54. Khatri P, Sirota M, Butte AJ (2012) Ten years of pathway analysis: current approaches and outstanding challenges. PLoS Comput Biol 8(2), e1002375. doi:10.1371/journal.pcbi.1002375
55. Subramanian A, Tamayo P, Mootha VK et al (2005) Gene set enrichment analysis: a knowledge-based approach for interpreting genome-wide expression profiles. Proc Natl Acad Sci U S A 102(43):15545–15550
56. Da Wei Huang BTS, Lempicki RA (2008) Systematic and integrative analysis of large gene lists using DAVID bioinformatics resources. Nat Protoc 4(1):44–57
57. Merico D, Isserlin R, Stueker O et al (2010) Enrichment map: a network-based method for gene-set enrichment visualization and interpretation. PLoS One 5(11), e13984
58. Alexa A, Rahnenfuhrer J (2010) topGO: enrichment analysis for gene ontology. R package version 28
59. Ideker T, Ozier O, Schwikowski B et al (2002) Discovering regulatory and signalling circuits in molecular interaction networks. Bioinformatics 18(Suppl 1): S233–S240
60. Mitra K, Carvunis A-R, Ramesh SK et al (2013) Integrative approaches for finding modular structure in biological networks. Nat Rev Genet 14(10):719–732
61. Iorio F, Saez-Rodriguez J, Di Bernardo D (2013) Network based elucidation of drug response: from modulators to targets. BMC Syst Biol 7(1):139

62. Vidal M, Cusick ME, Barabasi A-L (2011) Interactome networks and human disease. Cell 144(6):986–998
63. Franceschini A, Szklarczyk D, Frankild S et al (2013) STRING v9.1: protein-protein interaction networks, with increased coverage and integration. Nucleic Acids Res 41(Database issue):D808–D815. doi:10.1093/nar/gks1094
64. Croft D, O'Kelly G, Wu G et al (2011) Reactome: a database of reactions, pathways and biological processes. Nucleic Acids Res 39(Database issue):D691–D697. doi:10.1093/nar/gkq1018
65. Wu G, Feng X, Stein L (2010) A human functional protein interaction network and its application to cancer data analysis. Genome Biol 11(5):R53. doi:10.1186/gb-2010-11-5-r53
66. Calvano SE, Xiao W, Richards DR et al (2005) A network-based analysis of systemic inflammation in humans. Nature 437(7061): 1032–1037
67. Racine J, Gerber V, Feutz MM et al (2011) Comparison of genomic and proteomic data in recurrent airway obstruction affected horses using ingenuity pathway analysis®. BMC Vet Res 7(1):48
68. Go Y-M, Roede JR, Orr M et al (2014) Integrated redox proteomics and metabolomics of mitochondria to identify mechanisms of Cd toxicity. Toxicol Sci 139(1):59–73
69. Hoeng J, Talikka M, Martin F et al (2013) Case study: the role of mechanistic network models in systems toxicology. Drug Discov Today 19(2):183–192
70. Schlage WK, Westra JW, Gebel S et al (2011) A computable cellular stress network model for non-diseased pulmonary and cardiovascular tissue. BMC Syst Biol 5(1):168
71. Gebel S, Lichtner RB, Frushour B et al (2013) Construction of a computable network model for DNA damage, autophagy, cell death, and senescence. Bioinform Biol Insights 7:97
72. Westra JW, Schlage WK, Frushour BP et al (2011) Construction of a computable cell proliferation network focused on non-diseased lung cells. BMC Syst Biol 5(1):105
73. Westra JW, Schlage WK, Hengstermann A et al (2013) A modular cell-type focused inflammatory process network model for non-diseased pulmonary tissue. Bioinform Biol Insights 7:167
74. Slater T (2014) Recent advances in modeling languages for pathway maps and computable biological networks. Drug Discov Today 19(2):193–198
75. Smoot ME, Ono K, Ruscheinski J et al (2011) Cytoscape 2.8: new features for data integration and network visualization. Bioinformatics 27(3):431–432. doi:10.1093/bioinformatics/btq675
76. Saito R, Smoot ME, Ono K et al (2012) A travel guide to cytoscape plugins. Nat Methods 9(11):1069–1076. doi:10.1038/nmeth.2212
77. Nepusz T, Yu H, Paccanaro A (2012) Detecting overlapping protein complexes in protein-protein interaction networks. Nat Methods 9(5):471–472
78. Maere S, Heymans K, Kuiper M (2005) BiNGO: a cytoscape plugin to assess overrepresentation of gene ontology categories in biological networks. Bioinformatics 21(16):3448–3449. doi:10.1093/bioinformatics/bti551
79. Leung MC, Rooney JP, Ryde IT et al (2013) Effects of early life exposure to ultraviolet C radiation on mitochondrial DNA content, transcription, ATP production, and oxygen consumption in developing Caenorhabditis elegans. BMC Pharm Toxicol 14(1):9
80. McRedmond JP, Park SD, Reilly DF et al (2004) Integration of proteomics and genomics in platelets: a profile of platelet proteins and platelet-specific genes. Mol Cell Proteomics 3(2):133–144. doi:10.1074/mcp.M300063-MCP200
81. Buschow SI, Lasonder E, van Deutekom HW et al (2010) Dominant processes during human dendritic cell maturation revealed by integration of proteome and transcriptome at the pathway level. J Proteome Res 9(4): 1727–1737
82. Fisher RA (1925) Statistical methods for research workers. Genesis Publishing Pvt, Ltd, New Delhi
83. Stouffer SA, Suchman EA, DeVinney LC et al (1949) Studies in social psychology in world war II, vol 1. The American soldier: adjustment during army life. Princeton University Press, Princeton, NJ, US
84. Kaever A, Landesfeind M, Feussner K et al (2014) Meta-analysis of pathway enrichment: combining independent and dependent omics data sets. PLoS One 9(2), e89297. doi:10.1371/journal.pone.0089297
85. Balbin OA, Prensner JR, Sahu A et al (2013) Reconstructing targetable pathways in lung cancer by integrating diverse omics data. Nat Commun 4:2617
86. Hood LE, Omenn GS, Moritz RL et al (2012) New and improved proteomics technologies for understanding complex biological systems: addressing a grand challenge in the life sciences. Proteomics 12(18):2773–2783

87. van Vliet E (2011) Current standing and future prospects for the technologies proposed to transform toxicity testing in the 21st century. ALTEX 28(1):17
88. Martin SF, Falkenberg H, Dyrlund TF et al (2013) PROTEINCHALLENGE: crowd sourcing in proteomics analysis and software development. J Proteomics 88:41–46
89. Roemer E, Schramke H, Weiler H et al (2012) Mainstream smoke chemistry and in vitro and in vivo toxicity of the reference cigarettes 3R4F and 2R4F. Contribute Tobacco Res 25:316–335
90. Deutsch EW, Mendoza L, Shteynberg D et al (2010) A guided tour of the Trans-Proteomic Pipeline. Proteomics 10(6):1150–1159
91. Rauch A, Bellew M, Eng J et al (2006) Computational proteomics analysis system (CPAS): an extensible, open-source analytic system for evaluating and publishing proteomic data and high throughput biological experiments. J Proteome Res 5(1): 112–121
92. Cox J, Mann M (2008) MaxQuant enables high peptide identification rates, individualized ppb-range mass accuracies and proteome-wide protein quantification. Nat Biotechnol 26(12):1367–1372
93. Chen EY, Tan CM, Kou Y et al (2013) Enrichr: interactive and collaborative HTML5 gene list enrichment analysis tool. BMC Bioinformatics 14(1):128
94. Tarca AL, Draghici S, Khatri P et al (2009) A novel signaling pathway impact analysis. Bioinformatics 25(1):75–82
95. Martin F, Sewer A, Talikka M et al (2014) Quantification of biological network perturbations for mechanistic insight and diagnostics using two-layer causal models. BMC Bioinformatics 15(1):238. doi:10.1186/1471-2105-15-238
96. Dittrich MT, Klau GW, Rosenwald A et al (2008) Identifying functional modules in protein-protein interaction networks: an integrated exact approach. Bioinformatics 24(13):i223–i231. doi:10.1093/bioinformatics/btn161
97. Beisser D, Klau GW, Dandekar T et al (2010) BioNet: an R-Package for the functional analysis of biological networks. Bioinformatics 26(8):1129–1130
98. Aluru M, Zola J, Nettleton D et al (2013) Reverse engineering and analysis of large genome-scale gene networks. Nucleic Acids Res 41(1):e24
99. Komurov K, Dursun S, Erdin S et al (2012) NetWalker: a contextual network analysis tool for functional genomics. BMC Genomics 13(1):282
100. Komurov K, White MA, Ram PT (2010) Use of data-biased random walks on graphs for the retrieval of context-specific networks from genomic data. PLoS Comput Biol 6(8), e1000889
101. Petrochilos D, Shojaie A, Gennari J et al (2013) Using random walks to identify cancer-associated modules in expression data. BioData Min 6(1):17
102. Ulitsky I, Krishnamurthy A, Karp RM et al (2010) DEGAS: de novo discovery of dysregulated pathways in human diseases. PLoS One 5(10), e13367
103. McClellan EA, Moerland PD, van der Spek PJ et al (2013) NetWeAvers: an R package for integrative biological network analysis with mass spectrometry data. Bioinformatics 29(22):2946–2947. doi:10.1093/bioinformatics/btt513
104. Saha A, Tan AC, Kang J (2014) Automatic context-specific subnetwork discovery from large interaction networks. PLoS One 9(1), e84227
105. Eddy JA, Hood L, Price ND et al (2010) Identifying tightly regulated and variably expressed networks by differential rank conservation (DIRAC). PLoS Comput Biol 6(5), e1000792

Chapter 12

Xenobiotic Metabolism Activation as a Biomarker of Cigarette Smoke Exposure Response

Anita R. Iskandar

Abstract

The recent advances in "omics" technologies have generated various in silico approaches for toxicity assessment. In silico-based toxicity predictions can overcome certain major drawbacks of laboratory experiments, including the limitation of conducting experiments in a chemical-by-chemical basis that can be expensive. This chapter discusses some recent applications of in silico approaches utilizing xenobiotic metabolism that can be used to assess the impact of cigarette smoke (CS). We first outline recent studies using quantum mechanics/molecular modeling and quantitative structure–activity relationships that focus on smoking-relevant cytochrome P450 (CYP) enzymes. Subsequently, we describe several network-based approaches for toxicity assessment and relevant use cases leveraging a xenobiotic metabolism network model for a quantitative assessment of CS impact.

Key words Xenobiotics, Cytochrome P450, Risk assessment, Systems biology, Computational toxicology, Structure–activity relationship studies, Biological networks, Biological pathways, Omics

1 Introduction

In recent years, in silico approaches for predictions of metabolism-related property have been developed and reported. These efforts are aligned with the reduction of toxicity testing in animals, which has been encouraged by various regulations, including the European Union's regulation REACH (Registration, Evaluation, and Authorization of Chemicals) and the USA's Toxicology in the 21st Century program (Tox21) [1]. In the realms of drug development, in silico predictions of metabolism-related property become valuable to generate ADMET data (Absorption, Distribution, Metabolism, Excretion, and Toxicity) of drug candidates. Furthermore, in silico predictions are expected to escalate the numbers of hits coming out from high-throughput screenings as compared to those from in vitro or in vivo screenings [2].

Xenobiotic metabolism in particular becomes attractive for toxicity predictions as one of the most prominent defense mechanisms

Julia Hoeng and Manuel C. Peitsch (eds.), *Computational Systems Toxicology*, Methods in Pharmacology and Toxicology, DOI 10.1007/978-1-4939-2778-4_12, © Springer Science+Business Media New York 2015

in a living organism [3]. It comprises phase I, II, and phase III metabolizing enzymes. Phase I enzymes are responsible for the first step of xenobiotic metabolism mainly via oxidation reactions [3]. Phase I metabolism aims to generate less harmful and more soluble metabolites that can be excreted more quickly [4]. The cytochrome P450 (CYP) family is the major class of phase I enzymes. Phase I-mediated transformation can also generate toxic metabolites [4]. This situation is often utilized in drug development, for example, a CYP-mediated local metabolism of anticancer prodrugs in a specific target tissue becomes valuable for the drugs' effectiveness [4]. Various xenobiotics (e.g., cigarette smoke (CS) and its constituents, drugs, and air pollutants) can influence not only the levels of CYP gene expression but also its activity. Therefore, studying the interactions between xenobiotics and CYP enzymes becomes essential to avoid unknown side-effects of drug candidates [4]. Furthermore, phase II enzymes (mainly the transferases) catalyze conjugation reactions, such as glucuronidation, sulfation, methylation, and acetylation [5, 6]. Finally, phase III enzymes refer to active membrane transporters responsible for the translocation of xenobiotic metabolites across cellular membranes [6, 7]. The prominent member of this enzyme family is the ATP-Binding Cassette family of drug transporters [6]. Phase II and III enzymes typically lead to the elimination of xenobiotic metabolites [6].

Studies have demonstrated that CS exposure affects xenobiotic metabolism. In this chapter, we discuss the recent in silico approaches relevant for the impact assessment of CS that utilize xenobiotic metabolism activation as a tool for the exposure assessment, which includes (Fig. 1):

1. Quantum mechanics and molecular modeling (QM/MM).
2. Quantitative structure–activity relationship (QSAR).
3. Biological network analysis.

2 Methods

2.1 QM/MM Studies for CYP-Mediated Reactions

QM approach has been used to study CYP-mediated reactions. However, to evaluate more complex situations, such as the transition states of CYP enzymes and substrate specificities to CYP enzymes, a hybrid approach of QM/MM is more suitable [8–11]. QM approach alone describes the electronic structure of a molecule and attempts to solve the Schrödinger equation [10, 12]. A good quality QM model is computationally demanding; therefore, QM approach is hardly applied to a large system (e.g., to study a complete enzymatic reaction) and is often applied for small and isolated model systems (e.g., to reveal the physical chemistry of nucleobases, charge distributions, dipole moments, polarizabilities, protein affinities, and amino group hybridization) [4, 12, 13].

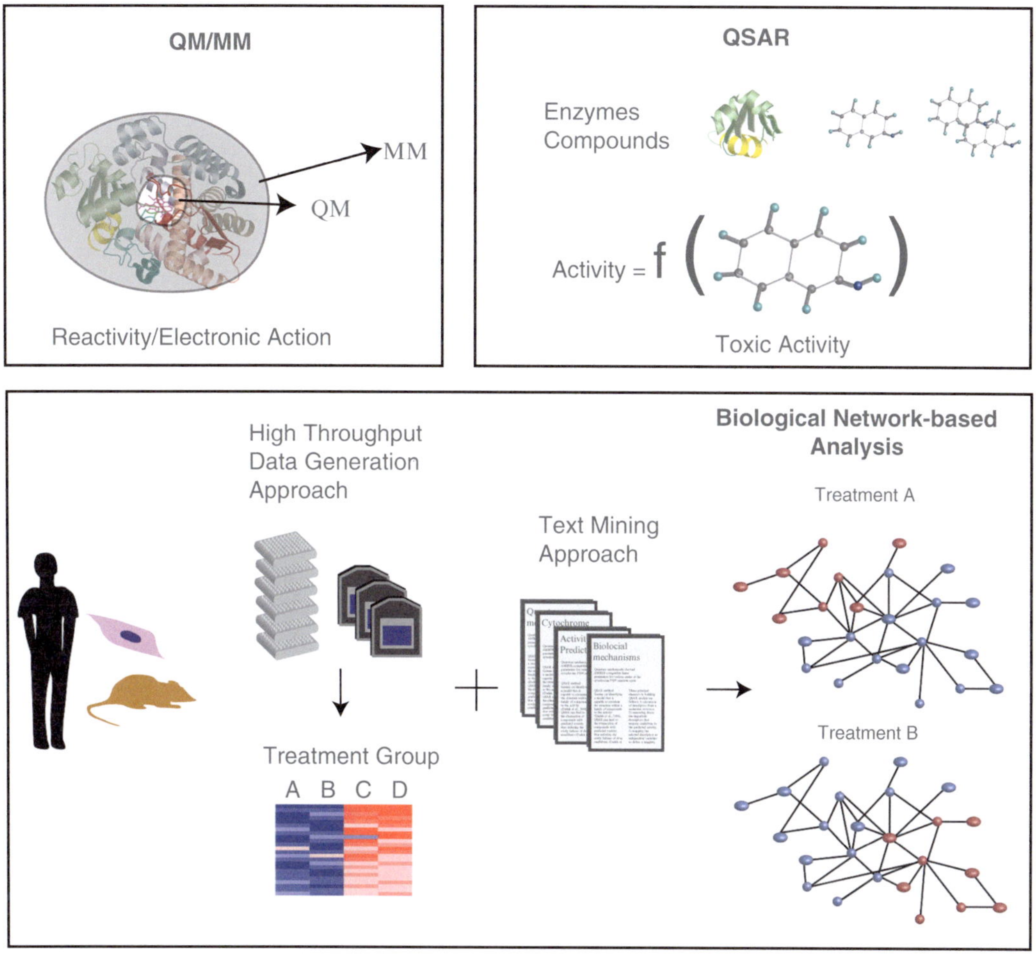

Fig. 1 Schematic representation describing in silico approaches for the study of xenobiotic metabolism. The three approaches discussed in this chapter are shown: QM/MM, QSAR and biological network-based approaches. QM/MM and QSAR rely on the physiochemical factors (e.g., topological, structural, electronic) of a particular protein/compound for reactivity and toxicity prediction. Moreover, network-based analysis relies on the compilation of data (e.g., from high throughput "omics" experiment and/or text-mining approaches) to deduce the activity/toxicity of a compound(s) of interest. Network-based evaluation can also be conducted at the levels of gene/protein/metabolite pattern, pathway/reaction, or sub-network. Integration of these approaches could be implemented and are expected to advance the value of in silico toxicology. *QM/MM* quantum mechanics/molecular modeling, *QSAR* quantitative structure–activity relationships

This limitation can be solved by using the hybrid QM/MM approach [12].

Principally, in QM/MM studies, the region of interest (i.e., the catalytic core/active site of the enzyme) is defined using an electronic structure method, i.e., the QM method [10, 12, 14, 15]. The rest of the protein is defined using the MM force field method capturing the contribution of the enzymatic structure and the dynamic of the catalytic reaction [10, 12]. The MM element expresses the total energy of the molecular system as a sum of the bonded energy, electrostatic, and van der Waals interactions [8]. In this manner, QM/MM calculations can represent the specific reactivity/electronic events of the protein of interest, thus they are capable to handle more extensive systems [4, 8]. Enzymatic studies using QM/MM generally involve the following elements: (1) determination of the QM and MM methods; (2) partitioning of the QM and MM regions; (3) simulation and/or calculation of the QM/MM approach; and (4) preparing the molecular structures of the protein of interest [10]. We briefly describe these components in the following paragraphs.

Several QM methods are available starting from the quick but semiempirical methods (e.g., Austin Model 1 (AM1), Parametric Method 3 (PM3), and Self-Consistent Charge Density-Functional Tight-Binding (SCC-DFTB), Modified Neglect of Diatomic Overlap (MNDO)), to the more accurate but computationally expensive methods, such as those using Density Functional Theory (DFT) and molecular orbital ab initio theory (e.g., Hartree-Fock, MP2) [8, 10]. In the DFT-based studies of CYP enzymes (and other heme-containing enzymes), the B3YLP method works well and by far has become the most widespread QM module used [8]. The structural models for QM calculations often focus on the active region only, which for CYP enzymes, it includes the active site buried inside the protein (located below the regions *vi* and *vii*, Fig. 2) and a heme group (located between the helixes *ix* and *xii*, Fig. 2). The active region can also be modeled to include the heme group with or without the surrounding entities (e.g., the cysteinated residue below the heme group, water molecules near the

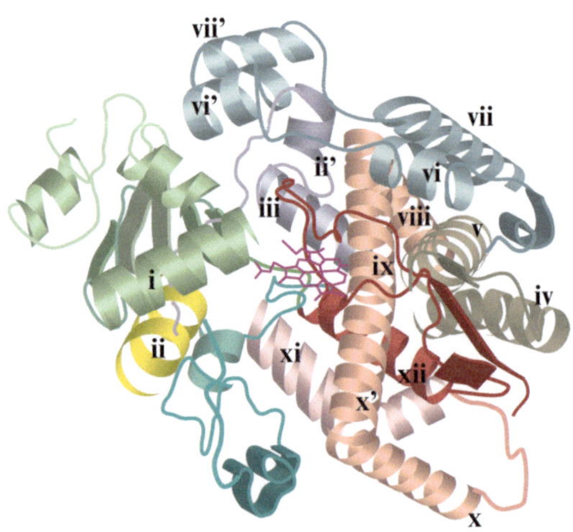

Fig. 2 Cartoon schematic showing the structure of cytochrome P450 (CYP). The heme moiety is present in all CYP enzymes and is illustrated in the center of the enzyme (fuchsia). The helical structures are labeled *i–xii*. The structure is drawn based on the CYP structure illustrated in a previous publication [14]

active site, or other residues of the protein environment) [8, 15]. In addition, one must define the protonation states of the associated residues in the QM region, establish how to handle the electrostatic interactions, and determine the covalent bonds crossing the boundary between the QM and MM regions [10].

For the MM method, the Chemistry at HARvard Macromolecular Mechanics (CHARMM) is the most widely used method in CYP enzyme studies [8]. Other commonly used MM calculations are the Assisted Model Building with Energy Refinement (AMBER), GROMOS and Optimized Potentials for Liquid Simulations (OPLS) [8, 10]. Although the use of static MM force fields rarely includes the effects of polarization, polarizable force field models are increasingly being developed to improve the prediction accuracy [8].

In typical QM/MM studies, the partitioning of the QM/MM regions is often fixed and defined once [8]. However, the choice of which should not be rigid but should be tested to determine whether the approach is reliable [10]. Pezeshki and Lin recently described "on-the-fly" adaptive partitioning for the QM/MM boundary during a dynamics simulation [16]. In addition, an adjustment to minimize the computational demand can be done by condensing the QM region [8, 9].

We outline recent QM/MM-based investigations of xenobiotic metabolism enzymes in Table 1, which indicates that most studies have focused on the phase I enzymes. In regard to CS assessment, QM/MM methods have been used to evaluate the reactivity and/or interaction of constituents of CS (e.g., nicotine) to CYP enzymes. In particular, the evaluation of CYP2A6 would be important in regard to smoking because it is the primary enzyme responsible for the metabolism of nicotine [17]. The calculations of free energy of CYP enzymes tends to become computationally expensive, thus they are rarely performed [8–10]. However, Li and colleagues have demonstrated that based on the free energy barrier calculations for CYP2A6-(−)-nicotine binding reactions, the regioselectivity for CYP2A6-mediated 5′-hydroxylation was higher than the *N*-methylhydroxylation. Such studies may open new avenues in deciphering detailed mechanisms of the metabolism of nicotine and other CS toxicants.

2.2 QSAR Studies for CYP-Activity Predictions

In a laboratory setting, the evaluation of CYP activity is done in microsomes obtained from the specific tissue of interest [18]. In actual smoking situation, the lung is the primary target organ, which is also capable in metabolizing xenobiotics [18]. However, the expression and activity of human pulmonary CYP enzymes are considerably lower than those in the liver or in other species (e.g., rodents), making the study of human pulmonary xenobiotic metabolism difficult [18]. In this regard, QSAR-based studies can be implemented to facilitate the investigation of pulmonary

Table 1
Recent QM/MM-based evaluation of human CYP enzymes

Enzyme	Method	QM Region	Endpoint	Ref
CYP2A6	AMBER ff03; B3LYP B-31G* (for the iron atom and the 6-atoms coordinated to the iron atom) and 6-31G for other atoms in Gaussian 03	The (*S*)-(−)-nicotine, the porphyrin-iron-complex, and the side chain of Cysteine-439	Mechanism of N-methylhydroxylation of (*S*)-(−)-nicotine and the free energy profile	[64]
	AMBER ff03 (for CYP2A6 residues) and AMBER gaff (for heme and inhibitor); B3LYP (6-31G(d) in Gaussian 03)	The heme group and the inhibitor molecule	Binding affinity of nicotine analogs to CYP2A6	[65]
CYP2B4	CHARMM; B3LYP (LACVP and TZVP)	The QM region was minimized using CHARMM force field before being subjected to subsequent QM/MM study. For the protein delivery pathways Conf S included Compound 0, the crucial Thr302, Glu301, and three water molecules that may be involved in the proton shuttle. In all other snapshots, the Phe/His429 residues and additional water molecules were also included. For the investigation of electronic structural features of Compound 0, only the Compound 0 complex was considered	Impact of the single mutation F429M on CYP2B4 activity	[66]
CYP2C9	CHARMM 22ff, B3LYP (6-31G(d) in Jaguar 5.0)	Compound I, modeled as a truncated heme with all ring substituents replaced by hydrogen atoms	Mechanism of the oxidation reactions of S-ibuprofen, diclofenac, and S-warfarin	[67]
CYP2D6	CHARMM 27ff; UB3LYP in Jaguar 6.0; LACVP	The heme group without substituents on the porphyrin ring, the SCH2 group of the cysteinyl ligand, and the benzene molecule	Preference of O-demethylation for dextromethorphan metabolism by CYP2D6 as opposed to aromatic carbon oxidation	[4]
	AMBER ff03 (for the protein) and gaff (for the ligand); B3LYP 6-311G+(2d,p) (for the non-Fe atom) and LACVP (for the Fe atom) in Gaussian 09	The QM part contains 79 atoms, including the heme ring, as illustrated in the original publication	Conformational of the aripiprazole-CYP2D6 complex for the 4-hydroxylation reaction	[68]

CYP2C9, CYP2D6, and CYP3A4	CHARMM 27ff; B3LYP LACVP for the iron and 6-31G* (for all other atoms)	The QM region includes Compound I, modeled as a truncated heme with all ring substituents replaced by hydrogen atoms, as described in the original publication, with the proximal cysteine represented by methyl mercaptide	Calculation of electronic structure and estimation of the Fe–O bound enthalpy for Compound I	[69]
CYP3A4	AMBER (ff9X and ff10), B3LYP/LACVP (6-31G* for the organic atoms and LANL2DZ for the frozen core electrons) in Gaussian 03 for the high spin ferric species	The QM coordinates comprise various force constant parameters: Fe–S, Fe–N and Fe–O bond; and Fe–N–C, N–Fe–N, S–Fe–N, N–Fe–O, and Fe–O–O	Docking prediction of substrates to CYP3A4	[70]
GSTA3-3	AMBER ff99 and gaff, B3LYP/DZVP in Gaussian 03 and Gaussian 09	The QM region is the innermost layer (referred to as the "H" region) including the molecular fragments directly involved in the enzyme-catalyzed isomerization process, the side chain of the cysteine residue belonging to the tripeptide SH < the tyrosine residue (Y9) including the phenolic ring fragment, two rings of the steroid nucleus, and (the saturated) dangling bonds with hydrogen atoms	Catalytic mechanism of GSTA3-3	[71]
GSTP1-1	AMBER; DFT (Gaussian and plane-wave formalism); TZV2P to describe the wave function; 320Ry/GTH pseudopotentials to describe the core electrons; GGA for the exchange and correlation energies using PBE functional	The platinum center with the ligands directly coordinated to it, including both Cys101 side chains up to the β-carbon atom	Mechanism of the inhibition activity of GSTP1-1 by ethacraplatin, a platinum(IV)-based anticancer compound	[72]
NQO2	CHARMM 22, SCC-DFTB	The dimethylisoalloxazine (i.e., flavin) ring atoms of FAD that can account for the change in the electronic structure upon redox transition.	Computation of redox potential and pKa of the enzyme-bound flavin	[73]

ff force field, *GGA* generalized gradient approximation, *GTH* Goedecker Teter Hutter, *GSTP* glutathione S-transferase, *NQO* NAD(P)H:quinone acceptor oxidoreductase, *FAD* flavin adenine dinucleotide xenobiotic metabolizing enzymes. QSAR approach aims to develop a model that can associate the chemical structure of a protein/compound to its activity [19], thus they can be used to evaluate the toxicity of specific airborne toxicants, inhaled drugs, or constituents of CS. In Table 2, we outline recent QSAR studies on xenobiotic-metabolizing enzymes that are known to be expressed in the lung. Such studies could be leveraged to predict the potential interactions between inhaled drugs and cigarette smoke.

In an ideal situation, QSAR models should be developed from homogeneous datasets, such as those generated from experiments using identical protocol; however, this is rarely feasible as illustrated in Tables 2 and 3. Scientists should clearly reveal how the datasets were divided into the training and test sets, because imbalance datasets would bias the prediction toward the majority class and reduce the predictive accuracy for the minority class [20]. Recently, strategies have been developed to rebalance the training and test sets, such as to over-sampling the minority class and under-sampling the majority class [20]. Furthermore, efforts have been made toward open-source platforms for QSAR studies. For example, Table 2 includes the open-source online chemical modeling environment (OCHEM, http://ochem.eu [21]), which was used to develop QSAR models to assess the inhibition of CYP1A2 activity [22].

Building QSAR models generally requires the following elements [19, 23]: (1) drawing of the molecular structures; (2) extracting descriptors from a molecular structure; (3) narrowing down the important descriptors that contribute majorly to the activity prediction; and finally (4) mapping the selected descriptors as independent variables to define a mapping that correlates them with the activity in question. These elements are briefly discussed in the following paragraphs.

In building QSAR models, molecular descriptors are extracted and used to map the molecular structure into a set of numerical or binary values that represent various molecular properties [19, 23]. There are more than 30,000 known molecular descriptors, the choice of which are determined by the QSAR methods, e.g., 2D- or 3D-QSAR [19, 23]. The descriptors for 2D-QSAR can be comprised of ordinary descriptors (e.g., molecular weights, number of atoms, type of bonds, and number of aromatic rings); electrostatic descriptors (e.g., atomic, quantum-chemical, and polarizability); and topological descriptors (e.g., molecular surface, and topological pharmacophores) [19, 23]. For 3D-QSAR, the descriptors can be alignment-dependent or alignment-independent [23]. Alignment-dependent descriptors include the comparative molecular field analysis (CoMFA) and comparative molecular similarity indices analysis (CoMSIA) [23]. Furthermore, alignment-independent descriptors are invariable to molecular rotation and transformation in space, such as Comparative Molecular Moment Analysis (CoMMA), Weighted Holistic Invariant Molecular (WHIM),

Table 2
Recent QSAR-based evaluation of CYP enzymes known to be expressed in the lung

Enzyme	Endpoint	Method/model	Dataset size	Ref
CYP1A1	Docking analysis	3D-QSAR with CoMFA descriptors was performed using PLS	36 ligands (training set) 12 ligands (test set)	[74]
	Metabolic activation	QSAR using MLR and ANN methods, following a docking study	24 ligands (training set) 8 ligands (test set)	[75]
	Inhibition of CYP1A1 activity	QSAR project manager module of the Tripos SYBYL-X 1.3 program was used for model creation	21 ligands[a]	[76]
CYP1A2	Inhibition of CYP1A2 activity	80 in silico QSAR models from OCHEM	3745 ligands (training set) 3741 ligands (test set)	[21]
	Inhibition of CYP1A2 activity	CoMFA, comparative molecular similarity analysis, and hologram QSAR approaches	36 ligands (Model 1) 34 ligands (Model 2)	[77]
	Inhibition of CYP1A2 activity	QSAR project manager module of Tripos SYBYL-X 1.3 program was used for model creation	21 ligands[a]	[76]
	Metabolic activation	QSAR using SVM modules	14226 ligands (training set) 2011 ligands (test set)	[20]
CYP1B1	Metabolic activation	QSAR using MLR and ANN methods, following a docking study	24 ligands (training set) 8 ligands (test set)	[75]
CYP2A6	Inhibition of CYP2A6 activity	3D-QSAR model with CoMFA descriptors in modeling package SYBYL (v. 8.1)	86 (training set)[a]	[78]
	Inhibition of CYP2A6 activity	QSAR using SVM and least square-LSM, following a molecular docking and molecular dynamics simulation	30 (training set) 8 (test set)	[79]
CYP2B6	Inhibition of CYP2B6 activity	QSAR using MLR	12 ligands[a]	[80]
	Inhibition of CYP2B6 activity	Pharmacophore mapping, molecular docking followed by QSAR were used	22 ligands (training set) 26 ligands (test set)	[81]
CYP2C8	Inhibition of CYP2C8 activity	2D-QSAR model using MLR	20 ligands (training set) 9 ligands (test set)	[82]

(continued)

Table 2 (continued)

Enzyme	Endpoint	Method/model	Dataset size	Ref
CYP2D6	Inhibition of CYP2D6 activity	QSAR modeling using PLS in CODESSA software	51 ligands[a]	[83]
	Metabolic activation and inhibition studies	QSAR models using PLS	746 ligands (2D6 substrates) 691 ligands (2D6 inhibitors)	[84]
	Ligand binding assay	QSAR models using MLR	3 ligands (2D6 substrates) for initial generation and 5 ligands for the final QSAR model[a]	[85]
	Metabolic activation	QSAR using SVM modules	15428 ligands (training set) 1945 ligands (test set)	[20]
CYP2E1	Inhibition of CYP2E1 activity	QSAR with CoMFA descriptors using PLS in modeling package SYBYL (v. 8.1), following a docking study	9 ligands for the open form of the enzyme and 29 ligands for the closed form of the enzyme (the training set) 6 ligands (test set)	[86]
CYP3A4	Inhibition of CYP3A4 activity	QSAR using GALAS methodology and logistic PLS	190 compounds (internal test set) and 8528 compounds (external test set)	[87]
	Inhibition of CYP3A4 activity	QSAR using GA	359 ligands (training set) 301 ligands (test set)	[88]
	Inhibition of CYP3A4 activity	3D-QSAR models with CoMFA descriptors using PLS in modeling package SYBYL (v. 8.1)	11–14 ligands (training set) and 3–6 ligands (test set)	[89]
	Metabolic activation	QSAR using SVM	13433 ligands (training set) 1803 ligands (test set)	[20]
UGT2B7	Glucuronidation of substrates	3D-QSAR with VolSurf descriptors using PLS in modeling package SYBYL (v. 8.1)	36 ligands (training set) 17 ligands (test set)	[90]

ANN artificial neural network, *CoMFA* comparative molecular field analysis, *GA* genetic algorithms, *GALAS* global, adjusted locally according to similarity model, *MLR* multiple linear regression, *OCHEM* online chemical modeling environment (http://ochem.eu), *PLS* partial least square, *SVM* support vector machine.

[a]Detailed information on the number of ligands for training and test datasets were not stated in the original paper

Molecular Surface WHIM, VolSuf, and Grid-Independent Descriptors (GRIND) [19, 23].

The descriptors will be further selected based on their contribution to the activity prediction [19, 23]; many techniques are available for this selection step. The selected techniques should be applicable for large datasets containing structurally diverse

Table 3
Recent QSAR-based evaluation of PAHs

Endpoint	Descriptor	Method	Statistics		Dataset size		Ref.
			Training	Test	Training	Test	
Toxicity of PAHs to earthworms	Ionization potential, Kow, soil pore-water concentration	MLR and PLS	$R^2 = 0.81$	NR	9	9	[91]
Covalent binding of hydroxylated PAHs to calf thymus DNA	Molecular docking; hydrophobic interactions; molecular size; polarizability; electrostatic potential	PLS	$R^2 = 0.751$	$R^2 = 0.656$	17	7	[92]
Phototoxicity	Electronic structure (HOMO, LUMO)	MLR	$R^2 = 0.824$	$R^2 = 0.909$	14	12	[93]
Estrogenic activity; binding affinity to estrogen receptor	Molecular size, Van der Waals volumes, polarizabilities, electrotopological	CoMFA, PLS	$R^2 = 0.94$	$R^2 = 0.58$	29	7	[94]
Biodegradation activity of PAHs	Physical descriptors, such as HOMO and LUMO orbital energy difference; molecular dipole moment; molar volume	Stepwise MLR	$R^2 = 0.856$	$R^2 = 0.905$	17	17	[95]
Calculation of soot-water partition coefficients	Surface area, electronic (HOMO, LUMO), various electrostatic potentials	Stepwise linear regression	$R^2 = 0.9797$	$R^2 = 0.9811$	25	11	[96]
Cross reactivity in an ELISA kit (cross reaction potential)	Molar solubility and valence molecular connectivity index	MLR, principal component regression and PLS regression	RaPID kit $R^2 = 0.5943$ Ris kit $R^2 = 0.6383$	NR	14	NR	[97]

(continued)

Table 3
(continued)

Endpoint	Descriptor	Method	Statistics		Dataset size		Ref.
			Training	Test	Training	Test	
Cross-reactivity in immunoassays used to detect PAHs in the environment	CoMFA, electrostatic fields	PLS	$R^2 = 0.772$ regression analysis $R^2 = 0.926$ CoMFA	NR	23–37	NR	[98]
Disappearance of PAHs from soil, sediments and mangrove leaves	Log solubility, water diffusivity, HOMO-LUMO, molecular volume, aspect ratio	nonlinear models	R^2 was presented individually for each descriptor from soil, sediment or leaf separately but was not combined		15	ND	[99]
Bioconcentration on fish, mutagenicity, carcinogenicity, developmental toxicity, skin sensitization	Analysis of applicability domains, and similarity sets, which consist of the similar compounds from the training sets. These results were further analyzed with chemometric methods (hierarchical clustering, principal component analysis, and self-organizing maps, which were used for clustering and to define the cluster indicators)	QSAR models using CAESAR program	NR	NR	16	NR	[100]

Half-life (environmental persistence) of pharmaceuticals in aqueous environments	Biodegradation, sorption, co-metabolic oxidation in addition to standard biodegradation rate constants	QSAR using EPI suite	2000 pharmaceuticals	NR	NR	NR	[101]
Photodegradation of PAHs	HOMO, LUMO, molecular dipole moment, molar volume, number of six-carbon benzenoid ring, number of all type of rings, most negative charge, most positive charge of hydrogen atoms, in-plane bending vibration frequency of the conjugated ring of PAHs	Stepwise MLR	Various	NR	12	NR	[102]

CoMFA comparative molecular field analysis, *HOMO* highest occupied molecular orbital, *LUMO* lowest unoccupied molecular orbital, *MLR* multiple linear regression, *NR* not reported, *OCHEM* online chemical modeling environment (http://ochem.eu), *PAH* polycyclic aromatic hydrocarbon, *PLS* partial least square compounds and should optimize a quick and automatic selection [23]. The selection methods can generally be divided into two major categories: the filtering and the wrapper methods [19, 23]. Both of the methods are not exclusive and can be combined [19, 23]. The filtering methods can be based on correlation-based methods (e.g., Pearson's correlation), information theory-based methods (i.e., using entropy of descriptor that is treated as a random variable), or the statistical-based methods (e.g., Fisher's ratio, Kolmogorov-Smirnov) [19, 23]. Moreover, the wrapper methods run together during a mapping procedure because the error of the mapping algorithm determines the choice of the best set of descriptors [19, 23]. Examples of the wrapper methods include principal component analysis (PCA), Genetic Algorithms (GA), Simulated Annealing (SA), Sequential Forward Feature Selection, and Sequential Backward Feature Elimination [19, 23].

Having selected the descriptors, the last step in building QSAR models is to map the features of the molecular structure to the activity. Various mapping approaches in QSAR studies are available. Frequently, the mapping uses straightforward linear models (i.e., the linear combination of the descriptors), such as multiple linear regression (MLR), partial least squares (PLS), and linear discriminant analysis (LDA) [19, 23]. Multiple linear regression (MLR) can be used to determine the most relevant features of the model; however, it relies on the assumption that the relationship between descriptors and activity is linear [24]. Table 2 indicates that many of the CYP enzyme studies implement MLR and PLS. Although linearity relationship might exist in the data, other possible nonlinearity association should not be disregarded; especially if, a priori knowledge is not available for the set of compounds being tested [24]. In such situations, nonlinear approaches would be more suitable, such as Bayes classifier, machine learning k-Nearest neighbor, Artificial Neural Network (ANN), Decision Trees (DT), Support Vector Machines (SVM), Evolutionaly Algorithms (EA), and Ensemble methods [23].

Furthermore, QSAR models have been used to predict the toxicity of PAHs. PAHs are formed from the incomplete combustion and pyrolysis of organic materials, and are known to have carcinogenic activity [25]. In addition to their presence in CS, PAHs can be generated from automobile emissions, domestic heating, asphalt production, and agricultural burning [25]. Table 3 lists recent QSAR studies on PAHs. In most cases, the number of the training and test datasets are small. Future studies should attempt to cover a greater number of PAHs being tested, thus expanding the applicability of the models. Nevertheless, QSAR-based evaluations could overcome some of the difficulties of studying PAHs in laboratory settings, which mostly related to their diverse structures that influence their activities and rates of degradation [25]. The structures of PAH can be a simple two-ringed molecule to a more complex molecule with ten aromatic rings [25]. Therefore, in laboratory-based toxicity studies, researchers must either focus only on a small number of PAHs or screen a large number of PAHs simultaneously to comprehensively evaluate the associated toxicities, which are costly and time-consuming [25].

Because QSAR studies are increasingly important in public health practices, several principles for the Organization for Economic Co-operation and Development QSAR modeling and validation have been made available. They aim to facilitate the regulatory acceptance of QSAR models. These principles include: (1) a defined endpoint; (2) an unambiguous algorithm; (3) a defined domain of applicability; (4) an appropriate of goodness-of-fit, robustness, and prediction; and (5) if feasible, a mechanistic interpretation [26]. In addition to the before-mentioned principles, Fourches and colleagues further emphasized that QSAR models must be trained and validated using comprehensively curated chemical datasets, from which the chemical structure and associated characteristic values are known [26]. Unfortunately, the guidance on the optimal size of the datasets and how to rationally divide them into the training and validation sets (e.g., randomly chosen or divided based on a certain criteria) are currently not available [27, 28].

2.3 Network-Based Analyses of Transcriptomics Data for Exposure Assessment

Disease development is influenced by environmental factors (e.g., xenobiotics) and often originates not only from a disruption of a single gene but also from perturbed regulation of genetics and epigenetics [1]. Molecular signals interact with one another, forms pathways, and regulate normal cellular functions. In reality, biological pathways overlap and interact with each other; therefore, models of biological pathways can be extended into nonlinear molecular networks, which would be a better representation of biological systems [1].

Nodes of molecular networks typically represent molecular entities, such as genes, proteins, mRNAs, and metabolites [1]. Edges in networks connect the nodes and embody the functional relationship among the entities, for example, the edges may represent protein–protein interactions, gene modifications, transcriptional regulations, DNA methylation, pretranslational and posttranslational modifications, metabolic reactions, or siRNA–mRNA interactions [1]. The impact of exposure (or xenobiotics) can be perceived as a network perturbation, which is a reflection of pathophysiological alterations from a normal state.

Publications on pathway/network-based analyses keep escalating (Fig. 3). This growth in publication numbers reflects the advances in "omics" technologies and systems biology approaches, as well as "the recent boom in network methods" [29]. In this section, we briefly describe recent studies on the assessment of exposure-related toxicity using network-based approaches and

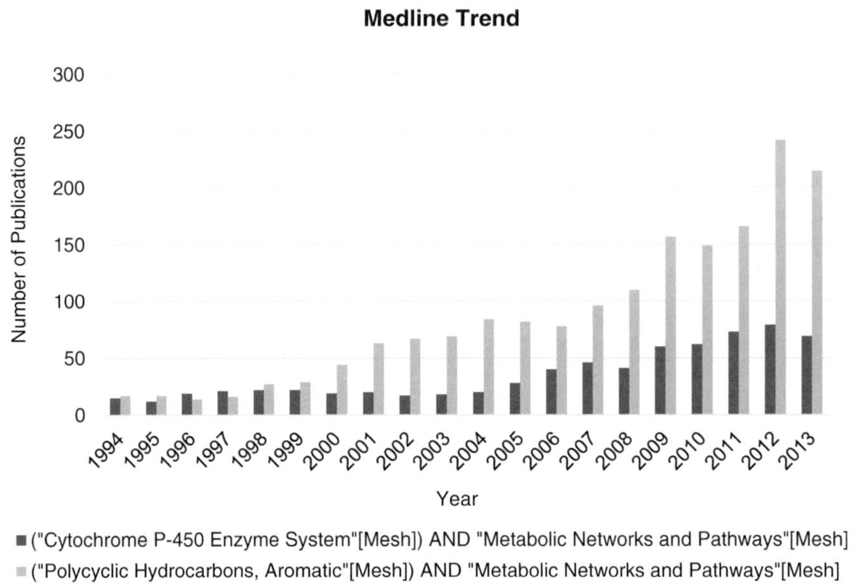

Fig. 3 Number of publications in Medline on cytochrome P450 (CYP) enzymes and polycyclic aromatic hydrocarbons using network/pathway analysis (between the years of 1994–2013). The data were obtained from http://dan.corlan.net/medline-trend.html using the specific query-terms stated in the legend below the graph

our recent approach on the quantification of network perturbations. Readers are also referred to a recent comprehensive review for a general applications of molecular network analysis in drug discovery [29].

In 2010, Van Delft and colleagues generated transcription factor networks to investigate relevant transcription factors affected benzo[a]pyrene (BaP), a well-known PAH [30]. Exposure of cells to BaP (and other carcinogens) is known to alter various molecular responses, including changes in transcription factor activations. To investigate the most relevant transcription factors (which were responsible for the changes of the gene transcription), transcriptomics data were generated from the human hepatoma HepG2 cells that were treated with BaP for various exposure durations. A network-building algorithm on transcription regulation from MetaCore (GeneGo) [31] was used to evaluate whether the modulated genes were connected to any transcription factors; transcription regulation networks were subsequently built using 233 genes. The networks were constructed dynamically by connecting all genes via direct mechanistic connection based on manually curated evidence from scientific literature. The analysis revealed 30 transcription factors that were mostly proto-oncogenes and tumor suppressor genes. Similar approach could be used to assess the impact of other known CS toxicants, whole CS exposure, as well as reduced risk products on the regulation of transcription factors.

In addition, Kiyosawa and colleagues in 2010 generated networks of toxicologically relevant gene set level and networks of phenotype-level [32]. The study was designed to identify the chemical-associated changes in gene transcript levels and phenotypes. The transcriptomics and phenotype data (for example, blood chemistry, body weight, and organ weight) were retrieved from the database of Toxicogenomics Project-Genomics Assisted Toxicity Evaluation system (TG-GATEs) in Japan. The chosen datasets were generated from rats treated with different chemicals. The relationships among the gene sets and phenotype information were evaluated using a partial correlation coefficient using a Gaussian graphical model algorithm. Subsequently, the relationships (edges) with significant correlation coefficients were used for the network constructions using Cytoscape. From the constructed networks, subnetworks could be further extracted, for example, a toxicity subnetwork could be generated from a particular gene set, e.g., toxicity-related gene set, with its neighboring nodes containing the phenotypic information. However, the interpretation is only applicable and relevant for the specific context of the systems/datasets under investigation. This approach envisions a better toxicity profiling and biomarker discovery [32]. Likewise, the approach can be applied for xenobiotic metabolism-related gene set and applied to datasets of CS-exposed samples.

In 2013, Yang and colleagues described a proof-of-concept approach leveraging "differential networks" that can be used for risk assessment and identify "toxicological mode of action" [33]. Briefly, the approach starts by constructing networks from differentially expressed genes generated from experiments. Subsequently, the networks were compared to identify differential edges, which are edges that were presently different (presence or absence) between the networks being compared. These differential edges were used to construct "differential networks." Finally, the differentially expressed genes possessing the differential edges were mapped to biological pathways in various databases, including KEGG (Kyoto Encyclopedia of Genes and Genome) database, ecoCyc, and RegulonDB. The pathway annotations were then used to infer biological processes-associated with the specific treatment condition [33].

Moreover, to facilitate the evaluation of exposure impact and compare the impacts of various exposures, our group has developed an algorithm—the "Network Perturbation Amplitude" (NPA) approach. We describe this approach in the following section and present several use cases, in which the approach was implemented to assess the impact of CS exposure.

2.3.1 Quantification of Network Perturbation Amplitude

Our group have built comprehensive network models capturing physiological and biological responses to stressors [34]. The biological network models can be utilized to qualitatively deduce the biological impact of a given stimulus (stimuli), treatment(s), or drug(s). Our network models are encoded in the Biological Expression Language (BEL, http://www.openbel.org/), a computable language that allows in silico modeling and analyses [35]. Specifically, the "Xenobiotic Metabolism Network" (XMN) model captures the biological mechanisms in response to various xenobiotics. The network model is built based on evidence from scientific literature in the non-diseased human lung tissue context [34]. The XMN model was first published in 2011 [34] and subsequently improved to capture a more comprehensive xenobiotic metabolism process [36, 37], shown in Fig. 4. The yellow oval nodes (referred to as "backbone nodes") represent biological entities, including protein activities, protein abundances, and biological entities (e.g., reactive oxygen species, oxidative stress). The edges connecting them reflect the relationships between the nodes based on published literature [34]. The central backbone node in the model is the transcriptional activity of aryl hydrocarbon receptor (depicted as "taof(AHR)"). AHR is a transcription factor activated by xenobiotic compounds [38] and ubiquitously expressed in most cells and organs [38]. Binding of diverse toxicants to AHR leads to its activation, eventually leading to transcription of genes, including *CYP1A1* and *CYP1B1* [38].

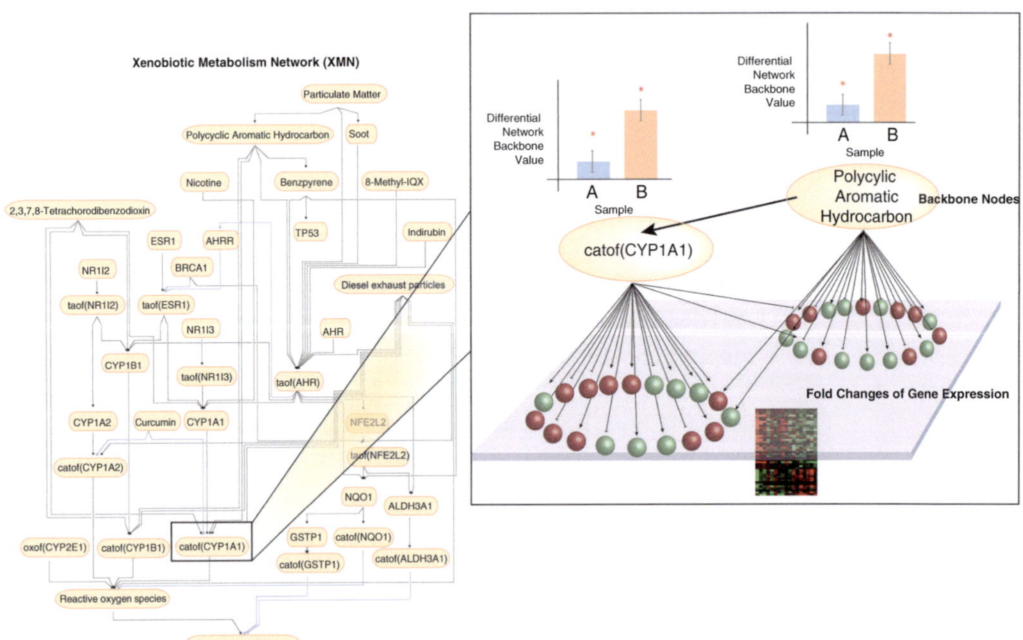

Fig. 4 Xenobiotic metabolism network (XMN) model representing the associated biological processes and relevant mechanisms in the lung tissue context. Inset, an illustration of the NPA approach for the quantification of exposure impact. * indicates the significance perturbation of the XMN model computed using the NPA approach; the details on the NPA approach are described in Chap. 15 of this book and in our previous publication [40]. *NPA* network perturbation amplitude

Quantitatively, these network models can be overlaid using transcriptomics data as an input to computationally predict the activity/functionality of the backbone nodes [34, 39]. The transcriptomics data can be generated, for example, from exposed vs. non-exposed samples or treated vs. control samples. The NPA algorithm, which we have developed, aims to score the activity of the backbone nodes: the "differential network backbone values" (Fig. 4, inset). The details of the NPA approach are described in Chap. 15 of this book and in our previous publication [40]. The NPA approach is based on backward reasoning, by which changes in gene expression are considered as the consequence of an activity of an upstream biological entity, i.e., a backbone node (Fig. 4, inset). Alteration of the differential network backbone value thus reflects a perturbation of the specific biological entity captured in the model. In the case of the XMN model, the differential network backbone values illustrate the specific alterations in the pulmonary xenobiotic metabolism upon airborne exposure.

Using the NPA approach, we used the XMN model to deduce potential biological mechanisms associated with CS exposure. The impact of CS was studied in in vivo tissue samples, from which the transcriptomics data were generated from the bronchial and nasal epithelial cells of smokers and nonsmokers [37]. Furthermore, the same approach was used to assess the impact of CS in vitro, i.e., in bronchial and nasal organotypic 3D tissue culture models that were exposed to whole-smoke of mainstream cigarettes [37]. The following use cases utilize the XMN model to probe the tissue responses to CS exposure.

Use Case 1: Profiling Potential Surrogate Tissues in the Context of Tissue Responses Following CS Exposure

Monitoring the impact of toxic exposure in a more accessible tissue is often conducted to estimate the damage occurring in a target tissue [41]. In the case of CS exposure and lung cancer, the levels of DNA adducts in white blood cells are measured to predict the risk of lung cancer [41]. The issue remains whether alterations in surrogate tissues correctly reflects those in a target tissue. The damage in the target tissue is likely to be more pronounced than in surrogate tissues [41]. Acknowledging this limitation, we describe in the following paragraphs that our NPA approach using the XMN model could facilitate the comparison of tissue responses between a target tissue and its surrogate tissue(s) following CS exposure.

Studies have suggested that the nasal tissue can be a suitable surrogate for the lung tissue in regard to tissue response following smoking [42, 43]. We leveraged our XMN model and the NPA algorithm to compare the responses of the nasal and bronchial tissues to CS exposure (described in greater detail in our previous publication [37]). The public dataset GSE16008 [44] was used for this analysis; the dataset contains a collection of bronchial and nasal epithelial gene expressions of non-diseased smokers and nonsmokers.

The bronchial epithelial samples were collected by bronchoscopy and the nasal epithelial samples were collected by brushing the inferior turbinates. We hypothesized that the CS-related effects on xenobiotic metabolism in the in vivo nasal samples would be similar to those in the in vivo bronchial samples. Changes in the differential network backbone value for each of the backbone nodes in the XMN model would indicate a perturbation of the specific biological process (i.e., represented by the node entity) in response to CS exposure.

Figure 5a shows each of the differential network backbone values in the XMN model. These values reflect the biological activities of the important mechanisms (i.e., the nodes) in xenobiotic metabolism. The inset is a cartoon illustrating the derivation of the differential network backbone value of the AHR node, computed from the transcriptomics data. Figure 5b shows the significantly positive correlation between the differential network backbone values obtained from the bronchial (x-axis) and nasal samples (y-axis). Each data point in the graph represent one of the backbone nodes of the XMN model (a linear regression line is shown in blue). In general, the differential network backbone values were lower in the nasal as compared to the bronchial samples, which was also reflected in the overall NPA score of the network (Fig. 5c). This weaker response of the nasal samples could be attributed to the proposed lower detoxification capacity of the nasal tissue as compared to the bronchus [44]. Zhang and colleagues detected lower fold-changes of the phase II enzymes gene expression, i.e., *ALDH3A1*, *GPX2*, and *AKR1C2*, in nasal tissue (~1.5-fold) than in the bronchus (~fourfold) [44].

Figure 5b also showed that the data point for the AHR backbone node fell outside the regression line. However, the differential network backbone values for the AHR node for both the nasal bronchial tissues were equivalent (approximately 0.15, *y*-

Fig. 5 A comparison between tissue responses to CS exposure generated from in vivo human bronchial and nasal public dataset GSE16008 in the context of the XMN model. The differential network backbone values were generated from bronchial and nasal samples ((**a**), *left* and *right*, respectively), illustrating the impact of smoking on the activity of each node in the XMN model. Inset, a representative illustration of the backward reasoning-based computation to predict the activity of the AHR node. The color gradation represents the NPA score for the activity of the nodes: blue indicates decreasing activity and darker red indicates increasing activity of the node. The correlation between the differential network backbone values generated from the in vivo bronchial and nasal samples is shown in (**b**). Quantifications of the entire network perturbation in the bronchial and nasal tissues are shown in (**c**) (**O*, *, *K** indicate the significance perturbation of the XMN model computed using the NPA approach; the details on the NPA approach are described in Chap. 15 of this book and in our previous publication [40]). The correlation at the level of gene expression fold-changes is shown in (**d**). *AHR* aryl hydrocarbon receptor, *CS* cigarette smoke, *NPA* network perturbation amplitude, *XMN* xenobiotic metabolism network

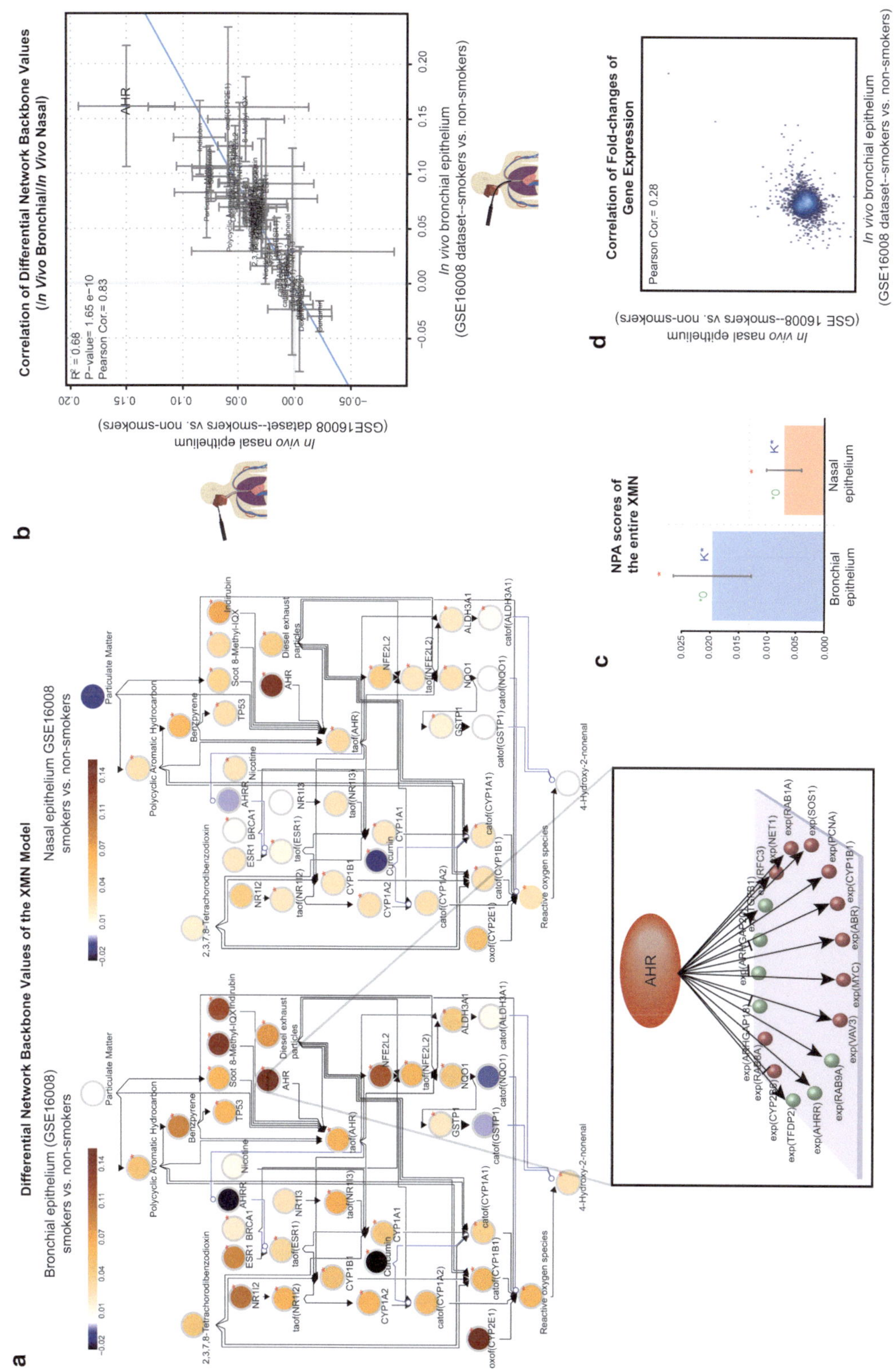
a
Differential Network Backbone Values of the XMN Model
Bronchial epithelium (GSE16008) smokers vs. non-smokers
Nasal epithelium GSE16008 smokers vs. non-smokers
b
Correlation of Differential Network Backbone Values
(In Vivo Bronchial/In Vivo Nasal)
R^2 = 0.68
P-value= 1.65 e-10
Pearson Cor.= 0.83
In vivo bronchial epithelium
(GSE16008 dataset--smokers vs. non-smokers)
In vivo nasal epithelium
(GSE16008 dataset--smokers vs. non-smokers)
c
NPA scores of the entire XMN
Bronchial epithelium
Nasal epithelium
d
Correlation of Fold-changes of Gene Expression
Pearson Cor.= 0.28
In vivo bronchial epithelium
(GSE16008 dataset--smokers vs. non-smokers)
In vivo nasal epithelium
(GSE 16008--smokers vs. non-smokers)
AHR

and x-axis, respectively). This observation suggested that CS exposure similarly affected the activity of AHR in both tissues. This was not surprising, because AHR is known to be ubiquitously expressed in various cells and tissues [38]. Moreover, the comparability of the tissue responses at the network level (i.e., differential network backbone values) was more robust than that at the gene level. Figure 5d shows that when we correlated the fold-changes of the gene expressions of the nasal samples to those of the bronchial samples, no significant correlation was obtained.

Use Case 2: Comparing In Vitro to In Vivo Tissue Responses Upon CS Exposure

Recently, organotypic culture models from human cells have been developed and increasingly used for mechanistic studies [45–49]. In particular, for toxicology assessment of airborne toxicants, pollutants, or inhaled drugs, the in vitro organotypic tissue models become truly valuable because they can be cultured at the air-liquid interphase. An exposure of a test item can be administered at the apical side of the tissue, which mimics the actual in vivo exposure situation. Using the NPA approach and the XMN model, we tested the impact of CS exposure on the organotypic tissue models and we probed the comparability of the tissue responses against CS exposure in vitro and in vivo.

The in vitro tissue responses to CS were assessed in the human bronchial tissue model (i.e., MucilAir™-human fibroblasts-bronchial full thickness from Epithelix Sàrl, Geneva, Switzerland) and human nasal tissue model (i.e., MucilAir™-human fibroblast-nasal full-thickness, Epithelix Sàrl). The experimental design is shown in Fig. 6a (the study was described in greater detail in our previous publication [37]). Briefly, the transcriptome profile was generated from the tissue models following a repeated CS exposure procedure (16 % vol/vol mainstream smoke generated from the reference cigarette 3R4F purchased from the University of Kentucky). The tissues were harvested immediately after the last CS exposure. The repeated exposure to CS was administered to mimic the actual smoking situation.

To determine whether the in vivo responses (i.e., in smokers) could be captured in in vitro tissue models, correlations between the in vitro/in vivo differential network backbone values of the XMN model were generated. Figure 6 shows that the in vitro and in vivo differential network backbone values were significantly correlated for the bronchial (Fig. 6b) and nasal samples (Fig. 6c). These results suggested that the in vitro tissue models could capture similar perturbation in the xenobiotic metabolism following the actual smoking situation. This similar response was not observed if the correlation was performed at the level of gene expression fold-changes (insets), suggesting that the network-based approach was more robust to detect the impact of CS exposure.

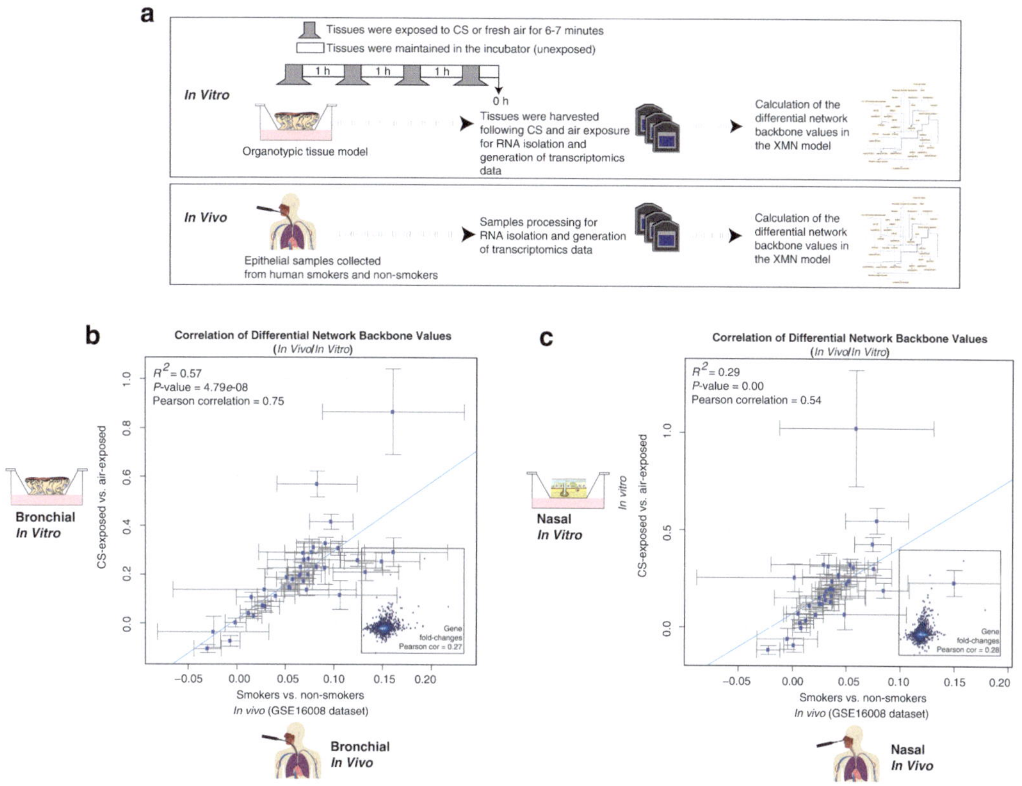

Fig. 6 Comparability between in vivo and in vitro tissue responses upon CS exposure. Illustrations of the sample generation from the in vitro and in vivo systems are shown in (**a**). In vivo/in vitro correlations between the differential network backbone values of the XMN model generated from the bronchial (**b**) and nasal samples (**c**). Pearson correlations for each of the post-exposure timepoints are shown. *Insets* show the correlations at the level of gene expression fold-changes. *CS* cigarette smoke, *XMN* xenobiotic metabolism network

Use Case 3: Assessing the Comparability of Tissue Responses at Various Post-exposure Timepoints of CS Exposure

Using the in vitro tissue models, we investigated whether the duration of post-exposure affected the comparability of the tissue responses following repeated CS exposure by leveraging the XMN model and NPA approach. The experimental design is shown in Fig. 7a (described in a greater detail in our previous publication [37]). We hypothesized that the longer the post-exposure duration, the weaker the tissue responses would be.

Figure 7b shows that the differential network backbone values generated from the bronchial and nasal tissues remained correlated at the 4, 24, and 48 h post-exposure timepoints. However, the correlations were reduced as the post-exposure time increased. This was also reflected by the perturbation scores for the entire XMN model (Fig. 7c). The results remain consistent when a commercially available data-analysis and interpretation tool, i.e., Ingenuity Pathways Analysis (IPA) was used for the analysis. The same datasets were uploaded to IPA; the two canonical pathways within the IPA's knowledge base, "Aryl Hydrocarbon Receptor

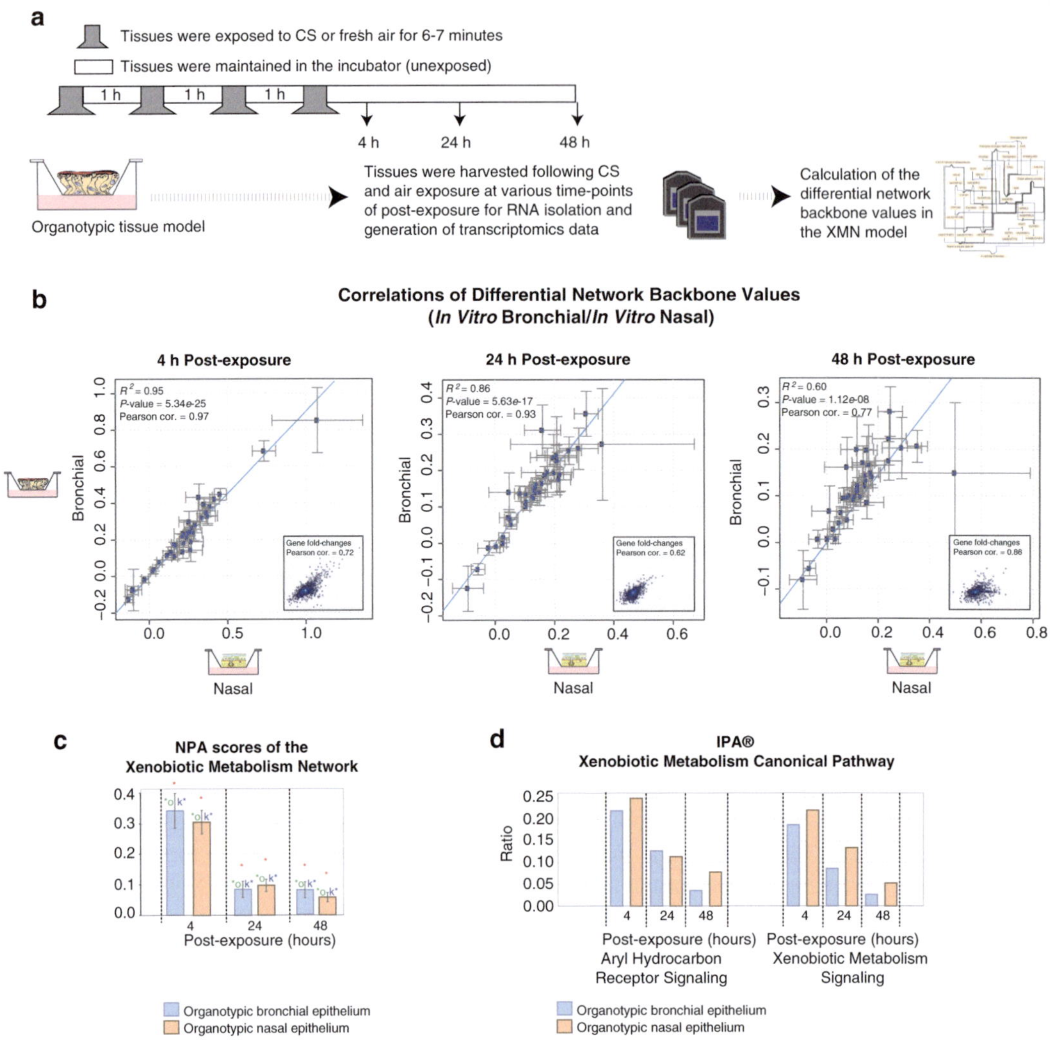

Fig. 7 Comparability between tissue responses at different post-exposure timepoints. The experimental design is shown in (**a**). The responses of the bronchial and nasal in vitro organotypic tissue models at the various post-exposure timepoints were compared at the level of backbone level the XMN model (**b**). Pearson correlations for each of the post-exposure timepoints are shown. Quantifications of the entire network perturbation are shown in (**c**) (**O*, *, *K** indicate the significance perturbation of the XMN model computed using the NPA approach; the details on the NPA approach are described in Chap. 15 of this book and in our previous publication [40]). Enrichment scores (ratio-scores) between the xenobiotic metabolism-related pathways and the in vitro dataset analyzed using IPA are shown in (**d**). *CS* cigarette smoke, *IPA* ingenuity pathway analysis, *NPA* network perturbation amplitude, *XMN* xenobiotic metabolism network

Signaling" and the "Xenobiotic Metabolism Signaling" were significantly associated with the datasets (Fig. 7d). At the later post-exposure timepoints, the ratio-scores for these pathways were reduced as compared to the earlier timepoint (Fig. 7d). The declining perturbation of xenobiotic metabolism was in agreement with a previous publication that reported that the activation of phase I enzyme was transient [50].

3 Notes

The outcomes from in silico approaches are aligned with the goal of the US Agency for Toxic Substances and Disease Registry (ATSDR), which is to minimize risk levels. A minimal risk level is used as a screening benchmark to select potential contaminants (e.g., CS) and identify their potential toxicity. QSAR, for example, is used by the Agency for Toxic Substances and Disease Registry to conduct scientific evaluation when laboratory and other computational methods are inadequate for such assessment. Future improvement of in silico predictions is likely to be driven by resources from public domain and open-source software, such as the OCHEM platform, which has been developed to automate and simplify QSAR modeling [21]. OCHEM was successfully used to study the activity inhibition of CYP1A2 [21], CYP2D6 [51], and CYP3A4 [51].

The reactivity of particular CS constituents to CYP enzymes (e.g., nicotine to CYP2A6) has been successfully studied using QM/MM. Moreover, QSAR-based evaluations have been used to predict the toxicity of PAHs (as shown in Table 3). Toxicity of other "harmful and potentially harmful constituents in tobacco products and tobacco smoke," which are specified by the US Food and Drug Administration, could be evaluated using similar approaches [20]. However, prediction models that largely depend on the chemical structures of the proteins/receptors-ligands/compounds of interest cannot be applied to predict the toxicity of other compounds that are structurally different [52]. This becomes a challenge for assessment of whole mixtures, e.g., herbal medicines, air pollutants, and CS. Additional challenges in QM/MM and QSAR studies arise from limited knowledge about the precise mechanisms of the enzyme–substrate binding, rates of the chemical reaction, and the existence of possible competing reactions that initially could be overlooked during the model building [52]. Furthermore, the lack of consensus methodology in QM/MM and QSAR studies adds another challenge during the initial method selection. Finally, choosing a validation method that can confirm the robustness, performance, and applicability of the model, is also not an easy task but fundamental to demonstrate the reliability of the models [53].

Moreover, network-based approaches become more advanced along with the expanding systems biology field. This has become evident in the field of herbal medicine because of their complex mixtures [54, 55], which evaluation has increasingly relied on "omics" approaches [56]. These approaches are similarly attractive for the assessment of a complex mixture of CS. A comparison at the network levels should be more reliable to investigate systems-level differences of exposure impact. It should be acknowledged that variability in the metabolism of xenobiotics can occur at any point in the process (i.e., from exposures to endpoints) or at the systems level (i.e., organs, tissues, and cells). Nonetheless, we have discussed that using a network-based approach, the tissue responses upon CS exposure could be evaluated (i.e., the responses between in vivo and in vitro, or those between potential surrogate and target tissues). Furthermore, in favor for xenobiotic metabolism studies, interspecies sequence-variations among the gene orthologs are mostly well documented and understood [1]. Although still limited, future studies may leverage network-based approaches to evaluate species-specific responses.

Kell and Goodacre recently stated that metabolomics network (more than transcriptomics and proteomics networks) could be more relevant for finding disease biomarkers because the change in metabolite concentration is likely to be more robust than the change of the enzyme that controls the metabolic fluxes [57]. Although we do not discuss the evaluation and interpretation of metabolomics data in this chapter, the relevant tools are increasingly available. Several tools are available for enrichment and overrepresentation analyses of metabolite data, such as, the MSEA (Metabolite Set Enrichment Analysis, http://www.msea.ca) [58], which is now part of the MetaboAnalyst (http://www.metaboanalyst.ca); MBRole database (Metabolite Biological Role, http://csbg.cnb.csic.es/mbrole/) [59]; IMPaLA web-tool (http://impala.molgen.mpg.de/) [60]; Consensus PathDB (http://cpdb.molgen.mpg.de/) [61]; and the recently developed web-tool 3Omics (http://3omics.cmdm.tw) that is able to integrate transcriptomics, proteomics, and metabolomics data in a single platform [62]. 3Omics correlation analysis can also be used generate a correlation network, which can be downloaded from the web interface [62]. Furthermore, a community-driven generation of human metabolic network (Recon 2) was recently developed; Recon2 allows mapping available information from DrugBank and identify the connection among metabolomic, genomic, and eventually phenotypic information [63]. These tools could potentially be used to predict toxicity associated with CS exposure, for example, to understand CYP-derived metabolite data generated from CS-exposed samples.

In silico prediction of metabolic transformation remains important for the first-step prediction of how exposure will impact the human body [52]. Xenobiotic metabolism in particular becomes an attractive tool for exposure assessment because of its role as a "biosensor" against CS toxicants and other toxic compounds. In pharmaceutical research, evaluating xenobiotic metabolism activation becomes interesting mainly because of the need to deduce unwanted drug–drug interactions and to recognize individuals' variability in metabolizing xenobiotics [10]. Furthermore, toxicity networks should consider not only the toxicity-promoting mechanisms (phase I enzymes), but also the detoxification mechanisms (phase II xenobiotic metabolism-related pathways) and the drug transporters (phase III enzymes), as well as pro-survival and pro-apoptosis pathways [29, 57]. Future studies could further integrate QM/MM or QSAR evaluation to a network-based analyses, for example, to evaluate a more comprehensive perturbation of xenobiotic metabolism network based on the QSAR-predicted CYP activation/inhibition. Integration of various approaches is expected to advance the value of in silico toxicology.

Acknowledgements

The in vitro experiments used for the use cases in Sect. 2.3 were conducted at and funded by Philip Morris International R&D and has been published previously in Biomed Research International [37]. We are grateful for the valuable comments from Elyette Martin, Marja Talikka, and Pavel Pospisil during the preparation of this manuscript. We thank the support from Edanz Group Ltd. for the assistance in editing and generating tables summarizing the recent QSAR studies.

References

1. Burgess-Herbert SL, Euling SY (2013) Use of comparative genomics approaches to characterize interspecies differences in response to environmental chemicals: challenges, opportunities, and research needs. Toxicol Appl Pharmacol 271(3):372–385
2. Peach ML, Zakharov AV, Liu R et al (2012) Computational tools and resources for metabolism-related property predictions. 1. Overview of publicly available (free and commercial) databases and software. Future Med Chem 4(15):1907–1932. doi:10.4155/fmc.12.150
3. Courcot E, Leclerc J, Lafitte J-J et al (2012) Xenobiotic metabolism and disposition in human lung cell models: comparison with in vivo expression profiles. Drug Metab Dispos 40(10):1953–1965
4. Oláh J, Mulholland AJ, Harvey JN (2011) Understanding the determinants of selectivity in drug metabolism through modeling of dextromethorphan oxidation by cytochrome P450. Proc Natl Acad Sci 108(15):6050–6055
5. Shimada T (2006) Xenobiotic-metabolizing enzymes involved in activation and detoxification of carcinogenic polycyclic aromatic hydrocarbons. Drug Metab Pharmacokinet 21(4):257–276
6. Omiecinski CJ, Vanden Heuvel JP, Perdew GH et al (2011) Xenobiotic metabolism, disposition, and regulation by receptors: from biochemical phenomenon to predictors of major toxicities. Toxicol Sci 120(Suppl 1): S49–S75. doi:10.1093/toxsci/kfq338
7. Croom E (2012) Metabolism of xenobiotics of human environments. Prog Mol Biol Transl Sci 112:31–88. doi:10.1016/B978-0-12-415813-9.00003-9
8. Shaik S, Cohen S, Wang Y et al (2009) P450 enzymes: their structure, reactivity, and selectivity modeled by QM/MM calculations. Chem Rev 110(2):949–1017
9. Groenhof G (2013) Introduction to QM/MM simulations, Biomolecular simulations. Springer, New York, NY, pp 43–66
10. van der Kamp MW, Mulholland AJ (2013) Combined quantum mechanics/molecular mechanics (QM/MM) methods in computational enzymology. Biochemistry 52(16): 2708–2728
11. Yang Y, Wong SE, Lightstone FC (2014) Understanding a substrate's product regioselectivity in a family of enzymes: a case study of acetaminophen binding in cytochrome P450s. PLoS One 9(2), e87058. doi:10.1371/journal.pone.0087058
12. Banáš P, Jurečka P, Walter NG et al (2009) Theoretical studies of RNA catalysis: hybrid QM/MM methods and their comparison with MD and QM. Methods 49(2):202–216

13. Sponer J, Leszczynski J, Hobza P (2001) Electronic properties, hydrogen bonding, stacking, and cation binding of DNA and RNA bases. Biopolymers 61(1):3–31. doi:10.1002/1097-0282(2001)61:1<3::aid-bip10048>3.0.co;2-4
14. Kirchmair J, Williamson MJ, Tyzack JD et al (2012) Computational prediction of metabolism: sites, products, SAR, P450 enzyme dynamics, and mechanisms. J Chem Inf Model 52(3):617–648
15. Rydberg P, Olsen L, Ryde U (2012) Quantum-mechanical studies of reactions performed by cytochrome P450 enzymes. Curr Inorg Chem 2(3):292–315
16. Pezeshki S, Lin H (2011) Adaptive-partitioning redistributed charge and dipole schemes for QM/MM dynamics simulations: on-the-fly relocation of boundaries that pass through covalent bonds. J Chem Theory Comput 7(11):3625–3634. doi:10.1021/ct2005209
17. Mishra NK (2011) Computational modeling of P450s for toxicity prediction. Expert Opin Drug Metab Toxicol 7(10):1211–1231
18. Castell JV, Donato MT, Gomez-Lechon MJ (2005) Metabolism and bioactivation of toxicants in the lung. The in vitro cellular approach. Exp Toxicol Pathol 57(Suppl 1): 189–204
19. Dudek AZ, Arodz T, Galvez J (2006) Computational methods in developing quantitative structure-activity relationships (QSAR): a review. Comb Chem High Throughput Screen 9(3):213–228
20. Sun H, Veith H, Xia M et al (2012) Prediction of cytochrome P450 profiles of environmental chemicals with QSAR models built from drug-like molecules. Mol Inform 31(11–12): 783–792
21. Sushko I, Novotarskyi S, Korner R et al (2011) Online chemical modeling environment (OCHEM): web platform for data storage, model development and publishing of chemical information. J Comput Aided Mol Des25(6):533–554.doi:10.1007/s10822-011-9440-2
22. Novotarskyi S, Sushko I, Körner R et al (2011) A comparison of different QSAR approaches to modeling CYP450 1A2 inhibition. J Chem Inf Model 51(6):1271–1280
23. Aguiar-Pulido V, Gestal M, Cruz-Monteagudo M et al (2013) Evolutionary computation and QSAR research. Curr Comput Aided Drug Des 9(2):206–225
24. Gertrudes JC, Maltarollo VG, Silva RA et al (2012) Machine learning techniques and drug design. Curr Med Chem 19(25): 4289–4297
25. Andrada MF, Duchowicz PR, Castro EA (2013) QSAR applications on polycyclic aromatic hydrocarbons and some derivatives. Curr Org Chem 17(23):2872–2879
26. Fourches D, Muratov E, Tropsha A (2010) Trust, but verify: on the importance of chemical structure curation in cheminformatics and QSAR modeling research. J Chem Inf Model 50(7):1189–1204
27. Leonard JT, Roy K (2006) On selection of training and test sets for the development of predictive QSAR models. QSAR Comb Sci 25(3):235–251
28. Roy PP, Leonard JT, Roy K (2008) Exploring the impact of size of training sets for the development of predictive QSAR models. Chemom Intell Lab Syst 90(1):31–42
29. Csermely P, Korcsmáros T, Kiss HJ et al (2013) Structure and dynamics of molecular networks: a novel paradigm of drug discovery: a comprehensive review. Pharmacol Ther 138(3):333–408
30. van Delft JH, Mathijs K, Staal YC et al (2010) Time series analysis of benzo[A]pyrene-induced transcriptome changes suggests that a network of transcription factors regulates the effects on functional gene sets. Toxicol Sci 117(2):381–392. doi:10.1093/toxsci/kfq214
31. Vastrik I, D'Eustachio P, Schmidt E et al (2007) Reactome: a knowledge base of biologic pathways and processes. Genome Biol 8(3):R39
32. Kiyosawa N, Manabe S, Sanbuissho A et al (2010) Gene set-level network analysis using a toxicogenomics database. Genomics 96(1): 39–49
33. Yang Y, Maxwell A, Zhang X et al (2013) Differential reconstructed gene interaction networks for deriving toxicity threshold in chemical risk assessment. BMC Bioinformatics 14(Suppl 14):S3
34. Schlage WK, Westra JW, Gebel S et al (2011) A computable cellular stress network model for non-diseased pulmonary and cardiovascular tissue. BMC Syst Biol 5:168. doi:10.1186/1752-0509-5-168, 1752-0509-5-168 [pii]
35. Slater T (2014) Recent advances in modeling languages for pathway maps and computable biological networks. Drug Discov Today 19(2):93–98
36. Hoeng J, Talikka M, Martin F et al (2013) Toxicopanomics: applications of genomics, transcriptomics, proteomics and lipidomics in predictive mechanistic toxicology. In: Hayes AW (ed) Principle and methods on toxicology. Taylor & Francis, London, In press
37. Iskandar AR, Martin F, Talikka M et al (2013) Systems approaches evaluating the perturba-

tion of xenobiotic metabolism in response to cigarette smoke exposure in nasal and bronchial tissues. Biomed Res Int 2013:512086. doi:10.1155/2013/512086
38. Gonzalez FJ, Fernandez-Salguero P (1998) The Aryl hydrocarbon receptor: studies using the AHR-null mice. Drug Metab Dispos 26(12):1194–1198
39. Thomson TM, Sewer A, Martin F et al (2013) Quantitative assessment of biological impact using transcriptomic data and mechanistic network models. Toxicol Appl Pharmacol 272(3):863–878
40. Martin F, Thomson TM, Sewer A et al (2012) Assessment of network perturbation amplitudes by applying high-throughput data to causal biological networks. BMC Syst Biol 6(1):54
41. Demetriou CA, Raaschou-Nielsen O, Loft S et al (2012) Biomarkers of ambient air pollution and lung cancer: a systematic review. Occup Environ Med 69(9):619–627
42. Peluso M, Neri M, Margarino G et al (2004) Comparison of DNA adduct levels in nasal mucosa, lymphocytes and bronchial mucosa of cigarette smokers and interaction with metabolic gene polymorphisms. Carcinogenesis 25(12):2459–2465. doi:10.1093/carcin/bgh259
43. Sridhar S, Schembri F, Zeskind J et al (2008) Smoking-induced gene expression changes in the bronchial airway are reflected in nasal and buccal epithelium. BMC Genomics 9(1):259
44. Zhang X, Sebastiani P, Liu G et al (2010) Similarities and differences between smoking-related gene expression in nasal and bronchial epithelium. Physiol Genomics 41(1):1–8. doi:10.1152/physiolgenomics.00167.2009
45. Bosse Y, Postma DS, Sin DD et al (2012) Molecular signature of smoking in human lung tissues. Cancer Res 72(15):3753–3763. doi:10.1158/0008-5472.CAN-12-1160, 0008-5472.CAN-12-1160 [pii]
46. Karp PH, Moniger T, Weber SP et al (2002) An in vitro model of differentiated human airway epithelia. Methods Mol Biol 188: 115–137
47. Mathis C, Poussin C, Weisensee D et al (2013) Human bronchial epithelial cells exposed in vitro to cigarette smoke at the air-liquid interface resemble bronchial epithelium from human smokers. Am J Physiol Lung Cell Mol Physiol 304(7):L489–L503
48. Maunders H, Patwardhan S, Phillips J et al (2007) Human bronchial epithelial cell transcriptome: gene expression changes following acute exposure to whole cigarette smoke in vitro. Am J Physiol Lung Cell Mol Physiol 292(5):L1248–L1256. doi:10.1152/ajplung.00290.2006, 00290.2006 [pii]
49. Pezzulo AA, Starner TD, Scheetz TE et al (2011) The air-liquid interface and use of primary cell cultures are important to recapitulate the transcriptional profile of in vivo airway epithelia. Am J Physiol Lung Cell Mol Physiol 300(1):L25–31. doi:10.1152/ajplung.00256.2010, ajplung.00256.2010 [pii]
50. Gebel S, Gerstmayer B, Kuhl P et al (2006) The kinetics of transcriptomic changes induced by cigarette smoke in rat lungs reveals a specific program of defense, inflammation, and circadian clock gene expression. Toxicol Sci 93(2):422–431
51. Novotarskyi S, Sushko I, Koerner R et al (2013) Chemogenomic approach to increase accuracy of QSAR modeling of inhibition activity against five major P450 isoforms. J Cheminform 5(Suppl 1):P23
52. Bugrim A, Nikolskaya T, Nikolsky Y (2004) Early prediction of drug metabolism and toxicity: systems biology approach and modeling. Drug Discov Today 9(3):127–135. doi:10.1016/S1359-6446(03)02971-4
53. Roy K (2007) On some aspects of validation of predictive quantitative structure-activity relationship models. Expert Opin Drug Discov 2(12):1567–1577
54. Buriani A, Garcia-Bermejo ML, Bosisio E et al (2012) Omic techniques in systems biology approaches to traditional Chinese medicine research: present and future. J Ethnopharmacol 140(3):535–544
55. Tao W, Xu X, Wang X et al (2013) Network pharmacology-based prediction of the active ingredients and potential targets of Chinese herbal Radix Curcumae formula for application to cardiovascular disease. J Ethnopharmacol 145(1):1–10
56. Leung EL, Cao Z-W, Jiang Z-H et al (2013) Network-based drug discovery by integrating systems biology and computational technologies. Brief Bioinform 14(4):491–505
57. Kell DB, Goodacre R (2014) Metabolomics and systems pharmacology: why and how to model the human metabolic network for drug discovery. Drug Discov Today 19(2): 171–182
58. Nam D, Kim SY (2008) Gene-set approach for expression pattern analysis. Brief Bioinform 9(3):189–197. doi:10.1093/bib/bbn001
59. Chagoyen M, Pazos F (2011) MBRole: enrichment analysis of metabolomic data. Bioinformatics 27(5):730–731. doi:10.1093/bioinformatics/btr001
60. Kamburov A, Cavill R, Ebbels TMD et al (2011) Integrated pathway-level analysis of transcriptomics and metabolomics data with IMPaLA. Bioinformatics 27(20):2917–2918. doi:10.1093/bioinformatics/btr499

61. Kamburov A, Pentchev K, Galicka H et al (2011) ConsensusPathDB: toward a more complete picture of cell biology. Nucleic Acids Res 39(suppl 1):D712–D717
62. Kuo TC, Tian TF, Tseng YJ (2013) 3Omics: a web-based systems biology tool for analysis, integration and visualization of human transcriptomic, proteomic and metabolomic data. BMC Syst Biol 7:64. doi:10.1186/1752-0509-7-64
63. Thiele I, Swainston N, Fleming RM et al (2013) A community-driven global reconstruction of human metabolism. Nat Biotechnol 31(5):419–425
64. Li D, Huang X, Lin J et al (2013) Catalytic mechanism of cytochrome P450 for N-methylhydroxylation of nicotine: reaction pathways and regioselectivity of the enzymatic nicotine oxidation. Dalton Trans 42(11): 3812–3820
65. Lu H, Huang X, AbdulHameed MDM et al (2014) Binding free energies for nicotine analogs inhibiting cytochrome P450 2A6 by a combined use of molecular dynamics simulations and QM/MM-PBSA calculations. Bioorg Med Chem 22(7):2149–2156
66. Usharani D, Zazza C, Lai W et al (2012) A Single-site mutation (F429H) converts the enzyme CYP 2B4 into a heme oxygenase: a QM/MM study. J Am Chem Soc 134(9): 4053–4056
67. Lonsdale R, Houghton KT, Zurek J et al (2013) Quantum mechanics/molecular mechanics modeling of regioselectivity of drug metabolism in cytochrome P450 2C9. J Am Chem Soc 135(21):8001–8015
68. Shi R, Li W, Liu G et al (2013) Catalytic mechanism of cytochrome P450 2D6 for 4-hydroxylation of aripiprazole: A QM/MM study. Chin J Chem 31(9):1219–1227. doi:10.1002/cjoc.201300427
69. Lonsdale R, Olah J, Mulholland AJ et al (2011) Does compound I vary significantly between isoforms of cytochrome P450? J Am Chem Soc 133(39):15464–15474. doi:10.1021/ja203157u
70. Shahrokh K, Orendt A, Yost GS et al (2012) Quantum mechanically derived AMBER-compatible heme parameters for various states of the cytochrome P450 catalytic cycle. J Comput Chem 33(2):119–133. doi:10.1002/jcc.21922
71. Calvaresi M, Stenta M, Garavelli M et al (2012) Computational evidence for the catalytic mechanism of human glutathione S-transferase A3-3: a QM/MM investigation. ACS Catal 2(2):280–286
72. Parker LJ, Italiano LC, Morton CJ et al (2011) Studies of glutathione transferase P1-1 bound to a platinum (IV)-based anticancer compound reveal the molecular basis of its activation. Chemistry 17(28):7806–7816
73. Mueller RM, North MA, Yang C et al (2011) Interplay of flavin's redox states and protein dynamics: an insight from QM/MM simulations of dihydronicotinamide riboside quinone oxidoreductase 2. J Phys Chem B 115(13):3632–3641
74. Pan J, Liu G-Y, Cheng J et al (2010) CoMFA and molecular docking studies of benzoxazoles and benzothiazoles as CYP450 1A1 inhibitors. Eur J Med Chem 45(3):967–972
75. Gonzalez J, Marchand-Geneste N, Giraudel J et al (2012) Docking and QSAR comparative studies of polycyclic aromatic hydrocarbons and other procarcinogen interactions with cytochromes P450 1A1 and 1B1. SAR QSAR Environ Res 23(1-2):87–109
76. Sridhar J, Ellis J, Dupart P et al (2012) Development of flavone propargyl ethers as potent and selective inhibitors of cytochrome P450 enzymes 1A1 and 1A2. Drug Metab Lett 6(4):275–284
77. Sridhar J, Foroozesh M, Stevens CK (2011) QSAR models of cytochrome P450 enzyme 1A2 inhibitors using CoMFA, CoMSIA and HQSAR. SAR QSAR Environ Res 22(7-8):681–697
78. Rahnasto MK, Raunio HA, Wittekindt C et al (2011) Identification of novel CYP2A6 inhibitors by virtual screening. Bioorg Med Chem 19(23):7186–7193
79. Gharaghani S, Khayamian T, Keshavarz F (2012) Docking, molecular dynamics simulation studies, and structure-based QSAR model on cytochrome P450 2A6 inhibitors. Struct Chem 23(2):341–350
80. Lewis DF, Ito Y, Lake BG (2010) Quantitative structure-activity relationships (QSARs) for inhibitors and substrates of CYP2B enzymes: importance of compound lipophilicity in explanation of potency differences. J Enzyme Inhib Med Chem 25(5):679–684
81. Roy PP, Roy K (2010) Pharmacophore mapping, molecular docking and QSAR studies of structurally diverse compounds as CYP2B6 inhibitors. Mol Simul 36(11):887–905
82. Taxak N, Bharatam P (2013) 2D QSAR study for gemfibrozil glucuronide as the mechanism-based inhibitor of CYP2C8. Indian J Pharm Sci 75(6)
83. Saraceno M, Massarelli I, Imbriani M et al (2011) Optimizing QSAR models for predicting ligand binding to the drug-metabolizing cytochrome P450 isoenzyme CYP2D6. Chem Biol Drug Des 78(2):236–251
84. Jónsdóttir SÓ, Ringsted T, Nikolov NG et al (2012) Identification of cytochrome P450 2D6 and 2C9 substrates and inhibitors by QSAR analysis. Bioorg Med Chem 20(6):2042–2053

85. Mo S-L, Liu W-F, Li C-G et al (2012) Pharmacophore, QSAR, and binding mode studies of substrates of human cytochrome P450 2D6 (CYP2D6) using molecular docking and virtual mutations and an application to Chinese herbal medicine screening. Curr Pharm Biotechnol 13(9):1640–1704
86. Martikainen LE, Rahnasto-Rilla M, Neshybova S et al (2012) Interactions of inhibitor molecules with the human CYP2E1 enzyme active site. Eur J Pharm Sci 47(5): 996–1005
87. Didziapetris R, Dapkunas J, Sazonovas A et al (2010) Trainable structure–activity relationship model for virtual screening of CYP3A4 inhibition. J Comput Aided Mol Des 24(11): 891–906
88. Hamon V, Horvath D, Gaudin C et al (2012) QSAR modelling of CYP3A4 inhibition as a screening tool in the context of drug-drug interaction studies. Mol Inform 31(9): 669–677
89. Handa K, Nakagome I, Yamaotsu N et al (2012) Three-dimensional quantitative structure-activity relationship analysis of inhibitors of human and rat cytochrome P4503A enzymes. Drug Metab Pharmacokinet 28(4):345–355
90. Ako R, Dong D, Wu B (2012) 3D-QSAR studies on UDP-glucuronosyltransferase 2B7 substrates using the pharmacophore and VolSurf approaches. Xenobiotica 42(9): 891–900
91. Kobeticova K, Simek Z, Brezovsky J et al (2011) Toxic effects of nine polycyclic aromatic compounds on Enchytraeus crypticus in artificial soil in relation to their properties. Ecotoxicol Environ Saf 74(6):1727–1733. doi:10.1016/j.ecoenv.2011.04.013
92. Li F, Li X, Liu X et al (2011) Noncovalent interactions between hydroxylated polycyclic aromatic hydrocarbon and DNA: molecular docking and QSAR study. Environ Toxicol Pharmacol 32(3):373–381. doi:10.1016/j.etap.2011.08.001
93. Al-Fahemi JH (2012) The use of quantum-chemical descriptors for predicting the photoinduced toxicity of PAHs. J Mol Model 18(9):4121–4129.doi:10.1007/s00894-012-1417-0
94. Li F, Wu H, Li L et al (2012) Docking and QSAR study on the binding interactions between polycyclic aromatic hydrocarbons and estrogen receptor. Ecotoxicol Environ Saf 80:273–279. doi:10.1016/j.ecoenv.2012.03.009
95. Xu X, Li X-G, Sun S-W (2012) A QSAR study on the biodegradation activity of PAHs in aged contaminated sediments. Chemometr Intell Lab Syst 114:50–55, http://dx.doi.org/10.1016/j.chemolab.2012.03.002
96. Xu HY, Zou JW, Min JQ et al (2012) A quantitative structure-property relationship analysis of soot-water partition coefficients for persistent organic pollutants. Ecotoxicol Environ Saf 80:1–5. doi:10.1016/j.ecoenv.2012.02.002
97. Zhang Y-F, Zhang L, Gao Z-X et al (2012) Investigating the quantitative structure-activity relationships for antibody recognition of two immunoassays for polycyclic aromatic hydrocarbons by multiple regression methods. Sensors 12(7):9363–9374
98. Zhang YF, Ma Y, Gao ZX et al (2010) Predicting the cross-reactivities of polycyclic aromatic hydrocarbons in ELISA by regression analysis and CoMFA methods. Anal Bioanal Chem397(6):2551–2557.doi:10.1007/s00216-010-3785-6
99. Tansel B, Lee M, Tansel DZ (2013) Comparison of fate profiles of PAHs in soil, sediments and mangrove leaves after oil spills by QSAR and QSPR. Mar Pollut Bull 73(1):258–262
100. Vračko M, Bobst S (2013) Performance evaluation of CAESAR–QSAR output using PAHs as a case study. J Chemom 28(2):100–107
101. Lim SJ, Fox P (2014) Effects of halogenated aromatics/aliphatics and nitrogen (N)-heterocyclic aromatics on estimating the persistence of future pharmaceutical compounds using a modified QSAR model. Sci Total Environ 470:348–355
102. Xu X, Li X (2014) QSAR for photodegradation activity of polycyclic aromatic hydrocarbons in aqueous systems. J Ocean Univ China 13(1):66–72

Chapter 13

High-Content Screening: Understanding and Managing Mechanistic Data to Better Predict Toxicity

Paul Walker, Tim Smith, Kerry Frost, Sarah Kelly, and Ignacio Gonzalez Suarez

Abstract

An increased understanding of the cellular pathways involved in toxicity responses, coupled with a simultaneous advance in technology, has allowed for a shift in the way that the safety assessment of novel chemicals is performed. The development of assays that offer a high-throughput and low-cost option in comparison to more traditional approaches has been a focus of recent years.

Early identification of compounds which have the potential to cause Drug Induced Liver Injury (DILI) remains a major challenge for the pharmaceutical industry. Improvements in the mechanistic understanding of the cellular processes involved in this complex response have allowed for models to be generated and more reliable predictions to be made.

High-Content Screening (HCS) describes an approach whereby multiple end points can be monitored in a single assay. The focus of this chapter is to introduce HCS with relation to using automated fluorescence microscopy in order to assess phenotypic changes within cells and, more specifically, how this can be incorporated into the drug discovery process.

We discuss the advantages that HCS can offer, whilst highlighting important factors to take into account when considering establishing the approach within a laboratory. Four papers whereby HCS has been used to highlight the potential of compounds to cause DILI are reviewed and compared. In addition, an option for including HCS as part of a wider workflow to identify environmental toxicants is introduced.

Key words High-content screening, HCS, Mechanistic, Data interpretation, Cell culture, Image analysis, Toxicity, Exposure, Dose response, DILI, Cytotoxicity, Systems biology, Biological pathways

1 Introduction

This chapter reviews high-content screening (HCS) as a useful tool in assessing toxicities observed in drug development and also applied other industry sectors. In recent years, the technology of HCS has gained significant traction as a useful tool in the in vitro predictive assessment of toxicity. HCS refers to any technique that can generate multiple readouts from a cellular assay simultaneously; however, HCS or high-content analysis (HCA) is more

Julia Hoeng and Manuel C. Peitsch (eds.), *Computational Systems Toxicology*, Methods in Pharmacology and Toxicology, DOI 10.1007/978-1-4939-2778-4_13, © Springer Science+Business Media New York 2015

often used as the term referring specifically to the monitoring of different phenotypic changes as measured by automated fluorescence microscopy [1, 2]. Multiparametric cellular analysis allows mechanistic insights to be gained, and therefore HCS has proven to be very useful in deconvoluting complex biological processes in a number of important areas, including drug development and toxicological assessment. Any new drug on average costs approximately two billion dollars between discovery and registration; thus, it is clear that elucidating any potential safety concerns in the early phases of drug development is of prime importance [3]. For decades, drug-induced liver injury (DILI) has been the most frequent single cause of safety-related drug marketing withdrawals [4] and consequently, the assessment of risk of DILI is a top priority in the pharmaceutical industry [5, 6]. Current approaches to the assessment of DILI that employ HCS are described later in this chapter and used as examples of HCS capabilities and potential.

1.1 Technology, Advantages, and Challenges of HCS

Historically, toxicological or safety assessment of new chemical entities has been built around in vivo animal studies. These types of approaches have been, and still are, valuable, providing important information on systemic toxic effects and long-term exposures. However, there are a number of major drawbacks that continuously challenge the suitability of in vivo studies in modern-day toxicological assessment. First, animal studies are time-consuming, labor-intensive and rather limited in the number of doses and end points that can be assessed simultaneously. As a result, they offer little mechanistic understanding and limited insight into the mode of action by which chemicals affect cellular processes, leading to adverse outcomes [7–9]. This is particularly the case for idiosyncratic hepatic toxicity, which has caused several late-stage and post-launch drug withdrawals [10]. Second, most studies rely on the development of phenotypic responses observed only at high doses that do not represent relevant exposure scenarios in the human population. Third, relevant aspects of human and rodent biology are quite different (e.g., xenobiotic metabolism), which sometimes makes data interpretation elusive. Finally, they require a large number of animals, which raises ethical concerns. Clearly, in vitro assays that offer high levels of sensitivity and specificity in predicting in vivo toxicity are of great utility. With HCS techniques being of relatively low cost and high throughput they become an attractive tool—especially given the relative ease with which assays can be multiplexed.

HCS involves cells or tissues in well plate format being typically dosed with test compound and then incubated for a set period of time (typically 4–72 h). After the initial incubation period, the relevant fluorescent dyes or antibodies are added depending on the desired end point to be determined. This is followed by fixing and washing steps if required, and then the plates are ready for analysis and data interpretation [11] *see* Fig. 1 for an example HCS workflow. Automated image capture and processing hardware and software is

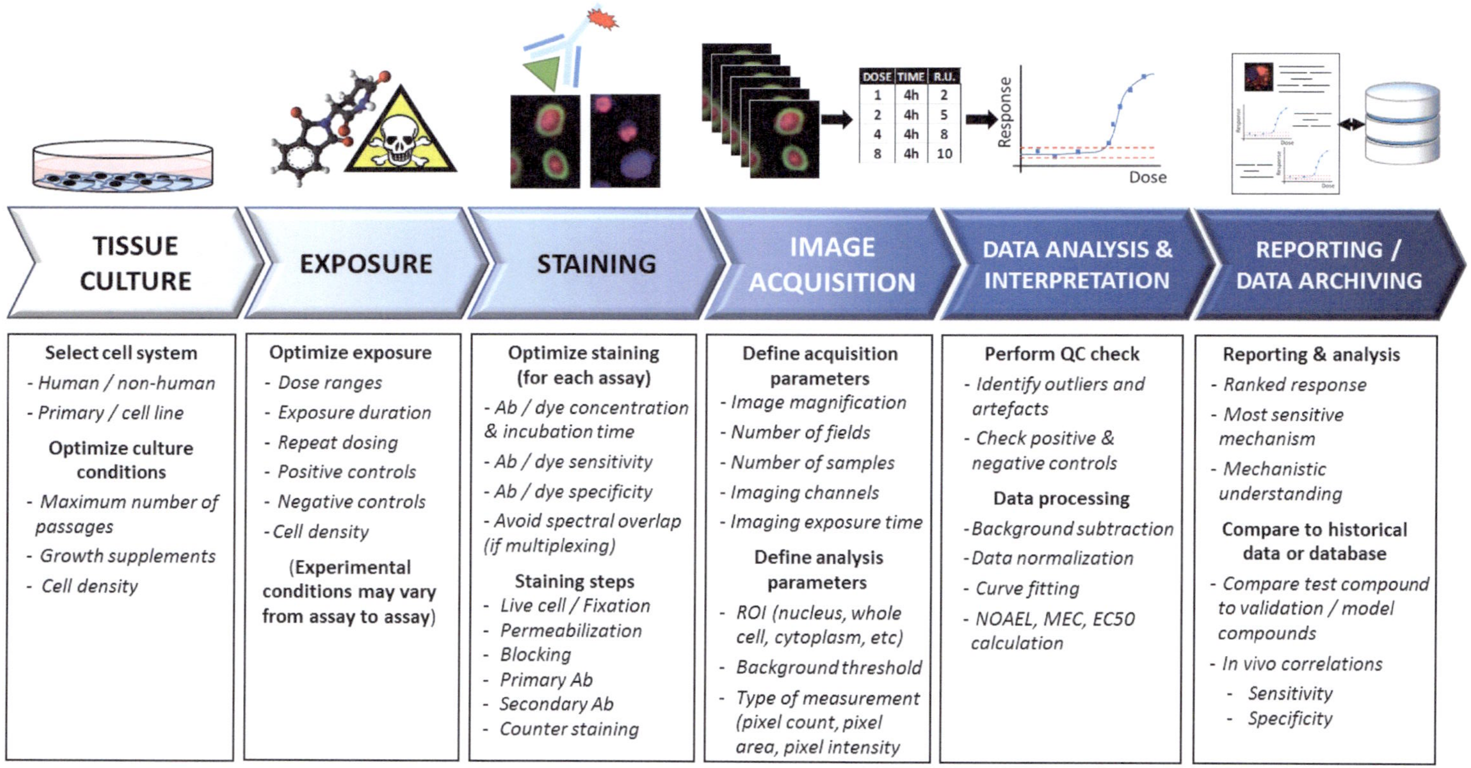

Fig. 1 The key steps in HCS-based assays. The first step is the selection of an appropriate cellular model, which will depend on the type of assay(s) to be performed, as well as the optimization of the tissue culture conditions. The second step is the optimization of the exposure conditions (dose, exposure time, etc.) so that a dose response is observed. Exposure conditions will vary depending of the final end point and, therefore, need to optimized independently for each HCS assay. The use of positive and negative controls is highly recommended. The third step is detection, which involves the use of live cell dyes and/or fluorescent-labelled antibodies as well as standard immunofluorescence procedures. The fourth is image acquisition and involves not only the generation of the images but also the measurement and quantification of the fluorescent signals. In order to generate robust data, the image acquisition and analysis parameters need to be carefully selected. The use of appropriate positive and negative controls during the optimization of this step is essential. The fifth step is data analysis and interpretation, which usually involves quality check of the data using appropriate controls. Data interpretation also takes place at this step. The results in dosed cells are compared to the controls and the non-exposed cells and relevant parameters (e.g., MEC) are calculated. During the final step, analyzed data needs to be formatted and organized into a report. The use of predefined templates for automatic report generation is recommended when a large number of samples are analyzed. During this step, the data is often compared to a database of validated compounds or to historical data generated in-house. *ROI* region of interest, *MEC* minimum effective concentration, *NOAEL* no observed adverse effect level

capable of capturing fluorescence from discrete areas of the cell (e.g., the nucleus), and thus specific areas of dye accumulation or loss are quickly determined. HCS is a technique that relies on fluorescent imaging, and hence the only barrier to multiplexing of assays lies in the spectral interference of the chosen dyes or antibodies. This allows for the creation of customized assays that can be tailored to highly specific targets or mechanisms of action. The choice of cell type can also be tailored to particular needs and in fact, the selection of an appropriate experimental model, which will depend on the type of question that needs to be addressed, becomes a critical step in HCS.

HCS offers the ability to detect cellular pathways associated with sublethal toxicities allowing a sensitive early indication. Many conventional cytotoxicity assays (e.g., LDH release, trypan blue influx, and ATP measurement) demonstrate low sensitivity because the end points are typically only measured at a point very close to or at cell death [12]. HCS-based assays are much more flexible, with a huge variety of end points that can be assessed. Subtle changes in single cells can be picked up using automated imaging algorithms, allowing for rapid data turnaround. With perhaps thousands of cells being imaged per well, whole population behavior is determined easily. Not only does the combination of multiple end points allow for a much better understanding of the mechanisms involved, but, with well-chosen drug exposure times (and potentially the use of live cell imaging), it also permits an insight into the sequence of events and mechanisms leading to toxicity. This is not always possible with conventional cytotoxicity assays, most of which are terminal in nature.

The possibility to measure multiple end points simultaneously with HCS using live cellular dyes in a high-throughput manner dramatically decreases experiment time and therefore cost. In turn, this has resulted in a higher proportion of compounds being studied (and therefore meaningful information about their toxicity profile gained) at an earlier phase of the drug discovery program. HCS represents a modern approach to toxicity screening that can provide timely go/no-go decisions with adequate selectivity and sensitivity. Nevertheless, HCS-based approaches face some challenges if they are to be integrated fully into the drug discovery process.

Numerous cell types have been employed in published strategies, with HepG2 hepatocarcinoma cells being a popular choice in the assessment of DILI [5, 13]. Other cell types including HEK293, MCF7, HCT116, and HeLa have been used [1]. Apart from HEK293 (human embryonic kidney), all these cell types are human cancer cell lines. Indeed, Gasparri [1] suggested that of the 12 most common cell types used in HCS assays up to that point, approximately 75 % were derived from cancer cell lines. These immortalized cell lines offer ease of culture and handling, and allow for effects on cell cycling to be assessed. Furthermore, with the capacity for HCS to elucidate specific pathways of action, this method has a particular application in early potency screening in the oncology field. To that end, numerous assays have been developed that use the relevant cell types e.g., the adenocarcinomic human cell line A549 [14]. Immortalized cell lines, however, are not always physiologically relevant, with no such thing as the "perfect cell"; therefore, the selection will depend ultimately on the questions that need to be addressed. Immortalized cell lines are characterized by a dividing phenotype, which is often stark contrast to non-dividing primary cells. In addition, a number of such cell lines (for example HepG2) have been reported to be metabolically deficient [15]. This potentially reduces their ability to elucidate pathways of toxicity that may be related to drug metabolites. Primary cells are inherently more heterogeneous in their biological responses when compared to immortalized cell lines, so it is generally more challenging to optimize the experimental conditions to ensure reproducibility. Nevertheless, primary human cells are a better representation of in vivo physiology. Because both adherent primary cells and cell lines are compatible with HCS techniques there are few barriers to cell type choice. The high level of flexibility in both cell type and dye/antibody choice renders HCS a very powerful tool in the assessment of both toxicity and efficacy.

1.2 Data Storage and Automization

Because HCS is a high-throughput image capturing technique, thousands of images can be generated from each run, it is critical that image capture and storage are well managed. It is important to determine how many images are required to represent the whole well in order to reduce associated costs of image storage. Assuming a 60 well assay, with 4 dyes to be imaged and 20 fields of view per well, then it is not uncommon for image data to account for over 1 gigabyte per plate. Multiplying this value by a reasonable throughput of 50 plates per month and the need for high capacity storage becomes clear. High volume server storage coupled with strict image management is of prime importance. Image storage is the major but not the only part of the data management issue, because of the derived data (e.g., the cell features recorded) and the metadata (e.g., the protocol information) there is a need for database management too. Indeed, storage issues are major drivers of the cost of setting up a fully functioning HCS screening facility; however, standards such as the minimum information standard (originally set up to support genomics and proteomics, *see* [16]) have allowed data to be stored in XML format, which confers the ability to analyze it remotely using a wide variety of tools [17]. It is important to consider not only the server space requirement but also the need for careful raw data archive management because it is unlikely that most research facilities will have the ability to store data endlessly. Thus, data backup solutions are likely to be necessary.

As dyes and antibodies can be multiplexed using HCS, and because each dye/antibody can be localized to defined areas of the cell, image analysis is a complex computational step that has been addressed by software offered by HCS instrument manufacturers and third party software developers. These solutions provide user-friendly software and protocols that can support rapid data processing. After data have been created and stored, the complex issue of data analysis arises. Because the creation of dose–response curves for each end point is the likely output of HCS-based toxicity assays, the best ways to process the data still have to be addressed. For certain mechanisms or for ease of analysis, it may be necessary to determine the minimum effective concentration, defined as the lowest concentration of test compound that produces a response outside of vehicle control responses (*see* [5, 6, 13]). Other approaches use an AC_{50} based analysis that determines the concentration at which half maximal response is observed (*see* [18]), and/or incorporating exposure information such as Cmax (i.e., the observed maximum concentration of absorbed drug) in the data analysis (*see* [5, 11]). Recently, early physicochemical data and modeling approaches have been used to enhance the predictivity of HCS approaches (*see* [18]). The length of exposure to the test compound can also be varied and may be important to determine a certain mechanistic response; for example, some end points may respond early to test compound insult but adaptive responses may dominate at longer exposure times, and different classes of compounds may have different temporal responses. In addition, hormetic responses can be followed; for instance, in the reactive oxygen species (ROS)-driven toxicity to pancreatic ß-cells, low levels of ROS are protective, whereas high levels are often toxic [19], showing that an understanding of the mechanism of toxicity and cellular response is essential in understanding toxicity.

In this chapter, variables such as mechanisms of toxicity, cell type selection, incubation time, and test compound concentration, and data interpretation—all of which are essential in understanding what approaches are most relevant in determining in vivo toxicity from in vitro HCS data—are discussed. All such variables, however, only optimize the in vitro experiments. Where HCS experiments need to add real power is in the extrapolation of the end point responses into the in vivo situation. This is where a clear impact can be made on cost and timelines of drug development. If in vivo experiments can be either better informed in their direction, or compounds can be dropped from in vivo testing altogether because robust and predictive in vitro HCS data are available, then significant efficiencies can be gained. HCS methods have given toxicologists a uniquely sensitive, data-rich, and highly customizable platform on which to base decisions on drug development. As discussed later in this chapter, there is a natural fit with omics- and systems biology-based approaches. HCS techniques can be customized to follow a variety of biological pathways or mechanisms over time, and coupled with advances in modeling techniques HCS is becoming a key tool in many industries including pharmaceutical, agrochemical, cosmetic, and chemical for the better prediction and understanding of the potential for new chemical entities to be toxic.

2 Methods

2.1 Mechanistic HCS Assays

The techniques of HCS can be used to investigate many individual or combined mechanistic end points of toxicity. With the introduction of different dyes to monitor cellular organelles and antibodies to specific signals within pathways, complex cellular pathways and organ-specific toxicity can be monitored [20]. For assays that are used to determine a mechanism of action, it is important that the relevance to the in vivo situation is understood. This can be achieved by determining, if possible, the sensitivity, specificity, and accuracy of each responding mechanism. Here, sensitivity is defined as the number of true positives that tested positive in the assay divided by the total number of true positives; specificity is defined as the number of true negatives that tested negative in the assay divided by the total number of true negatives; and accuracy is defined as the number of correctly assigned compounds in the assay divided by the total number of compounds tested.

Various studies have tried to determine the predictivity of certain HCS end points when HCS is performed in different cell types and at different time points (e.g., [21]). For each mechanistic end point, sufficient data are required from both positive and negative compounds that have known effects in vivo. The set of compounds chosen for the in vitro validation must also be sufficiently large to give confidence in the prediction and ideally reevaluated with an independent test set.

One of the difficulties of validating such assays is a lack of in vivo data to test the hypothesis. Generally, the development of compounds can be stopped because of gross toxicity; however, the underlying mechanisms are not always completely understood or determined. Furthermore, a number of HCS assays reported in the literature appear to predict in vivo toxicity well; however, data sets tend to be small and often there is a lack of known truly negative compounds that could be tested in these assays. If chemical entities are tested to a high enough concentration, generally they will cause some cytotoxicity. Therefore, any in vitro data that are generated are often evaluated in context with free circulating plasma or tissue levels to realistically compare them to the in vivo situation.

A mechanistic end point ideally would allow clear interpretation of a positive or negative response. Traditionally, from dose–response curves combinations of AC_{50}, minimum effective concentration, and/or fold increase above vehicle control have been used to determine the effect of a particular drug or chemical (e.g., [13, 21]).

Mechanistic end points should also be considered alongside general cytotoxicity assays. It may be that observed effects are due to cytotoxicity. Furthermore, often cell loss may mask a mechanistic response due to temporal effects of the test compound.

Appropriate cell lines or cellular models must be considered when approaching any in vitro assay. Generally, it is recommended that cells from tissues/species of interest are used (i.e., liver cells to investigate hepatotoxicity and renal cells to predict nephrotoxicity), assuming the cell line or model best represent the in vivo situation. Some cell lines can influence assay outputs as shown by Fowler et al. [22] who suggested that the use of human p-53 competent cell lines may reduce the incidence of false positives in genotoxicity assays. Cell lines with reduced or no metabolic capability (e.g., HepG2 cells) may reduce the ability to detect pro-toxicants that undergo metabolic transformation to produce toxic metabolites [23]. As such, metabolically competent cell lines may be preferable, although this can also be achieved by approaches such as co-incubation techniques with rat liver S9 extract, to allow metabolic transformations to occur within the experiment. These techniques have been used routinely in genotoxicity in vitro studies to study metabolic activation; however, they are now beginning to be used in HCS assays (e.g., [24]).

2.2 Mechanistic Understanding Using HCS

Many mechanistic end points can be investigated using HCS. Here, three examples are discussed from their individual contexts of assay validation and prediction to in vivo outcomes.

2.2.1 Steatosis

Steatosis is the accumulation of lipid droplets in the liver, which can lead to steatohepatitis, potentially resulting in liver failure. It is therefore important to screen for drugs for steatotic potential. Amacher [25] described several non-HCS in vivo and in vitro approaches to investigate steatosis (i.e., histological evaluation of biopsy or postmortem tissue, colorimetric or fluorescent end point in vitro assays, and flow cytometry assays) and discussed the need to develop novel and improved assays because of the unavailability, unreliability, and/or subjectivity of histological data from in vivo animal studies to correlate in vitro studies with in vivo data. Donato et al. [26] presented data on a HCS end point assay for drug-induced steatosis in HepG2 cells to replace a flow-cytometry assay reported earlier [26] to avoid having to detach the cells for analysis, which was labor intensive and limited the application of the assay to high-through-put screening. Steatosis was determined by measuring the staining of intracellular fat deposits by BODIPY 493/503 fluorescent dye. Twenty-two compounds were used (16 positive and 6 negative for steatosis in vivo). Evaluation criteria were based on a simple fold increase over vehicle control (i.e., 130 % or higher increase in intracellular BODIPY 493/503 fluorescence compared with untreated cells), and sensitivity and specificity were both calculated to be 100 % with steatosis generally observed at sub-cytotoxic concentrations for positive compounds. However, to increase confidence in the data and to be certain of the usefulness of this assay within a screening program, a larger data set of compounds would need to be tested using this assay and compared with in vivo data. As mentioned previously, one of the difficulties in validating such assays is the lack of in vivo data and small in vitro data sets. Having a larger number of compounds tested through the assays would add more weight and statistical significance to the usefulness of this assay as an early toxicity screen. Interestingly, in Donato et al.'s study [26], additional end points of reactive oxygen species (ROS) and mitochondrial membrane potential were also monitored and the different response of these end points were reported. This allowed these data to be used to pin down the underlying mechanisms of action of the steatotic effect of the compounds tested, highlighting the added benefit of multiplexing end points in a single experiment. Some compounds that were positive for steatosis showed changes in both ROS and mitochondrial potential while others showed alterations in only one of these additional end points. These findings gave some insights in to the mechanisms of action of the compounds; for example, combined effects of oxidative stress and mitochondrial dysfunction (e.g., tetracyclines), and mitochondrial DNA depletion (e.g., antiviral agents) [26].

2.2.2 Histone 2AX Phosphorylation

Histone 2AX phosphorylation (H2aX) is known to be a marker of DNA damage (double-strand breaks) and has been evaluated as a marker for genotoxicity. H2aX has been investigated in vitro using flow cytometry end points as an early screen prior to in-depth and more expensive in vitro and in vivo studies [27]. Tsamou et al. (2012) showed that the flow cytometry measurement of H2aX in conjunction with other genotoxicity in vitro assays (e.g., bacterial reverse mutation assay (Ames), mouse lymphoma assay (MLA), and micronucleus test (MNT)) could be a useful addition to early screen genotoxicity tests. Garcia-Canton et al. [24] presented data on a novel HCS assay where an increase in H2aX phosphorylation was detected by HCS as a marker for DNA double-strand breaks. This assay was performed in a metabolically limited cell-line (BEAS-2B), and therefore rat liver S9 extract was included in the incubations to try to improve predictions for metabolically activated toxicants. Evaluation criteria were based on previous data from flow cytometry experiments [28] using a simple 1.5-fold increase over vehicle control to indicate positive compounds. Twenty-two compounds (12 positive, 8 negative, and two aneugenic compounds that were equivocal) were tested in an assay based on ECVAM recommendations that have been used in previous assessments of in vitro genotoxicity such as Ames, MLA, and in vitro chromosome aberration (CA) or MNT [29]. Aneugenic compounds affect the cell cycle process and are not necessarily linked to direct DNA damage; therefore, they were classed as non-genotoxic. However, to calculate sensitivity and specificity, the data were analyzed by considering that these compounds were in both the positive and negative groups. When the aneugenic compounds were considered positive, the sensitivity, specificity, and accuracy of the assay were 86, 88, and 86 %, respectively, and when the aneugenic compounds were considered negative, the sensitivity, specificity, and accuracy remained high at 92, 80, and 86 %, respectively. On comparison with the data from other in vitro assays, it was found that Ames was more specific and the other in vitro tests (MNT and MLA) were more sensitive. However, in general, the assay performed well and out-performed flow cytometry testing based purely on specificity and sensitivity data reported by Tsamou et al. [27], making the HCS measurement of H2aX a promising addition to early screening in this area. Because of current regulatory guidelines, considerably more data with a larger data set (Kirkland et al. [29] tested >800 compounds) would be required before this assay could replace any of the current in vitro genotoxicity assays; however, its use as a supporting assay alongside these other assays would be a powerful addition to early screens in the genotoxicity field.

2.2.3 Phospholipidosis

Phospholipidosis, the drug-induced accumulation of phospholipids in lysosomes, is an adverse drug response that can delay or prevent drug registration. Phospholipidosis can be investigated using HCS approaches through direct measurement of phospholipid accumulation. Van de Water et al. [21] developed a HCS assay that compared two cell lines (CHO-K1 and HepG2 cells) to detect phospholipidosis using a fluorescent phospholipid biomarker (NBD-PE). In this assay, a compound was classified as positive when it showed a dose dependent and statistically significant ($P<0.025$) increase of fluorescence at non-cytotoxic concentrations compared with the vehicle control. In their study, 56 compounds were analyzed, 25 of which were known to induce phospholipidosis in vivo and 31 that were classified as negative. Based on their HCS response alone, the accuracy, specificity, and sensitivity were reported to be greater in CHO-K1 cells (89, 87, and 92 %, respectively) than in HepG2 cells (84, 81, and 88 %, respectively). Both these assays out-performed the in silico model for phospholipidosis developed by Ploemen et al. [30], which was reported to have accuracy, specificity, and sensitivity of 80, 84, and 76 %, respectively, highlighting the benefit of using HCS screens early in drug development to investigate drug induced phospholipidosis. A study by Bhandari et al. [31] found that species differences between rhesus monkey and rat in vivo phospholipidosis could be predicted using species-specific primary hepatocytes. The study of 15 compounds was performed in HepG2, primary rat, and primary rhesus monkey hepatocytes, and phospholipidosis was measured by monitoring for NBD-PE accumulation. Data were analyzed by normalizing the response to the maximal response observed with amiodarone as the most potent compound tested, and any concentrations where cytotoxicity was observed (LDH release) were excluded. The data showed that, with the exception of thioridazine, NDB-PE accumulation was similar in rat hepatocytes and HepG2 cells; however, four compounds (tamoxifen, perhexiline, clomipramine, and haloperidol) that caused potent NBD-PE accumulation in HepG2 and rat hepatocytes showed little effect in rhesus monkey hepatocytes. While these data suggest that this assay could be used to investigate species differences in in vivo phospholipidosis, the lack of published confirmatory in vivo data and the small number of compounds that have been tested, means more data would likely be needed to confirm these findings. This observation highlights the need to test appropriate cell lines or cellular models.

As the above examples demonstrate, HCS is a powerful tool to investigate mechanisms of toxicity and species/organ specific toxicities. In addition, the potential power of HCS comes from multiplexing different end points and mechanisms of cellular response to investigate the underlying causes of toxicity and to use this data to predict more complex toxicological pathways. Attempts have been made to predict DILI using multi-mechanistic HCS in both academic and pharmaceutical situations. These approaches are discussed in relation to how difficult it can be to determine the best approach when comparisons are made across different studies (see Sect. 3).

3 Results

3.1 Using HCS to Predict DILI in Drug Discovery

There are multiple applications for HCS technologies within drug development, some of which have been discussed earlier. However, here the focus will be on applying the technology to understand the mechanisms behind and to predict DILI. DILI is a type of adverse drug reaction and, partly because of the lack of good concordance between animals and man, is a leading cause of failed drug development, failed registration, withdrawal of licensed drugs, and cautionary labelling, which restricts drug usage [32, 33]. More than 50 approved drugs have been withdrawn from domestic markets worldwide because of DILI [34, 35]. The underlying mechanisms of DILI are complex and the clinical manifestations are diverse [36].

Often the mechanisms that result in DILI are unrelated to the pharmacology of the drug. It has been proposed that idiosyncratic DILI may involve an immune component, where a drug may induce an immune response to a reactive metabolite acting as a hapten or to a native protein. The difference in susceptibility to this immune response, which may have delayed onset, offers an explanation for the inter-individual differences observed in human DILI [37].

The hepatotoxicity that occurs as a result of the immune response may present itself through changes in a variety of cellular processes and, ultimately, in many cases, cell death. More traditional approaches to predicting human liver damage by drugs have relied on high dose testing in animals with extrapolations to relate the results to lower-dose human studies. More recently, testing has progressed from using these expensive, low-throughput approaches where high usage of laboratory animals is required towards developing well-designed in vitro assays to aid in understanding the innate cellular processes, termed toxicity pathways, that may be perturbed by chemicals [8].

Although such toxicity pathways are not always implicit in the early mechanisms that result in DILI, by assessing the downstream effects of compounds on these pathways, an in vitro assessment of DILI potential can be obtained. Because HCS is an in vitro approach whereby multiple cellular markers can be assessed in a single, high-throughput assay, the potential strength of HCS in the early identification of drug candidates with the potential to cause DILI has been investigated in a number of publications [5, 6, 11, 13, 38–40].

3.2 Assessment of the Current Progress of HCS Approaches to Predict DILI

Four papers Persson et al. [11], Xu et al. [6], O'Brien et al. [5], and Tolosa et al. [13], that discuss HCS as a strategy for predicting compounds that pose a risk of DILI were selected for further analysis and discussion. The general approach and assay conditions reported in these papers are summarized in Table 1.

3.2.1 Similarities and Differences Across the Four Selected DILI Publications

Although, in general, the methodologies used in the four publications were the same, clearly there are many variables (Table 1) that can be adjusted within a complex approach such as HCS. For example, in three of the four papers, HepG2 cells were the cell model of choice, whereas in one group human hepatocytes were used [6]. The seeding density also varied from 5 to 15 K cells per well, although in all four studies a 24-h equilibration time was used once the cells were seeded. There were additional differences across the compound treatment regimens that were applied. In three of the four papers a 24-h time point was used for the treatment time; however, in the Xu et al. [6] study, this was the only time point studied, while in the Tolosa et al. [13] and Persson et al. [11] studies, additional times of 3 and 72 h were used, respectively. In the paper by O'Brien et al. [5] the compounds were applied only for a time period of 72 h.

Similarly, the compound treatment concentrations varied across the four studies. Persson et al. [11] and Tolosa et al. [13] used a fixed concentration range across all compounds (0.01–100 and 1–1000 μM, respectively), while Xu et al. [6] and O'Brien et al. [5] selected the concentrations based on physiological concentrations (measured or predicted) of a compound (100-fold or 30-fold the efficacious concentration or Cmax respectively).

Table 1
Summary of HCS approaches used in the four DILI publications selected for review

Literature source (group)	Compounds tested	Cell type	Exposure time (hours)	Test concentration used (μM)	HCS cell parameters	Dyes combined	Sensitivity reported (calculated from data)	Specificity reported (calculated from data)
Xu et al. [6]	344	Human hepatocytes	24	100 × Cmax[a] or 100 μM	Nuclei count Nuclear area ROS intensity Lipid intensity MMP[b] GSH[c] content GSH area	DRAQ5 DRAQ5 CM-H2DCFDA DRAQ5 TMRM MCBI MCBI	50–60 (51)	95–100 (100)
O'Brien et al. [5]	243	HepG2	72	30 × Cmax or 100 μM	Nuclei count DNA content Calcium MMP PMP[d]	Hoechst Hoechst Fluo-4 AM TMRM TOTO-3	90 (89)	98 (98)
Tolosa et al. [13]	78	HepG2	3 and 24	1–1000 μM	Nuclear morphology PMP MMP Calcium Oxidative stress	Hoechst Propidium Iodide TMRM Fluo-4 AM BODIPY	89 (94)	92 (92)
Persson et al. [11]	102	HepG2	24 and 72	0.01–100 μM	Nuclei count Nuclear area PMP Lysosomal activity MMP Mitochondrial area	Hoechst Hoechst TOTO-3 LysoTracker MitoTracker MitoTracker	50 (52)	90 (91)

[a] *Cmax* rate of bioavailability
[b] *MMP* mitochondrial membrane potential
[c] *GSH* glutathione
[d] *PMP* plasma membrane permeability

The cellular mechanisms that were monitored in the four studies show some agreement but there are also some important differences. An assessment of nuclear area was reported in all four papers. Three of the four studies used Hoechst 33342 dye to measure this area. In all four papers, mitochondrial area, function, and/or membrane potential were assessed using the TMRM and MitoTracker Orange dyes. Plasma membrane integrity was monitored in three of the publications [5, 11, 13], while intracellular calcium concentration was measured in only two studies [5, 13]. Assessments of glutathione content [6], ROS intensity [6], oxidative stress [13] or lysosomal activity [11] were each reported in only one of the papers.

3.2.2 Comparison of Assignment of DILI Scores and Classification Across Reference Compounds

A list of compounds known to cause DILI has been characterized by the FDA as part of the Liver Toxicity Knowledge Database (LTKB). They were classified according to concern levels 1–3, where 1 is no DILI concern, 3 is of most DILI concern, and 2 is "less" DILI concern [41].

To make comparisons across the four selected papers, all the studied compounds were rated according to this scale. In the Xu et al. [6] and Tolosa et al. [13] papers, the compounds were assigned either a "+" or a "–," depending on their predicted DILI potential, and so were assigned only a 1 or 3 on the FDA scale. However, in the O'Brien et al. [5] and Persson et al. [11] papers, a three-tier scale (severe, moderate, or nontoxic) was used. The numerical ratings assigned by the FDA and how these have been aligned to the approaches used in the four selected studies are shown in Table 2.

Building and having access to a database of reference compounds, which have been assessed and ranked in this manner, is an essential tool when developing and validating appropriate predictive assays.

Table 2
Numerical rating from FDA and their relationship to the classifications used in the four DILI publications selected for review

	Publication			
FDA classification	Xu et al. [6]	Tolosa et al. [13]	O'Brien et al. [5]	Persson et al. [11]
3	+	+	3	3
2			2	2
1	–	–	1	1

3.2.3 Comparison of Compounds Studied Across the Four Selected DILI Publications and the FDA DILI Listing

In the four papers plus the FDA listing, a total of 231 compounds appeared in more than one group. The overlap of these compounds between groups is shown in bubble diagrams (Fig. 2). These diagrams show a simple two-way comparison between the source (FDA or cited publication in the center) and each of the other groups in isolation. When the crossover of chemicals is considered between any two groups, in general there is good correlation between the compound choices; therefore, there is plenty of scope for comparisons on a group-by-group basis. To allow a more complex visual comparison and to illustrate the complexity involved in making comparisons, the FDA group was not considered and only those compounds that were studied in more than one of the DILI HCS publications were included, and a four-way Venn diagram was constructed (Fig. 3) to compare the overlap among the four published studies. A total of 79 compounds were present in one or more of the four papers. The highest overlap was 51 compounds between O'Brien et al. [5] and Xu et al. [6], followed by 31 compounds between Persson et al. [11] and O'Brien et al. [5], and 14

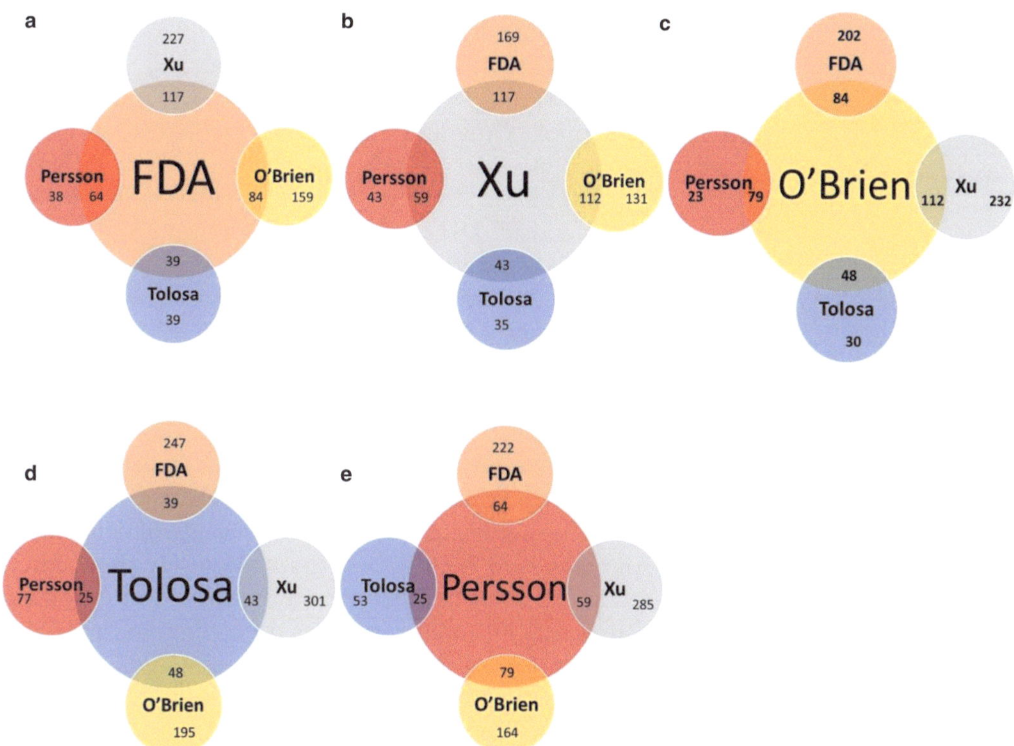

Fig. 2 Bubble diagrams of pair-wise comparisons and five-way overlaps of compounds between the FDA DILI list and the four DILI publications selected for review (**a**) and between and the four DILI publications and the FDA LTKB DILI list (**b**–**e**). The total number of compounds was FDA = 286 [41], Xu [6] = 344, O'Brien [5] = 243, Tolosa [13] = 78, and Persson [11] = 102

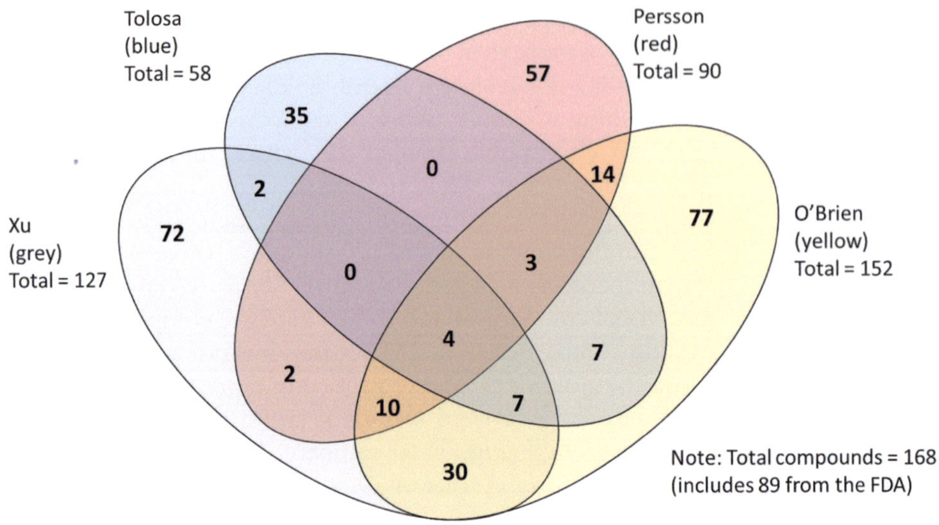

Fig. 3 Four-way Venn diagram showing the overlap of compounds across the four selected DILI publications [5, 6, 11, 13]

compounds when all these groups were compared. Only 7 or 13 compounds overlapped between Tolosa et al. [13] and Persson et al. [11] or Tolosa et al. [13] and Xu et al. [6], respectively. Between Tolosa et al. [13] and O'Brien et al. [5] only 21 compounds overlapped. Only four compounds overlap across all four studies, as such in order to adequately compare the difference between these studies a larger set of compounds is desirable, indicating that a standard DILI compound reference compound set such as the LTKB is essential to allow a true cross comparison and understanding of the value of different mechanisms and cell type choices across studies.

Most of the selected publications predate the release of the FDA LTKB; therefore, different approaches were taken to categorize the reference compounds. For example, Xu et al. [6], classified hepatotoxicants as: (a) withdrawal from the market due to hepatotoxicity; (b) not marketed in the US because of hepatotoxicity; and (c) received a black box warning from the FDA because of hepatotoxicity. Any compounds in the database that fell outside these criteria were classed as DILI negative. Conversely, O'Brien et al. [5] classified hepatotoxicants into four categories depending on the severity of human hepatotoxicity they produced as: (a) severe; i.e., >1 % increase of serum of alanine aminotransferase (ALT) plus 2 % increase of either jaundice, >3 % reports of liver failure, or containing a warning for adverse liver events on the FDA label; (b) moderate, i.e., producing 0.1–1 % frequency of ALT plus either jaundice or a warning for adverse liver events on the FDA label; (c) non-hepatotoxic or minimally toxic, i.e., <1 % frequency of increase in ALT with no signs of clinical hepatotoxicity; and (d) compounds known to be toxic in organs other than the liver. Specific compounds known to be nontoxic in the clinic, known to cause idiosyncratic hepatotoxicity, and known to cause toxicity through the formation of reactive metabolites were also included in the set [5]. These different approaches to selecting and classifying compounds in relation to their potential clinical hepatotoxicity will obviously have an effect not only on the selectivity and specificity calculated for each approach but also on comparisons that are made later in this chapter (see Table 3).

3.2.4 What Understanding Can Be Obtained at the Individual Compound Level?

Differences observed between the four DILI publications [5, 6, 11, 13] are in Table 3 (a total of 33 compounds). When the sensitivities of HepG2 cells [5, 11, 13] and human hepatocytes [6] were compared, it was clear that 18 of the compounds that responded in HepG2 cells did not respond in human hepatocytes (i.e., chloramphenicol, furazolidone, alloxan, amitriptyline, bezafibrate, bupivacaine, fluvastatin, memantine, nocodazole, orphenadrine citrate, pamidronate, temozolomide, tetracaine, vidarabine, diethylcarbamazine, etoposide, metoclopramide, and p-bromophenol). Interestingly, 12 of these compounds were characterized differently with regard to in vivo DILI potential and therefore would not be considered as false positives or negatives in either approach. HepG2 are cycling cells and primary hepatocytes are not, therefore, the potential differences observed may not be due to cytotoxicity, but reduced cell number as several of these compounds are known cell cycle inhibitors. For example, amitriptyline was reported to inhibit D-cyclin transactivation and arrest cells at the G_0/G_1 phase of the cell cycle [42], bupivacaine was found to arrest cell growth by inactivating ribosomal protein S6 kinase 1 [43], nocodazole is an antimicrotubule agent that was shown to arrest cells in the G2/M phase of the cell cycle [44] and etoposide is a topoisomerase II (topo II) inhibitor [45].

Alloxan, a known inhibitor of mitochondrial aconitase and other enzymes associated with the citric acid cycle [46], was tested at the same concentration in two studies [5, 6]. However, in the O'Brien et al. [5], study, alloxan showed changes only in the mitochondrial potential, suggesting that HepG2 cells may be more sensitive than primary human hepatocytes. However, a longer incubation time of 72 h was used than the incubation time used in Xu et al.'s study [6], which could explain the observed difference. For fluvastatin, calcium dyshomeostasis was the most sensitive marker of toxicity used by O'Brien et al. [5], and fluvastatin has been shown previously to modify calcium homeostasis in rat skeletal muscle fibers by impairing the sarcoplasmic reticulum/mitochondria Ca^{2+} release system [47]. Calcium homeostasis was not determined by Xu et al. [6], and because other end points also changed in the O'Brien et al. [5] assay, the importance of the calcium homeostasis end point in detecting fluvastatin is not clear.

Table 3
Comparison of in vivo DILI scores, response in HCS assay, and overall predictions across the four DILI studies

	DILI score					Positive					
Compound name	**A**	**B**	**C**	**D**	**I. DILI score disagrees**	**A**	**B**	**C**	**D**	**II. Response disagrees**	**III. Prediction disagrees**
Differences in DILI score classification, assay response and assay prediction											
Acetylsalicylic acid	1	2	3	2	✓	−ve	+ve	+ve	−ve	✓	✓
Chloramphenicol	3	2	3		✓	−ve	+ve	+ve		✓	✓
Furazolidone	3	2			✓	−ve	+ve			✓	✓
Nimodipine	3			2	✓	+ve			−ve	✓	✓
Differences in DILI score classification and assay response											
Alloxan	1	3			✓	−ve	+ve			✓	
Amitriptyline	1	3	3	3	✓	−ve	+ve	+ve	+ve	✓	
Bezafbrate	1	3			✓	−ve	+ve			✓	
Bupivacaine	1	3			✓	−ve	+ve			✓	
Caffeine		3	1	1	✓		+ve	−ve	−ve	✓	
Fluvastatin	1	3			✓	−ve	+ve			✓	
Gentamycin		3	1		✓		+ve	−ve		✓	
Lidocaine	1	3		1	✓	−ve	+ve		−ve	✓	
Memantine	1	3			✓	−ve	+ve			✓	
Nicotine		3		1	✓		+ve		−ve	✓	
Nocodazole	1	3			✓	−ve	+ve			✓	
Orphenadrine Citrate	1		3		✓	−ve		+ve		✓	
Pamidronate	1	3			✓	−ve	+ve			✓	
Temozolomide	1	3			✓	−ve	+ve			✓	
Tetracaine HCl	1		3		✓	−ve		+ve		✓	
Topiramate		3		1	✓		+ve		−ve	✓	
Vidarabine	1	3			✓	−ve	+ve			✓	
Differences in DILI score and assay prediction											
Astemizole	1	3			✓	−ve	−ve				✓
Cisapride	1	3		1	✓	−ve	−ve		−ve		✓
Terfenadine	1	2			✓	−ve	−ve				✓
Zomepirac	1	3			✓	−ve	−ve				✓
Differences in assay response and assay prediction											
d-galactosamine	3	3	3			−ve	+ve	−ve		✓	✓
Diethylcarbamazine	2	2				−ve	+ve			✓	✓
Etoposide	3	3	3	3		−ve	+ve	+ve	+ve	✓	✓
Ketoprofen		2		2			+ve		−ve	✓	✓
Ketotifen	1	1	1			−ve	−ve	+ve		✓	✓
Metoclopramide	3	3				−ve	+ve			✓	✓
Novobiocin	3	3		3		+ve	+ve		−ve	✓	✓
p-Bromophenol	3	3				−ve	+ve			✓	✓

A: 100 × C_{max} [6]. Where C_{max} was not available it was assigned as 1 μM

B: 30 × maximal efficacious concentration or 100 μM [5]

C: 1, 10, and 100 μM and, if not toxic, then at 100 μM and 1,000 μM, or the maximum possible concentration was looked at [13]

D: 0.01–100 μM [11]

Therefore, it is not possible to conclude why fluvastatin was not identified as hepatotoxic in the Xu et al. [6] study. Here again, it would seem that the cell type and the incubation time adopted are critical factors in determining the results of HCS assays.

It is not evident how the specific cut-off values used to determine DILI compounds described in each publication were adhered to. For example, in the O'Brien et al. [5] study, for memantine (30× Cmax is 12.3 μM) the parameter most sensitive to toxicity was cell number, which was detected at a concentration of 25 μM or higher, and although this is above the concentration highlighted for DILI, it was classified as positive. A similar situation was evident for diethylcarbamazine (cell number was the most sensitive toxicity marker at 100 μM) and p-bromophenol (cell number, calcium dyshomeostasis, membrane permeability, and nuclear area were sensitive toxicity markers at 630 μM), which were classified as positive although both these concentrations were above the concentration highlighted for DILI. Inclusion of compounds at concentrations outside the range described will have an impact on the reported sensitivity and specificity values for that study, which makes comparisons between approaches more complicated (Table 3). Initially, it appeared that detection of the toxicity of these compounds was dependent on the cell type and incubation times used but, in the absence of data for the exposure of hepatocytes to higher concentrations of these compounds, it is difficult to ascertain that this is correct.

By identifying a group of compounds which gave a negative response when studied using the Xu et al. [6], approach but a positive response in the other three assays even at lower concentrations, it could be seen that HepG2 cells were more sensitive than primary human hepatocytes for detecting potentially hepatotoxic compounds. However, compared with primary human hepatocytes, HepG2 cells are known to express low levels of phase I and phase II drug metabolizing enzymes [48, 49]. Therefore, HepG2 cells may under-predict the DILI potential of compounds where metabolic activation plays a role in their toxicity or, conversely, over-predict where compounds with intrinsic DILI potential are metabolized rapidly in hepatocytes. In a recent study the effect of 144 drugs on HepG2 cells cultured in the absence or presence of rat liver S9 extract and on cryopreserved human hepatocytes was investigated [38]. By adopting a clustering analysis approach, the authors estimated a sensitivity of 58 % and a specificity of 75 % and noted that these values were lower than reported by Xu et al. [6] and O'Brien et al. [5], where the specificity values were greater than 90 %. Other than for a single compound, cyclophosphamide, inclusion of S9 and cofactors did not potentiate the cell toxicity to HepG2 cells. Garside et al. [38] proposed that, due to the large variability in cytochrome P450 enzymes between species, inclusion of Aroclor-induced S9 from rat may not be optimal, having an impact on the sensitivity and specificity values observed. However they also note that, although many drugs are metabolized by P450 enzymes to form toxic metabolites, for most it results only in a moderate increase in toxicity that can also be detected in the absence of P450 dependent metabolism [38, 50, 51].

Here, the strengths of HCS approaches in predicting compounds with the potential to cause DILI have been discussed; however, it is also interesting to consider the attempts that have been made to combine HCS with other end points to improve the HCS predictions. Iskar et al. [52] proposed that integration of in silico approaches with HCS may improve the predictive power of HCS. Recent efforts to build predictive models to reduce the risk of progressing candidate drugs with DILI risk have been categorized into three groups based on the type of data used; i.e., chemical structure, in vitro assay, and toxicogenomics [18]. Recently, Chen et al. [53] coupled a rule-of-two (RO2) model with a HCS assay using primary rat hepatocytes (CellCiphr™) to investigate the DILI potential of a set of 70 drugs. The RO2 model is defined by a daily dose of ≥100 mg/day and logP ≥3 and is based on the observation that high daily doses and lipophilicity for oral medication show high correlation with significant DILI risk. In brief, this approach reduced in the number of drugs that required an experimental assessment by 20 % and improved the accuracy of predicting DILI compounds by 10 % [53].

The inclusion of cytokines in the cell media when performing in vitro HCS assays to predict idiosyncratic hepatotoxicity may be a beneficial approach. A number of idiosyncratic hepatotoxic drugs (ranitidine, trovafloxacin, nefazodone, nimesulide, and clarithromycin) were shown to synergistically induce hepatocellular death in vitro when co-administered with a cytokine mix that contained the LPS-upregulated cytokines TNF, IFNγ, and IL-1α, and LPS [54]. This approach allowed the efficient in vitro investigation of inflammation-associated idiosyncratic drug hepatotoxicity using HCS as an end point with both primary hepatocytes and HepG2 cells.

4 Integration of HCS and Omics Technologies in Toxicological Assessment

Traditional approaches using in vivo models pose serious limitations and ethical concerns that challenge their use in toxicity assessment. Therefore, the need for alternative approaches in toxicity testing has forced both the scientific community and regulatory bodies to consider incorporating recent advances in molecular and systems biology, high-throughput screening methods, computational toxicology, and bioinformatics [8, 55–63].

The European Union has launched the "Safety Evaluation Ultimately Replacing Animal Testing" initiative (http://www.seurat-1.eu/), which aims to replace traditional animal studies by predictive toxicology based on computational and in vitro experimental models. This initiative is also aligned with the European 7th Amendment to the Cosmetics Directive [64], which banned any type of animal testing on cosmetic ingredients after 2013. Several other initiatives have also been established; e.g., the Next Generation Risk Assessment program [63] and, particularly, the Toxicology in the 21st Century (Tox21) program [9, 65], a multidisciplinary effort to screen 10,000 chemicals using a battery of 150 HTS assays.

Most in vitro toxicity approaches share a common working hypothesis, namely the existence of a finite number of molecular pathways that are perturbed under stress conditions (i.e., chemical exposure) and determine cell-fate decisions [8, 66]. The number and magnitude of the perturbations are dose-dependent. Beyond a certain level of exposure, such perturbations become excessive, thereby leading to adverse health effects [8]. HCS technologies can provide valuable mechanistic understanding. Despite the advantages and case studies already discussed in this chapter, there are some limitations to the use of HCS technologies in toxicological assessment. The number of biological assays is limited by the availability of appropriate dyes/antibodies and how easily these reagents can be multiplexed in a single assay. In addition, the experimental conditions need to be optimized for each cellular model. Furthermore, reporter assays for transcriptional activation end points are only available in transformed cell lines and not in physiologically relevant cells [9]. Moreover, even when appropriate assays are available, the number of end points that can be measured simultaneously is generally restricted from four to ten per assay plate. As a systems biology approach, this number is relatively low and can be overcome only by increasing the number of assays. Unfortunately, the development of the potentially hundreds of mechanistic end points that would be needed to generate a comprehensive mechanistic understanding on all possible molecular pathways affected by a toxicant would require too many resources to be viable. A more efficient and less costly alternative may be the integration of HCS assays with high-throughput profiling techniques. Omics technologies, namely transcriptomics, genomics, proteomics, and metabolomics, can measure up to thousands of end points in one single experiment and thus could complement and/or be complemented by HCS approaches.

A potential workflow for the integration of HCS and omics technologies is shown in Fig. 4. This is just one example of many possible workflows, which illustrates how relevant mechanistic information on the effects of a compound of interest in a biological system could be extracted. High-throughput profiling generates an overwhelmingly high amount of information, which often requires the use of complex and time-consuming computational and/or bioinformatics tools for pathway analysis or network model development, for example, to predict mechanistic information [67–71].

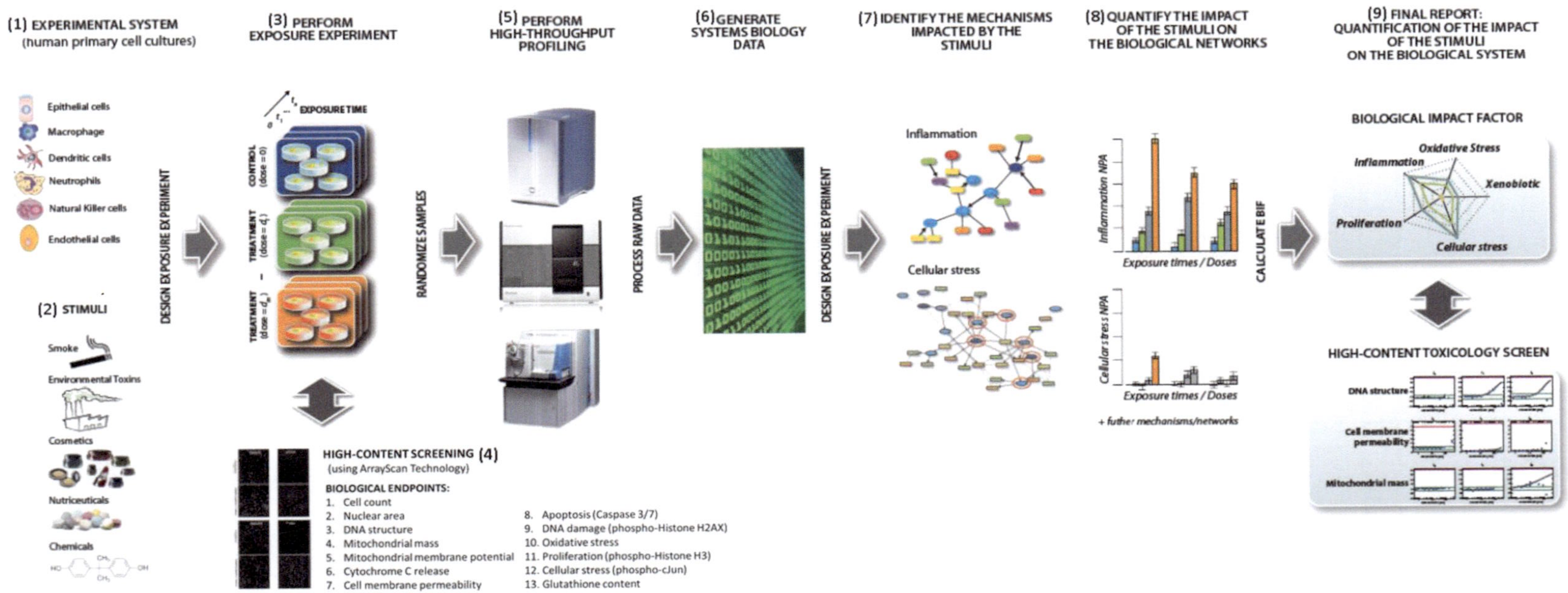

Fig. 4 Systems biology approach for the assessment of environmental toxicants. One or more experimental systems (*1*) are exposed to different stimuli (*2*) at multiple doses and time points (*3*). The biological impact of each stimulus on the cells is evaluated initially by a number of HCS-based toxicity assays (*4*). Based on the results from the HCS end points, relevant doses and time-points are selected for further analysis by high-throughput profiling techniques (metabolomics, proteomics, genomics, or transcriptomics) (*5*). The raw data are processed and replicate samples are grouped and compared with the vehicle control to identify differences in expression (*6*). The data are processed further using computational tools and network biology to identify molecular pathways perturbed by the different stimuli (*7*) and to quantify the level of perturbation under the different experimental conditions (*8*). The quantitative data are then computed and plotted together into a Biological Impact Factor, which provides a straightforward, yet comprehensive overview of the overall impact on the biological system. Finally, the results from the HCS assays can be used to support the gene expression data (*9*). (Adapted (with permission) from [76])

HCS assays generate robust and reproducible data, and also provide a level of flexibility and speed that allows testing of multiple experimental designs in a cost-effective manner. Thus, a small number of HCS-based end points can be used as an initial step to help select the optimum experimental conditions for further investigation using gene expression or other profiling methods. As discussed in Sect. 3 for the analysis of HCS data, it is important to consider that with omics technologies it is difficult to discriminate between true, relevant biological signals and irrelevant or erroneous signals caused by technical and biological variations—otherwise known as noise—[72]. Often, the only thing that differentiates a true signal from the noise is the adoption of an arbitrary threshold, and the establishment of an appropriate threshold can be challenging because there is a risk of either erroneously rejecting true signals if the statistical power is low or incorporating noise as true signals if false discovery rates are high [72]. In this regard, HCS assays can be used to validate the omics-derived predictions, thus providing additional support to the results. A specific end point, for example, can be used as a functional indicator of a particular cellular response (e.g. increased nuclear levels of the transcription factor NRF2 as an indicator of oxidative stress response (*see* [73]).

The number and type of assays will be determined by many factors, such as whether or not there is existing knowledge about the biological impact of the compound being tested, the experimental system of choice, and importantly, the specific aims of the study. Another critical aspect in these types of studies is the selection of an appropriate in vitro system. With no such thing as the "perfect cell", the selection will depend ultimately on the questions that need to be addressed. Primary cells are inherently more heterogeneous in their biological responses than are immortalized cell lines, so it is generally more challenging to optimize the experimental conditions to ensure reproducibility. However, primary human cells are a better representation of in vivo physiology, and using them helps to minimize the risk of genetic mutations (always present in immortalized cell lines) that could later affect data interpretation. This aspect is highly relevant when the aim is to extract mechanistic information using whole genome transcriptomics [74, 75].

In a recent study the workflow shown in Fig. 4. was used, with a combination of HCS end points and whole-genome transcriptomics to assess acrolein and other well-known toxicants that are present in cigarette smoke [76]. Primary human bronchial epithelial (NHBE) cells were selected to minimize the risk of genetic mutations [74, 75] and because lung epithelial cells make an efficient barrier against pathogens and aggressive molecules. After exposure to cigarette smoke and pathogens, these cells have been shown to release a variety of pro-inflammatory mediators, including chemokines, cytokines, and growth factors through various signaling pathways. More importantly, NHBE cells express

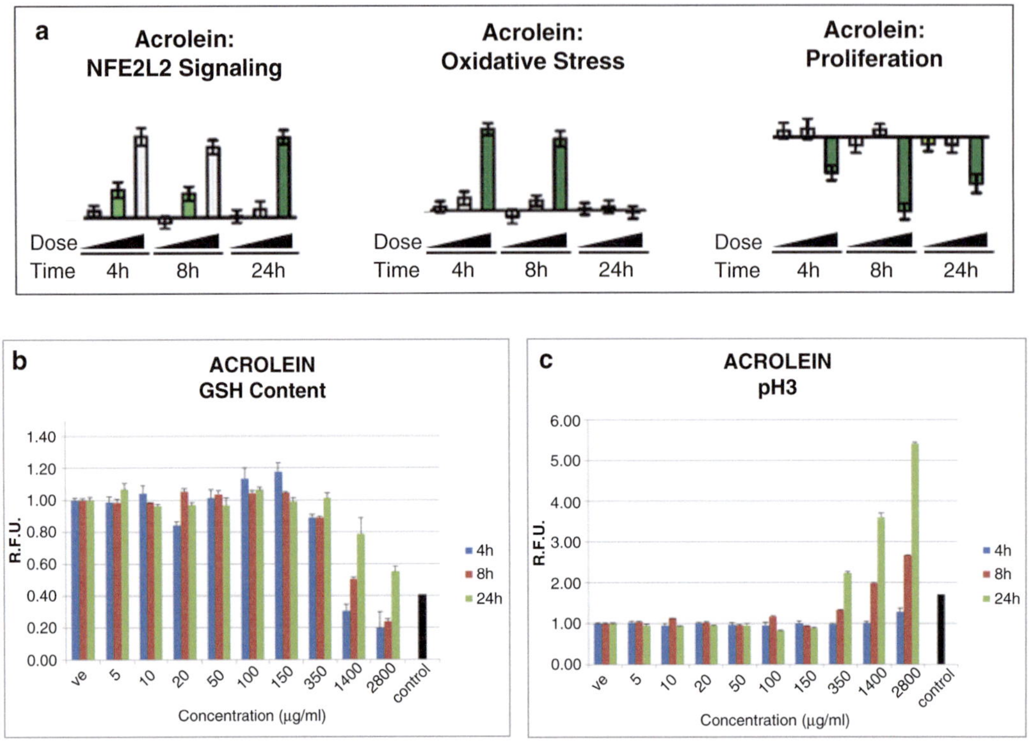

Fig. 5 Comparison of gene expression and HCS results. (**a**) Gene expression profiling predicted the activation of oxidative stress-response pathways and the inhibition of cell proliferation pathways in cells exposed to acrolein. (**b**) HCS results show a dose-dependent decrease in glutathione (GSH) content in exposed cells. This is indicative of the presence of oxidative stress and is consistent with the gene expression results. (**c**) HCS results show the accumulation of the mitosis market phospho-histone H3. This is indicative of growth arrest and is consistent with the gene expression results. *R.F.U.* Relative fluorescence units. (Adapted with permission from Gonzalez-Suarez et al. [76]; Copyright 2014 American Chemical Society)

adequate levels of the phase I enzymes, cytochrome P450s [77, 78], which allows the chemicals that need metabolic activation to be studied. NHBE cells were exposed to acrolein under different experimental conditions and a total of 13 multiparametric indicators of toxicity were measured using HCS-based assays. Relevant doses and exposure times were selected for whole genome transcriptomics based on the results of the assays. After the generated data were processed, gene expression analysis predicted the activation of cellular and oxidative stress responses, as well as the inhibition of cell proliferation (Fig. 5). These predictions were supported by HCS. In particular, a dose-dependent decrease was observed in glutathione, a major endogenous antioxidant and a sensitive indicator of oxidative stress [79]. In addition, the accumulation of the mitosis marker phospho-histone 3 was observed, which is consistent with G2/M cell cycle arrest [80].

The combination of HCS and omics technologies could increase dramatically the amount of mechanistic information that can be extracted from each experiment, and overcome some of the individual limitations of each approach. The use of an omics approach within the context of a relevant in vitro model can provide detailed mechanistic insights into the mode of action of a compound. At the same time, the use of HCS-based end points can help substantiate omics predictions in a timely and cost-effective manner. Although the example discussed here was focused on the assessment of well-known environmental toxicants for which much information was available [81], a similar workflow could be applied to the toxicological assessment of compounds for which little or no knowledge is available.

5 Summary

The cellular processes behind DILI are complicated and still relatively poorly understood; therefore, using a technique such as HCS to detect multiple outcomes with many potential pathways involved seems appropriate. Here we have discussed the use of HCS as a screening approach to detect DILI and highlighted that, because it requires complex data analysis, the comparison of data across different experimental approaches is a major challenge. Choices relating to details of the HCS methodology are important in understanding the sensitivities and specificities of the different approaches. Choices around test concentration ranges and relationship to in vivo exposure can have a large influence on the sensitivity and specificity of each approach in determining whether a compound has DILI potential. Furthermore, there is a need for a standard DILI reference compound list such as the LTKB to truly understand the mechanistic and predictive power of differing cell types, incubation times, and HCS end points.

When adopting or developing a HCS assay to determine compounds with a liability to cause DILI, extensive validation is critical. The selection of compounds is obviously essential; therefore, a database of how these compounds have performed in other assays is a desirable approach to collate data sets and improve our understanding of HCS assay results. It is also important to consider the physiological relevance of adapting assay conditions and how relevant any response achieved under in vitro conditions will be to an in vivo or clinical situation. In summary, HCS is a very useful tool in in vitro toxicity testing and, as has been highlighted here, in drug development where approaches to predict DILI risk are essential tools. Not only can HCS be used to gain detailed information about potential hepatotoxicity for later-stage drug development candidates, but because of its high-throughput nature, it can also be used to rank compounds early in the development of both the experimental and data analysis phases. The combination of HCS with other technologies or as part of a wider workflow can add further to the value and future development of these approaches.

Acknowledgments

The research described in this chapter of the book was funded in part by Philip Morris International.

References

1. Gasparri F (2009) An overview of cell phenotypes in HCS: limitations and advantages. Expert Opin Drug Discov 4(6):643–657
2. Persson M et al (2014) High-content analysis/screening for predictive toxicology: application to hepatotoxicity and genotoxicity. Basic Clin Pharmacol Toxicol 115(1):18–23
3. Morgan S et al (2011) The cost of drug development: a systematic review. Health Policy 100(1):4–17
4. U.S. Department of Health and Human (2009) Services Food and Drug Administration Center for Drug Evaluation and Research and Center for Biologics Evaluation and Research. Guidance for Industry Drug-Induced Liver Injury: Premarketing Clinical Evaluation
5. O'Brien PJ et al (2006) High concordance of drug-induced human hepatotoxicity with in vitro cytotoxicity measured in a novel cell-based model using high content screening. Arch Toxicol 80(9):580–604
6. Xu JJ et al (2008) Cellular imaging predictions of clinical drug-induced liver injury. Toxicol Sci 105(1):97–105
7. Trosko JE, Upham BL (2010) A paradigm shift is required for the risk assessment of potential human health after exposure to low level chemical exposures: a response to the toxicity testing in the 21st century report. Int J Toxicol 29(4):344–357
8. Bhattacharya S et al (2011) Toxicity testing in the 21 century: defining new risk assessment approaches based on perturbation of intracellular toxicity pathways. PLoS One 6(6), e20887
9. Tice RR et al (2013) Improving the human hazard characterization of chemicals: a Tox21 update. Environ Health Perspect 121(7):756–765
10. Andrade RJ et al (2006) Outcome of acute idiosyncratic drug-induced liver injury: long-term follow-up in a hepatotoxicity registry. Hepatology 44(6):1581–1588
11. Persson M et al (2013) A high content screening assay to predict human drug-induced liver injury during drug discovery. J Pharmacol Toxicol Methods 68(3):302–313
12. O'Brien PJ (2014) High-content analysis in toxicology: screening substances for human toxicity potential, elucidating subcellular mechanisms and in vivo use as translational safety biomarkers. Basic Clin Pharmacol Toxicol 115(1):4–17
13. Tolosa L et al (2012) Development of a multiparametric cell-based protocol to screen and classify the hepatotoxicity potential of drugs. Toxicol Sci 127(1):187–198
14. Wilson CJ et al (2006) Identification of a small molecule that induces mitotic arrest using a simplified high-content screening assay and data analysis method. J Biomol Screen 11(1): 21–28
15. Gerets HH et al (2012) Characterization of primary human hepatocytes, HepG2 cells, and HepaRG cells at the mRNA level and CYP activity in response to inducers and their predictivity for the detection of human hepatotoxins. Cell Biol Toxicol 28(2):69–87
16. Brazma A et al (2001) Minimum information about a microarray experiment (MIAME)-toward standards for microarray data. Nat Genet 29(4):365–371
17. Collins MA (2009) Generating 'omic knowledge': the role of informatics in high content screening. Comb Chem High Throughput Screen 12(9):917–925
18. Chen M et al (2014) A testing strategy to predict risk for drug-induced liver injury in humans using high-content screen assays and the 'rule-of-two' model. Arch Toxicol 88(7): 1439–1449
19. Li N, Stojanovski S, Maechler P (2012) Mitochondrial hormesis in pancreatic beta cells: does uncoupling protein 2 play a role? Oxid Med Cell Longev 2012:740849
20. Zock JM (2009) Applications of high content screening in life science research. Comb Chem High Throughput Screen 12(9):870–876
21. van de Water FM et al (2011) High content screening analysis of phospholipidosis: validation of a 96-well assay with CHO-K1 and HepG2 cells for the prediction of in vivo based phospholipidosis. Toxicol In Vitro 25(8):1870–1882
22. Fowler P et al (2012) Reduction of misleading ("false") positive results in mammalian cell genotoxicity assays. II Importance of accurate toxicity measurement. Mutat Res 747(1):104–117
23. Westerink WM, Schoonen WG (2007) Cytochrome P450 enzyme levels in HepG2

cells and cryopreserved primary human hepatocytes and their induction in HepG2 cells. Toxicol In Vitro 21(8):1581–1591
24. Garcia-Canton C, Anadon A, Meredith C (2013) Assessment of the in vitro gammaH2AX assay by High Content Screening as a novel genotoxicity test. Mutat Res 757(2):158–166
25. Amacher DE (2011) Strategies for the early detection of drug-induced hepatic steatosis in preclinical drug safety evaluation studies. Toxicology 279(1-3):10–18
26. Donato MT et al (2012) High-content imaging technology for the evaluation of drug-induced steatosis using a multiparametric cell-based assay. J Biomol Screen 17(3):394–400
27. Tsamou M et al (2012) Performance of in vitro gammaH2AX assay in HepG2 cells to predict in vivo genotoxicity. Mutagenesis 27(6):645–652
28. Smart DJ et al (2011) Genotoxicity screening via the gammaH2AX by flow assay. Mutat Res 715(1–2):25–31
29. Kirkland D et al (2005) Evaluation of the ability of a battery of three in vitro genotoxicity tests to discriminate rodent carcinogens and non-carcinogens I. Sensitivity, specificity and relative predictivity. Mutat Res 584(1-2):1–256
30. Ploemen JP et al (2004) Use of physicochemical calculation of pKa and CLogP to predict phospholipidosis-inducing potential: a case study with structurally related piperazines. Exp Toxicol Pathol 55(5):347–355
31. Bhandari N et al (2008) Phospholipidosis assay in HepG2 cells and rat or rhesus hepatocytes using phospholipid probe NBD-PE. Assay Drug Dev Technol 6(3):407–419
32. Kaplowitz N (2005) Idiosyncratic drug hepatotoxicity. Nat Rev Drug Discov 4(6):489–499
33. Kola I, Landis J (2004) Can the pharmaceutical industry reduce attrition rates? Nat Rev Drug Discov 3(8):711–715
34. Fung M et al (2001) Evaluation of the characteristics of safety withdrawal of prescription drugs from worldwide pharmaceutical markets-1960 to 1999*. Drug Inform J 35(1):293–317
35. Chen M et al (2011) FDA-approved drug labeling for the study of drug-induced liver injury. Drug Discov Today 16(15-16):697–703
36. Park BK, Pirmohamed M, Kitteringham NR (1998) Role of drug disposition in drug hypersensitivity: a chemical, molecular, and clinical perspective. Chem Res Toxicol 11(9):969–988
37. Uetrecht J (2009) Immunoallergic drug-induced liver injury in humans. Semin Liver Dis 29(4):383–392
38. Garside H et al (2014) Evaluation of the use of imaging parameters for the detection of compound-induced hepatotoxicity in 384-well cultures of HepG2 cells and cryopreserved primary human hepatocytes. Toxicol In Vitro 28(2):171–181
39. Hornberg JJ et al (2014) Exploratory toxicology as an integrated part of drug discovery. Part II: screening strategies. Drug Discov Today 19(8):1137–1144
40. Hornberg JJ et al (2014) Exploratory toxicology as an integrated part of drug discovery. Part I: why and how. Drug Discov Today 19(8):1131–1136
41. Chen M et al (2013) The liver toxicity knowledge Base: a systems approach to a complex end point. Clin Pharmacol Therapeutics 93(5):409–12
42. Mao X et al (2011) The tricyclic antidepressant amitriptyline inhibits D-cyclin transactivation and induces myeloma cell apoptosis by inhibiting histone deacetylases: in vitro and in silico evidence. Mol Pharmacol 79(4):672–680
43. Beigh MA et al (2014) Growth inhibition by bupivacaine is associated with inactivation of ribosomal protein S6 kinase 1. Biomed Res Int 2014:831–845
44. Chou CF, Omary MB (1994) Mitotic arrest with anti-microtubule agents or okadaic acid is associated with increased glycoprotein terminal GlcNAc's. J Cell Sci 107(Pt 7): 1833–1843
45. Osheroff N (1989) Effect of antineoplastic agents on the DNA cleavage/religation reaction of eukaryotic topoisomerase II: inhibition of DNA religation by etoposide. Biochemistry 28(15):6157–6160
46. Boquist L, Ericsson I (1984) Inhibition by alloxan of mitochondrial aconitase and other enzymes associated with the citric acid cycle. FEBS Lett 178(2):245–248
47. Liantonio A et al (2007) Fluvastatin and atorvastatin affect calcium homeostasis of rat skeletal muscle fibers in vivo and in vitro by impairing the sarcoplasmic reticulum/mitochondria Ca2+-release system. J Pharmacol Exp Ther 321(2):626–634
48. Riley RJ, Workman P (1992) DT-diaphorase and cancer chemotherapy. Biochem Pharmacol 43(8):1657–1669
49. Wilkening S, Stahl F, Bader A (2003) Comparison of primary human hepatocytes and hepatoma cell line Hepg2 with regard to their biotransformation properties. Drug Metab Dispos 31(8):1035–1042
50. Gustafsson F et al (2014) A correlation between the in vitro drug toxicity of drugs to cell lines that express human P450s and their propensity to cause liver injury in humans. Toxicol Sci 137(1):189–211
51. Thompson RA et al (2012) In vitro approach to assess the potential for risk of idiosyncratic

adverse reactions caused by candidate drugs. Chem Res Toxicol 25(8):1616–1632
52. Iskar M et al (2012) Drug discovery in the age of systems biology: the rise of computational approaches for data integration. Curr Opin Biotechnol 23(4):609–616
53. Chen M et al (2014) Toward predictive models for drug-induced liver injury in humans: are we there yet? Biomark Med 8(2):201–213
54. Cosgrove BD et al (2009) Synergistic drug-cytokine induction of hepatocellular death as an in vitro approach for the study of inflammation-associated idiosyncratic drug hepatotoxicity. Toxicol Appl Pharmacol 237(3):317–330
55. Andersen ME, Krewski D (2009) Toxicity testing in the 21st century: bringing the vision to life. Toxicol Sci 107(2):324–330
56. Hartung T (2009) A toxicology for the 21st century–mapping the road ahead. Toxicol Sci 109(1):18–23
57. Hubal EA (2009) Biologically relevant exposure science for 21st century toxicity testing. Toxicol Sci 111(2):226–232
58. Andersen ME, Krewski D (2010) The vision of toxicity testing in the 21st century: moving from discussion to action. Toxicol Sci 117(1):17–24
59. Boekelheide K, Campion SN (2010) Toxicity testing in the 21st century: using the new toxicity testing paradigm to create a taxonomy of adverse effects. Toxicol Sci 114(1):20–24
60. Krewski D et al (2011) New directions in toxicity testing. Annu Rev Public Health 32:161–178
61. Davis M et al (2013) The new revolution in toxicology: the good, the bad, and the ugly. Ann N Y Acad Sci 1278:11–24
62. Dourson M et al (2013) Advancing human health risk assessment: integrating recent advisory committee recommendations. Crit Rev Toxicol 43(6):467–492
63. U.S.E.P. Agency (2013) Next generation risk assessment: incorporation of recent advances in molecular, computational, and systems biology (external review draft)
64. European Commission (2009) Alternative testing strategies progress report 2009. Replacing, reducing and refining use of animals in research
65. Collins FS, Gray GM, Bucher JR (2008) Toxicology. Transforming environmental health protection. Science 319(5865):906–907
66. Simmons SO, Fan CY, Ramabhadran R (2009) Cellular stress response pathway system as a sentinel ensemble in toxicological screening. Toxicol Sci 111(2):202–225
67. Lee E et al (2008) Inferring pathway activity toward precise disease classification. PLoS Comput Biol 4(11):e1000217
68. Hoeng J et al (2012) A network-based approach to quantifying the impact of biologically active substances. Drug Discov Today 17(9-10):413–418
69. Martin F et al (2014) Quantification of biological network perturbations for mechanistic insight and diagnostics using two-layer causal models. BMC Bioinformatics 15:238
70. Kramer A et al (2014) Causal analysis approaches in Ingenuity Pathway Analysis. Bioinformatics 30(4):523–530
71. Bourdon JA et al (2013) Gene expression profiling to identify potentially relevant disease outcomes and support human health risk assessment for carbon black nanoparticle exposure. Toxicology 303:83–93
72. Ideker T, Dutkowski J, Hood L (2011) Boosting signal-to-noise in complex biology: prior knowledge is power. Cell 144(6):860–863
73. Ma Q (2013) Role of nrf2 in oxidative stress and toxicity. Annu Rev Pharmacol Toxicol 53:401–426
74. Fields WR et al (2005) Gene expression in normal human bronchial epithelial (NHBE) cells following in vitro exposure to cigarette smoke condensate. Toxicol Sci 86(1):84–91
75. Mathis C et al (2013) Human bronchial epithelial cells exposed in vitro to cigarette smoke at the air-liquid interface resemble bronchial epithelium from human smokers. Am J Physiol Lung Cell Mol Physiol 304(7): L489–L503
76. Gonzalez-Suarez I et al (2014) Systems biology approach for evaluating the biological impact of environmental toxicants in vitro. Chem Res Toxicol 27(3):367–376
77. Ding X, Kaminsky LS (2003) Human extrahepatic cytochromes P450: function in xenobiotic metabolism and tissue-selective chemical toxicity in the respiratory and gastrointestinal tracts. Annu Rev Pharmacol Toxicol 43: 149–173
78. Iskandar AR et al (2013) Systems approaches evaluating the perturbation of xenobiotic metabolism in response to cigarette smoke exposure in nasal and bronchial tissues. Biomed Res Int 2013:512086
79. Slater AF et al (1995) Signalling mechanisms and oxidative stress in apoptosis. Toxicol Lett 82–83:149–153
80. Rapino F, Naumann I, Fulda S (2013) Bortezomib antagonizes microtubule-interfering drug-induced apoptosis by inhibiting G2/M transition and MCL-1 degradation. Cell Death Dis 4:e925
81. Kehrer JP, Biswal SS (2000) The molecular effects of acrolein. Toxicol Sci 57(1):6–15

Chapter 14

A Systems Toxicology Approach to Investigating the Cardiovascular Effects of Cigarette Smoke and Environmental Pollutants in *ApoE*-Deficient Mice

Héctor De León, Stephanie Boue, Manuel C. Peitsch, and Julia Hoeng

Abstract

Epidemiological evidence indicates that exposure to combustion-derived particles is linked to an increased risk of cardiovascular disease. Despite this strong association, there remains a lack of data that can be used to identify the molecular mechanisms through which exposure to particulate matter (PM) leads to adverse cardiovascular events. The biological complexity of the responses generated by exposure to PM is compounded by the inherent multidimensional nature of the chemical mixtures associated with PM emitted from a range of sources, including diesel and gasoline exhausts, cigarette smoke, and ambient particles. The current challenges in this field of toxicology include the development and adoption of standardized in vitro and in vivo methods for testing multiple types of PM as individual entities or as mixtures. In this chapter, we focused on studies that have been published on the effects of combustion-derived PM in the *ApoE*-deficient mouse, a model of atherosclerosis that has proven to be a powerful in vivo surrogate for vascular disease in humans. A number of studies have been conducted in which the effects of diesel and gasoline exhausts, cigarette smoke, and concentrated ambient particles have been investigated, however, the exposure times, routes of PM delivery, and end points used in those studies have varied widely, making it challenging to compare the results of the studies. Despite these limitations, a few key findings on the development of vascular lesions in $ApoE^{-/-}$ mice can consistently be extracted from the published data. Future efforts should be aimed at systematizing the experimental designs and functional and molecular end points in studies using $ApoE^{-/-}$ mice so that the effects of different sizes and types of PM can be compared reliably. Functional end points and molecular data generated using 'omics' approaches, in which the transcriptome, lipidome, and proteome of $ApoE^{-/-}$ mice exposed to a diverse range of PM sources are examined, will be invaluable components in a systems biology approach to achieving a full understanding of the underlying biology of exposure to combustion-derived PM.

Key words Systems toxicology, $ApoE^{-/-}$ mice, Cigarette smoke, Cardiovascular risk, Diesel exhaust, Gasoline exhaust, Particulate matter, Environmental tobacco smoke

Julia Hoeng and Manuel C. Peitsch (eds.), *Computational Systems Toxicology*, Methods in Pharmacology and Toxicology,
DOI 10.1007/978-1-4939-2778-4_14, © Springer Science+Business Media New York 2015

1 Introduction

The complex chemical mixtures present in particulate matter (PM) from tobacco smoke and ambient air are common pollutants that have harmful effects on the cardiovascular system, ranging from endothelial dysfunction to the development of atherosclerotic plaques. Ultimately, these effects cause increased cardiovascular morbidity and mortality [1]. Although the biological mechanisms linking PM to cardiovascular outcomes are not entirely clear, key mechanisms whereby cigarette smoke (CS) and traffic-related PM e.g., diesel engine exhaust (DE) cause vascular dysfunction have been characterized in the past few years using the $ApoE^{-/-}$ mouse model. *ApoE*-deficiency is associated with a number of clinical conditions, including atherosclerosis and Alzheimer's disease [2]. Subjects with *APOE* deficiency or a defective APOE gene may develop type III hyperlipidemia (dysbetalipoproteinemia according to the Fredrickson lipid disorder classification) [3, 4], which is characterized by elevated total cholesterol and triglyceride concentrations. Similarly to humans harboring a defective *APOE* isoform (*ApoE2*), the β-very low density lipoproteins (VLDL) are the main plasma lipoproteins in $ApoE^{-/-}$ mice [5]. The $ApoE^{-/-}$ mouse model may also be used to investigate the effects of airborne exposure to toxic compounds and lung disease because $ApoE^{-/-}$ mice develop emphysematous changes prematurely compared with wild-type animals [6].

Pollutants derived from CS and ambient air cause various common functional and molecular effects that are associated with morphological vascular end points indicative of disease progression (e.g., atherosclerotic plaque size). Given the clinical and epidemiological relevance of cardiovascular morbidity and mortality associated with exposure to CS and other environmental pollutants, a review of the currently available functional and molecular evidence gathered from exposure studies using $ApoE^{-/-}$ mice, a widely used model for experimentally studying atherosclerosis [7] (Fig. 1), is presented in this chapter. The systemic inflammatory status of *ApoE–/–* mice and their susceptibility to premature atherosclerosis brands this strain as one of the best models for investigating cardiovascular morbidity associated with exposure to CS and environmental pollutants.

Fig. 1 (continued) substances (TBARS), nitric oxide and superoxide production), endothelial dysfunction (e.g., impaired vasodilation and/or vasoconstriction), dyslipidemia, and thrombosis. Standardization of biomarkers of disease progression should be expanded to include high-throughput methodologies to assess the mouse lipidome and transcriptome. *iNOS* inducible nitric oxide synthase, *oxLDL* oxidized low-density lipoproteins, *EDN1* endothelin 1, *MMP*, matrix metalloproteinase

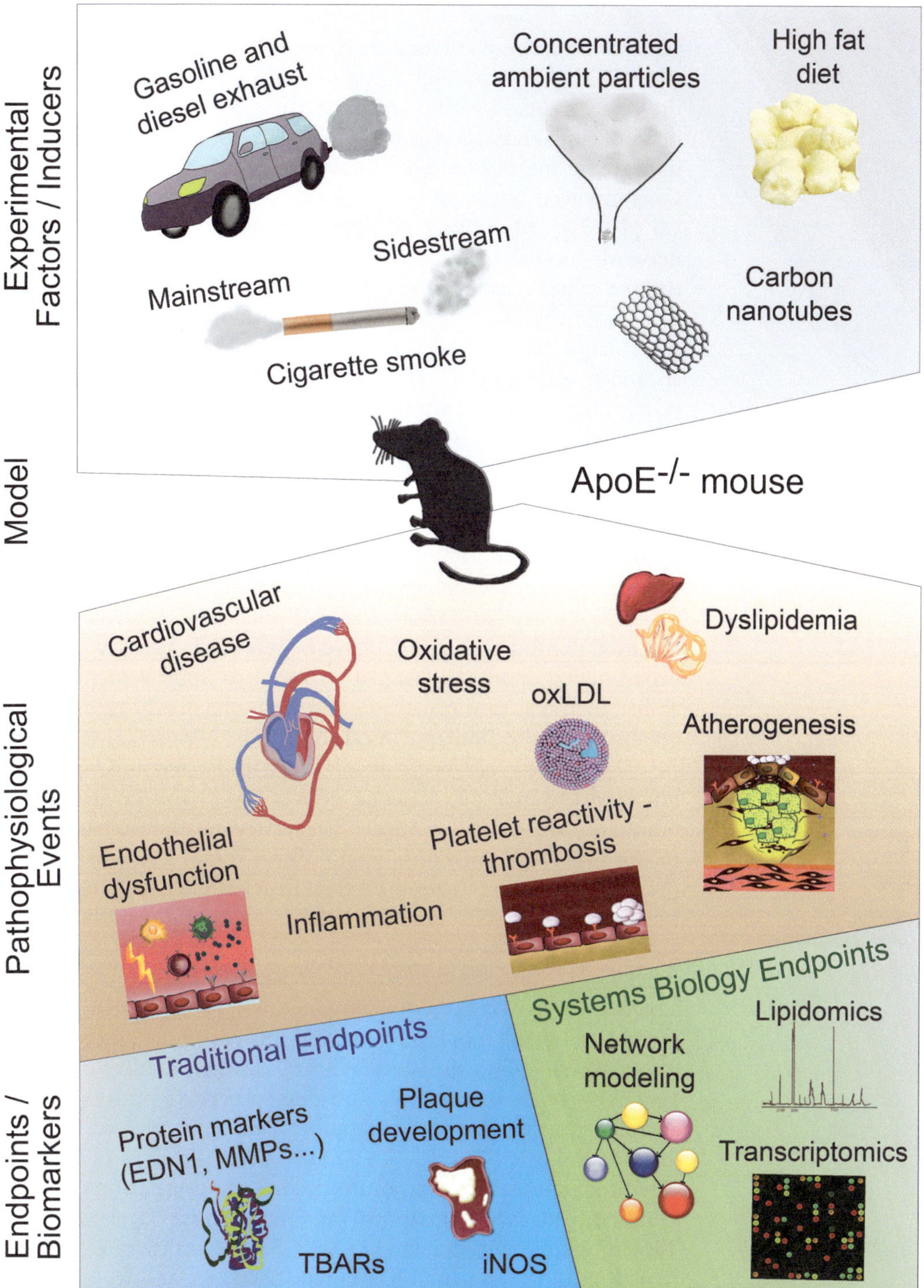

Fig. 1 Exposure studies using the ApoE$^{-/-}$ mouse model of dyslipidemia and atherosclerosis. Studies have been conducted in ApoE$^{-/-}$ mice exposed to PM from a variety of sources including combustion-derived particles from diesel and gasoline exhaust, ambient particles, and particles present in cigarette smoke. Surrogates of cardiovascular disease include conventional end points for oxidative stress (e.g., thiobarbituric acid reactive

Recent advances in high-throughput techniques have made it possible to analyze datasets from experiments using cells and tissues [8–10]. For example, retention of individual lipid species in the vessel wall and within atherosclerotic plaques can now be conducted using mass-spectrometry-based methods [11–13]. We have used transcriptomics and lipidomics to identify the molecular pathways involved in plaque progression in *ApoE*$^{-/-}$ mice exposed to CS [14–16]. Moreover, we have developed a suite of biological network models of vascular inflammatory processes, which use reverse causal reasoning analysis of large gene expression datasets to explore the mechanistic variability that underlies atherogenesis and plaque destabilization [17]. A similar systems toxicology approach using *ApoE*$^{-/-}$ mice exposed to gasoline engine exhaust (GE), DE, hardwood smoke, biogenic secondary organic aerosol, and individual gases would help to unravel the mechanisms underlying the complex pathophysiological responses that occur. The *ApoE*$^{-/-}$ model for hypercholesterolemia and atherosclerosis could be used as a benchmark to assess the effects of environmental, motor engine-, and CS-derived pollutants (Fig. 1). Efforts to use such an approach to determine the contributions to vascular toxicity made by exposure to a variety of pollutants, regardless of the reported combustion source, have begun [18]. This diverse set of experimental approaches is likely to lead to improvements in our ability to translate mechanisms through which exposure leads to effects in mice to human clinical settings.

The chapter has been divided into three sections. First, the reported systemic and vascular effects of CS and environmental pollutants are described (Sect. 2). Particular attention is paid to motor vehicle emissions from diesel and gasoline engines (Sect. 2.1). Studies focused on the effects of environmental pollution using the *ApoE*$^{-/-}$ mouse model are discussed including a subsection of studies reporting the actions of carbon nanotubes (Sect. 2.2). Studies reporting the impact of mainstream and sidestream tobacco smoke complement this first section of environmental pollutants (Sect. 2.3). Second, papers reporting the various experimental approaches conducted to discern whether the vascular effects of cigarette smoke are due to particle translocation in alveoli or to spillover of inflammatory mediators released from the lungs is discussed in Sect. 2.4. Third, the wide array of reported experimental designs, exposure items, exposure times, tissues examined, analytical methods, and end points is presented in Tables 1 and 2 and discussed in Sect. 3. Last, an example of a systems toxicology approach to examine the vascular effects of CS in *ApoE*$^{-/-}$ mice is presented in Sect. 4. The implementation of such systems toxicology methodologies requires method validation and standardization processes, which would facilitate comparative analyses of datasets generated by different laboratories worldwide.

Table 1
Particulate matter exposure conditions from acute studies conducted in ApoE$^{-/-}$ mice

Exposure category	Exposure route	Exposure item	Exposure item concentration	Exposure protocol	Exposure time (day)	Diet	Mouse strain	
							ApoE$^{-/-}$	C57BL/6
MVE	IP	DE	50–5000 μg DEP/kg BW	Single dose	0.25, 1	Standard	17602821	17602821
			500–5000 μg DEP/kg BW	Single dose	0.04	Standard	17234226	17234226
	IV	DE	7.5–15 μg	Single dose	0.04	Standard		24255950
EAP	Inhaled	CB	60 mg/m^3 CB	0.5–1.5 h	1	NA	19138394	
	IP	C60 fullerenes	0.05–0.5 mg/kg of C60 fullerenes	Single dose	0.04	Standard	19243580	
	IT	Au	0.54 μg of gold	Single dose	0.125–1	NA	19138394	
		C60 fullerenes	54 μg of C60 fullerenes					
		CB	18, 54 μg CB					
		QD	– and + charged CdTe quantum dots (63 μg Cd)	Single dose	0.125–1	NA	19138394	
		SWCNT	54 μg of SWCNT	Single dose	0.125–1	NA	19138394	

CB carbon black, *BM* body weight, *d* days, *DE* diesel engine, *EAP* environmental air particles, *HFD* high-fat diet, *IP* intraperitoneal, *IT* intratracheal, *IV* intravenous, *MVE* motor vehicle emissions, *SWCNT* single-walled carbon nanotubes, *w* weeks. Mouse strain columns show PubMed IDs (PMID)

Table 2
Particulate matter exposure conditions from chronic studies conducted in $ApoE^{-/-}$ mice

							Mouse strain				
Exposure category	Exposure route	Exposure item	Exposure item concentration	Exposure protocol	Exposure time (d)	Diet	$ApoE^{-/-}$	C57BL/6	$ApoE^{-/-}$ $LDLr^{-/-}$	$ApoE^{-/-}$ $SOD2^{+/-}$	$LDLr^{-/-}$
MVE	Nasal	DE	1.7 mg/m³ DE ± 9 ppm nanosized cerium oxide	0.3–1.3 h/d, 5 d/w	28	HFD	22507957				
			2 µg DEP/d	1 h/d	21	Standard	23584878	23584878			
						HFD	23584878				
	OPA	DE	35 µl DEP (1 mg/ml)	2 d/w	28	Standard		24330719			
						HFD	24330719				
	WBE	DE	100 µg/m³ PM	6 h/d, 7 d/w	50	HFD	19891982				
			200 µg/m³ PM	6 h/d, 5 d/w	49	Standard	21435644, 21722660, 22746401				
			300 µg/m³ PM	4 h/d, 7 d/w	30	Standard	21222557	21222557			
			300 µg/m³ PM	6 h/d, 7 d/w	50	HFD	19891982				
			436 µg/m³	5 h/d, 4 d/w	150	Standard	20462391				
			1000 µg/m³ PM	4 h/d, 7 d/w	30	Standard	21222557	21222557			
				6 h/d, 7 d/w	50	HFD	19891982				
			DEG—equivalent to DEP = 436 µg/m³	5 h/d, 4 d/w	150	Standard	20462391				
			PM-filtered DE with gaseous concentrations ~1000 µg/m³ PM	6 h/d, 7 d/w	50	HFD	19891982				
		DE + GE	250 mg PM/m³ from DE +150 mg PM/m³ from GE	6 h/d	50	HFD	21493736				
			MVE 300 µg/m³ PM: 50 µg/m³ from GE +250 µg/m³ from DE MVE-G (300 µg/m³ PM with gases removed) MVE-PM (MVE 300 µg/m³ PM with PM filtered)	6 h/d, 7 d/w	50	HFD	24730681				
		GE	8, 40, 60 µg/m³ PM MVE-PM (MVE 60 µg/m³ PM with PM filtered)	6 h/d	50	HFD	17065432				

EAP		MWCNT	38, 128 μg/mouse	1 d/w	35	HFD	24431218	24431218		
	OPA	SWCNT	20 μg/mouse	1/2 w	56	Standard	1849906			
	WBE	CAPs	61.0 μg/m^3 PM 2.5 and 99.45 μg/m^3 PM10 in PM chamber	24 h/d, 7 d/w	60	HFD	24045146			
			85 μg/m^3 PM 2.5	6 h/d, 5 d/w	180	Standard	18236227			
						HFD	16414948, 17107850, 18236227			
			110 μg/m^3 PM	6 h/d, 5 d/w	150	Standard			15804939	
						Standard/HFD	15804939			
			110 μg/m^3 PM 2.5	6 h/d, 5 d/w	180	Standard	15804943		15804943	
			113 μg/m^3 PM 2.5	6 h/d, 5 d/w	150	Standard	16263514	16263514		
			120 μg/m^3 PM	6 h/d, 5 d/w	150	Standard/HFD	19496694			
			134 μg/m^3 CAP	6 h/d, 5 d/w	60, 180	Standard	20235771			
			138.4 μg/m^3 PM 2.5	6 h/d, 5 d/w	120	HFD	19182107			
			CAPs (from Los Angeles freeway) <0.18 μm (UFPs)	5 h/d, 3 d/w	40	Standard	18202315			
			CAPs (from Los Angeles freeway) <2.5 μm (PM)							
			Concentrated ambient PM 2.5 (CAP; 105 μg/m^3)	5 h/d, 4 d/w	150	Standard	20462391			
		UAA	100.3 μg/m^3 PM 2.5 in the PM chamber	24 h/d, 7 d/w	60	HFD	24523034			
MVE+ EAP	WBE	DE + CAPs	CAPs PM 2.5 = 113 μg/m^3 + DEG equivalent to DEP 436 μg/m^3	5 h/d, 4 d/w	150	Standard	20462391			
MTS	WBE	MCS	ETS (10 mg/m^3 TSP)	6 h/d, 5 d/w	28	Standard	21169125			21169125
			MCS (600 μg/l TPM)	3 h/d, 5 d/w	90	Standard	23669247			
					180	Standard	23669247, 24096154, 23102783			
			Up to 1 cigarette per day	1 h/d	35	HFD	15186941			
		MCS + Ang II	– 1.44 or 0.7 mg/kg/d Ang II for 4 w – ISO conditions (TPM: 600 μg/l)	4 h/d, 5 h/d	28	Standard	20937366			
ETS	WBE	SSCS	25 mg/m^3 SP	6 h/d, 5 d/w	49, 70,98	HFD	11368996			
			35 mg/m^3 SP	4 h/d, 5 d/w	42, 105	HFD +/- Q10	20227489			
			40 mg/m^3 SP	4 h/d, 5 d/w	105	HFD	22659316			22659316
			450 μg/m^3 SP	6 h/d, 5 d/w	60, 120, 180	Standard/HFD	20235771			
			Not reported	4 h/d, 5 d/w	84	HFD	20451958	20451958		

CB carbon black, *BM* body weight, *CAPs* concentrated ambient particles, *d* days, *DE* diesel engine, *EAP* environmental air particles, *ETS* environmental tobacco smoke, *GE* gasoline engine, *HFD* high-fat diet, *IP* intraperitoneal, *IT* intratracheal, *IV* intravenous, *MCS* mainstream cigarette smoke, *MTS* mainstream tobacco smoke, *MVE* motor vehicle emissions, *MWCNT* multi-walled carbon nanotubes, *OPA* oropharyngeal aspiration, *PM* particulate matter, *SSCS* sidestream cigarette smoke, *SP* smoke particulates, *SWCNT* single-walled carbon nanotubes, *TPM* total particulate matter, *UAA* untreated ambient air, *w* weeks, *WBE* whole body exposure. Mouse strain columns show PubMed IDs (PMID)

2 Systemic and Vascular Effects of Environmental Contaminants on ApoE$^{-/-}$ Mice

Exhausts from motor vehicles are major pollutants in the industrialized world and they exemplify public health environmental hazards. CS contributes not only to environmental pollution but also to the development of cardiovascular and respiratory disease. Research on diesel and gasoline exhaust, as well as CS, using the *ApoE*$^{-/-}$ mouse model will be described in this section followed by a discussion of potential mechanisms by which exposure to these environmental contaminants can lead to systemic cardiovascular effects.

2.1 Diesel and Gasoline Motor Vehicle Emissions

Epidemiological data suggest that air pollution caused by motor vehicle emissions distresses the cardiovascular system [19, 20]. Using the *ApoE*$^{-/-}$ mouse model, several research groups have shown that emissions from diesel and gasoline engines influence the development of vascular disease [21, 22]. Increased levels of markers of inflammation (e.g., lipid peroxides) and vascular remodeling (e.g., endothelin 1 [EDN1], and matrix metallopeptidase [MMP] 9) have been found in aortas from *ApoE*$^{-/-}$ mice exposed to DE [21]. The robust expression of enzymes involved in vascular remodeling, including MMP7, MMP3, MMP9, tissue inhibitor of metalloproteinase (TIMP) 2, EDN1, and heme oxygenase 1 (HMOX1), has been found in aortas from *ApoE–/–* mice exposed to GE [22]. The vascular effects of pollutants derived from motor vehicles may be mediated, at least partly, through the oxidized low-density lipoprotein (oxLDL) and its receptor on endothelial cells, oxLDL lectin-like receptor 1 [23]. It has been postulated that the impairment of vascular function by DE particles is mediated by the decreased bioavailability of endothelium-derived nitric oxide [24]. The question of whether vascular toxicity induced by vehicular traffic is mediated by PM or pollutants in the gaseous phase is particularly relevant from a regulatory perspective because decreasing PM concentrations (e.g., by filtering the gases emitted from engines) can be addressed more easily than decreasing gas phase pollutants. Campen and colleagues reported that removing PM from DE by filtering the DE did not alter the vascular toxicity of the DE to *ApoE*$^{-/-}$ mice, assessed by the expression of remodeling markers (e.g., MMPs), suggesting that the gaseous phase of the DE is key to the toxicological effects [25]. The same group of investigators conducted a head-to-head comparison of the effects of chemical mixtures and individual pollutants [25], and found that the effects of whole combustion emissions on EDN1 and MMP9 levels could be recreated by exposure to CO and NO but not to NO_2 or biogenic secondary organic aerosol. Interestingly, unlike whole emissions from diesel and gasoline engines, neither the separate gases nor the PM induced vascular lipid peroxidation (determined by measuring the concentration of thiobarbituric acid reactive substances [TBARS]) suggesting that important interactions may occur between the gases and the PM on the reactive surface of the PM. Since the study was designed as an acute exposure (6 h/day for 7 days), the plaque size was not determined. Adding cerium oxide (CeO_2) improves the burning efficiency of fuel and has been shown to decrease the number and surface area of DE particles [26]. Atherosclerotic plaques of *ApoE*$^{-/-}$ mice exposed to DE for 4 weeks showed a trend towards an increased size, whereas the plaques of mice exposed to DE produced with 9 ppm CeO_2 added to the diesel showed no increases in size [26]. The authors of that study postulated that the difference may have been caused by the CeO_2 decreasing the number and surface area of the particles produced. The plaque composition in the two groups was also different. Further studies with larger sample sizes are required to verify the reported trend toward a greater and more complex plaque promoted by DE than DE with added CeO_2.

Acute effects of exposure to DE particles (DEP) have been found in livers from *ApoE*$^{-/-}$ mice. The hepatic expression of inducible nitric oxide synthase (iNOS) has been detected as early as 6 h after the intraperitoneal administration of DEP [27]. Serum concentrations of NO and iNOS, and DNA damage in lungs and aortas, remained unchanged 6 and 24 h after the intraperitoneal administration of DEP, indicating that the DEP did not cause vascular or respiratory inflammation within the 24-h observation period [27]. However, the inhalation of DEP caused significant transcriptional changes in endothelial cells from *ApoE*$^{-/-}$ mice, indicating endothelial dysfunction [28]. Moreover, the direct exposure of aortic rings from rats to DEP has been shown to decrease the bioavailability of endothelium-derived NO because of the generation of reactive oxygen species (ROS) [29]. The exposure of *ApoE*$^{-/-}$ mice to DEP has been found to decrease the number of circulating endothelial progenitor cells (colony-forming units) and to decrease their ability to migrate [30]. Brief exposure to inhaled non-particulate DE was found to induce electrocardiographic changes (bradycardia and T wave depression) in *ApoE–/–* mice implanted with telemetry devices [31]. Coronary artery rings from the *ApoE*-deficient animals exposed to non-particulate DE exhibited enhanced constriction and reduced dilatation. Taken together, these findings indicate that the particulate and volatile fractions of common pollutants promote vascular changes that may lead to plaque development.

It is unclear whether chronic exposure to DEP or to the volatile fractions of DE promotes the growth of atherosclerotic plaques and may, therefore, contribute to an increased risk of atherothrombotic events (e.g., myocardial infarction or stroke). No atherosclerotic plaque growth was detected in a 50-day study in which *ApoE*$^{-/-}$ mice were exposed to DE, even though plaque inflammation, collagen deposition, the expression of MMPs and EDN1, and increased aortic lipid peroxidation (TBARS) were found [21]. Recent morphological data on the effect of DEP exposure on the development of atherosclerosis have shown that *ApoE*$^{-/-}$ mice fed a high-fat diet had more brachiocephalic plaques than did animals fed a standard diet [32]. These data suggest that, regardless of potential DEP-mediated changes in the plaque area, DEP activates molecular pathways involved in plaque formation and vessel remodeling, which may lead to the destabilization of the plaque and its eventual rupture.

2.2 Environmental Pollution

2.2.1 Concentrated Ambient Particles (CAPs)

Toxicological studies in animals have enhanced our mechanistic understanding of the biological responses to $PM_{2.5}$, however, large knowledge gaps remain about the underlying cell-, tissue, and organ-specific pathobiology and the risks associated to exposure to larger and smaller PM and its constituents [33]. Lippmann et al have recently conducted a comprehensive review of epidemiological and toxicological studies linking exposure to ambient air fine particles ($PM_{2.5}$) and cardiovascular effects [33]. Other reviews have examined published literature on the association of particulates from air pollution and cardiovascular disease [34–36]. Unlike previous works, the present review is focused on studies that used the ApoE$^{-/-}$ mouse model of dyslipidemia and atherosclerosis to better understand the cardiovascular pathobiology of PM exposure.

Continuously exposing *ApoE*$^{-/-}$ mice to untreated ambient PM (24 h/day, 7 days/week, for 2 months) was found to cause accelerated plaque development and trigger a systemic inflammatory response, indicating that non-concentrated exposure to ambient PM is sufficient to trigger systemic responses that contribute to vascular lesion formation in susceptible mice [37]. Abnormalities in cardiac function and mildly increased atherosclerotic plaque areas have been found in *ApoE*$^{-/-}$ mice exposed to CAPs ($PM_{2.5}$) for 6 h/day, 5 days per week, for up to 5 months [38–40]. These results are similar to those reported for the exposure of the same animal model to sidestream cigarette smoke (environmental tobacco smoke; ETS) [41]. However, higher PM concentrations were used in the ETS study than in the CAPs studies, so it has been suggested that atherosclerosis may be accelerated more strongly by CAPs than by ETS [38]. Similar results to those observed in plaques from *ApoE*-deficient mice, and increased lipid pools in coronary and aortic lesions, have been reported in hypercholesterolemic rabbits exposed to PM_{10} twice a week for 4 weeks [42]. Functional changes in vascular tissues, associated with exposure to $PM_{2.5}$, have been described. The aortic rings of *ApoE*$^{-/-}$ mice exposed to CAPs ($PM_{2.5}$) for 6 months (6 h/day, for 5 days per week) were found to exhibit exaggerated vasoconstrictor responses to phenylephrine and serotonin and to exhibit attenuated relaxation responses to acetylcholine [43]. In that study, plaques from *ApoE*$^{-/-}$ mice fed a standard diet or a high-fat diet exhibited higher expression levels of iNOS and 3-nitrotyrosine if the mice were exposed to $PM_{2.5}$ than if the mice were exposed to filtered air. The in situ detection of reactive oxygen species (ROS) and macrophages in aortas from *ApoE*$^{-/-}$ mice revealed that animals exposed to CAPs had higher ROS concentrations and higher macrophage infiltration than did animals not exposed to CAPs [43]. Most studies investigating the vascular effects of CAPs have focused on PM with aerodynamic diameters of 2.5 and 10 μm, however, experimental evidence suggests that PM with diameters of less than 2.5 μm may be more atherogenic than PM with larger diameters [44]. *ApoE*$^{-/-}$ mice exposed to an atmosphere containing ultrafine particles (<0.18 μm) for 75 h over a period of 40 days had larger aortic lesions than mice that had been exposed to fine particles (<2.5 μm) or filtered air. Interestingly, that study also demonstrated that exposure to CAPs results in the loss of the anti-inflammatory properties that high-density lipoproteins provide. Exposure to particles with diameters of less than 0.18 μm has been found to cause activation of hepatic enzymes related to xenobiotic metabolism, including NFE2-related transcription factor 2 (NFE2L2), activating transcription factor 4 (ATF4), glutathione S-transferase (GSTA1), SOD2, and NAD(P)H-quinone oxidoreductase 1 (NQO1) [44]. These data are consistent with the hypothesis that the adverse effects of PM on the cardiovascular system are mediated by the activation of oxidative stress and inflammation mechanisms in which the regulation of the expression of protective antioxidant enzymes by *Nfe2l2* plays a crucial role [45].

2.2.2 Carbon Nanotubes (CNTs)

The unique physicochemical properties (including size, shape, surface area, and surface chemistry) of nanomaterials have triggered great enthusiasm for their use in biomedical applications [46]. However, these same physicochemical characteristics may lead to unidentified biohazards that need to be toxicologically characterized using rigorous screening strategies [47–49]. The intratracheal instillation or pharyngeal aspiration of CNTs in rats and mice lead to a number of deleterious pulmonary effects, ranging from inflammation to fibrosis and granuloma formation [50–52]. Acute pulmonary exposure (for 24–72 h) of spontaneously hypertensive rats to single walled CNTs (SWCNTs) delivered by intratracheal instillation resulted in lung injury and increased plasma concentrations of endothelin-1 (EDN1) and angiotensin-I converting enzyme (ACE), suggesting that systemic endothelial dysfunction is caused [53]. Exposure to intravenously injected multi-walled CNTs (MWCNTs) induces platelet aggregation and thrombosis in rat carotid arteries [54], whereas pulmonary exposure to MWCNTs impairs coronary arteriolar vasodilation [55]. Although none of the studies described above were conducted using *ApoE*$^{-/-}$ mice, they indicate that CNTs are involved in various crucial biological responses that are ultimately associated with vascular disease and atherogenesis.

Repeatedly (every other week for 8 weeks) exposing *ApoE*$^{-/-}$ mice to 20 μg of SWCNTs by oropharyngeal aspiration results in acceleration of plaque formation in the aortic arches and brachiocephalic arteries, but no change in the lipid profile [56]. In that study, oxidative stress, determined by the activation of heme oxygenase-1 (HMOX1), and mitochondrial DNA damage were increased. Administering *ApoE*$^{-/-}$ mice with between 32 and 128 μg of MWCNT per mouse for 5 weeks has been shown to accelerate atherosclerosis [57]; the highest dose that was used (128 μg) was chosen because it corresponds to workplace exposure levels (400 μg/m^3) that have been determined for humans [58]. Jacobsen and colleagues studied the acute pulmonary effects of administering *ApoE*$^{-/-}$ mice with a single intratracheal instillation of five different types of nanoparticles, which were carbonaceous particles, C_{60} fullerenes, SWCNTs, gold particles, and quantum dots [59]. Increased recruitment of neutrophils in bronchoalveolar lavage fluid (BALF) and the increased expression of inflammatory molecules in lung tissue were observed 3–24 h following SWCNTs, carbonaceous particles, and quantum dots. Weaker responses were observed for the gold and C_{60} fullerene particles. *ApoE*$^{-/-}$ mice exhibited faster and stronger responses than wild-type mice to such stimuli [59]. These data confirmed that the *ApoE*$^{-/-}$ mouse strain is particularly sensitive to the acute CNT-mediated effects and is suitable for the toxicological screening of these effects. Exposing old (40–42 weeks old) and young (11–13 weeks old) *ApoE*$^{-/-}$ mice to C_{60} fullerenes through intraperitoneal injection for 1 h resulted in the slight impairment of endothelium-dependent (acetylcholine-dependent) and endothelium-independent (sodium nitroprusside-mediated) relaxation [60].

2.3 Cigarette Smoke

2.3.1 Mainstream Tobacco Smoke (MTS)

Cigarette smokers display a clinical state characterized by hypercoagulability [61, 62]. The increased expression of tissue factor observed in atherosclerotic plaques from *ApoE*$^{-/-}$ mice is a mechanism that could explain the increased risk of atherothrombotic events in smokers [63]. *ApoE*$^{-/-}$ mice exposed to MTS exhibit increased rates of plaque development [14, 64] and aortic aneurysm formation [65] compared with animals exposed to filtered air. The accumulation of aortic cholesterol and increased plasma concentrations of substances related to oxidative stress (including TBARS, oxLDL, and 3-nitrotyrosine) have been suggested as evidence of the MTS-dependent acceleration of atherogenesis [66].

2.3.2 Environmental Tobacco Smoke (ETS)

Epidemiologic studies have shown an association between exposure to ETS (also known as passive smoking, involuntary smoking, or exposure to second-hand smoke) and cardiovascular [67, 68] and periodontal disease [69], as well as an increased risk of stroke among never-smokers [70]. No head-to-head comparisons of the effects of MTS and ETS on cardiovascular end points have been reported, however, $ApoE^{-/-}$ mice on a high-fat diet exhibited significantly increased aortic plaque areas when they were exposed to MTS [41, 71]. Changes in plaque size were accompanied by increased free and esterified cholesterol concentrations in the aortas of the exposed animals even though no significant changes in plasma cholesterol concentrations were observed [41, 71]. Using the ferric chloride injury model in carotid arteries of $Apoe^{-/-}$ mice that had been fed a high-fat diet, Dong and colleagues demonstrated that exposure to ETS increased platelet reactivity and accelerated thrombosis [72]. In summary, studies that have been conducted using $ApoE^{-/-}$ mice have shown that exposure to MTS results in the aggravation of vascular lesion formation. Although there are few studies aimed at identifying potential mechanistic differences between exposure to MTS and ETS, Harrison and colleagues generated *ApoE–/–* mice with decreased superoxide dismutase 2 activity ($ApoE^{-/-}$, SOD2$^{+/-}$) to investigating the roles of oxidative stress and impaired mitochondrial function in animals exposed to CS [73]. $ApoE^{-/-}$, SOD2$^{+/-}$ animals that were exposed to ETS exhibited larger aortic plaques than did $ApoE^{-/-}$ animals also exposed to ETS. $ApoE^{-/-}$, SOD2$^{+/-}$ animals that were unexposed to ETS also exhibited larger aortic plaques than did $ApoE^{-/-}$ animals that were unexposed to ETS. A key observation in that study was that the plasma levels of cholesterol were the same in both mice genotypes, indicating that a decrease in the local mitochondrial antioxidant capacity may affect the development of vascular disease in the absence of systemic changes. In this context, using lipidomics to analyze the molecular lipid species in the subcellular organelles affected by increased oxidant concentrations would support the notion that oxidative stress associated with inflammation leads to mitochondrial damage in arteries that are prone to develop atherosclerosis.

2.4 Vascular Effects of Pollutants and Cigarette Smoke in $ApoE^{-/-}$ Mice: Alveolar Translocation of Particles Versus Spillover of Lung Inflammation

The mechanisms underlying the systemic effects of inhaled toxicants remain uncertain. It has been postulated that the impairment of vascular function by inhaled environmental pollutants is mediated by a variety of mechanisms that fall into three categories: (1) direct vascular effects of inhaled PM; (2) direct effects of inhaled gases derived from emissions and organic constituents and transition metals derived from PM that can readily cross the alveolar barrier; and (3) indirect vascular effects mediated by spillover to the systemic circulation of pulmonary inflammatory molecules released in response to inhaled pollutants [44]. Although a combination of these mechanisms may occur simultaneously, the relative contributions and potential additive or synergistic effects of the individual mechanisms on vascular dysfunction and disease progression have not yet been fully elucidated.

Several studies have demonstrated that PM induces endothelial dysfunction in humans [74, 75]. Healthy volunteers exposed to DE for 2 h have shown increased systolic blood pressure and decreased relaxation in response to bradykinin, acetylcholine, and sodium nitroprusside, whereas exposure to pure carbon nanoparticles and filtered DE had no effect on vascular end points [76]. These results indicate that combustion-derived nanoparticles are responsible for the adverse vascular effects caused by inhaling DE. Although carbon nanoparticles induce inflammatory effects in vitro [77] and between 3 and 24 h after intratracheal instillation into *ApoE*$^{-/-}$ mice [59], carbon nanoparticles cause only minor changes in vascular reactivity 1 h after intraperitoneal injection [60]. Furthermore, exposure to CAPs, which are low in combustion-derived particles, did not affect vascular function in healthy subjects or coronary artery disease patients [78]. Taken together, data from acute exposure studies in humans and animals suggest that combustion-derived PM is a prominent factor in vascular dysfunction driven by environmental pollution. A limitation of the studies described above is that they followed short-term exposure protocols; therefore, no inferences regarding long-term effects could be drawn from them. A recent study of *ApoE*$^{-/-}$ mice exposed to vehicle emissions for 50 days demonstrated that removing the gaseous components from mixed vehicle emissions abolished the vascular toxicity of the whole vehicle emissions, underlining the relevance of the interactions between fresh gases and PM [79].

Inhaled particles must gain access to the systemic circulation by penetrating the alveolar barrier to exert vascular effects in organs other than the lungs. The rather complex biology involved in the translocation of particles of different sizes and different physicochemical properties has been addressed in a number of studies [80–83]. The target organs of inhaled PM include not only the lungs but also the liver, spleen, kidneys, heart, and brain [80]. The translocation of inhaled carbon and iridium nanoparticle aggregates has been shown to be size-dependent, and larger 24-h fractional retention has been found for small particles than for large particles in all organs examined [80]. Furthermore, the size of the inhaled PM seems to correlate with the intensity of the microvascular dysfunction observed [24].

Experiments designed to circumvent the pulmonary inflammatory responses triggered by inhaled PM have indicated that PM elicits inflammatory responses (increased TNFα) in C57BL/6 mice once it has entered the systemic circulation, within 24 h of being intraperitoneally injected [84]. Similarly, the intraperitoneal administration of DE particles into *ApoE*$^{-/-}$ mice was found to result in increased hepatic iNOS expression [27]. However, no changes in iNOS expression were observed in the lungs and aortas of *ApoE*$^{-/-}$ mice [27], and the vascular reactivity (e.g., acetylcholine-induced relaxation) remained unmodified in C57BL/6 animals [84]. The latter finding may be accounted for by the rapid sequestration of particles by the spleen and liver [84], however, in a previous study, endothelial dysfunction occurred as early as 1 h after mice were intraperitoneally injected with DE particles [85]. Intraperitoneal administration of C_{60} fullerenes (nanoparticles with diameters of about 1 nm) into *ApoE*$^{-/-}$ mice elicited moderate vasomotor impairment in aortas harvested and evaluated 1 h after exposure [60]. The exclusively acute nature of these studies suggests that full time-course experiments using non-pulmonary administration routes should be conducted. Multiple surrogates of tissue inflammation and oxidative stress should be also examined to investigate whether the acute administration of DE particles and nanoparticles has a direct impact on molecular inflammation pathways consistent with the development of vascular lesions.

Further substantiation of the notion that ultrafine particles are rapidly translocated by biological membranes has been provided by short-term (<24 h) post-exposure studies that have shown that ultrafine particles are actively transferred not only to the interstitial space in the respiratory tract but also to non-respiratory organs including the liver [86] and to the central nervous system via olfactory nerve axons [81]. Nanoparticles have also been shown to be translocated across non-respiratory cellular membranes, including endothelial cell monolayers and atherosclerotic plaques in rabbit aortas [87]. These data suggest that circulating particles derived from toxicants have the potential to influence plaque development, and this has been demonstrated experimentally [54, 56]. However, the data also show the potential therapeutic benefits of using nanotechnology to effectively deliver drugs to atherosclerotic plaques [87–89].

Cigarette smoke is a complex chemical mixture of thousands of compounds, and the effects of these compounds on vascular biology remain largely uncharacterized. Future efforts should be aimed at measuring the particle size distributions in cigarette aerosols for use in in vitro and in vivo tests in systems relevant to vascular disease.

3 Assessment of the Comparability of Data from PM Exposure Studies Conducted Using ApoE$^{-/-}$ Mice

We have constructed two tables in which most, if not all, published PM exposure studies conducted so far in *ApoE*$^{-/-}$ mice are summarized (Table 1 and 2). The number of end points (spanning cardiovascular, inflammatory, and respiratory diseases) varied greatly between the 43 studies included in Table 1. Between one and six types of exposure (and conditions) were reported in most of the studies, 12 were reported in two studies [27, 90], and 15 were reported in one study [59]. The exposure categories, items administered, routes of administration, and detailed exposure protocols were compiled. The exposure periods were divided into acute (≤24 h, Table 1) and chronic (8–180 days, Table 2) according to the reported exposure times. Four exposure categories were defined: motor vehicle emissions (MVE), environmental air particles (EAP), MTS, and ETS. A few general observations can be made from the experimental conditions depicted in Tables 1 and 2. First, more end points were studied and more studies were performed using whole body exposure (WBE) protocols than using other exposure routes. The four exposure categories were well represented in the studies using WBE protocols, whereas they were poorly represented in the studies using other PM exposure routes and items (e.g., DE and CAPs). Chronic effects of PM exposure, as defined in this review (>8 days) were investigated in the majority of the studies (Table 2). Acute effects were investigated in a few studies (Table 1), and the effects of PM exposure for 2–7 days were not investigated in any of the studies. Overall, the shorter the exposure time, the more often the reported end point remained unchanged or statistically nonsignificant. A single PM dose was administered when the intraperitoneal, intravenous, or intratracheal PM exposure routes were used, except in one intratracheal study in which the chronic effects of MWCNTs were investigated [57]. In the latter study, animals were fed a high-fat diet, whereas the animals were fed a standard diet in all of the other intraperitoneal, intravenous, and intratracheal exposure studies (Table 1). Therefore, inferences regarding vascular effects from these acute studies should be taken with caution when compared with subchronic or chronic protocols using animals fed a high-fat diet. The exposure item concentrations, particle types, and particle sizes were relatively consistent across the studies that used similar administration routes (e.g., WBE) for each exposure item (e.g., CAPs), but varied greatly between exposure categories (e.g., MVE versus EAP). The biological significance of these differences is unknown and warrants further investigation.

Several functional categories related to atherogenesis were created (e.g., dyslipidemia, inflammation, and plaque development) according to the specific end point that was investigated to facilitate the analysis of such dissimilar sets of data (Table 3). For example, macrophage infiltration and TNFα production were categorized as inflammatory end points, and the plaque area and volume determinations were included in the plaque development category. Plaque development has been investigated after exposure to environmental and combustion-derived PM (EAP, MVE + EAP, MTS and ETS). With few reported exceptions (listed in Table 3), the chronic WBE exposure of *ApoE*$^{-/-}$ mice to PM resulted in aortic lesions that progressed faster than in the unexposed control or sham-exposed groups. Chronic WBE and nasal exposure to environmental or combustion-derived PM showed less consistent effects on inflammatory markers in the studies. Significant increases,

Table 3
Functional end point categories related to atherogenesis in studies conducted in ApoE$^{-/-}$ mice

Exposure Category	Exposure Route	PMID	Dyslipidemia	Inflammation	Oxidative Stress	Plaque Development	Vascular Function (constriction)	Vascular Function (relaxation)	Vascular Remodeling
MVE	IP	17234226					↓ →	↓ →	
		17602821			→ ↗				
	Nasal	22507957	→	→ ↘		→ ↘ ↗			
		23584878				↑			
	OPA	24330719				↑	→	→	
	WBE	17065432			↑ ↗		↑ → ↗		↑ → ↘ ↗
		19891982		↑ →	↑ → ↗	↓ → ↘ ↗	↑ →		↓ ↑ → ↘ ↗
		20462391				↑ → ↗	↗	→	
		21435644	→		↑	↑			
		21722660			↑	→	↓ ↑		
		22746401		↑ →		↑	↓		
		24730681		↑ → ↗	↑ →			↓ →	↑ ↗
EAP	IP	19243580					→	↓ ↘	
	OPA	1849906				↑			
	WBE	15804939				↑			
		16414948				↑	↓	↑	
		18202315				↑ ↗			
		18236227		↑		↑ ↗			↑ ↗
		19182107		↑	↑ ↑ →	↑	↓		
		20235771				↑			
		20462391				↑ ↑	↗	→	
		24045146				↑			
		24523034				↑			
MVE+EAP	WBE	20462391				↑ ↗	↗	→	
ETS	WBE	11368996				↑ → ↗			
		20227489				↑			
		20235771				↑ → ↗			
		22659316				↑			

Functional categories related to plaque development were created according to the specific endpoint investigated (see main text). This approach facilitated the analysis of dissimilar sets of data. The table is broken down by Exposure Category and Exposure Route. PubMed IDs (PMID) for each route and category are shown. EAP: environmental air particles, ETS: environmental tobacco smoke, IP: intraperitoneal, MTS: mainstream tobacco smoke, MVE: motor vehicle emissions, OPA: oropharyngeal aspiration, WBE: whole body exposure. Arrows indicate increase (upwards), decrease (downwards), no change (rightwards), trend to increase (north east), trend to decrease (south east).

increasing trends, decreases, and a lack of change were observed using different inflammatory end points, including cytokine production (IL-4, IL-10, and IFN-γ) [22] and macrophage recruitment [91]. Given the systemic inflammatory status that the *ApoE*$^{-/-}$ mouse model displays, it would be worthwhile to develop a standardized test for vascular inflammation and oxidative stress biomarkers that can be interrogated using plasma and vascular tissues from *ApoE*$^{-/-}$ mice that have been chronically exposed to a variety of noxious ambient and combustion-derived particles (Fig. 2). Despite the fact that the dyslipidemic profile of *ApoE*$^{-/-}$ mice is modified by exposure to MTS [15, 16], few studies investigated the effects of exposure on plasma lipids and two studies showed no effects of MVE exposure (nasal and WBE) on plasma lipids (Table 3). In vitro evaluation of vascular function illustrated primarily as changes in endothelium-dependent (acetylcholine), or

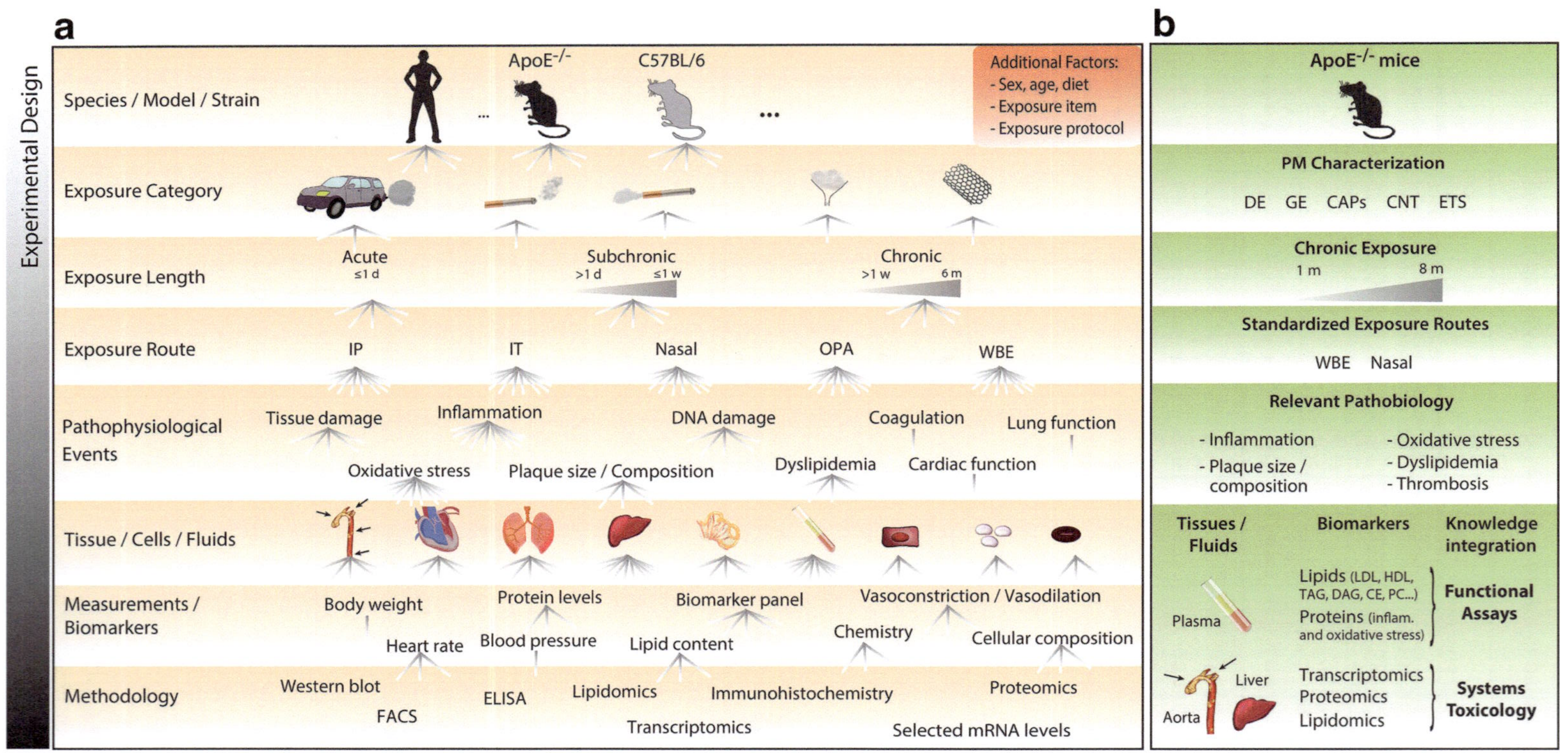

Fig. 2 Experimental approaches to assess the biology of exposure to combustion-derived pollutants. (**a**) Exposure studies conducted so far lack the systematic use of uniform experimental exposure parameters (e.g., time, route of administration, exposure items and animal models) to compare end points and datasets from different studies. Acute experiments were defined as those with exposure times of 1 day or less, whereas chronic studies were considered to be those reporting exposure times of 1 week or longer. Sub-chronic exposure times (>1 day, ≤1 week) were rarely reported. (**b**) The *ApoE*-deficient mouse model of atherosclerosis can be utilized as a benchmark to assess the effects of environmental, motor engine, and CS-derived pollutants. Efforts to standardize the experimental conditions should include exposure times that would also require the development of exposure protocols to clearly distinguish short- and long term effects on vascular toxicity using a combination of conventional biomarkers and systems toxicology approaches. *CAPs* concentrated ambient particles, *CE* cholesteryl esters, *CNT* carbon nanotubes, *d* day *DAG* diacylglycerols, *DE* diesel exhaust, *ELISA* enzyme-linked immunosorbent assay, *ETS* environmental tobacco smoke, *FACS* fluorescence-activated cell sorting, *GE* gasoline exhaust, *HDL* high-density lipoproteins, *LDL* low-density lipoproteins, *m* month, *TAG* triacylglycerols, *PC* phosphatidylcholine, *w* week, *WBE* whole body exposure

endothelium-independent (sodium nitroprusside) vasorelaxation and vasoconstriction (phenylephrine- or K^+-induced) were conducted in a few studies, but the results were mixed (Table 3). Differences in PM exposure times and tissue collection times, a lack of standardized methods for isolating vessels, different aortic segments being used (e.g., thoracic aorta versus aortic arch), and different organ chamber incubation and stimulation conditions may all have played a role in the conflicting results reported in different studies (Table 2). In summary, clear knowledge gaps exist regarding the vascular biological effects of PM from a variety of sources. Standardized, simple and reliable methods should be developed to generate scientific data using a variety of functional and molecular end points. Systems toxicology approaches and in silico modeling would provide the required tools to integrate knowledge derived from large datasets (Fig. 2, panel b).

4 Systems Toxicology of Cigarette Smoking: Transcriptomic- and Lipidomic-Based Analysis of Plasma and Vascular Tissues from ApoE$^{-/-}$ Mice

The CS-dependent acceleration of aortic plaque development is accompanied by the accumulation of all lipid classes determined by using mass spectrometry, including cholesteryl esters (CE), diacylglycerol (DAG), triacylglycerol (TAG), and sphingolipids [15] (Fig. 3). Smoking has also been found to induce compositional changes in the aortic lipidomes, in which the relative proportions of several phospholipid classes decreased while the proportion of CE increased (Fig. 3). The effects of CS on hepatic lipid metabolism are reflected in plasma, with the concentrations of several lipid classes being increased. Free cholesterol (FC), lysophosphatidylethanolamine (LPE), and sphyingosine-1-phosphate (S1P) were found to increase more than other lipid classes in response to CS (Fig. 3). Interestingly, the lipidomics profiles of *ApoE–/–* mouse aortas and human plaques from patients undergoing carotid, femoral, or radial endarterectomy exhibited a number of similarities. Lipids that were enriched in human arteries were also detected in the aortas of *ApoE–/–* mice. Specifically, CE16:1, CE18:3, and CE18:1 were found to be enriched in CS-exposed murine aortas, plasma, and liver, and were also enriched in the human atherosclerotic plaques [13]. Other lipid classes that were found to be enriched in murine and human plaques included phosphatidylserine (PS), phosphatidylcholine (PC), lysophosphatidylcholine (LPC), and sphingomyelin (SM) [13, 15]. Transcriptomics analysis of livers of $ApoE^{-/-}$ mice exposed to CS revealed expression upregulation of genes related to lipid metabolism [14, 92].

The relevance of lipidomics-based approaches to cardiovascular research has been demonstrated by molecular lipid species being identified that can only be achieved using these methodological

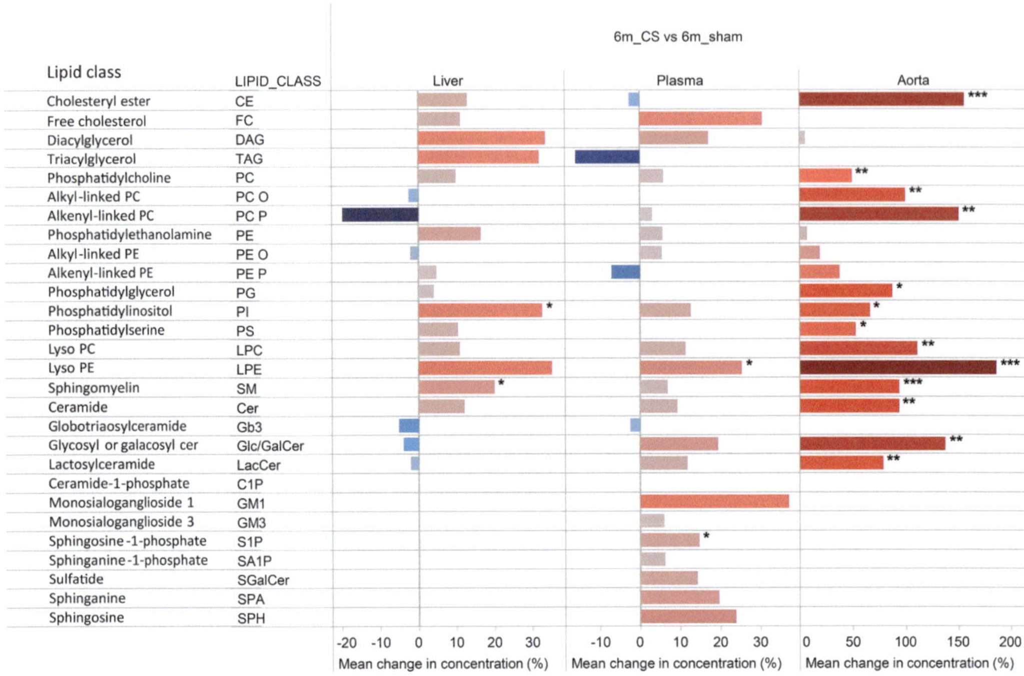

Fig. 3 Lipidomics datasets contrasts using tissues from CS-exposed and sham-exposed ApoE$^{-/-}$ mice. *Bars* indicate percentage differences in lipid class concentrations in liver, plasma, and aorta between the CS group and the sham group. *$p<0.05$, **$p<0.01$, ***$p<0.001$. The largest relative changes in lipid class concentrations were observed in aortic tissues. Reproduced with permission from [15]

approaches. For instance, changes in polyunsaturated fatty acids contained in plasma CE and PC species (20:4 and 22:6) have shown the effects of smoking on eicosanoid metabolism, because arachidonic acid (20:4) is a primary substrate for eicosanoid synthesis. Using a smoking protocol that switches from smoking exposure to smoking cessation (fresh air), we have shown that most lipid species that were elevated in the livers of CS-exposed animals decreased in concentration when smoking ceased [92]. The aortas of hamsters fed a high-fat diet have also been shown to undergo specific changes in the tissue content of lipid classes and molecular species present [93]. However, while some lipid classes were similarly upregulated or downregulated compared with *ApoE*$^{-/-}$ mice exposed to CS, most molecular species were regulated differently, supporting the power of lipidomics analysis to discriminate between diet-induced changes and changes driven by *ApoE*-deficiency. To address this matter, and the challenge of analyzing thousands of molecular species and genes in complex lipidomics and whole-genome gene expression datasets, we have developed a vascular biology model that recreates the inflammatory processes that take place during atherogenesis, from endothelial dysfunction to atheroma formation and plaque destabilization

[17]. Using this systems toxicology tool, we have identified fundamental molecular differences in plaque development between murine and human atherosclerotic plaques. It is reasonable to expect that additional models aimed at integrating the concentrations of molecular lipid species enriched in vascular tissues with the activities of enzymes involved in the synthesis and catabolism of modulated lipids would be an invaluable resource for the identification of novel therapeutic targets.

5 Conclusion

Control of environmental pollution due to motor vehicle emissions, cigarette smoke, and other combustion-derived PM are daunting regulatory challenges. Health problems derived from exposure to combustion-derived pollutants are the result of the effects of numerous chemical entities on various cell types, tissues, organs and entire biological systems in humans and animals. Current scientific approaches aimed at expanding our understanding of the molecular mechanisms underlying the actions of combustion-derived pollutants face unique challenges due to the inherent chemical complexity of the testing material (e.g., motor vehicle emissions, ambient particles and tobacco smoke fractions). The *ApoE*-deficient mouse has greatly contributed to the understanding of cardiovascular function, dyslipidemia and atherosclerosis since its development more than two decades ago [7, 94]. Over the past decade, several research groups have utilized the *ApoE*$^{-/-}$ mouse model to investigate the effects of exposure to PM from a variety of sources. Although some of the results of those studies are relatively unambiguous (e.g., effects on plaque development), others are more difficult to interpret and reconcile due primarily to differences in exposure times, study end points, exposure items, particle size distribution, and routes of administration between studies. Numerous questions, including the effects of filtering PM and the actions of particles and nanoparticles on oxidative stress, inflammation, plasma lipids and vascular lipid content remain unanswered. A conceptual framework designed to support the generation of such information would be highly valuable not only from a human health perspective but to drive and sustain an evidence-based system to regulate combustion-derived emissions. The inclusion of multidisciplinary systems toxicology approaches and methodologies (e.g., transcriptomics and lipidomics), as well as the development and implementation of guidelines and standards to investigating the effects of combustion-derived chemicals and their interactions in animal models of disease, would vastly enhance our knowledge and thrust academic, industrial, and regulatory efforts at controlling and reducing the toxic effects induced by chemicals and chemical mixtures.

References

1. Peters A (2005) Particulate matter and heart disease: evidence from epidemiological studies. Toxicol Appl Pharmacol 207(2 Suppl):477–482. doi:10.1016/j.taap.2005.04.030
2. Moghadasian MH, McManus BM, Nguyen LB et al (2001) Pathophysiology of apolipoprotein E deficiency in mice: relevance to apo E-related disorders in humans. FASEB J 15(14):2623–2630. doi:10.1096/fj.01-0463com
3. Ghiselli G, Schaefer EJ, Gascon P et al (1981) Type III hyperlipoproteinemia associated with apolipoprotein E deficiency. Science 214(4526):1239–1241
4. Zannis VI, Breslow JL (1980) Characterization of a unique human apolipoprotein E variant associated with type III hyperlipoproteinemia. J Biol Chem 255(5):1759–1762
5. Plump AS, Breslow JL (1995) Apolipoprotein E and the apolipoprotein E-deficient mouse. Annu Rev Nutr 15:495–518. doi:10.1146/annurev.nu.15.070195.002431
6. Arunachalam G, Sundar IK, Hwang JW et al (2010) Emphysema is associated with increased inflammation in lungs of atherosclerosis-prone mice by cigarette smoke: implications in comorbidities of COPD. J Inflamm 7:34. doi:10.1186/1476-9255-7-34
7. Plump AS, Smith JD, Hayek T et al (1992) Severe hypercholesterolemia and atherosclerosis in apolipoprotein E-deficient mice created by homologous recombination in ES cells. Cell 71(2):343–353
8. Puri R, Duong M, Uno K et al (2012) The emerging role of plasma lipidomics in cardiovascular drug discovery. Expert Opin Drug Discov 7(1):63–72. doi:10.1517/17460441.2012.644041
9. Stahlman M, Ejsing CS, Tarasov K et al (2009) High-throughput shotgun lipidomics by quadrupole time-of-flight mass spectrometry. J Chromatogr B Analyt Technol Biomed Life Sci 877(26):2664–2672. doi:10.1016/j.jchromb.2009.02.037
10. Vihervaara T, Suoniemi M, Laaksonen R (2013) Lipidomics in drug discovery. Drug Discov Today. doi:10.1016/j.drudis.2013.09.008
11. Didangelos A, Stegemann C, Mayr M (2012) The -omics era: proteomics and lipidomics in vascular research. Atherosclerosis 221(1):12–17. doi:10.1016/j.atherosclerosis.2011.09.043
12. Ekroos K, Janis M, Tarasov K et al (2010) Lipidomics: a tool for studies of atherosclerosis. Curr Atheroscler Rep 12(4):273–281. doi:10.1007/s11883-010-0110-y
13. Stegemann C, Drozdov I, Shalhoub J et al (2011) Comparative lipidomics profiling of human atherosclerotic plaques. Circ Cardiovasc Genet 4(3):232–242. doi:10.1161/CIRCGENETICS.110.959098
14. Boue S, De Leon H, Schlage WK et al (2013) Cigarette smoke induces molecular responses in respiratory tissues of ApoE(-/-) mice that are progressively deactivated upon cessation. Toxicology 314(1):112–124. doi:10.1016/j.tox.2013.09.013
15. Boue S, Tarasov K, Janis M et al (2012) Modulation of atherogenic lipidome by cigarette smoke in apolipoprotein E-deficient mice. Atherosclerosis 225(2):328–334. doi:10.1016/j.atherosclerosis.2012.09.032
16. Lietz M, Berges A, Lebrun S et al (2013) Cigarette-smoke-induced atherogenic lipid profiles in plasma and vascular tissue of apolipoprotein E-deficient mice are attenuated by smoking cessation. Atherosclerosis 229(1):86–93. doi:10.1016/j.atherosclerosis.2013.03.036
17. De Leon H, Boue S, Schlage WK et al (2014) A vascular biology network model focused on inflammatory processes to investigate atherogenesis and plaque instability. J Transl Med 12(1):185. doi:10.1186/1479-5876-12-185
18. Seilkop SK, Campen MJ, Lund AK et al (2012) Identification of chemical components of combustion emissions that affect pro-atherosclerotic vascular responses in mice. Inhal Toxicol 24(5):270–287. doi:10.3109/08958378.2012.667455
19. Hoffmann B, Moebus S, Mohlenkamp S et al (2007) Residential exposure to traffic is associated with coronary atherosclerosis. Circulation 116(5):489–496. doi:10.1161/CIRCULATIONAHA.107.693622
20. Peters A, von Klot S, Heier M et al (2004) Exposure to traffic and the onset of myocardial infarction. N Engl J Med 351(17):1721–1730. doi:10.1056/NEJMoa040203
21. Campen MJ, Lund AK, Knuckles TL et al (2010) Inhaled diesel emissions alter atherosclerotic plaque composition in ApoE(-/-) mice. Toxicol Appl Pharmacol 242(3):310–317. doi:10.1016/j.taap.2009.10.021
22. Lund AK, Knuckles TL, Obot Akata C et al (2007) Gasoline exhaust emissions induce vascular remodeling pathways involved in atherosclerosis. Toxicol Sci 95(2):485–494. doi:10.1093/toxsci/kfl145
23. Lund AK, Lucero J, Harman M et al (2011) The oxidized low-density lipoprotein receptor

mediates vascular effects of inhaled vehicle emissions. Am J Respir Crit Care Med 184(1):82–91. doi:10.1164/rccm.201012-1967OC

24. Nurkiewicz TR, Porter DW, Hubbs AF et al (2011) Pulmonary particulate matter and systemic microvascular dysfunction. Res Rep Health Eff Inst 164:3–48
25. Campen MJ, Lund AK, Doyle-Eisele ML et al (2010) A comparison of vascular effects from complex and individual air pollutants indicates a role for monoxide gases and volatile hydrocarbons. Environ Health Perspect 118(7):921–927. doi:10.1289/ehp.0901207
26. Cassee FR, Campbell A, Boere AJ et al (2012) The biological effects of subacute inhalation of diesel exhaust following addition of cerium oxide nanoparticles in atherosclerosis-prone mice. Environ Res 115:1–10. doi:10.1016/j.envres.2012.03.004
27. Folkmann JK, Risom L, Hansen CS et al (2007) Oxidatively damaged DNA and inflammation in the liver of dyslipidemic ApoE$^{-/-}$ mice exposed to diesel exhaust particles. Toxicology 237(1-3):134–144. doi:10.1016/j.tox.2007.05.009
28. Maresh JG, Campen MJ, Reed MD et al (2011) Hypercholesterolemia potentiates aortic endothelial response to inhaled diesel exhaust. Inhal Toxicol 23(1):1–10. doi:10.3109/08958378.2010.535572
29. Miller MR, Borthwick SJ, Shaw CA et al (2009) Direct impairment of vascular function by diesel exhaust particulate through reduced bioavailability of endothelium-derived nitric oxide induced by superoxide free radicals. Environ Health Perspect 117(4):611–616. doi:10.1289/ehp.0800235
30. Poss J, Lorenz D, Werner C et al (2013) Diesel exhaust particles impair endothelial progenitor cells, compromise endothelial integrity, reduce neoangiogenesis, and increase atherogenesis in mice. Cardiovasc Toxicol 13(3):290–300. doi:10.1007/s12012-013-9208-0
31. Campen MJ, Babu NS, Helms GA et al (2005) Nonparticulate components of diesel exhaust promote constriction in coronary arteries from ApoE$^{-/-}$ mice. Toxicol Sci 88(1):95–102. doi:10.1093/toxsci/kfi283
32. Miller MR, McLean SG, Duffin R et al (2013) Diesel exhaust particulate increases the size and complexity of lesions in atherosclerotic mice. Part Fibre Toxicol 10:61. doi:10.1186/1743-8977-10-61
33. Lippmann M (2014) Toxicological and epidemiological studies on effects of airborne fibers: coherence and pubic health implications. Crit Rev Toxicol 44(8):643–695. doi:10.3109/10408444.2014.928266
34. Lee BJ, Kim B, Lee K (2014) Air pollution exposure and cardiovascular disease. Toxicol Res 30(2):71–75. doi:10.5487/TR.2014.30.2.071
35. Grahame TJ, Klemm R, Schlesinger RB (2014) Public health and components of particulate matter: the changing assessment of black carbon. J Air Waste Manage Assoc 64(6):620–660
36. Grunig G, Marsh LM, Esmaeil N et al (2014) Perspective: ambient air pollution: inflammatory response and effects on the lung's vasculature. Pulm Circ 4(1):25–35. doi:10.1086/674902
37. Wan Q, Cui X, Shao J et al (2014) Beijing ambient particle exposure accelerates atherosclerosis in ApoE knockout mice by upregulating visfatin expression. Cell Stress Chaperones. doi:10.1007/s12192-014-0499-2
38. Chen LC, Nadziejko C (2005) Effects of subchronic exposures to concentrated ambient particles (CAPs) in mice. V. CAPs exacerbate aortic plaque development in hyperlipidemic mice. Inhal Toxicol 17(4-5):217–224. doi:10.1080/08958370590912815
39. Lippmann M, Gordon T, Chen LC (2005) Effects of subchronic exposures to concentrated ambient particles in mice. IX. Integral assessment and human health implications of subchronic exposures of mice to CAPs. Inhal Toxicol 17(4-5):255–261. doi:10.1080/08958370590912941
40. Lippmann M, Hwang JS, Maciejczyk P et al (2005) PM source apportionment for short-term cardiac function changes in ApoE$^{-/-}$ mice. Environ Health Perspect 113(11):1575–1579
41. Gairola CG, Drawdy ML, Block AE et al (2001) Sidestream cigarette smoke accelerates atherogenesis in apolipoprotein E$^{-/-}$ mice. Atherosclerosis 156(1):49–55
42. Suwa T, Hogg JC, Quinlan KB et al (2002) Particulate air pollution induces progression of atherosclerosis. J Am Coll Cardiol 39(6):935–942
43. Sun Q, Wang A, Jin X et al (2005) Long-term air pollution exposure and acceleration of atherosclerosis and vascular inflammation in an animal model. JAMA 294(23):3003–3010. doi:10.1001/jama.294.23.3003
44. Araujo JA, Barajas B, Kleinman M et al (2008) Ambient particulate pollutants in the ultrafine range promote early atherosclerosis and systemic oxidative stress. Circ Res 102(5):589–596. doi:10.1161/CIRCRESAHA.107.164970
45. Li N, Nel AE (2006) Role of the Nrf2-mediated signaling pathway as a negative regulator of inflammation: implications for the impact of particulate pollutants on asthma. Antioxid Redox Signal 8(1-2):88–98. doi:10.1089/ars.2006.8.88

46. Farokhzad OC, Langer R (2009) Impact of nanotechnology on drug delivery. ACS Nano 3(1):16–20. doi:10.1021/nn900002m
47. Kagan VE, Bayir H, Shvedova AA (2005) Nanomedicine and nanotoxicology: two sides of the same coin. Nanomedicine 1(4):313–316. doi:10.1016/j.nano.2005.10.003
48. Shvedova AA, Kisin ER, Porter D et al (2009) Mechanisms of pulmonary toxicity and medical applications of carbon nanotubes: two faces of Janus? Pharmacol Ther 121(2):192–204. doi:10.1016/j.pharmthera.2008.10.009
49. Oberdorster G, Maynard A, Donaldson K et al (2005) Principles for characterizing the potential human health effects from exposure to nanomaterials: elements of a screening strategy. Part Fibre Toxicol 2:8. doi:10.1186/1743-8977-2-8
50. Lam CW, James JT, McCluskey R et al (2004) Pulmonary toxicity of single-wall carbon nanotubes in mice 7 and 90 days after intratracheal instillation. Toxicol Sci 77(1):126–134. doi:10.1093/toxsci/kfg243
51. Muller J, Huaux F, Moreau N et al (2005) Respiratory toxicity of multi-wall carbon nanotubes. Toxicol Appl Pharmacol 207(3):221–231. doi:10.1016/j.taap.2005.01.008
52. Shvedova AA, Kisin ER, Mercer R et al (2005) Unusual inflammatory and fibrogenic pulmonary responses to single-walled carbon nanotubes in mice. Am J Physiol Lung Cell Mol Physiol 289(5):L698–L708. doi:10.1152/ajplung.00084.2005
53. Ge C, Meng L, Xu L et al (2012) Acute pulmonary and moderate cardiovascular responses of spontaneously hypertensive rats after exposure to single-wall carbon nanotubes. Nanotoxicology 6(5):526–542. doi:10.3109/17435390.2011.587905
54. Radomski A, Jurasz P, Alonso-Escolano D et al (2005) Nanoparticle-induced platelet aggregation and vascular thrombosis. Br J Pharmacol 146(6):882–893. doi:10.1038/sj.bjp.0706386
55. Stapleton PA, Minarchick VC, Cumpston AM et al (2012) Impairment of coronary arteriolar endothelium-dependent dilation after multi-walled carbon nanotube inhalation: a time-course study. Int J Mol Sci 13(11):13781–13803. doi:10.3390/ijms131113781
56. Li Z, Hulderman T, Salmen R et al (2007) Cardiovascular effects of pulmonary exposure to single-wall carbon nanotubes. Environ Health Perspect 115(3):377–382. doi:10.1289/ehp.9688
57. Cao Y, Jacobsen NR, Danielsen PH et al (2014) Vascular effects of multiwalled carbon nanotubes in dyslipidemic ApoE$^{-/-}$ mice and cultured endothelial cells. Toxicol Sci 138(1):104–116. doi:10.1093/toxsci/kft328
58. Mercer RR, Scabilloni JF, Hubbs AF et al (2013) Extrapulmonary transport of MWCNT following inhalation exposure. Part Fibre Toxicol 10(1):38. doi:10.1186/1743-8977-10-38
59. Jacobsen NR, Moller P, Jensen KA et al (2009) Lung inflammation and genotoxicity following pulmonary exposure to nanoparticles in ApoE$^{-/-}$ mice. Part Fibre Toxicol 6:2. doi:10.1186/1743-8977-6-2
60. Vesterdal LK, Folkmann JK, Jacobsen NR et al (2009) Modest vasomotor dysfunction induced by low doses of C60 fullerenes in apolipoprotein E knockout mice with different degree of atherosclerosis. Part Fibre Toxicol 6:5. doi:10.1186/1743-8977-6-5
61. Benowitz NL, Fitzgerald GA, Wilson M et al (1993) Nicotine effects on eicosanoid formation and hemostatic function: comparison of transdermal nicotine and cigarette smoking. J Am Coll Cardiol 22(4):1159–1167
62. Miller GJ, Bauer KA, Cooper JA et al (1998) Activation of the coagulant pathway in cigarette smokers. Thromb Haemost 79(3):549–553
63. Matetzky S, Tani S, Kangavari S et al (2000) Smoking increases tissue factor expression in atherosclerotic plaques: implications for plaque thrombogenicity. Circulation 102(6):602–604
64. von Holt K, Lebrun S, Stinn W et al (2009) Progression of atherosclerosis in the Apo E$^{-/-}$ model: 12-month exposure to cigarette mainstream smoke combined with high-cholesterol/fat diet. Atherosclerosis 205(1):135–143. doi:10.1016/j.atherosclerosis.2008.11.031
65. Stolle K, Berges A, Lietz M et al (2010) Cigarette smoke enhances abdominal aortic aneurysm formation in angiotensin II-treated apolipoprotein E-deficient mice. Toxicol Lett 199(3):403–409. doi:10.1016/j.toxlet.2010.10.005
66. Kunitomo M, Yamaguchi Y, Kagota S et al (2009) Biochemical evidence of atherosclerosis progression mediated by increased oxidative stress in apolipoprotein E-deficient spontaneously hyperlipidemic mice exposed to chronic cigarette smoke. J Pharmacol Sci 110(3):354–361
67. Jousilahti P, Patja K, Salomaa V (2002) Environmental tobacco smoke and the risk of cardiovascular disease. Scand J Work Environ Health 28(Suppl 2):41–51
68. Munteanu I, Mihaltan FD (2014) Second-hand smoking and CV risk. Curr Treat Options in cardiovasc Med 16(12):348. doi:10.1007/s11936-014-0348-8
69. Javed F, Bashir Ahmed H, Romanos GE (2014) Association between environmental

tobacco smoke and periodontal disease: a systematic review. Environ Res 133:117–122. doi:10.1016/j.envres.2014.05.008

70. Nishino Y, Tsuji I, Tanaka H et al (2014) Stroke mortality associated with environmental tobacco smoke among never-smoking Japanese women: a prospective cohort study. Prev Med 67:41–45. doi:10.1016/j.ypmed.2014.06.029
71. Han SG, Howatt DA, Daugherty A et al (2012) Atherogenic and pulmonary responses of ApoE- and LDL receptor-deficient mice to sidestream cigarette smoke. Toxicology 299(2-3):133–138. doi:10.1016/j.tox.2012.05.015
72. Dong A, Caicedo J, Han SG et al (2010) Enhanced platelet reactivity and thrombosis in Apoe$^{-/-}$ mice exposed to cigarette smoke is attenuated by P2Y12 antagonism. Thromb Res 126(4):e312–e317. doi:10.1016/j.thromres.2010.03.010
73. Harrison CM, Pompilius M, Pinkerton KE et al (2011) Mitochondrial oxidative stress significantly influences atherogenic risk and cytokine-induced oxidant production. Environ Health Perspect 119(5):676–681. doi:10.1289/ehp.1002857
74. Brook RD, Brook JR, Urch B et al (2002) Inhalation of fine particulate air pollution and ozone causes acute arterial vasoconstriction in healthy adults. Circulation 105(13):1534–1536
75. LeBlanc AJ, Moseley AM, Chen BT et al (2010) Nanoparticle inhalation impairs coronary microvascular reactivity via a local reactive oxygen species-dependent mechanism. Cardiovasc Toxicol 10(1):27–36. doi:10.1007/s12012-009-9060-4
76. Mills NL, Miller MR, Lucking AJ et al (2011) Combustion-derived nanoparticulate induces the adverse vascular effects of diesel exhaust inhalation. Eur Heart J 32(21):2660–2671. doi:10.1093/eurheartj/ehr195
77. Barlow PG, Clouter-Baker A, Donaldson K et al (2005) Carbon black nanoparticles induce type II epithelial cells to release chemotaxins for alveolar macrophages. Part Fibre Toxicol 2:11. doi:10.1186/1743-8977-2-11
78. Mills NL, Robinson SD, Fokkens PH et al (2008) Exposure to concentrated ambient particles does not affect vascular function in patients with coronary heart disease. Environ Health Perspect 116(6):709–715. doi:10.1289/ehp.11016
79. Campen M, Robertson S, Lund A et al (2014) Engine exhaust particulate and gas phase contributions to vascular toxicity. Inhal Toxicol 26(6):353–360. doi:10.3109/08958378.2014.897776
80. Kreyling WG, Semmler-Behnke M, Seitz J et al (2009) Size dependence of the translocation of inhaled iridium and carbon nanoparticle aggregates from the lung of rats to the blood and secondary target organs. Inhal Toxicol 21(Suppl 1):55–60. doi:10.1080/08958370902942517
81. Oberdorster G, Sharp Z, Atudorei V et al (2004) Translocation of inhaled ultrafine particles to the brain. Inhal Toxicol 16(6-7):437–445. doi:10.1080/08958370490439597
82. Nemmar A, Vanbilloen H, Hoylaerts MF et al (2001) Passage of intratracheally instilled ultrafine particles from the lung into the systemic circulation in hamster. Am J Respir Crit Care Med 164(9):1665–1668. doi:10.1164/ajrccm.164.9.2101036
83. Shimada A, Kawamura N, Okajima M et al (2006) Translocation pathway of the intratracheally instilled ultrafine particles from the lung into the blood circulation in the mouse. Toxicol Pathol 34(7):949–957. doi:10.1080/01926230601080502
84. Bai N, van Eeden SF (2013) Systemic and vascular effects of circulating diesel exhaust particulate matter. Inhal Toxicol 25(13):725–734. doi:10.3109/08958378.2013.844749
85. Hansen CS, Sheykhzade M, Moller P et al (2007) Diesel exhaust particles induce endothelial dysfunction in apoE$^{-/-}$ mice. Toxicol Appl Pharmacol 219(1):24–32. doi:10.1016/j.taap.2006.10.032
86. Oberdorster G, Sharp Z, Atudorei V et al (2002) Extrapulmonary translocation of ultrafine carbon particles following whole-body inhalation exposure of rats. J Toxicol Environ Health A 65(20):1531–1543. doi:10.1080/00984100290071658
87. Kim Y, Lobatto ME, Kawahara T et al (2014) Probing nanoparticle translocation across the permeable endothelium in experimental atherosclerosis. Proc Natl Acad Sci U S A 111(3):1078–1083. doi:10.1073/pnas.1322725111
88. Chou LY, Ming K, Chan WC (2011) Strategies for the intracellular delivery of nanoparticles. Chem Soc Rev 40(1):233–245. doi:10.1039/c0cs00003e
89. Petros RA, DeSimone JM (2010) Strategies in the design of nanoparticles for therapeutic applications. Nat Rev Drug Discov 9(8):615–627. doi:10.1038/nrd2591
90. Chen LC, Quan C, Hwang JS et al (2010) Atherosclerosis lesion progression during inhalation exposure to environmental tobacco smoke: a comparison to concentrated ambient air fine particles exposure. Inhal Toxicol 22(6):449–459. doi:10.3109/08958370903373845
91. Sun Q, Yue P, Kirk RI et al (2008) Ambient air particulate matter exposure and tissue

factor expression in atherosclerosis. Inhal Toxicol 20(2):127–137. doi:10.1080/08958370701821482

92. De Leon H, Boue S, Peitsch MC et al (2013) Modulation of the hepatic lipidome and transcriptome of Apoe$^{-/-}$ mice in response to smoking cessation. J Liver 2:132. doi:10.4172/2167-0889.1000132

93. Jove M, Ayala V, Ramirez-Nunez O et al (2013) Lipidomic and metabolomic analyses reveal potential plasma biomarkers of early atheromatous plaque formation in hamsters. Cardiovasc Res 97(4):642–652. doi:10.1093/cvr/cvs368

94. Vasquez EC, Peotta VA, Meyrelles SS (2012) Cardiovascular autonomic imbalance and baroreflex dysfunction in the apolipoprotein E-deficient mouse. Cell Physiol Biochem 29(5–6):635–646. doi:10.1159/000277623

Chapter 15

Computational Data Integration in Toxicogenomics

Simona Constantinescu, Shana J. Sturla, Giancarlo Marra, Bernd Wollscheid, and Niko Beerenwinkel

Abstract

Toxicogenomics is an emerging field defined by the adaptation and application of functional genomics techniques to toxicology. Recent advances in generating and analyzing multi-omics data have facilitated the development of toxicogenomics to provide novel answers to many toxicology-related questions. In this chapter, we discuss five recent toxicogenomics studies presenting complementary strategies for mining genome-wide molecular profiling data after exposure of cells to chemicals in vitro. The case studies cover various areas of systems toxicology and pharmacogenomics and illustrate the rapid evolution of toxicogenomics, including computational methods for the analysis, integration, and interpretation of omics data.

Key words Toxicogenomics, Data integration, Data analysis, Cancer, Pharmacogenomics, Systems toxicology, Cell line, Case studies

1 Introduction

The ultimate goal of toxicology is to provide guidelines that ensure the safety of humans and the environment. To this end, repeated and rigorous assessments of toxicity dose levels following exposures to potentially harmful substances, such as chemical and drugs, are essential. Potentially harmful effects need to be evaluated in particular tissues as well as systemically, on the whole organism. Technological advancements in genomic sciences, as well as the development and availability of computational methods to address toxicology-related questions, have greatly increased recently and given rise to the new field of toxicogenomics [1]. The concept of toxicogenomics dates back to at least 1999 and was used initially to refer to microarray-based gene expression studies for identifying biomarkers of toxicity [2, 3]. With time, it evolved to describe multiple omics technologies used to assess toxic effects, and can now be defined as the combination of functional genomics, namely genome, transcript, protein, and metabolite profiling, with toxicology. Based on omics data, toxicogenomics includes the generation of novel

Julia Hoeng and Manuel C. Peitsch (eds.), *Computational Systems Toxicology*, Methods in Pharmacology and Toxicology, DOI 10.1007/978-1-4939-2778-4_15,

toxicological models that aim to provide a comprehensive view of modulations of cellular machineries in response to compounds at various dose ranges [1, 4]. From a broader perspective, toxicogenomics can be regarded as a central element of systems toxicology, which seeks to analyze and predict systemic cellular responses to toxicants across multiple layers of biological organization, ultimately seeking a mechanistic understanding of adverse responses [5].

In therapeutic contexts, for example cancer, the fruitful application of pharmacogenomics approaches has led to the discovery of biomarkers for personalized medicine. Thus, the identification of tumor subtypes susceptible to the intended cytotoxic effects of a particular drug because of specific perturbations within targeted pathways has become one of the current driving forces in improving anticancer therapies [6]. In this regard, high-throughput screening is used to assess the perturbation effects of large numbers of drugs on large numbers of samples representing different conditions, usually cell lines, by measuring indicators such as proliferation or death [7]. The measured effects are potentially associated with specific characteristics of the cell lines, such as expression or mutational profiles, with the aim of revealing new drug biomarkers, as well as to provide an in-depth understanding of the mechanism of action of the drugs.

In this chapter, we discuss five toxicogenomics studies, illustrating complementary strategies for mining genome-wide data after exposure to a toxicant. Section 2 highlights some of the widely used systems toxicology techniques of simplifying the unspecific, heterogeneous, and noisy input data into intuitive representations, namely network and pathway analyses, and pharmacokinetic-pharmacodynamic (PK-PD) modeling. In Sect. 3, we discuss two recent large-scale pharmacogenomics studies in which the impact of a large panel of anticancer drugs was assessed within a large panel of cancer cell lines. Additionally, we present a systematic way of evaluating modeling choices for pharmacogenomics datasets.

2 Systems Toxicology

The increasing amount of available genomic data concerning cellular changes after chemical exposures has expanded our understanding of how perturbations impact biological networks. However, in addition to their inherent complexity, such datasets typically capture only a snapshot of specific cellular activities at a given point in time. Therefore, detecting robust cellular signals and, ultimately, elucidating toxicity mechanisms, remain largely unsolved problems. To this end, large-scale genomic datasets are often transformed to reduced representations, which contain the underlying signal from the data, otherwise hard to detect because of biological heterogeneity and technical noise. The transformed genomic data can then serve as basis for a generalizable model [8, 9]. In principle, any classical computational systems biology method, e.g., network analysis, modularization, or pattern recognition, may be applied to problems in systems toxicology, but special considerations must be accounted for. Following exposure to realistic doses of chemicals, the perturbations to the biological system are often not expected to occur genome-wide, but rather to target specific genes and pathways [10]. In many of these cases, changes at the level of single genes are difficult to detect statistically, especially if the system is inherently noisy. However, the detection power is often increased at the level of groups of genes [11]. Additionally, for some chemicals, a basic mechanistic biological understanding of their mode of action already exists [12], making mechanistic modeling of the kinetics of compounds a well-suited systems toxicology approach.

2.1 Network and Pathway Analyses

One of the most widely used ways to simplify the underlying biology of a complex process is to represent it as a network [13]. Biological networks model biological entities, usually genes or proteins, as nodes in a graph. Relationships between them, such as coexpression, belonging to the same protein family, or evolutionary relatedness, are modeled as edges [14]. Biological networks are an intuitive representation of large-scale omics data. Therefore, several academic and commercial software tools have been developed that perform reconstruction, analysis, and visualization of biological networks. For example, visualization tools [15] include Arena 3D [16], BioLayout Express3D [17], VANTED [18], CellDesigner [19], Medusa [20], Osprey [21], Cytoscape [22, 23], and others (Table 1). Specifically, Cytoscape is a popular tool used for both data integration and network visualization, and can be extended with various plugins.

Pathway analysis, also termed gene set analysis, involves the analysis of networks in which genes are connected by their related functionality. With pathway analysis methods, large gene expression datasets are mined on the basis of existing biological knowledge and enriched groups of genes that share common biological functions are extracted. Examples of databases containing functional information include Gene Ontology (GO) [30], the Kyoto Encyclopedia of Genes and Genomes (KEGG) [31], and the Molecular Signatures Database (MSigDB) [32]. Available pathway analysis methods can roughly be divided into three groups [33]: classical methods that use tests on contingency tables, such as Fisher's exact test or the χ^2 test, to detect whether a predefined gene set is overrepresented in a list of significantly differentially expressed genes, e.g., BINGO [34], DAVID [35], GOstat [36], and GOEAST [37]; multivariate methods in which multivariate tests are performed with raw expression data rather than starting

Table 1
Selected tools performing visualization of biological networks

Tool	Type	Description
Arena 3D [16]	Stand-alone	3D visualization of biological data on multiple levels via multilayered graphs
BioLayout Express 3D [17]	Stand-alone	Integration via clustering, 3D visualization, and analysis of very large network graphs
BioTapestry [24]	Stand-alone	Visualization of complex, spatially and temporally distributed genetic regulatory networks
CellDesigner [19]	Stand-alone	Modeling, simulation, and visualization of gene-regulatory and biochemical networks
Cytoscape [22, 23]	Stand-alone	Complex analysis and visualization of biological networks, extendable with various plug-ins performing specific modeling and visualization tasks
Medusa [20]	Stand-alone	Visualization and clustering of large-scale biological networks from multiple sources
Osprey [21]	Stand-alone	Visualization and manipulation of complex interaction networks
VANTED [18]	Stand-alone	Visualization and analysis of networks with related experimental datasets
iPath [25]	Web-based	Visualization, analysis, and customization of various KEGG pathways
KEGG Atlas [26]	Web-based	Interactive visualization of KEGG pathways
VisANT [27]	Web-based	Analysis and visualization of integrated biological datasets
GeneMania [28]	Cytoscape plugin	Biological network integration for gene prioritization and gene function prediction
Cerebral [29]	Cytoscape plugin	Visualization of separate layers of an interaction network, according to subcellular localization annotation

from a list of differentially expressed genes, e.g., Hotelling's T [38], analysis of covariance (ANCOVA) [39], and multivariate analysis of variance (MANOVA) [40]; and methods that involve starting from a list of gene-level statistics and calculating a gene set-level statistic for each analyzed gene set, e.g., the popular gene set enrichment analysis method (GSEA) first introduced by Mootha et al. [41] and subsequently improved by Subramanian et al. [11].

With GSEA, an enrichment score is assigned to each analyzed gene set, reflecting its degree of overrepresentation in a ranked list of genes, either at the top (corresponding to overexpression), or at the bottom (corresponding to underexpression). The enrichment score is computed as a weighted Kolmogorov–Smirnov-like statistic, and its statistical significance is assessed using an empirical permutation test. The R package *piano* [33] provides a unified and simplified framework for testing and evaluating various pathway approaches. Pathway analysis methods are particularly useful for extracting biological signals from microarray datasets, because it can happen that a cellular process jointly performed by a large number of genes is more significantly perturbed than the individual genes, especially after multiple testing correction.

2.1.1 Case Study: Transcriptomic Effects of Sulforaphane and Selenium on Colon Cancer and Noncancerous Cell Lines

In a recent network study by Constantinescu et al. [10], the genome-level responses of human cells in vitro were examined after administering diet-relevant doses of sulforaphane and selenium, two food components with anticancer properties (Fig. 1a). Sulforaphane has a well-characterized transcriptional activation mechanism, and is known to induce detoxifying and antioxidant enzymes, as well as selenoproteins. The enzyme induction is associated with the nuclear factor (erythroid-derived 2)-like 2 (NRF2)-regulated transcriptional activation of genes encoding carcinogen detoxification and antioxidant enzymes. Selenium is an essential redox-active trace element that dictates the intracellular levels of selenoproteins. Given the complementary profiles of these two food compounds as well as their connections to the regulation of selenoproteins and to the NRF2 pathway, their effects on the colon cancer cell line HT29 and on the noncancerous human colon

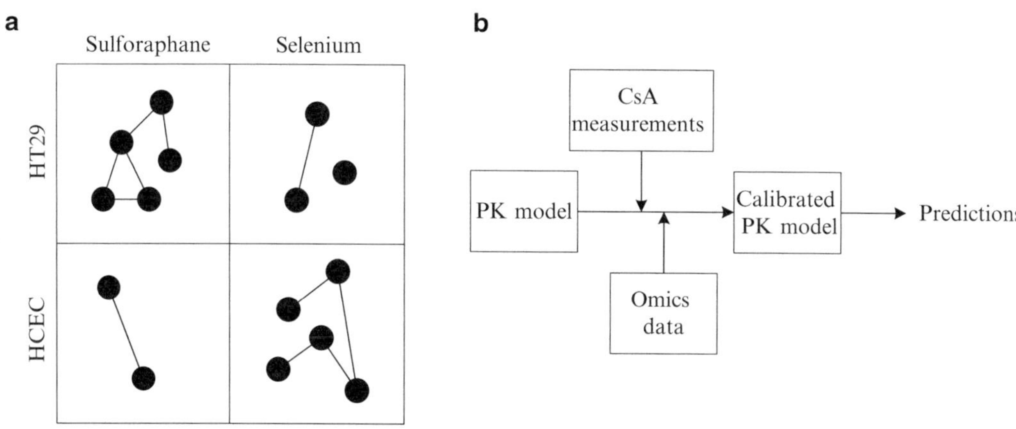

Fig. 1 Systems toxicology case studies. (**a**) Schematic representations of the network study by Constantinescu et al. [10]. The impact that sulforaphane and selenium had on the colon cancer cell line HT29 and on the noncancerous colon cell line HCEC is schematically shown by network representations. The analysis was performed using transcriptomic changes between each cell line treated with each chemical and the corresponding control samples, as explained in the main text. The different number of nodes and edges of the four graphs corresponding to the four comparisons symbolically represent different complexity levels of the inferred gene networks. (**b**) Schematic representation of the modeling of repeated CsA exposure on an in vitro system. The modeling was performed by coupling a three-compartment PK model with a systems biology model of the NRF2 pathway, and calibrating it with both omics measurements and measurements of CsA concentrations [48]. Adapted from Hamon et al. [53] with permission from BioMed Central

epithelial cell line (HCEC) [42] were compared. Since HCECs share many characteristics with normal colon epithelial cells, including a diploid karyotype and no mutations in genes known to be important in carcinogenesis, this immortalized cell line was used as an in vitro model for noncancerous colon tissue. This factorial design allowed the examination of perturbation effects in response to different agents in both cell lines, and provided a way to assess the significance of cell type in interpreting human cellular transcriptome-level responses to food components.

At the single gene level, differential expression between each cell line treated with each food component and the corresponding controls was quantified by fitting a linear model for which the standard errors of the estimated log-fold changes were moderated with an empirical Bayes method. This resulted in more stable inference and improved power [43]. Only a few significant changes were detected at the transcriptomic level in all four comparisons, i.e., the two test compounds in the two cell types, probably because the two compounds were applied in diet-relevant doses. In HT29 cells, four members of the aldo-keto reductase (AKR) family 1 (AKR1C2, AKR1C1, AKR1C3, and AKR1B10) were significantly overexpressed after treatment with sulforaphane, reinforcing the notion that AKR family 1 is inducible by sulforaphane under NRF2 mediation. Additionally, the protein-level and mRNA-level fold changes of the four AKR genes were very highly correlated.

Because the changes at the level of single genes were moderate, network and pathway analyses were highly effective for extracting additional information about the induced perturbations. Network analysis was performed with the GeneMania [28] plugin of Cytoscape, using as inputs for each of the four comparisons the differentially expressed genes with absolute log-fold changes higher than a chosen threshold. The resulting network contained a number of genes functionally similar to the input genes, which were identified using transcriptomics, genomics, and proteomics databases. The similarity criteria were evidence of coexpression from previous studies, shared protein domains, and participation in the same functional pathway. In HT29 cells treated with sulforaphane, most of the detected related genes were either members of the AKR superfamily or known NRF2-regulated genes.

Functional pathway-level analysis was applied to the complete list of normalized gene expression values in each of the four comparisons. As described by Väremo et al. [33], gene sets from the complete GO database were tested for either upregulated or downregulated significant enrichment. The gene-level statistics consisted of both log-fold ratios and corrected *p*-values following differential expression analysis, and the group-level statistic of each GO term was computed as the mean of the gene-level statistics of the member genes. Using the mean as the group-level statistic has been shown to be more powerful than the modified Kolmogorov–Smirnov statistic from GSEA [44]. The statistical significance of the group-level enrichment was assessed with a permutation test, which is particularly suitable when gene-level changes are moderate because it does not require any a priori significance cutoff.

To further analyze the large number of significantly overexpressed or underexpressed GO terms, the space of pathways was narrowed by mapping the identified GO terms to more general supersets named GO Slim (GOS) terms. GOS terms contain multiple conceptually related GO terms, e.g., terms that describe similar aspects of the same biological process. Each GOS term was assigned upregulation and downregulation scores in each comparison by normalizing the count of its upregulated and downregulated members in that comparison to the total count of members detected across all four comparisons. This strategy addressed the commonly encountered problem of extracting meaningful biological signal when the number of differentially expressed pathways is large. Moreover, this pathway analysis provided a higher-resolution understanding of the impact that the two compounds had on the two cell lines. Following exposure to low doses of sulforaphane and selenium, transcriptomic changes that were expected but undetected at the single-gene level were detected at the pathway level. An example is the sulforaphane-induced pathway-level upregulation of the GOS term aldo-keto reductase (NADP) activity in both cell lines, as opposed to the single gene level. Individual AKRs were significantly upregulated only in HT29 cells, but not in HCECs. Compared with a number of previous studies, the work of Constantinescu et al. [10] suggests a potentially more realistic way of evaluating the impact of diet components on cells, and is an example of how to mine data that display few perturbations after exposure to realistic doses of compounds.

2.2 Pharmacokinetics and Pharmacodynamics (PK-PD) Modeling

In systems toxicology, PK-PD modeling represents a major class of models that aim to mechanistically characterize the response of a system after exposure to a drug. Pharmacokinetics (PK) concerns how drugs are absorbed, distributed, metabolized, and excreted, and adds the time dimension to the dose–effect relationship, incorporating dynamic changes of the effect [45]. By contrast, pharmacodynamics (PD) concerns the physiological impacts of a drug. PK-PD integrates these two facets by creating a common representation, which describes the intensity of the effect over time in response to the administration of a drug dose [46, 47]. Bassingthwaighte et al. [48] has reviewed PK-PD models, including model types, topologies, equation types, as well as useful software and data analysis indications.

Non-compartmental analysis (NCA) techniques are used in the analysis of PK data whenever a scalable modeling technique with few underlying assumptions is preferred. Usually, NCA is performed if the main goal is to determine the degree of exposure following administration of a drug, together with the associated PK parameters of the drug [49]. To estimate the temporal dynamics of drug concentration in plasma, either the linear or the log-linear trapezoidal rules methods, which effectively approximate the integral of a function, are used. In compartmental models, biological entities of interest are represented as interconnecting homogeneous and instantly mixed chambers with uniform compound concentrations. Under simplifying assumptions, chemical reactions, binding processes, and transmembrane transport are represented by systems of ordinary differential equation. Compartmental models are commonly used to describe the concentration-time curve of a drug concentration following its administration. The shape of the curve provides a measure of how long the drug concentration remains at a therapeutic level, and supports the understanding of expected effects following administration [48]. Physiology-based PK models are a more realistic representation of PK processes because they describe the actual physiology of the organism under investigation, and physical organs and tissues are modeled as compartments. In addition, instead of composite rate constants defined by fitting data, measured physicochemical and biochemical constants of the compound can be used [50, 51]. Compared with non-compartmental models, compartmental models provide a more realistic picture of the underlying biological process. However, mechanistic biological information is needed to construct a compartmental model, reducing their applicability to cases where this knowledge is reasonably reliable. A review of advantages and disadvantages of NCA versus compartmental models was provided by Gillespie [52].

2.2.1 Two Case Studies of Integrated PK-PD Modeling and Omics Analysis of In Vitro Cells Exposed to Cyclosporine A

Integrating large-scale omics data and PK-PD modeling is a powerful way to study in-depth the cellular response following drug exposure. Wilmes et al. [12] and Hamon et al. [53] characterized compound-induced stress at the cellular level, using omics data to fit either a mechanistic PK model [12] or a mechanistic coupled PK-NRF2 pathway model [53] (Fig. 1b). The experimental setup of both studies involved the repeated exposure of in vitro cells to cyclosporine A (CsA). CsA is a well-known immunosuppressive drug that is widely used in clinical applications, especially in organ transplantation. Its mechanism of action is based on its interference with the function of T cells, leading to activity reduction of the immune system. CsA has been shown to be nephrotoxic, causing disturbances to the vasculature, glomerulus, and proximal tubule, as well as to increase the risk of various infections [54].

Study 1: To understand how cells deal with stress induced by repeated exposure to CsA, Wilmes et al. [12] exposed cultured human renal epithelial cells (RPTEC/TERT1) [55], a human renal proximal tubular cell line, to two different concentrations of

CsA. The low dose was expected to have no significant effect, and the high dose was nonlethal but expected to trigger significant cellular perturbations. Two daily doses of CsA were applied to RPTEC cells for up to 14 days, and the cells were harvested for transcriptomics, proteomics, and metabolomics analyses at days 1, 3, and 14. Every day, CsA was measured in supernatants, cell lysates, and on cell culture walls, and supernatant was collected for metabolomic measurements.

Perturbations at the transcriptomic, proteomic, and metabolomic levels were determined as fold changes between each time point and the corresponding initial controls. Exposure to the higher dose of CsA only was found to trigger significant perturbations at all three omics levels, with many NRF2-related genes and pathways induced at all time points. Pearson's correlation coefficient between mRNA and protein levels increased over time; however, it never attained a very high value. In addition to the omics analyses, the CsA concentrations, which where monitored daily and externally reset to either the low or high dose value every 24 h, and supernatant metabolomics measurements were modeled with a three-compartment PK model of CsA exchange between supernatant, cells, and vial walls. According to the model design, CsA distributed inside the cell, adhered to the plastic vial walls, and was metabolized. The best fit for the form of the exchange rates was obtained using a Michaelis–Menten exit rate for first-order entry into cells, and Michaelis-Menten metabolism.

In a Bayesian framework, a Markov chain Monte Carlo (MCMC) sampling algorithm was used to estimate the model parameters on the basis of non-informative priors. Posterior distributions were computed to characterize the probability of the parameters taking any possible value, given the input data. The daily measured CsA concentrations in the supernatant and cell lysates were superimposed with the best maximum posterior fit of the PK model. After parameter estimation, the PK model was used to predict how the system evolved over time. A low dose of CsA triggered minor alterations, and a steady state behavior was quickly reached after 3 days. Conversely, after exposure to a high dose of CsA, an approximate steady state was reached after 7 days, and the intracellular concentration was much higher than the concentration following a low exposure. The model predicted rapid cellular uptake and accumulation of high doses of CsA, as well as a nonlinear relationship between the accumulation of CsA and extracellular concentration.

Study 2: In a follow-up study by Hamon et al. [53], researchers investigated in more detail one of the main conclusions of the study by Wilmes et al. [12], namely that CsA strongly activated the NRF2 pathway in RPTECs in vitro. The previously described three-compartment PK model of CsA exchange was coupled to a quantitative differential equations mechanistic model of the toxicity of the

NRF2 pathway, adapted from a model developed for a general xenobiotic chemical by Zhang et al. [56]. The in vitro cellular CsA measurements together with a subset of the omics data for specific markers were used to calibrate the coupled model in a Bayesian framework, similarly to the way it was performed by Wilmes et al. [12]. The parameters directly controlling CsA kinetics and part of the three-compartment model were set to the mode of the joint posterior distribution reported in the previous MCMC calibration by Wilmes et al. [12]. The other structural parameters were calibrated using omics measurements of several species thought to be influential in modeling the NRF2 pathway. Some of the prior distributions of these parameters were uninformative, while the others were log-normal distributions, centered on the values used by Zhang et al. [56] in the general xenobiotic model. Posterior distributions for each species were obtained by MCMC sampling.

After calibration, the relative differences between data and predictions were considered to be, on average, in the range of the expected measurement error. The calibrated model then was used for prediction. At the low CsA dose, periodic oscillations were pervasive, while at the high dose, the concentration of almost all species of interest reached a plateau after 3 days. The model was used to predict dose-dependent responses after simulating the exposure of the cells to a wide range of concentrations. Consequently, the time profiles of the concentrations of several species of interest, including the nuclear NRF2 protein, were either oscillating or plateauing, depending on the extracellular CsA concentration.

The experimental setup and modeling techniques employed by Wilmes et al. [12] and Hamon et al. [53] provide ways of characterizing compound-induced stress on an in vitro system at various levels. For example, the nonlinear accumulation of CsA was detectable only after repeated exposure to CsA. Moreover, these two cases studies are examples of fruitfully combining data-driven and mechanistic modeling approaches.

3 Pharmacogenomics

Pharmacogenomics is qualitatively different from systems toxicology approaches in that it specifically assesses how both single genes and gene groups influence drug responses. A translational aim of using pharmacogenomics is to identify genetic biomarkers predictive of drug sensitivity that may be directly integrated in clinical trial design. In early-phase clinical trials, among the large number of compounds under development, pharmacogenomics high-throughput screening accelerates the selection of patient-drug combinations for effective targeted therapy [57, 58]. Examples of targetable tumor subtypes discovered following high-throughput analysis of cancer cell lines include BCR-ABL positive chronic myelogenous leukemia, BRAF-mutant melanoma, and EGFR-mutant-positive lung cancer [59]. Despite these rapid developments, robustly assessing drug impacts on cell lines and deriving reliable biomarkers for personalized medicine remain complex and challenging tasks. Biological variation and technical noise have made it difficult to measure the perturbations induced by drugs in cell lines in a highly reproducible manner [58, 60, 61]. To address these issues, performing experiments in biological replicates and assessing their variability is essential.

3.1 Large-Scale Pharmacogenomics Studies

In two recent large-scale pharmacogenomics studies, the effects of a large panel of anticancer drugs on a large panel of cancer cell lines have been assessed on the basis of gene expression, copy number, and point mutation information. While linkages between genomic or transcriptomic markers and drug function had been characterized in a number of earlier studies [62–66], the numbers of tested cell lines and compounds were not high enough to allow systematic analyses. The Cancer Cell Line Encyclopedia (CCLE) [67] and the Cancer Genome Project (CGP) [59] have provided valuable resources for many follow-up investigations, in both clinical and basic research.

3.1.1 The Cancer Cell Line Encyclopedia (CCLE)

The CCLE contains the expression quantity, copy number, and mutational profiles of 947 human cancer cell lines representing 36 tumor types. The effects of 24 anticancer drugs (both targeted and cytotoxic) were evaluated in 479 of these cell lines, with the aim of identifying gene-based predictors of drug sensitivity. To assess how well the characterized cell lines represented primary tumors, several of their genetic profiles, such as chromosomal copy number alterations, gene expression patterns, and mutation landscapes, were compared with the profiles of a large collection of primary tumors of the same cancer type. Chromosomal copy number data were obtained from Tumorscape [68], a comprehensive portal that contains high-resolution copy number alteration data for multiple cancer types. Gene expression data were obtained from the expO and MILE transcriptomic data collections [69], and mutation data were downloaded from COSMIC [70], a database of somatic mutations related to multiple human cancers. Most cell lines were highly correlated with primary tumor data in all analyzed data types, indicating that the cell line models included in this study were representative genetic proxies for primary tumors.

Genetic markers that significantly influenced drug sensitivity were identified from both continuous and discrete drug sensitivity values. Continuous values included the concentration at which the drug response reached an inhibition of 50 % (IC_{50}), the concentration at half-maximal activity of the drug (EC_{50}), and the area under the dose response curve (AUC). Discretized drug sensitivity values were separated into sensitive, intermediate, or refractory.

Continuous data were modeled with the elastic net [71], a regularized linear regression method that linearly combines an L_1 penalty, amounting to the sum of the absolute values of all predictors and corresponding to the least absolute shrinkage and selection operator (LASSO) [72], and an L_2 penalty, amounting to the sum of squares of all predictors and corresponding to ridge regression [73] on regression coefficients. Drug sensitivity was the response variable, and its variation was explained by measured predictors (or features) such as gene expression values, copy number profiles, or pathway activity scores. Regularization was necessary in order to avoid solving an ill-posed problem or overfitting, as the number of predictors was much larger than the number of samples, a widely encountered situation in genomic applications. In practice, elastic net was shown to often outperform LASSO or ridge [71] because it aims to balance the effects of the two penalties by obtaining a parsimonious model through the L_1 term, while retaining groups of correlated features through the L_2 term. To compute confidence intervals of the detected gene–drug associations, 200 bootstrap datasets were generated at the cell-line level; i.e., each cell line was sampled with replacement and the elastic net optimization was performed on each of the 200 datasets. For each predictive feature, its final contribution to drug sensitivity was calculated as the percentage of the bootstrap estimates in which the feature was inferred as significantly different from zero. The prediction performance of the model was computed using Pearson's correlation coefficient or Kendal's τ rank coefficient between the average predicted response, computed as the mean of cross-validated predictions, and the observed response.

Discrete data were modeled with a naive Bayes classifier, a classification method based on the assumption that, given the class variable, the features are independent of each other. Therefore, only a small amount of training data is needed to estimate the parameters necessary for classification, and the classifier works efficiently on datasets with much larger numbers of features than samples. Following the shape of their corresponding dose–response curves, the drug sensitivity values were discretized into three classes: sensitive, intermediate, and refractory. The classifier included as predictive features mainly the top features significantly different between the sensitive and refractory classes, identified after using either Wilcoxon sum rank test for continuous features or Fisher's exact test for discrete features, and ranked by their corrected p-value. To reduce the number of correlated features, the remaining significant features were clustered using a message-passing algorithm [74], and the cluster representatives were additionally included as predictors in the naive Bayes. The area under the receiver operator curve described the performance of the binary classifier with increasing discrimination threshold, by computing the false positive and the true positive rates for each value of the threshold.

3.1.2 The Cancer Genome Project (CGP)

In the CGP, the expression quantity, copy number, and mutational profiles of 639 human tumor cell lines were assessed after exposure to 30 drugs, both targeted agents and cytotoxic chemotherapeutics, as IC_{50} and AUC values. To identify associations between single mutations and drug sensitivity, the authors used a fixed effects MANOVA without interaction terms. Tissue type, mutation status, chromosomal rearrangement, and microsatellite instability status were modeled as factors, and drug sensitivity was modeled as response. Cooperative interaction between genes, on the genomic and transcriptomic levels, was analyzed using the elastic net with cross-validated penalty parameters. For each drug, associations between groups of features and drug sensitivities were tested for stability by performing the optimization problem 100 times. In most cases, transcriptional features were more strongly correlated with drug sensitivity than point mutations.

3.1.3 Comparisons of the CCLE and CGP Datasets

The two studies described above have generated, mined, and shared valuable information for the entire cancer community by making most of their data available and by confirming old associations and proposing novel ones between biomarkers and drugs. Recently, Haibe-Kains et al. [60] compared the measured effects after treatment of the 471 common cell lines with the 15 common drugs in the two studies (Fig. 2a). Transcription profiles of replicate samples within the same study, as well as between identical cell lines in the two studies, were highly correlated, despite the fact that different microarray platforms were used in these two studies [75]. The correlation between point mutations from exome sequencing was also high, but lower than for gene expression, probably because of the differences in sequencing platforms and variant callers. By contrast, the sensitivities of most of the drugs, measured by either IC_{50} or AUC, were poorly correlated between the two studies, as assessed by Spearman's rank correlation. Gene–drug associations were predicted by fitting a linear regression model for every gene, including either IC_{50} or AUC as response and the gene expression feature as a predictor. The identified associations were also poorly correlated between the two studies. Following a pre-ranked GSEA [11] performed on the set of genes ranked by the significance of their associations to drugs, overrepresented GO terms (pathways) were associated with drug sensitivity. Although the pathway-based correlations were the best obtained throughout the entire study, many drugs still displayed very low correlations.

Drug sensitivity measurements were identified as the main source of inconsistency, since the gene expression values of identical cell lines were highly correlated between the two datasets. Even if the two studies used the same terminology for reporting drug sensitivity, namely IC_{50}, EC_{50}, and AUC, they tested different ranges of concentrations, and employed different procedures to fit dose–response curves and to generate summary statistics.

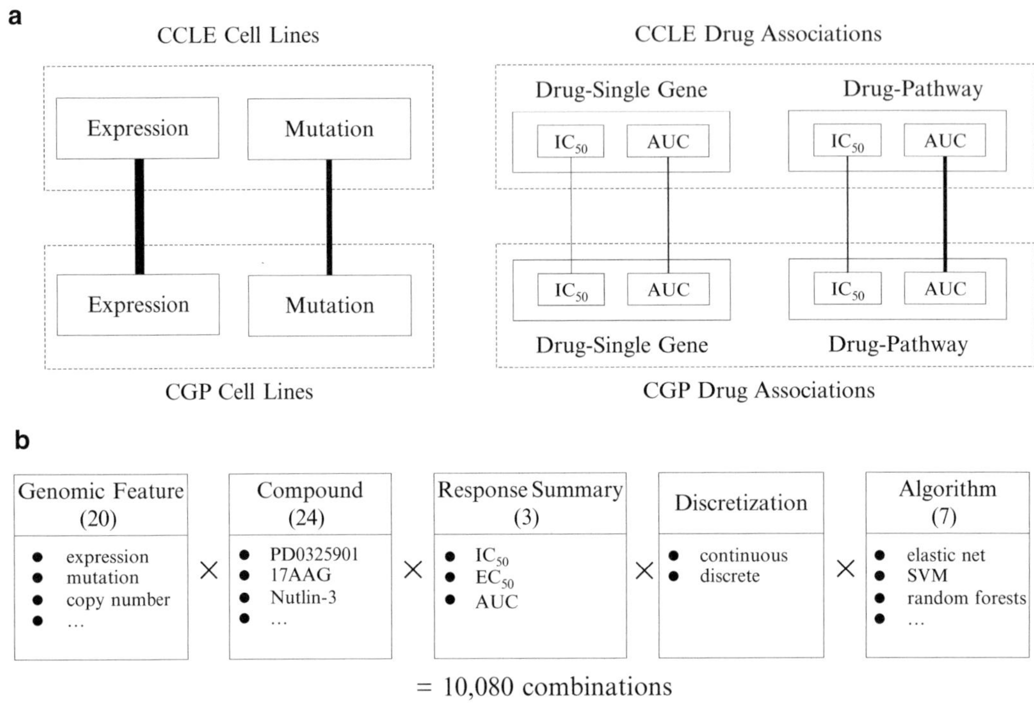

Fig. 2 Pharmacogenomics case studies. (**a**) Inconsistencies in gene–drug associations between two large pharmacogenomics studies, CCLE and CGP, as assessed by Haibe-Kains et al. [54]. Correlations between expression data of identical cell lines in the two studies were very high, and correlations between point mutation data were satisfactory. However, correlations between the associations of single genes or pathways and drugs, as measured by fitting a linear regression model, were only moderate. Higher correlations were obtained when using AUC drug sensitivity measurements as response and GO terms as predictors, rather than using IC_{50} drug sensitivity measurements as response and single gene as predictors. The thickness of the *black vertical lines* indicates strength of correlation (the thicker the line, the higher the correlation) as reported by Haibe-Kains et al. [54]. (**b**) Factorial evaluation design of modeling categories performed by Jang et al. [70], and applied to the two large-scale pharmacogenomics datasets CCLE and CGP. This framework was used to statistically test the performance of the systematically generated modeling options, providing general modeling guidelines for pharmacogenomics datasets. The *numbers* in brackets represent the actual numbers of levels of the modeling category to which they correspond, in the case of the CCLE dataset and of continuous drug sensitivity measurements

For example, in many cases, IC_{50} could not be estimated because the drug concentration necessary to inhibit 50 % growth was not reached. While CGP used a Bayesian approach to extrapolate the IC_{50} value, CCLE only reported the maximum growth inhibition level. This aspect represented a source of variability in the reported drug sensitivity measurements. Although for each cell line the drugs were assayed at least in duplicate, the raw data were not made publicly available by either of the two studies.

In a follow-up discussion article, Hatzis et al. [58] suggested that high-throughput pharmacogenomics studies would only be used at their full potential when experimental protocols for measuring drug response and for generating summary statistics are standardized through a collective effort. At the moment, possible technical artifacts include the purity, integrity, and concentration of the compound stock solutions, variations in cell culture conditions, reference source and passage numbers of the cell lines, as well as the type of cytotoxicity assay used.

3.2 Systematic Modeling of Pharmacogenomics Data

In an independent study by Jang et al. [76], the CGP and CCLE datasets were mined and a model selection framework was proposed that was built on a multifactorial experimental design. The authors tested more than 100,000 systematic combinations of modeling factors within several categories of modeling choices, such as type of molecular data used or algorithm employed for prediction (Fig. 2b). A large multifactorial experiment with crossed factors was performed, with experimental factors represented by the modeling choices and responses represented by the predictive performance measures. The resulting models were evaluated in an automated, statistically transparent and consistent fashion, which provided reliable modeling guidelines that may be applied for future pharmacogenomics studies.

3.2.1 Model Selection

Five categories of modeling choices were considered: genomic features, compound, response summary, discretization, and algorithm; and each category consisted of several modeling factors. Genomic features represented the input data types used, namely gene expression measurements, copy number variation measurements, cell line tumor type classification, mutation profiling, and mutation profiling using hybrid capture sequencing (available only for the CCLE dataset). All possible combinations of these data types were tested, except those with a small feature set. This strategy created 12 testing combinations for CGP and 20 testing combinations for CCLE. Compound represented the anticancer drug screened by the two studies, namely 24 drugs in CCLE and 130 in CGP. Response summary represented the statistic used to summarize the dose–response curve to a single number, namely AUC and IC_{50} for CGP; and IC_{50}, EC_{50}, and an inverse measure of AUC for CCLE. Discretization assessed whether predictions were based on continuous or discretized response summary measurements. Multiple discretization schemes were evaluated, including mean and deviation, Gaussian mixture models, and upper/lower third quartile thresholds. Among these schemes, upper/lower third quartile thresholds achieved the highest average classification accuracy. Finally, algorithm represented the predictive algorithms used to associate drug response with genomic features.

Most of the methods used to analyze continuous data were based on multiple linear regression and included principal component regression [77], partial least square regression [78], LASSO, ridge, elastic net, least squares support vector machine regression with linear kernels [79, 80], and random forests [81]. Both principal component regression and partial least square regression employ concepts from principal component analysis; least squares support vector machine regression with linear kernels is a class of kernel-based learning methods in which the objective function is linear, as opposed to quadratic optimization for classical support vector machines; and random forests detect associations based on multiple decision trees. For binary response data, the applied methods included least squares support vector machine classification with linear kernels, random forests, binomial LASSO, ridge binomial regression, and elastic net binomial regression [82].

All combinations resulting from the cross product of all factors of the five categories described above, amounting to more than 114,000 possible models, were tested on the data from each of the two studies. For each model, cross-validation was used for both parameter tuning and predictive performance assessment, and the predicted response vector was computed by concatenating the predictions for each cross-validation fold. The final performance measure between the predicted and observed responses was computed using Pearson correlation for continuous models and the area under the receiver operating characteristic curve for discrete models. The effects of each modeling category and their interactions on the predicted performance were evaluated using multi-way ANOVA with four crossed factors and a single observation per cell, detecting interactions of up to order three between the factors of the model. For example, to evaluate the performance of continuous models for the CCLE dataset, a four-factor ANOVA would be performed with the following factors: genomic features, composed of the 20 levels that represent the distinct data type combinations; compound, composed of the 24 levels that represent the anticancer drugs tested in the CCLE panel; response summary, composed of the three levels used by CCLE to measure drug sensitivity, namely EC_{50}, IC_{50}, and the inverse measure of AUC; and algorithm, composed of the seven levels that represent the chosen methods suitable for continuous data. For each of the possible 10,800 modeling choices, a model would be fitted to the CCLE data, a prediction would be made using cross validation, and a response variable would be assigned to measure the correlation between the observed and predicted outcomes.

3.2.2 Effects of Modeling Factors

Expectedly, the input data, namely the type of molecular feature and the predicted drug, were the dominant factors that influenced model performance, whereas the specific modeling techniques, i.e., type of response measure, discretization, and algorithm, were of secondary importance. The multi-way ANOVA also confirmed that the CGP and CCLE studies reported inconsistent drug sensitivity values, although they were reported under the same standard terminology. The response summary factor had variable effects in the two datasets, while all the other four factors had highly concordant effects. In the CCLE dataset, a strong interaction term existed between compound and response summary, suggesting that the performance of the model depended both on the drug being modeled, and the ability to summarize its dose–response measurements. Conversely, in the CGP dataset, the effect of the response summary factor was negligible.

In terms of modeling guidelines, the multifactorial experiment suggested that the best choices were the elastic net or ridge regression applied to continuous drug response data summarized using the area under the fitted dose–response curve and predicted on the basis on gene expression data. Using additional molecular feature types only modestly improved the predictive performance. In general, regression models trained using continuous response summary values significantly outperformed classification models trained using discretized response summary values. Moreover, discretizing data, e.g., into sensitive and resistant cells, was found to cause decreased model accuracy. Regarding the drug used, pathway-targeted compounds led to more accurate predictions than classical cytotoxic chemotherapies.

4 Conclusions and Outlook

In this chapter, we highlighted several applications of large-scale modeling strategies for omics datasets that span various areas of toxicogenomics. In the context of systems toxicology, we discussed network and pathway approaches, exemplified by a case study assessing the impact of diet-relevant doses of two food components on a colon cancer and a noncancerous cell line. Additionally, we discussed PK-PD modeling techniques, exemplified by two case studies in which the temporal impacts of high vs. low doses of extracellular CsA concentration on the evolution of the accumulated intracellular CsA concentration were examined. Finally, in the context of pharmacogenomics, we reviewed the two largest studies currently available, which assessed the impact of a large panel of anticancer drugs on a large panel of cell lines. Additionally, we discussed the observed inconsistencies in drug responses between the two studies, which mainly resulted from the lack of standardized drug sensitivity measurements. Furthermore, we presented a model selection framework that systematically evaluated combinations of modeling categories, statistically tested their performance, and on the basis on which general modeling guidelines for future pharmacogenomics datasets were proposed.

Despite its many successful applications, toxicogenomics is still a very young field. Initially, modeling predictions based on toxicogenomics data were regarded as inaccurate and in vitro cell line models were regarded as not representative of primary tumor tissues. Nevertheless, continuous advances in data generation and data analysis have allowed for the formulation and validation of important biological hypotheses now routinely used in clinics, such as the discovery of genetic biomarkers in targeted cancer therapy [83–87]. Although they are undoubtedly different, in vitro cell line models and primary tumors were found to be reasonably well correlated following a statistically rigorous large-scale comparison [67], which indicated that cell lines may well be informative of primary tumors. Additionally, despite differences in platforms [75], inherently noisy technologies such as microarrays have proven to produce robust results after data preprocessing, as exemplified by high correlations of replicates between the two large-scale pharmacogenomics studies discussed here. However, toxic effects are highly complex and dynamic phenomena that are not easily captured, even with genome-wide measurements. Thus, studies such as the one by Jang et al. [76], in which systematic and statistical evaluation of many modeling categories is performed, provide useful general guidelines for the large-scale analysis of toxicogenomics datasets. In the near future, the novel and powerful technique of sequencing single cells is anticipated to support the generation of datasets with higher resolution, ultimately providing a complete picture of cellular heterogeneity [88, 89]. These techniques are particularly appealing in the context of cancer therapy, where the main reason for treatment failure is the high tumor heterogeneity.

Finally, reproducibility is a basic requirement for any scientific study, and its importance is underscored when the results of a study may influence medical decision-making. In order to ensure the reproducibility of pharmacogenomics research, a necessary condition is the public availability of biological replicates for drug sensitivity testing, and of raw data rather than incomplete or extrapolated IC_{50} values. The opportunity to analyze such data would contribute to the separation and quantification of the biological and technical sources of inconsistencies between pharmacogenomics studies [58]. Moreover, making the source code of the implemented methods publicly available for full testing and improvement is crucial to increase reproducibility and quality standards. In conclusion, as data are continuously produced and the predictive accuracy of modeling techniques improves, the prevalence and impact of large-scale data generation and analysis in toxicology will increase dramatically.

Acknowledgements

This work was financially supported by the Swiss National Science Foundation (Sinergia project 136247).

References

1. Nigsch F, Macaluso NM, Mitchell JB, Zmuidinavicius D (2009) Computational toxicology: an overview of the sources of data and of modelling methods. Expert Opin Drug Metab Toxicol 5(1):1–14
2. Nuwaysir EF, Bittner M, Trent J et al (1999) Microarrays and toxicology: the advent of toxicogenomics. Mol Carcinog 24(3):153–159
3. Waters MD, Fostel JM (2004) Toxicogenomics and systems toxicology: aims and prospects. Nat Rev Genet 5(12):936–948
4. Gatzidou ET, Zira AN, Theocharis SE (2007) Toxicogenomics: a pivotal piece in the puzzle of toxicological research. J Appl Toxicol 27(4):302–309
5. Sturla SJ, Boobis AR, FitzGerald RE et al (2014) Systems toxicology: from basic research to risk assessment. Chem Res Toxicol 27(3):314–329
6. Ferté C, André F, Soria J-C (2010) Molecular circuits of solid tumors: prognostic and predictive tools for bedside use. Nat Rev Clin Oncol 7(7):367–380
7. Macarron R, Banks MN, Bojanic D et al (2011) Impact of high-throughput screening in biomedical research. Nat Rev Drug Discov 10(3):188–195
8. Arrell D, Terzic A (2010) Network systems biology for drug discovery. Clin Pharmacol Ther 88(1):120–125
9. Kreeger PK, Lauffenburger DA (2010) Cancer systems biology: a network modeling perspective. Carcinogenesis 31(1):2–8
10. Constantinescu S, Hecht K, Sobotzki N et al (2014) Transcriptomic responses of cancerous and noncancerous human colon cells to sulforaphane and selenium. Chem Res Toxicol 27(3):377–386
11. Subramanian A, Tamayo P, Mootha VK et al (2005) Gene set enrichment analysis: a knowledge-based approach for interpreting genome-wide expression profiles. Proc Natl Acad Sci U S A 102(43):15545–15550
12. Wilmes A, Limonciel A, Aschauer L et al (2013) Application of integrated transcriptomic, proteomic and metabolomic profiling for the delineation of mechanisms of drug induced cell stress. J Proteomics 79:180–194
13. Barabasi A-L, Oltvai ZN (2004) Network biology: understanding the cell's functional organization. Nat Rev Genet 5(2):101–113
14. Zhu X, Gerstein M, Snyder M (2007) Getting connected: analysis and principles of biological networks. Genes Dev 21(9):1010–1024
15. Gehlenborg N, O'Donoghue SI, Baliga NS et al (2010) Visualization of omics data for systems biology. Nat Methods 7:S56–S68
16. Pavlopoulos GA, O'Donoghue SI, Satagopam VP et al (2008) Arena3D: visualization of biological networks in 3D. BMC Syst Biol 2(1):104
17. Theocharidis A, Van Dongen S, Enright AJ, Freeman TC (2009) Network visualization and analysis of gene expression data using BioLayout Express3D. Nat Protoc 4(10):1535–1550
18. Junker BH, Klukas C, Schreiber F (2006) VANTED: a system for advanced data analysis and visualization in the context of biological networks. BMC Bioinformatics 7(1):109
19. Funahashi A, Morohashi M, Kitano H, Tanimura N (2003) Cell designer: a process diagram editor for gene-regulatory and biochemical networks. BioSilico 1(5):159–162
20. Hooper SD, Bork P (2005) Medusa: a simple tool for interaction graph analysis. Bioinformatics 21(24):4432–4433
21. Breitkreutz B-J, Stark C, Tyers M (2003) Osprey: a network visualization system. Genome Biol 4(3):R22
22. Shannon P, Markiel A, Ozier O et al (2003) Cytoscape: a software environment for integrated models of biomolecular interaction networks. Genome Res 13(11):2498–2504
23. Smoot ME, Ono K, Ruscheinski J et al (2011) Cytoscape 2.8: new features for data integration and network visualization. Bioinformatics 27(3):431–432
24. Longabaugh WJ, Davidson EH, Bolouri H (2009) Visualization, documentation, analysis, and communication of large-scale gene regulatory networks. Biochim Biophys Acta 1789: 363–374
25. Letunic I, Yamada T, Kanehisa M, Bork P (2008) iPath: interactive exploration of biochemical pathways and networks. Trends Biochem Sci 33:101–103
26. Okuda S, Yamada T et al (2008) KEGG atlas mapping for global analysis of metabolic pathways. Nucleic Acids Res 36:W423–W426
27. Hu Z, Hung J-H, Wang Y et al (2009) VisANT 3.5: multi-scale network visualization, analysis and inference based on the gene ontology. Nucleic Acids Res 37:W115–W121
28. Warde-Farley D, Donaldson SL et al (2010) The GeneMANIA prediction server: biological network integration for gene prioritization and predicting gene function. Nucleic Acids Res 38(suppl 2):W214–W220

29. Barsky A, Gardy JL, Hancock RE, Munzner T (2007) Cerebral: a Cytoscape plugin for layout of and interaction with biological networks using subcellular localization annotation. Bioinformatics 23:1040–1042
30. Ashburner M, Ball CA, Blake JA et al (2000) Gene ontology: tool for the unification of biology. Nat Genet 25(1):25–29
31. Kanehisa M, Goto S (2000) KEGG: Kyoto Encyclopedia of Genes and Genomes. Nucleic Acids Res 28(1):27–30
32. Liberzon A, Subramanian A, Pinchback R et al (2011) Molecular signatures database (MSigDB) 3.0. Bioinformatics 27(12):1739–1740
33. Väremo L, Nielsen J, Nookaew I (2013) Enriching the gene set analysis of genome-wide data by incorporating directionality of gene expression and combining statistical hypotheses and methods. Nucleic Acids Res 41(8):4378–4391
34. Maere S, Heymans K, Kuiper M (2005) BiNGO: a cytoscape plugin to assess overrepresentation of gene ontology categories in biological networks. Bioinformatics 21(16):3448–3449
35. Dennis G Jr, Sherman BT, Hosack DA et al (2003) DAVID: database for annotation, visualization, and integrated discovery. Genome Biol 4(5):P3
36. Beißbarth T, Speed TP (2004) GOstat: find statistically overrepresented gene ontologies within a group of genes. Bioinformatics 20(9):1464–1465
37. Zheng Q, Wang X-J (2008) GOEAST: a web-based software toolkit for gene ontology enrichment analysis. Nucleic Acids Res 36(suppl 2):W358–W363
38. Kong SW, Pu WT, Park PJ (2006) A multivariate approach for integrating genome-wide expression data and biological knowledge. Bioinformatics 22(19):2373–2380
39. Hummel M, Meister R, Mansmann U (2008) GlobalANCOVA: exploration and assessment of gene group effects. Bioinformatics 24(1):78–85
40. Tsai C-A, Chen JJ (2009) Multivariate analysis of variance test for gene set analysis. Bioinformatics 25(7):897–903
41. Mootha VK, Lindgren CM, Eriksson K-F et al (2003) PGC-1α-responsive genes involved in oxidative phosphorylation are coordinately downregulated in human diabetes. Nat Genet 34(3):267–273
42. Roig AI, Eskiocak U, Hight SK et al (2010) Immortalized epithelial cells derived from human colon biopsies express stem cell markers and differentiate in vitro. Gastroenterology 138(3):1012–1021
43. Smyth GK (2004) Linear models and empirical bayes methods for assessing differential expression in microarray experiments. Stat Appl Genet Mol Biol 3:3
44. Efron B, Tibshirani R (2007) On testing the significance of sets of genes. Ann Appl Stat 1(1):107–129
45. Madan A, Dureja H (2012) Prediction of pharmacokinetic parameters. In: Reisfeld B, Mayeno AN (eds) Computational toxicology, vol 929, Methods in molecular biology series. Springer, New York, NY, pp 337–357
46. Chien JY, Friedrich S, Heathman MA et al (2005) Pharmacokinetics/pharmacodynamics and the stages of drug development: role of modeling and simulation. AAPS J 7(3):E544–E559
47. Czock D, Markert C, Hartman B, Keller F (2009) Pharmacokinetics and pharmacodynamics of antimicrobial drugs. Expert Opin Drug Metab Toxicol 5(5):475–487
48. Bassingthwaighte JB, Butterworth E, Jardine B, Raymond GM (2012) Compartmental modeling in the analysis of biological systems computational toxicology. In: Reisfeld B, Mayeno AN (eds) Computational toxicology, vol 929, Methods in molecular biology series. Springer, New York, NY, pp 391–438
49. Gabrielsson J, Weiner D (2012) Non-compartmental analysis. In: Reisfeld B, Mayeno AN (eds) Computational toxicology, vol 929, Methods in molecular biology series. Springer, New York, NY, pp 377–389
50. Huisinga W, Telgmann R, Wulkow M (2006) The virtual laboratory approach to pharmacokinetics: design principles and concepts. Drug Discov Today 11(17):800–805
51. von Kleist M, Huisinga W (2007) Physiologically based pharmacokinetic modelling: a sub-compartmentalized model of tissue distribution. J Pharmacokinet Pharmacodyn 34(6):789–806
52. Gillespie WR (1991) Noncompartmental versus compartmental modelling in clinical pharmacokinetics. Clin Pharmacokinet 20(4):253–262
53. Hamon J, Jennings P, Bois FY (2014) Systems biology modeling of omics data: effect of cyclosporine a on the Nrf2 pathway in human renal cells. BMC Syst Biol 8(1):76
54. Emmel EA, Verweij CL et al (1989) Cyclosporin A specifically inhibits function of nuclear proteins involved in T cell activation. Science 246:1617–1620
55. Wieser M, Stadler G, Jennings P et al (2008) hTERT alone immortalizes epithelial cells of renal proximal tubules without changing their functional characteristics. Am J Physiol Renal Physiol 295(5):F1365–F1375

56. Zhang Q, Pi J, Woods CG, Andersen ME (2009) Phase I to II cross-induction of xenobiotic metabolizing enzymes: a feedforward control mechanism for potential hormetic responses. Toxicol Appl Pharmacol 237(3):345–356
57. Bredel M, Jacoby E (2004) Chemogenomics: an emerging strategy for rapid target and drug discovery. Nat Rev Genet 5(4):262–275
58. Hatzis C, Bedard PL, Birkbak NJ et al (2014) Enhancing reproducibility in cancer drug screening: how do we move forward? Cancer Res 74(15):4016–4023
59. Garnett MJ, Edelman EJ, Heidorn SJ et al (2012) Systematic identification of genomic markers of drug sensitivity in cancer cells. Nature 483(7391):570–575
60. Haibe-Kains B, El-Hachem N, Birkbak NJ et al (2013) Inconsistency in large pharmacogenomic studies. Nature 504(7480): 389–393
61. Malo N, Hanley JA, Cerquozzi S et al (2006) Statistical practice in high-throughput screening data analysis. Nat Biotechnol 24(2): 167–175
62. Lin WM, Baker AC, Beroukhim R et al (2008) Modeling genomic diversity and tumor dependency in malignant melanoma. Cancer Res 68(3):664–673
63. Neve RM, Chin K, Fridlyand J et al (2006) A collection of breast cancer cell lines for the study of functionally distinct cancer subtypes. Cancer Cell 10(6):515–527
64. Solit DB, Garraway LA, Pratilas CA et al (2005) BRAF mutation predicts sensitivity to MEK inhibition. Nature 439(7074): 358–362
65. Sos ML, Michel K, Zander T et al (2009) Predicting drug susceptibility of non–small cell lung cancers based on genetic lesions. J Clin Invest 119(6):1727–1740
66. Staunton JE, Slonim DK, Coller HA et al (2001) Chemosensitivity prediction by transcriptional profiling. Proc Natl Acad Sci U S A 98(19):10787–10792
67. Barretina J, Caponigro G, Stransky N et al (2012) The Cancer Cell Line Encyclopedia enables predictive modelling of anticancer drug sensitivity. Nature 483(7391):603–607
68. Beroukhim R, Mermel CH, Porter D et al (2010) The landscape of somatic copy-number alteration across human cancers. Nature 463(7283):899–905
69. Kohlmann A, Kipps TJ, Rassenti LZ et al (2008) An international standardization programme towards the application of gene expression profiling in routine leukaemia diagnostics: The Microarray Innovations in LEukemia study prephase. Br J Haematol 142(5):802–807
70. Forbes S, Bhamra G, Bamford S et al (2008) The catalogue of somatic mutations in cancer (COSMIC). Curr Protoc Hum Genet Chapter 10:Unit 10.11
71. Zou H, Hastie T (2005) Regularization and variable selection via the elastic net. J R Stat Soc B 67(2):301–320
72. Tibshirani R (1996) Regression shrinkage and selection via the lasso. J R Stat Soc B 58(1):267–288
73. Hoerl AE, Kennard RW (1970) Ridge regression: biased estimation for nonorthogonal problems. Technometrics 12(1):55–67
74. Frey BJ, Dueck D (2007) Clustering by passing messages between data points. Science 315(5814):972–976
75. Larkin JE, Frank BC, Gavras H et al (2005) Independence and reproducibility across microarray platforms. Nat Methods 2(5): 337–344
76. Jang IS, Neto EC, Guinney J et al (2013) Systematic assessment of analytical methods for drug sensitivity prediction from cancer cell line data. Pac Symp Biocomput 19:63–74
77. Jolliffe IT (1982) A note on the use of principal components in regression. Appl Stat 31:300–303
78. Wold S, Sjöström M, Eriksson L (2001) PLS-regression: a basic tool of chemometrics. Chemometr Intell Lab Syst 58(2):109–130
79. Balabin RM, Lomakina EI (2011) Support vector machine regression (SVR/LS-SVM)—an alternative to neural networks (ANN) for analytical chemistry? Comparison of nonlinear methods on near infrared (NIR) spectroscopy data. Analyst 136(8):1703–1712
80. Wu Y, Krishnan S (2011) Combining least-squares support vector machines for classification of biomedical signals: a case study with knee-joint vibroarthrographic signals. J Exp Theor Artif Intell 23(1):63–77
81. Breiman L (2001) Random forests. Mach Learn 45(1):5–32
82. Hilbe JM (2011) Negative binomial regression. Cambridge University Press, Cambridge
83. Chapman PB, Hauschild A, Robert C et al (2011) Improved survival with vemurafenib in melanoma with BRAF V600E mutation. N Engl J Med 364(26):2507–2516
84. Druker BJ, Guilhot F, O'Brien SG et al (2006) Five-year follow-up of patients receiving imatinib for chronic myeloid leukemia. N Engl J Med 355(23):2408–2417
85. King M-C, Marks JH, Mandell JB (2003) Breast and ovarian cancer risks due to inherited

mutations in BRCA1 and BRCA2. Science 302(5645):643–646
86. Kwak EL, Bang Y-J, Camidge DR et al (2010) Anaplastic lymphoma kinase inhibition in non-small-cell lung cancer. N Engl J Med 363(18): 1693–1703
87. McDermott U, Sharma SV, Dowell L et al (2007) Identification of genotype-correlated sensitivity to selective kinase inhibitors by using high-throughput tumor cell line profiling. Proc Natl Acad Sci U S A 104(50):19936–19941
88. Navin N, Kendall J, Troge J et al (2011) Tumour evolution inferred by single-cell sequencing. Nature 472(7341):90–94
89. Wang Y, Waters J, Leung ML et al (2014) Clonal evolution in breast cancer revealed by single nucleus genome sequencing. Nature 512(7513):155–160

Chapter 16

Aerosol Dosimetry Modeling Using Computational Fluid Dynamics

Markus Nordlund and Arkadiusz K. Kuczaj

Abstract

In this chapter, the current state of the art in modeling and prediction of aerosol deposition in respiratory systems using computational fluid dynamics (CFD) is presented and reviewed. First, the physical and chemical processes governing aerosol transport, evolution, and deposition are described followed by their coupling via fundamental conservation laws. The different ways to numerically model aerosol dynamics are then described and a brief overview of the different methods that can be used to obtain a realistic geometry, computational mesh, and simulation conditions of the respiratory system is given. A short review of available numerical algorithms to solve the systems of strongly coupled partial differential equations is also presented followed by a brief discussion of the importance of proper model verification and validation. To complete the path between the inhaled aerosol and the toxicological effects, attempts to couple deterministic CFD-based aerosol deposition modeling with biological effects via physiologically based pharmacokinetic (PBPK) models are described. Finally, conclusions and the current general trends in the application of CFD to toxicological science are given.

Key words Aerosol dosimetry, Modeling, Computational fluid dynamics, Physiologically based pharmacokinetic, Deposition mechanisms, Respiratory tract

1 Introduction

An aerosol is defined as a colloidal system of solid or liquid particles suspended in a gas, and can have different complexity in its chemical and physical compositions, ranging from single-species gases and mono-disperse solid particle mixtures to more complex multispecies gases and polydisperse liquid droplet mixtures. Aerosols are naturally present in various natural and industrial processes, with common aerosol particle sizes in the range of 0.01–100 μm [1]. For example, the atmosphere surrounding the Earth is composed of a mixture of gases and suspended solid and liquid particles, such as dust, water droplets, organic materials from biomass burning, sea salt crystals, pollen, bacteria, viruses, and other natural and man-made microscopic particles. This indicates that humans are continuously exposed to aerosols during inhalation of

Julia Hoeng and Manuel C. Peitsch (eds.), *Computational Systems Toxicology*, Methods in Pharmacology and Toxicology, DOI 10.1007/978-1-4939-2778-4_16, © Springer Science+Business Media New York 2015

air, which is a major potential entrance pathway into the human body [2]. Depending on the physical and chemical properties and the concentration of the particles, the exposure can be harmless, present health hazards, or be therapeutic. Respiratory aerosols are typically generated at high concentrations during the consumption of conventional and electronic cigarettes, with a particle number density in the order of 10^9–10^{11} particles per cubic centimeter [3]. High concentrations of aerosols are also produced for systemic delivery of therapeutic substances, because the lungs provide a large absorptive surface area for efficient drug delivery. Regardless of the origin and toxicological nature of an inhaled aerosol, it is important to know where and how much of the aerosol is deposited in the respiratory system to evaluate the toxicity and biological impact of the retained aerosol.

Deposition of aerosols in the respiratory system has been studied in various areas of dosimetry research, with the aim of relating the health effects to the exposure of inhaled and deposited aerosols. Historically, dosimetry research has focused on the health effects of inhaled toxic substances present in the atmosphere (e.g., radioactive chemicals) and in cigarette smoke. However, the focus has now widened to also include targeted deposition of biological agents for therapeutic applications (e.g., asthma inhalers). This new avenue of research has emerged since the success of inhalation therapy was found to not only depend on the pharmacology of the drugs being inhaled but also on the site and the deposited dose retained in the respiratory system [4]. The multidisciplinary scientific discipline of dosimetry combines the knowledge of the properties of aerosols with their transport, evolution, deposition, and translocation (clearance) in the airways to determine the dosage of the retained aerosol in the respiratory system. The retained aerosol exposes biological cells to the deposited compounds, which activates various types of biological processes. These biological processes may damage the functionality of the cells and cause disease when the deposited compound is toxic. Linking the local exposure in respiratory systems with measured dose is therefore the ultimate goal of dosimetry research.

1.1 Experimental Dosimetry

Aerosol deposition in the respiratory system has been investigated experimentally for certain aerosols both in vivo and in vitro. In vivo experimental work generally involves measurements of the *total lung deposition* (TLD) or *regional deposition*. In TLD experiments, the total deposition fraction is determined by measuring the number of inhaled and exhaled particles, which determines the number of aerosol particles retained in the respiratory system. In many studies, TLD experiments have been performed to investigate the effect of breathing patterns and particle sizes on the total retention in lungs [5, 6]. *Regional deposition* is typically measured indirectly using radioactive labeled aerosols and subsequent analyses of the clearance of particles by gamma scintigraphy [7]. Bolus delivery methods have also been pursued to avoid the use of radioactive labeled aerosol particles [8, 9]. The spatial resolution of regional deposition measurements is generally limited to a small number of regions, such as conventional three-compartment deposition [10] or extended ten-region deposition [8, 9]. However, the accuracy of experimentally determined regional deposition suffers from the indirectness of the method. In vivo inhalation experiments on animals can also give local deposition information [11, 12]. In these types of experiments, radioactively labeled aerosols can be detected by dissecting the animal, which is a labor-intensive and challenging process. Furthermore, the clearance mechanism of the respiratory system and biological effects may transform or transport the deposited matter away from the initial deposition location. Biokinetic effects and the available methods for macroscopic and microscopic analysis of particle deposition, retention, and clearance are reviewed in more detail by Geiser and Kreyling [13].

Another possible way to experimentally determine the deposition locations is to carry out in vitro deposition experiments on physical casts of the human respiratory tract. These types of experiments have the potential to determine some characteristics of the deposition in realistic geometries. However, the experiments are generally challenging for volatile, evolving aerosols and are limited to the number of branches of the physical lung cast model used. These experiments are therefore typically performed in casts of the upper respiratory tract or in the upper bronchial airways. In ref. [14] aerosol deposition experiments were performed on a human airway cast composed of an oral cavity down to the third-generation branches of the lung. Local deposition experiments in small airways were also performed by Phalen et al. [15] on a small part of a female lung replicate. In these experiments, the deposition of fluorescent particles was studied for generations 9–11 of the lung where the airway diameter is about 1 mm. A realistic head airway geometry was generated by Sosnowski et al. [16, 17] from computed tomography (CT) and magnetic resonance imaging (MRI) images, in which the influence of realistic breathing patterns on the deposition was measured for various particle sizes. Experiments on cigarette smoke deposition have also been performed on a human replica of the tracheobronchial airways generated from CT scans [18].

In addition to the in vitro studies on physical lung casts, experimental dosimetry is increasingly used to study the effect of direct deposition on living cell-cultures in exposure systems, such as the air–liquid interface (ALI) system [19]. In contrast to conventional in vitro exposure systems, ALI systems mimic the conditions of exposure via inhalation [20]. An overview of different ALI systems and deposition studies is given by Paur et al. [21].

1.2 Dosimetry Modeling

The experimental challenge and lack of reliable methods to determine the local deposition in human airways creates a need for predictive aerosol deposition models. There are different ways to model aerosol deposition in the respiratory tract. The two main groups of deposition models are *whole-lung models* and *local deposition models* [3]. *Whole-lung models* are typically based on one-dimensional transport equations applied to simple, idealized lung geometries to predict the total or regional particle deposition in the whole lung from the oronasal cavities to the pulmonary region. Whole-lung models can be divided into two sub-categories: *empirical/semiempirical* and *mechanistic models. Empirical models* are typically based on empirical correlations from mathematical multivariable regression of experimental deposition data from a series of anatomical compartments. Typical variables in such models are subject age, gender, breathing patterns, and aerosol type. Although *semiempirical models* are still based on the regression of experimental data for unknown parameters, they integrate information about the physical and chemical processes into the equations. Empirical and semiempirical models are based on experimental deposition data. On one hand, these models provide accurate predictions for the conditions and cases for which they were developed. On the other hand, they are limited to only those cases and cannot provide reliable information outside those ranges of variables and experimental data. *Mechanistic* whole-lung models aim to be more generic, because they are based on more realistic structures, physiologies, and mathematical representations of the physical and chemical processes governing aerosol transport and deposition. Different mechanistic whole-lung models incorporate different representations of the lung geometry, including *symmetric single-path models*, *one-dimensional "trumpet models"*, *asymmetric multiple-path models*, and *stochastic, asymmetric generation models* [22]. These mechanistic whole-lung models are all based on simplified analytical equations of aerosol transport and deposition by diffusion, sedimentation, and impaction. The current state-of-the-art *empirical/semiempirical* and *mechanistic* whole-lung models are reviewed and discussed in refs. [3, 22, 23], and are not covered in detail here. Despite mechanistic whole-lung models being more general than empirical/semiempirical models, they are still based on idealized geometries and physical representations, and thus lack the ability to accurately determine local deposition.

A more deterministic way of predicting deposition patterns based on the physical and chemical characteristics of the aerosol with realistic inhalation patterns, conditions, and geometries, is to use CFD simulations based on the established laws of physics with integrated aerosol transport, evolution, and deposition mechanisms. Because this approach is generic and based on physical and chemical mechanisms, it is advantageous over experimentally based models, which are geometry specific or condition constrained.

Numerical models of particle deposition in the lungs were initially developed to assess the toxicology of inhaled environmental and occupational pollutants, such as coal dust, cigarette smoke, and radionuclides [24], because information about local deposition patterns is important to assess toxicological effects.

The generic CFD approach offers detailed, noninvasive information about the aerosol physics of the formation, evolution, transport, and deposition of the aerosol, and can be used to identify the deposition locations in human lungs provided that the computational approach is properly validated. In addition, the amounts of deposited compounds and their compositions can also be quantified, and the effect of various aerosol properties and the influence of other properties, such as humidity and inhalation behavior, on the deposition can be investigated in detail with CFD. Although, in principle, the CFD approach permits a large part (many generations of branches) of the respiratory system to be studied provided that a digitalized geometry of the system and computational resources are available, it is mostly used for the upper part of the respiratory tract, including the nasal, oral, laryngeal airways, and upper lung bifurcations.

Despite the vast amount of information that deterministic CFD can give about aerosol transport, evolution, and deposition, it is still a challenging task to correctly model the dynamics of multicomponent, polydisperse aerosols with high concentrations of dispersed particles. Furthermore, incorporating the biokinetic effects of living bodies and tissues, such as translocation and clearance of deposited particles [25], in a deterministic way is highly complex, because biokinetic effects are not fully understood and described by first principles. Instead of trying to describe the biokinetic effects deterministically and integrate them into CFD-based models of the aerosol physics, attempts to couple PBPK models with CFD have been made to link the inhaled aerosol characteristics with the measured biomarkers in blood or urine [26]. Coupling deposition with biological effects, together with the challenge of correctly modeling complex aerosol dynamics using CFD, is a major focus of current research to determine predictive aerosol dosimetry models for local deposition in the respiratory system.

2 Aerosol Transport, Evolution, and Deposition

The transport and deposition of aerosols in the respiratory system are governed by physical and chemical mechanisms. During inhalation, an aerosol travels through different regions of the respiratory tract, which have different geometric shapes and dimensions. Starting from the mouth and nasal cavity, the aerosol flows through the pharynx (throat) and larynx (upper respiratory tract) down to the trachea and bronchial tubes before it finally reaches the alveoli.

During this journey, the aerosol experiences a variety of conditions and undergoes dynamic evolution. Its state and how it transports and deposits are dependent on the physical and chemical properties of the aerosol, the surrounding thermal and chemical environment, and the flow conditions. Because an aerosol is a gas mixture containing suspended solid or liquid particles, different physical and chemical mechanisms simultaneously govern the *transport*, *evolution*, and *deposition* of the different phases. To be able to predict aerosol deposition in the respiratory tract using a deterministic modeling approach such as CFD, these mechanisms need to be understood and correctly described.

2.1 Transport Mechanisms

The entrance of an aerosol into the respiratory tract is driven by the reduced pressure induced by the contraction of the diaphragm and the resulting expansion of the chest cavity and intrapleural space. This reduced pressure generates a flow of the aerosol between the mouth and the alveoli. In the upper airways, where the geometric dimensions are relatively large and the flow rate is high, the flow is mainly *convection* driven. The flow in the mouth, nasal cavity, and pharynx is also often complex, resulting in the onset of *turbulence* in locations where the Reynolds number is sufficiently high. Moreover, merging streams behind bifurcations in the upper part of the airways generate circulating vortices with prevailing secondary flows during exhalation. Further down in the respiratory system, the flow rate in each branch and the geometrics dimensions are smaller, which makes the effect of convection less pronounced, and *diffusion*-driven effects take over. During the inhalation and exhalation cycle, the flow rate varies depending on the breathing pattern. This *transient* flow behavior results in non-trivial flow dynamics in the respiratory tract, where different flow effects are important at different locations and times during the breathing cycle. As mentioned in ref. [27], the transport of aerosol in the respiratory tract is dependent on the geometric features of the airways, as well as transient effects, which may significantly influence particle deposition. In addition, turbulence in the upper airways also strongly affects particle deposition [28, 29]. The mechanisms governing the transport of aerosols in the respiratory system are shown in Fig. 1.

2.2 Evolution Mechanisms

The main physical mechanisms governing the evolution of aerosols during inhalation are *evaporation*, *condensation*, and *coalescence (coagulation)* [1], as well as gas phase *compressibility* and *dilution*, as shown in Fig. 2. The nucleation process is generally considered to be negligible in the respiratory system, but it is still included in Fig. 2 for completeness. *Evaporation* relates to the phase change of a compound from the liquid phase in the particles of the aerosol to the gas phase, whereas condensation is the reverse process. Evaporation of the liquid particles leads to a reduction of

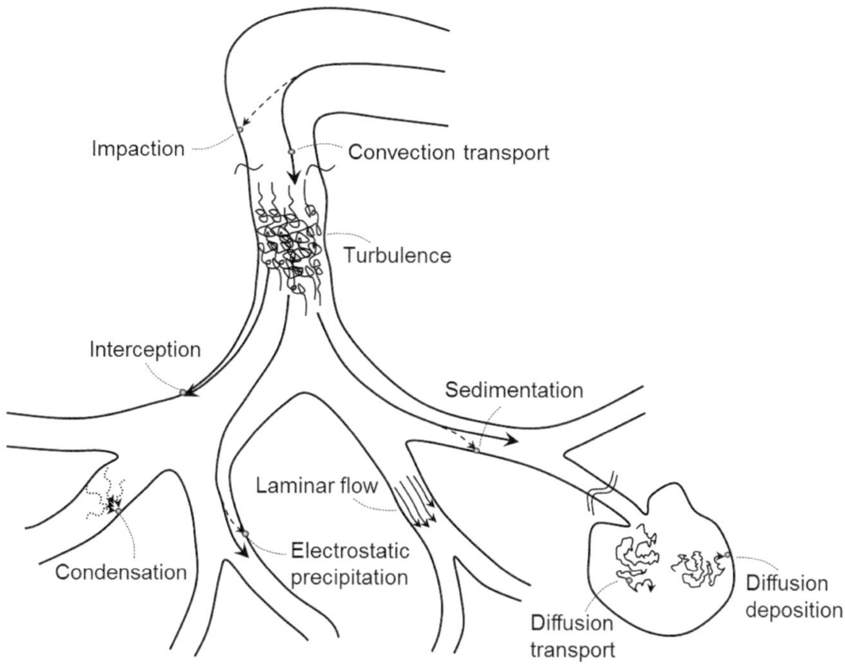

Fig. 1 Schematic diagram of the transport and deposition mechanisms in the respiratory tract

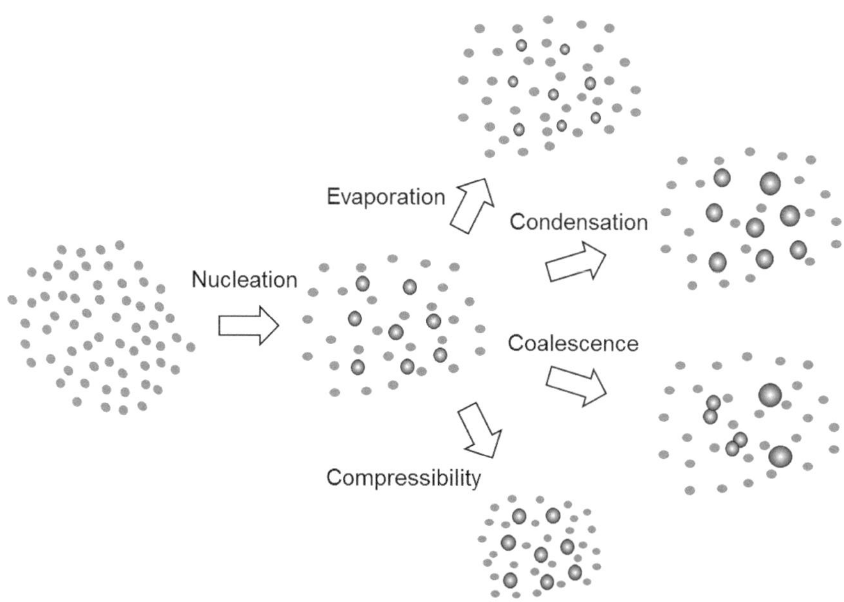

Fig. 2 Schematic diagram of the aerosol evolution mechanisms

the particle size and is dependent on the gas phase concentration, the partial vapor pressure of the compound, the Kelvin effect, and the temperature. *Condensation* of vapors onto existing droplets results in an increase of the particle size. Condensation is the main process behind hygroscopic growth of particles, and has been shown to enhance deposition in the upper respiratory tract [30, 31]. Condensation can also take place onto deposited liquid particles on the walls of the respiratory tract. During the evaporation and condensation processes, latent heat is consumed and released, respectively, which leads to changes of the thermal equilibrium and temperature-dependent thermophysical properties of the aerosol components. *Coalescence (or coagulation)* is the process where large particles are formed because of collisions between smaller aerosol particles. Coalescence is linked to the relative motion of the colliding particles, which can originate from either thermal or kinematic effects. *Thermal coalescence* relates to coalescence due to Brownian motion of the particles, whereas *kinematic coalescence* relates to coalescence induced by external forces or aerodynamic effects [1]. Coalescence is generally dominant at high concentrations of particles.

In addition to the evolution mechanisms related to the particulate phase of the aerosol, changes in the gas phase may also contribute to aerosol evolution. For instance, *compressibility* effects or *dilution* may contribute to the evolution of the aerosol. The aerosol can be compressed or expanded because of either pressure differences or density changes induced by temperature differences, evaporation, condensation, or particle deposition. During inhalation of the aerosol, the inhaled aerosol or vapor is also diluted with the residual air in the lung, which changes the physical characteristics of the aerosol [3]. All these relevant physical and chemical mechanisms contributing to the evolution of aerosols are strongly coupled and interdependent. This makes the aerosol dynamics in the airways nontrivial to study and predict accurately.

In many studies, aerosol evolution effects are excluded from the description of the aerosol dynamics and deposition in lungs, and it is assumed that the aerosol particle size remains constant during its residence time in the lung. This assumption can lead to incorrectly predicted deposition sites and amounts [32] because the aerosol evolution mechanisms change the aerosol properties (e.g., particle size distribution and particle number density), which influence the deposition mechanisms. Moreover, many biological effects depend not only on aerosol particle deposition but also on vapor uptake in the lung [33]. Because the evolution mechanisms described above determine the gas/particulate phase ratio, they are crucial to correctly describe the deposition of aerosols in predictive deterministic deposition models.

2.3 Deposition Mechanisms

Aerosol deposition is commonly associated with only the particulate phase. However, both the particulate phase and the gas phase contribute to the total deposited matter in the respiratory system. Particle deposition mechanisms include *impaction*, *diffusion*, *interception*, *sedimentation* (or *gravitational settling*), and *electrostatic precipitation* [1], as shown in Fig. 1. *Impaction* is related to the inability of particles to change directions because of inertial effects. Large particles suspended in a gas tend to not follow the motion of the gas but maintain their original direction when the flow suddenly changes direction. Impaction deposition is therefore associated with large particles in proximity to the deposition surface, where there is an abrupt change in the carrier gas flow direction. Deposition due to impaction typically occurs in the upper airways and at bifurcations. *Diffusion* is another mechanism that contributes to particulate phase deposition of aerosol constituents, mostly in the lower respiratory tract and in the alveolar region. Diffusion is directly related to stochastic Brownian motion of small particles suspended in the carrier gas. In contrast to impaction deposition, diffusion deposition is dominant for ultrafine particles, for which the mobility, and therefore the diffusion coefficients, are high. *Interception* is the deposition of particles because of collision with obstacles and walls of the respiratory tract. Interception is strongly dependent on the size and shape of particles and the surface characteristics of the obstacles or walls. For elongated particles, interception is an important deposition mechanism, whereas for spherical particles interception is less pronounced [34]. *Sedimentation* (or *gravitational settling*) is the deposition mechanism driven by the influence of gravity. The importance of sedimentation as a deposition mechanism is not only related to the particle size or density but also to the residence time and deposition surface orientation. Sedimentation typically occurs in the smaller airways, where the velocities and dimensions are small, as well as in locations where the deposition surface is horizontal. The last particle deposition mechanism is *electrostatic precipitation*, which relates to the deposition of electrically charged particles that are attracted to the surface by an induced electrical potential.

Along with particulate phase deposition, *gas phase condensation* onto the wall tissues of the respiratory tract may also occur depending on the surrounding conditions. Gas phase compounds may be absorbed by the alveolar–capillary membrane, where they can then enter the arterial blood flow and be carried to other organs of the body [33]. Respiratory vapor absorption generally depends on the chemical properties of the inhaled compounds, such as solubility, diffusivity, and metabolism/reactivity in the lung tissues [33, 35], but it also depends on the surrounding conditions, such as concentration and temperature.

Once the constituents are absorbed or deposited on the lung wall tissue, they are either taken up by the lungs by various uptake processes or eliminated from the body by removal or clearance processes. These biologically related processes are important for the ultimate therapeutic or toxic effects of the deposited matter.

3 CFD Modeling of Aerosol Dynamics

To simulate aerosol deposition in the respiratory tract using CFD, an appropriate computational model needs to be used, including all relevant physical and chemical mechanisms governing the aerosol dynamics. In addition to the mathematical description of the various mechanisms, information about the computational geometry, environmental and inhalation conditions, and physical characteristics of the aerosol is required. Once the physical processes, simulation conditions, and the case have been set up, the strongly coupled system of partial differential equations governing the aerosol transport, evolution, and deposition needs to be solved by numerical algorithms. Therefore, the essential elements of a CFD-based aerosol dynamics model for the respiratory system are

- Mathematical model of aerosol dynamics.
- Respiratory tract geometry and computational mesh.
- Simulation conditions.
- Numerical algorithms.

The accuracy of the results obtained from the simulations depends on how realistic and accurate all these elements are modeled and implemented. It is therefore of utmost importance to properly verify and validate the computational approach in controlled settings.

In the following subsections, the abovementioned elements of the CFD-based aerosol dynamics model will be described in detail to give an overview of the different computational approaches with various resolutions of the physics. The pros and cons of the different computational approaches and the state-of-the-art in applying CFD for evolving multicomponent aerosols at high particle number densities are also discussed. Finally, the important topics of *verification* and *validation* of CFD are covered.

3.1 Mathematical Model of Aerosol Dynamics

There are two main ways to model aerosol transport, evolution, and deposition in the respiratory tract, where the differences between the two approaches are related to how well the details of the aerosol system are captured. These two approaches are the *Eulerian–Lagrangian* and *Eulerian–Eulerian* methods. In both methods, the gas phase is considered as a continuum by solving the mass, momentum, and energy conservation equations in the Eulerian framework. The general set of equations describing the dynamics of a multicomponent gas mixture is formulated by the conservation laws for mass, momentum, and energy, as well as the concentration transport equations, expressed in terms of the mass fractions for each constituent in the n-component gas mixture according to [36]:

$$\partial_t \rho + \partial_j \left(\rho u_j\right) = 0 \tag{1}$$

$$\partial_t \left(\rho u_i\right) + \partial_j \left(\rho u_j u_i\right) = -\partial_i p + \partial_j \left(\tau_{ij}\right) + f_i;\ i = 1,\ldots,3 \tag{2}$$

$$\rho c_{\mathrm{p}} \left(\partial_t T + u_j \partial_j T\right) = -\partial_j q_j - \tau_{ij} \partial_i u_j + \left(\partial_{lnT}\left(\ln \hat{V}\right)\right)_{\mathrm{p},\chi_\alpha} \left(\partial_t p + u_j \partial_j p\right) + S_{\mathrm{h}} \tag{3}$$

$$\partial_t \left(\rho Y_\alpha\right) + \partial_j \left(\rho Y_\alpha u_j\right) = -\partial_j j_{j,\alpha} + S_\alpha;\ \alpha = 1,\ldots,n \tag{4}$$

where t is time, ρ is the total mass density of the mixture, u_i is the velocity, p is the pressure, and f_i is a momentum body source. The rate of strain tensor is defined as $\tau_{ij} = \mu\left(\partial_j u_i + \partial_i u_j\right) - \left(\frac{2}{3}\mu - \kappa\right)\delta_{ij}\partial_k u_k$, where μ is the dynamic viscosity of the mixture, δ_{ij} is the Kronecker delta tensor, and κ is the dilatational viscosity. The operator ∂_t is the temporal partial derivative and ∂_j is the partial derivative with respect to the spatial coordinate j. The Einstein summation convention is assumed for j and k. In the fluid energy equation, c_{p} is the specific heat of the mixture at constant pressure and T is the temperature. The heat flux $q_j = -\lambda \partial_j T + \sum_{\alpha=1}^{n} \bar{H}_\alpha M_\alpha^{-1} j_{j,\alpha} + q_j^{(x)}$, where λ is the thermal conductivity of the mixture. $\bar{H}_\alpha$ is the partial molar enthalpy of constituent α and $q_j^{(x)}$ is the usually negligible diffusion-thermo effect [36]. The molar mass is denoted M_α, the mass flux of constituent α is $j_{j,\alpha} = \rho Y_\alpha \left(u_{j,\alpha} - u_j\right)$, where $u_{j,\alpha}$ and Y_α are the velocity and mass fraction of component α in the gas phase, respectively. For a pure gas mixture with n gas species, the mass fractions agree with the following constraint:

$$\sum_{\alpha=1}^{n} Y_\alpha = 1 \tag{5}$$

This condition is only valid for pure gas mixtures. For aerosol systems, the mass fractions of the particulate phase also contribute to the total mass, and will therefore be part of the constraint, as will be shown later. To satisfy mass conservation, the diffusion fluxes, $j_{j,\alpha}$, obey the *zero-net flux condition* according to:

$$\sum_{\alpha=1}^{n} j_{j,\alpha} = 0 \tag{6}$$

Moreover, $\hat{V}$ is the volume per unit mass and χ_α is the molar fraction of constituent α. The energy source term $S_{\mathrm{h}} = \sum_{\alpha=1}^{n} \bar{H}_\alpha \left[\partial_j \left(M_\alpha^{-1} j_{j,\alpha}\right) - R_\alpha\right]$, where R_α is the molar rate of production of the species α by homogeneous chemical reactions or mass transfer from nucleation, evaporation, or condensation processes. S_α is the mass rate of production of species α from the mass transfer processes mentioned above. In the case of pure, non-reacting gas mixtures, $R_\alpha = S_\alpha = 0$. The transport equations for the particulate phase also need to be defined to fully describe the aerosol system. In the following subsections, the different ways to treat the particle phase are described.

3.1.1 Eulerian–Lagrangian Method

In the *Eulerian–Lagrangian* method, the gas phase is considered as a continuum by solving the mass, momentum, and energy conservation equations in the Eulerian frame, while the particulate phase is treated as a discrete phase. The equation of motion resulting from the various forces acting on an individual particle is solved for each particle to obtain individual trajectories for the particles. The equation of motion in the Lagrangian method can be expressed as [3]

$$m_\mathrm{p} d_t u_{i,\mathrm{p}} = F_{i,\mathrm{D}} + m_\mathrm{p} g_i + F_{i,\mathrm{Br}} + F_{i,a} \tag{7}$$

where $d_t = \partial_t + u_{j,\mathrm{p}} \partial_j$ denotes the differential operator with respect to time, m_p is the particle mass, $u_{i,\mathrm{p}}$ is the particle velocity, $F_{i,\mathrm{D}}$ is the drag or aerodynamic force, g_i is the gravitational acceleration, $F_{i,\mathrm{Br}}$ is the force owing to Brownian motion, and $F_{i,a}$ represents the additional forces per unit mass, such as electrostatic forces and thermophoresis. The forces depend on the particle shape, mixture viscosity, and the relative velocity between the particle and carrier gas. The exact expressions of the forces can be found in ref. [37] and will not be elaborated on here.

The coupling between the continuum and discrete phases can be performed at various levels of scrutiny, where the interactions between individual particles and their interaction with the gas phase depend on the physical conditions, such as the volume fraction of phases, particle number density, Reynolds number, and Stokes number [38]. For dilute aerosols, one-way coupling between the continuous and the discrete phases is usually assumed, i.e., the discrete particulate phase is affected by the continuous phase, but the continuous phase is not affected by the discrete phase [39]. Regardless of the assumption of one-way coupling between the particles and the carrier gas and evolution of the aerosol being generally ignored, Eulerian–Lagrangian methods are widely accepted to be the most accurate methods for aerosol deposition of dilute, low concentration aerosols. Eulerian–Lagrangian methods are commonly used for indoor aerosol transport and deposition investigations when the particle number density is low [40–42]. For deposition in the respiratory tract, the Eulerian–Lagrangian method has also been used in numerous studies to determine the local deposition of dilute aerosols. For example, [43, 44] used a Eulerian–Lagrangian method in simple bifurcation geometries to determine the local deposition patterns for various flow rates, branching angles, and dimensions for generations 3 and 4 of the Weibel airway model [45]. The Eulerian–Lagrangian approach has also been applied to track up to 10^5 particles in more anatomically realistic geometries [29, 46]. These studies found that a realistic geometrical representation is highly important for accurate local deposition information when using Lagrangian particle tracking for the aerosol particle transport.

The Lagrangian treatment can also be used for aerosols with high particle number densities if each particle in the Lagrangian treatment is considered to be a group or cloud of particles [3]. To provide a statistically stable representation of an aerosol system, the number of particles in such a particle cloud must be sufficiently large (in the order of 10^4) [3]. For smoke aerosols, which have a droplet concentration in the order of 10^9–10^{11} particles per cubic centimeter [3], the Eulerian–Lagrangian approach remains computationally challenging because the number of "particle clouds" to be tracked in the whole lung is in the order of 10^8–10^{10} (assuming that a particle cloud contains in the order of 10^4 particles). According to Longest and Xi [47], the disadvantages of the Eulerian–Lagrangian model include the requirement for an extremely large number of particles to establish convergent local deposition profiles and a stiff system-of-equations for fine and ultrafine particles. Therefore, for smoke aerosols with high particle number densities, in which evaporation, condensation, coalescence, and hygroscopic growth effects are not negligible, the Eulerian–Lagrangian approach is challenging and the Eulerian–Eulerian approach is more suitable.

3.1.2 Eulerian–Eulerian Method

The *Eulerian–Eulerian* method treats the particles as a continuum, similar to the continuous carrier gas phase. This method will hereafter be referred to as the *Eulerian* method for brevity. There are generally two different types of Eulerian methods: the *Eulerian chemical species model* and the *Eulerian drift-flux model.*

The *Eulerian chemical species model* treats the particle phase as a dilute chemical species in a multicomponent mixture. This approach ignores particle inertia and gravitational settling, and only considers deposition due to diffusion. The model is most often used for simulations of the transport and deposition of ultrafine and nanosized particles in situations where the Stokes number is below 5×10^{-5} and inertial effects can be safely ignored [48]. The governing equations for the *Eulerian chemical species model* include the gas phase Eqs. (1)–(4) for the mass, momentum, and energy conservation, as well as the mass fraction equations for the gas species. To complete the set of governing equations, the mass fraction equations for the dispersed particles of sizes $\beta \in \{1,\ldots,m\}$ and the related general dynamic equation (GDE) for the aerosol particle number concentration N_β in polydisperse aerosols can be written as

$$\partial_t\left(\rho Z_\beta\right)+\partial_j\left(\rho Z_\beta u_j\right)=-\partial_j j_{j,\beta}+S_\beta \ ; \ \beta=1,\ldots,m \qquad (8)$$

$$\partial_t\left(m_{p,\beta} N_\beta\right)+\partial_j\left(m_{p,\beta} N_\beta u_j\right)=-\partial_j j^*_{j,\beta}+S_\beta \ ; \ \beta=1,\ldots,m \qquad (9)$$

where the total mass fraction of particles of size β is denoted as Z_β. The aerosol particles are generally composed of multicomponent mixtures of liquid and/or solid components, where the mixture composition may differ between particle sizes. This implies that $Z_\beta = \sum_{\theta=1}^{w} Z_{\beta,\theta}$, where $Z_{\beta,\theta}$ is the mass fraction of component θ in the w-components liquid/solid mixture constituting the particles of size β. The diffusion mass flux is $j_{j,\beta} = \rho Z_\beta \left(u_{j,\beta} - u_j\right)$, where $u_{j,\beta}$ is the velocity of the particles of size β. The equivalent expression for the mass flux in the particle number concentration equation is $j^*_{j,\beta} = m_{p,\beta} N_\beta \left(u_{j,\beta} - u_j\right)$, where the mass of an individual particle of size β is denoted $m_{p,\beta}$. The mass transfer source term S_β includes the production of particle mass from the nucleation process, the coalescence as well as the increase and decrease of particle mass from evaporation and condensation, respectively. The GDE for the particle number density also allows information about the size distribution of the aerosol particles to be derived. More details about the expressions of the source terms, S_α and S_β, can be found in refs. [1, 49].

Mass conservation of an aerosol system with n gas or vapor species and m particle sizes imposes the following constraints on the mass fractions, diffusion fluxes, and the mass source terms:

$$\sum_{\alpha=1}^{n} Y_\alpha + \sum_{\beta=1}^{m} Z_\beta = 1 \tag{10}$$

$$\sum_{\alpha=1}^{n} j_{j,\alpha} + \sum_{\beta=1}^{m} j_{j,\beta} = 0 \tag{11}$$

$$\sum_{\alpha=1}^{n} S_\alpha + \sum_{\beta=1}^{m} S_\beta = 0 \tag{12}$$

The *Eulerian chemical species model* has been used in many studies to simulate the transport and deposition of aerosols in the respiratory system with particle diameters up to 0.1 μm [47, 48, 50–53]. For example, diffusional deposition of particles ranging from 5 nm to 1 μm in developing [48] and fully developed [47] tubular flow geometries showed good agreement with an analytical deposition expression in both cases. The *Eulerian chemical species model* has also been used to successfully predict the deposition in a 90° bend model compared with experiments for 12 nm particles [47]. However, when the *Eulerian chemical species model* was applied to a more complex airway bifurcation geometry, the model was shown to underpredict the total deposition fraction by one order-of-magnitude for 1 μm particles [53], for which the inertial effect is important.

The *Eulerian drift-flux model* is another Eulerian-based model that integrates gravitational settling and the inertial effects of particles into the concentration equation. As with the *Eulerian chemical species model*, the governing equations for the *Eulerian drift-flux model* include the gas phase Eqs. (1)–(3) for the mass, momentum, and energy conservation, respectively. The mass fractions of the various constituents in the multicomponent gas and particle mixture, and the GDE for the aerosol particle number concentration of polydisperse aerosols, can be written as

$$\partial_t\left(\rho Y_\alpha\right)+\partial_j\left(\rho Y_\alpha\left(u_j+v_{s,j}^{(g)}\right)\right)=-\partial_j j_{j,\alpha}+S_\alpha \quad ;\ \alpha=1,\ldots,n \quad (13)$$

$$\partial_t\left(\rho Z_\beta\right)+\partial_j\left(\rho Z_\beta\left(u_j+v_{s,j}^{(\beta)}\right)\right)=-\partial_j j_{j,\beta}+S_\beta \quad ;\ \beta=1,\ldots,m \quad (14)$$

$$\partial_t\left(m_{p,\beta}N_\beta\right)+\partial_j\left(m_{p,\beta}N_\beta\left(u_j+v_{s,j}^{(\beta)}\right)\right)=-\partial_j j_{j,\beta}^{*}+S_\beta;\ \beta=1,\ldots,m \quad (15)$$

where $v_{s,j}^{(g)}$ and $v_{s,j}^{(\beta)}$ are the slip velocities of the gas phase and particles of size β, respectively. The slip velocity of the particles of size β can be determined as a function of inertial and gravitational forces as [54]

$$v_{s,j}^{(\beta)}=\frac{C_c d_{p,\beta}^2}{18\mu}\left(\rho_{p,\beta}-\rho\right)\left[g_j-\partial_t u_j-u_k\partial_k u_j\right] \quad (16)$$

where C_c is the Cunningham correction factor, $d_{p,\beta}$ is the particle diameter, $\rho_{p,\beta}$ is the intrinsic density of the particles of size β, and g_j is the gravitational acceleration vector. The corresponding gas slip velocity can be written as

$$v_{s,j}^{(g)}=-\frac{\sum_{\beta=1}^{m}\left(Z_\beta v_{s,j}^{(\beta)}\right)}{\sum_{\alpha=1}^{n}Y_\alpha} \quad (17)$$

The *Eulerian drift-flux model* has been reported to predict the indoor particle distribution [55–60]. In addition, it has been applied to simulations of aerosol deposition owing to electrostatic forces [61] and for aerosol dispersion and deposition in the respiratory tract [53, 62, 63]. In ref. [62], a *drift-flux model* incorporating Brownian diffusion, gravitational settling, and electrostatic forces was validated by analytical results in a straight pipe prior to it being applied to a single bifurcation geometry to investigate the influence of the different deposition mechanisms for 1 nm and 10 μm particles. A *drift-flux model* was used to theoretically predict deposition due to inertia, diffusion, and turbulent dispersion in a nasal cavity, and showed good agreement with the results from high-resolution simulations [63]. Two velocity correction models for the *drift-flux model* have been developed to improve the near-wall particle velocity [53]. These extended *drift-flux models* were compared with the standard *drift-flux model* and an *Eulerian chemical species model* for particle deposition in a double bifurcation geometry for two particle sizes. It was concluded that *drift-flux models* with near-wall velocity corrections provide a highly effective solution for the deposition of fine respiratory aerosols.

In addition to the *Eulerian chemical species model* and the *Eulerian drift-flux model*, there are also other continuous two-phase methods available, such as the *Eulerian mixture model*. In the *Eulerian mixture model*, the gas and particle phases are treated as interpenetrating continua, and the two phases move at different velocities. A volume fraction differential equation coupled with the mass and momentum equations for the mixture is solved to obtain the volume fractions of the gas and particle phases. This method is also known as a volume-of-fluid (VOF) method for multiphase flows, and has been used to simulate the gas–liquid dynamics in cylindrical bubble reactors [64]. Because the model is based on volume fractions, it is not suitable for disperse aerosols, where the volume fraction of the particle phase is negligible despite the high number concentration. This was confirmed in ref. [37], where the model gave unacceptable results for the particle concentration in a ventilated model room simulation.

3.1.3 Turbulence Modeling

Once the modeling approach has been chosen for the aerosol system, the governing equations together with proper initial and boundary conditions are defined, and the mathematical problem can then be solved. However, for most mathematical problems, a numerical method must be used to find the solution, especially for transient cases and problems in complex geometries. As well as the problem of choosing appropriate numerical algorithms to solve the coupled system of equations, the inherent problems of discrete numerical representations need to be taken into account, such as spatial and temporal discretization. This is particularly important for simulations of aerosol dynamics in the respiratory system, because of transitional and turbulent flow behavior in the upper airways and the necessity to accurately capture wall effects (e.g., curvature [65] and deposition [66]) in realistic flows.

The most accurate, but most computationally demanding, way to numerically solve the system of governing equations is to satisfy the mesh resolution requirements dictated by the flow to avoid additional modeling. This "resolving flow features approach" for pure fluid flows is called direct numerical simulation (DNS). To resolve the flow features sufficiently well for turbulent flows, a mesh resolution one level lower than the typical size of the dissipation determined by the Kolmogorov scale [67, 68], needs to be adopted. This requirement is usually very difficult to achieve, particularly for cases of transitional flows where local, transient flow-features may be created. Such an approach leads to very dense mesh resolutions that are computationally unfeasible for complex airways geometries. To relax the mesh resolution constraint, the governing equations can be averaged in time using Reynolds decomposition [69], which separates the average and fluctuating quantities. For the momentum conservation equations (Navier–Stokes equations), the time averaging procedure leads to the well-known Reynolds-averaged Navier–Stokes (RANS) equations for steady flows, and the unsteady RANS (URANS) equations, also referred to as transient RANS (TRANS), for unsteady flows. The process of decomposition introduces unclosed terms, such as the Reynolds stress in the momentum equation and turbulent fluxes in the energy equation, which contain fluctuating parts of the velocity components and must be closed with additional equations. The closure of these additional terms is generally referred to as "turbulence modeling". There are many ways to close these terms, from *zero-equation models*, to *one-equation* and *two-equation models*, up to *algebraic* and *Reynolds stress models*. Each of these models makes certain assumptions, and they have strengths and weaknesses for different situations.

RANS models are widely used for many industrial applications to simulate average flow behavior, and they generally do not require the challenging mesh requirements of DNS to give indicative results. However, the lack of sufficient mesh resolution, especially close to the boundaries, generates unphysical behavior of flows in the vicinity of boundary layers. It is well-known that for turbulent flows the velocity profiles do not follow laminar/parabolic shapes [67]. In RANS, the so-called wall functions were introduced for both momentum and energy to simulate (usually algebraically) proper flow behavior near boundaries. Although widely accepted in engineering applications, the use of RANS models for unsteady transitional flows is questionable and the accuracy of the obtained numerical predictions needs to be carefully considered [67].

An alternative way to solve the conservation equations of turbulent flow is the *large-eddy simulation* (LES) approach. In this approach, a filtering operator is applied to the original set of governing equations. This operator decomposes the flow scales into those directly solvable on the computational mesh and those that are eliminated and therefore must be modeled/closed with appropriate models [70]. The most commonly used LES models are based on the eddy-viscosity hypothesis introduced by Smagorinsky [70]. Turbulent viscosity is added via the closure term (often called the subgrid-scale (SGS) model), which introduces additional dissipation into the flow to account for turbulent flow behavior. The filtering operation can also be introduced dynamically based on the local flow features [71]. The LES method gives much more detail and insight into the transitional turbulent flows generated in the upper airways than RANS, but is computationally demanding, lacks a near-wall treatment [72], and requires an a priori experimental investigation of the computational mesh resolution.

Various types of mixed/hybrid type models have been introduced over the past years, where RANS and LES models are applied dynamically depending on the flow conditions and the ability to resolve flow features on the computational mesh. This family of approaches is called the detached-eddy simulation (DES) approach [73].

To simulate aerosol dynamics in the upper respiratory tract, RANS models have been extensively used [46, 74–78]. For example, the k-ε model, where the closure term is modeled by solving two additional equations for turbulent kinetic energy and its dissipation, has been used to simulate the effect of turbulence on the overall deposition in the human larynx and trachea [74, 75]. These initial studies of turbulence effects was later followed by a number of attempts to investigate the influence of turbulence with more sophisticated RANS models, such as low-Reynolds k-ε, k-ω and k-ω Shear Stress Transport (SST) models, that are better suited to resolve transitional regions [46, 76–78]. The more accurate LES models were initially used to simulate simple geometries, such as bifurcation areas [79], but have gradually been used for more complex geometries, such as idealized mouth–throat geometries [80, 81]. For a detailed review of the various models used for simulations of different geometries and flow scenarios see [24].

3.1.4 Aerosol Modeling

Accurate and proper representation of the aerosol characteristics is necessary to accurately predict the transport, evolution, and deposition of aerosols in the respiratory system. One important characteristic of an aerosol is its particle size distribution, which is indirectly governed by the particle number concentration Eqs. (9) and (15) of the *Eulerian chemical species model* and *Eulerian drift-flux model*, respectively. Two of the most common approaches for describing particle size distributions of aerosols are *sectional methods* and *modal methods* [82].

A *sectional method* is a type of discrete particle size approach where the particle size range is split into a finite number of sections. All particles within a given size range are represented with the same composition. In principle, the size resolution of a sectional representation is only limited by the number of sections used, because the computational cost increases with the number of sections. An important feature of sectional methods is that they allow particles of different sizes to have different compositions, which is an attractive feature, especially for multicomponent aerosol systems. The original sectional method had fixed sectional size boundaries [83, 84]. The most prominent disadvantage of the fixed sectional method is that numerical diffusion occurs when simulating evolving aerosols. This numerical diffusion occurs when a particle grows or shrinks enough to fit in another size bin and its new size is artificially adjusted to the particle size in the new bin, leading to numerical diffusion of the size distribution. To address this shortcoming, the *moving sectional method* [85] and *moving center sectional method* [86] were developed.

In the *moving sectional method*, the boundaries of sections are not fixed and move with the evolving particle sizes rather than moving particles from one section to another. This method eliminates the numerical diffusion associated with condensational growth or evaporational shrinkage, but combining distributions with different size boundaries becomes extremely difficult and produces its own errors [86]. This limits its use for aerosol dynamics simulations in the airways.

In the *moving center sectional method*, the boundaries of sections are fixed but the diameter of particles is not limited to the middle of the size range, as in the original sectional method, but can vary within the bin. This type of sectional method has similar accuracy to the moving sectional method in representing the aerosol distribution of evolving aerosols and the fixed sectional method in combining distributions from different grid cells [87, 88]. Despite these advantages, it has been shown that the moving center sectional method may produce large numerical errors that result in artificial peaks in the distribution for situations when all particles in a bin are moved to another size range because of sufficiently large condensation [89].

Modal methods treat the aerosol size distribution as a set of individual, typically log-normal, distributions or modes [90]. In a log-normal modal distribution, the aerosol distribution is defined by the parameters of log-normal equations [91]. A modal representation of the particle size distribution assumes that components in each mode are well mixed such that all particles within a given mode have the same composition. However, different modes may have different compositions. Consequently, particles of the same size may belong to different modes and have different compositions. Aerosol particle size distributions often have multiple well-defined peaks. To resolve a multipeak particle size distribution, several modes are usually required. Each modal distribution is typically defined by three parameters: median diameter or geometric mean diameter, standard deviation, and total number (or mass, volume, surface area, depending on the particle characteristics of interest). Modal methods have much faster simulation speed and are less computational expensive than sectional methods, but the accuracy of the modal method is determined by how well the assumed analytical function fits the real size distribution profile [88]. Moreover, modal methods are not well-suited for the aerosol coalescence process when two particles from different modes merge [92], and it has difficulty handling the generation of new particles in an existing aerosol. The modal method is therefore not suitable for the detailed representation of a multicomponent system with highly diverse properties, and it is instead usually used in large-scale applications where computational efficiency is more important than the detailed treatment of aerosol chemical composition and the particle size distribution.

3.2 Respiratory Tract Geometry and Computational Mesh

To simulate the aerosol dynamics in the respiratory tract using numerical methods, the geometry needs to be defined and a computational mesh needs to be generated. In this section, the different ways to create the geometry are briefly reviewed and aspects of mesh generation are discussed.

3.2.1 Respiratory Tract Geometry

There are several ways that the respiratory system can be represented computationally from simple model geometries to more complex and realistic representations of the airways. The first step to generate a geometry model of the airways is to measure the physiological relevant data of the respiratory system. This was first done using light microscopy for generations 0 (trachea) to 15 and standard morphological techniques for generations 16–23 [45]. These data were then used to create empirical equations for the diameter and length of each generation (0–23). The branching in this symmetric model was based on Murray's law of optimal branching. Other models of the lung have also been developed where asymmetry [93] and randomness [94] were taken into account. In addition to morphometric measurements of the airways, a theoretical attempt to generate a complete three-dimensional (3D) lung geometry has been made based on nine rules [95].

In recent years, MRI and CT image processing have taken over as the conventional techniques to measure lung morphology. Both CT and MRI techniques generate stacks of two-dimensional (2D) images of the internal structure in the body. Different tissues have different intensities in the gray scale of the raw images. Image processing techniques are used to segment the stack of grayscale images and to extract the region of interest from the images. The stacks of 2D images are then used to render a 3D triangulated surface mesh of the respiratory tract. This surface mesh can then be used to create 3D solids, hollow geometries, or computational meshes. These imaging techniques have been extensively used to generate realistic representations of the respiratory tract. For example, replicas and computational geometries of tracheobronchial airways have been developed based on CT images [18, 96, 97], and MRI images were the basis for the realistic geometries of the oropharyngeal airways generated in refs. [98, 99]. Geometries of the respiratory system of animals, such as the replica of an average rat respiratory tract [100], have also been generated using CT images. Combinations of CT and MRI images have also been used to generate an average geometrical model of the mouth and throat [75], a realistic head airway geometry [16, 17], models of the human upper respiratory tract [101], and detailed geometrical models of rat, monkey and human respiratory tracts [102], including 17 ± 7, 19 ± 9 and 9 ± 2 airway generations, respectively.

Not all simulations of aerosol deposition require realistic geometries, and simulations in simplified model geometries and selected sections of the respiratory tract are often performed. For example, simple symmetric bifurcation geometries have been used to study particle deposition [43, 44, 103]. The alveolar region is another example where simplified model geometries are used to study aerosol deposition and gas absorption. The alveolar duct is typically represented by straight ducts with small cavities attached to represent the alveoli [104].

3.2.2 Computational Mesh

From the geometry model representing the respiratory tract, a computational mesh can be generated, for which numerical simulations of the aerosol dynamics in that geometry can be performed. A computational mesh is a discretized representation of the geometry in which the geometry is split into a finite number of control volumes or computational cells. These control volumes can be of many different types and shapes, which indicates that there are many different types of meshes that can be created. For airway modeling, body-fitted grids where the faces of the control volumes conform with the geometrical surface are preferable to preserve the geometrical features of the airways. Meshes that only consist of hexahedral control volumes are called structured meshes [105]. It is only possible to generate fully structured meshes for simple geometries. For more complex geometries, unstructured meshes consisting of, for example, tetrahedral or polyhedral control volumes, are required to obtain a body-conformed mesh. Unstructured meshes are known to generate large numerical diffusion, especially meshes based on tetrahedral elements [105], which may strongly alter the aerosol deposition results. Hexahedral dominant or polyhedral unstructured meshes are therefore preferable for simulations of aerosol dynamics in the respiratory system.

Regardless of the computational mesh that is chosen, controlling the mesh quality is a crucial aspect of CFD modeling. Attributes that have a negative influence on the mesh quality include highly skewed, high aspect ratio cells and cells with small angles or faces [106]. Therefore, controlling these mesh attributes is crucial to generate high-quality CFD results for aerosol deposition. The spatial resolution of the discretized mesh is also an important attribute to achieve predictive results. The spatial resolution needs to be sufficiently high to accurately capture the flow features, especially in the case of turbulence, as discussed previously. Computational meshes used for turbulence simulations usually include layers of control volumes in the vicinity to the boundaries of the mesh to accurately resolve the viscous boundary layer. This refinement of the mesh close to the boundaries is also beneficial for accurate prediction of particle deposition. Adaptive meshing techniques can also be used, in which the mesh resolution is locally increased depending on the gradients of certain quantities. The advantage of the adaptive grid method is that the total number of grid points is relatively small because the finest mesh occupies only a small part of the domain, so both the computational cost and the memory requirements are greatly reduced compared with if the mesh would have been refined uniformly to the same fine resolution [105].

3.3 Simulation Conditions

Before the aerosol dynamics can be simulated on a mesh representing a discretized version of the computational geometry of the respiratory tract, the simulation conditions need to be specified. There are three main simulation conditions that need to be defined prior to performing the simulation: (1) the *aerosol characteristics*, (2) the *environmental conditions*, and (3) the *inhalation pattern*.

1. *Aerosol characteristics* includes properties such as the *composition* of the multicomponent gas and particle mixture, the *particle size distribution*, the *particle number density* of the inhaled aerosol, and the *thermophysical properties* of the individual constituents of the mixture. The composition of the aerosol system is typically specified by the mass fractions of the individual constituents. The particle size distribution is specified in different ways depending on the aerosol model used, but typically involves the specification of the mass median diameter (MMD) or count median diameter (CMD), a standard deviation, the number of particles in different particle size ranges, or distribution functions. The thermophysical properties of the individual gas and particle components include their density, viscosity, thermal conductivity, specific heat, enthalpy of evaporation, vapor pressure curve, molar mass, surface tension, and compressibility. The aerosol system mixture properties are then characterized by mixture laws that are usually based on kinetic gas theory [107].
2. The *environmental conditions* need to be specified, and are typically ambient conditions, such as the gas pressure, temperature, relative humidity, and the initial gas composition in the respiratory tract. The attributes of the walls of the respiratory tract are also part of the environmental conditions. Environmental conditions are typically specified via boundary and initial conditions.
3. The *inhalation pattern* (or breathing pattern) describes how the aerosol enters the computational geometry. A typical human male breathes at a frequency of 12 breaths per minute and inhales 750 mL per breath while sitting [3]. Depending on the aim of the simulation, the inhalation pattern may follow measured inhalation patterns, idealized and cyclic breathing profiles [108], or continuous inhalation for steady-state studies. Inhalation can be induced by assigning a flow rate either at the inlet or the outlet of the computational geometry or by altering the pressure, which generates a non-zero pressure gradient driving the flow. A more complex, but less com-

mon, way of simulating inhalation is to move the boundaries of the geometry, which increases the volume of the computational geometry, similar to realistic breathing conditions.

3.4 Numerical Algorithms

Irrespective of the modeling approach chosen for the aerosol system and the computational geometry or simulation conditions, the coupled system of governing equations needs to be solved numerically using a suitable algorithm. To solve the set of continuous governing equations for the aerosol dynamics, the equations need to be formulated in a discretized form. Among the different methods for doing this, the finite difference (FD), finite element (FE), and finite volume (FV) methods are the most common [105]. For CFD applications, the collocated variable FV method is commonly used because of its inherent conservation properties and because it can be applied to non-conforming computational meshes.

Aerosol transport, evolution, and deposition involve physical and chemical processes acting on and within compressible gas mixtures with suspended liquid and/or solid particles. Temperature differences, evaporation, condensation, and deposition processes can result in density variations of the aerosol. This indicates that a compressible formulation of the numerical algorithm is required. There are many different numerical algorithms available, which all have their strengths and weaknesses. The two main categories of algorithms used to solve FV problems are *block-coupled* and *segregated* solvers. In a block-coupled algorithm, the dependent variables are largely implicitly coupled and the conservation equations are solved simultaneously in one block-matrix system. Despite the often fast convergence, such an approach is computationally expensive and convergence may be difficult to achieve in a strongly temperature-dependent system. In segregated algorithms, the conservation equations are solved one after the other, and the coupling between dependent variables is achieved in an iterative manner. The most common segregated algorithms for fluid flow problems are the semi-implicit method for pressure-linked equations (SIMPLE) algorithm [109] and the compressible pressure implicit with splitting of operators (PISO) algorithm [110]. There are many derivatives of the SIMPLE and PISO algorithms, such as the SIMPLE revised (SIMPLER) [111] and SIMPLE consistent (SIMPLEC) [112] algorithms, which have all been shown to have their advantages in certain cases. The original compressible PISO algorithm has been extended to reacting flows [113], and applied to combustion problems [113, 114]. Several studies of the applicability of the PISO algorithm for various applications have been carried out, which gave sufficient evidence that the PISO algorithm is computationally efficient, robust, accurate, and versatile. For example, it has been shown that the PISO algorithm has better convergence properties than the SIMPLE algorithm [115], as well as having better stability for large time steps and requiring less computational effort. Moreover, it has been shown that the PISO algorithm is suitable for low-Mach compressible and incompressible flow on collocated grids [116] and is accurate and robust for transient flows [117]. Steady turbulent flow problems were studied in ref. [118], where it was concluded that the PISO algorithm is computationally efficient and much more effective for time-accurate simulations than the SIMPLE algorithm. In the context of aerosol formation and transport, the compressible PISO algorithm for reacting flows [113] has been extended to incorporate the formation and transport of single-species aerosols and its numerical properties have been investigated [49].

The choice of a proper numerical algorithm is often overlooked when applying CFD models for aerosol deposition simulations in the human respiratory system. Robust algorithms with high accuracy are important for accurate prediction of local deposition, especially because many toxic substances deposits are relatively small but important quantities in the airways. In these cases, large errors from the numerical solution procedure can affect the deposition results in a similar way as insufficient spatial or temporal resolution.

There are a number of commercial and open source CFD software packages available (e.g., ANSYS Fluent®, ANSYS CFX®, STAR-CCM+®, and OpenFOAM®) that can be used for the simulation of the aerosol transport, evolution and deposition in the respiratory system. To consistently simulate all physical processes occurring during aerosol transport, evolution, and deposition in the airways, these software packages generally need to be adapted by user-defined functions or customized source code development.

3.5 Verification and Validation

Deterministic modeling of aerosol transport, evaporation, and deposition using CFD is based on mathematical formulations of physical and chemical processes. These formulations are implemented in a computer code that can be numerically solved on discretized computational geometries with specific initial and boundary conditions describing the simulation conditions. All of these steps may generate errors in the solutions, which may reduce the accuracy of the predicted quantities. The five major sources of error in CFD modeling are (1) *insufficient spatial discretization*, (2) *insufficient temporal discretization*, (3) *insufficient iterative convergence*, (4) *computer round-off errors*, and (5) *computer programming errors* [119]. The primary processes for assessing and quantifying the accuracy of computational results are through *verification* and *validation* [120].

Verification is the process of assessing the correctness and numerical accuracy of the solution to a given mathematical model [119] and provides substantiation that the conceptual (continuum mathematics) model is solved accurately by the discrete mathematics model embodied in the computer code. Highly accurate, reliable benchmark solutions are typically used to verify the computer code. *Validation* is the process of assessing the physical accuracy of a mathematical model based on comparisons between computational results and experimental data.

To ensure the quality of numerical simulations for the different physical and chemical mechanisms related to aerosol transport, evolution, and deposition, verification and validation of each mechanism should ideally be separately performed. This can be achieved within controlled experimental environments. For example, aerosol deposition in a bend pipe has been verified and validated with analytical and experimental data [121]. Because the mechanisms are strongly coupled and integrated into global conservation equations, relevant verification and validation cases integrating the coupling algorithm need to also be considered. This is important to guarantee correct predictions of the final aerosol deposition, especially for situations where the computational results are used qualitatively for comparison, as a basis for decisions, or when experimental data for validation are nonexistent or limited [119, 120], such as in the lower part of the respiratory tract. In the aerosol dosimetry modeling literature, little attention is generally paid to verification of the simulation results, such as controls of the temporal and spatial discretization errors, and iterative errors. However, comparison of model results with experimental data is relatively frequent in the literature of aerosol deposition studies in the respiratory system [14, 46, 53, 78, 101, 122], which strengthens the relevance and reliability of the simulation results despite the lack of verification.

4 Hybrid CFD-PBPK Models

In the previous sections, different ways to model aerosol dynamics in the respiratory system using deterministic CFD have been described together with their advantages and challenges. High-quality, state-of-the-art CFD modeling of aerosol transport, evolution, and deposition in the lung can provide accurate information about deposition locations, and the amount of the deposited matter and its composition. This information has the potential to predict where certain substances are retained in the lung, and can also be used for the comparison of the deposition behaviors of different aerosols, and inhalation conditions. More importantly, the amount and composition of chemical compounds deposited on the walls of the airways that living cells are exposed to can be estimated. This exposure may lead to inflammation in the tissues, or even to the onset of certain diseases. The biokinetic effects in the body related to the exposure of various chemical compounds can be modeled by PBPK models [123]. The combination of deterministic CFD models for aerosol deposition in the respiratory tract and PBPK models for the biokinetic effects taking place in the body in reaction to the exposure of a chemical substance has enormous potential in linking aerosol characteristics and inhalation conditions to the toxicity of an inhaled aerosol.

4.1 PBPK Models

PBPK models were initially developed for toxic substances and are therefore also known as physiologically based toxicokinetic (PBTK) models [124]. PBPK models typically involve simple mathematical transport/evolution equations and aim to predict the absorption, distribution, metabolism, and excretion (ADME) of chemical substances in living bodies using the circulating blood as the transporting media to reach various organs. The models attempt to be mechanistic in their way of describing anatomical, physiological, physical, and chemical phenomena involved in the ADME processes of the body. Because of the complexity of the biokinetic processes, the PBPK models rely on simplifications, where empirical data is used to determine the unknown coefficients and relationships, making the PBPK models semiempirical. PBPK models are typically based on the anatomical and physiological structure of the body. Because the anatomy and physiology of the body is different for different organs and regions, PBPK models are generally multicompartment models with specific systems of differential equations describing the active biokinetic processes in each compartment. For the concentration or quantity of a substance, the system of differential equations generally describes the functioning of various organs in relation to the exposure of deposited or metabolized matter using simple first-order time-evolution expressions. The various compartments are then coupled by flow rates that parallel the circulating blood system. PBPK models typically provide estimates of common pharmacokinetic (PK) parameters, such as clearance, volume of distribution, and effective half-life [125].

PBPK models have been developed for various substances and biomarkers. It has been demonstrated in physiologically based modeling strategy evaluation that PBPK models can lead to reasonable predictions of human pharmacokinetics [126]. PBPK models are therefore becoming an integrated part of the drug discovery and development processes of pharmaceutical molecules [127], and they are a valuable aid in toxicology safety assessment [128]. More extensive overviews of different PBPK models can be found in refs. [123, 125], where the fundamentals of PBPK models are described, including their components, structures and common system-related input parameters. For example, a PBPK model describing the inhalation of styrene in rats was developed [129], which was shown to predict the concentration of styrene in various tissues in rats. The same model was also used with scaled-up parameters to predict the styrene concentration in the blood of humans. Another seminal PBPK model was developed to describe the metabolism of dichloromethane in mice, rats, hamsters and humans [130]. The model structures provided by Andersen and coworkers [129, 130] have been replicated by many other groups over the years. For example [131] extended the styrene model developed by Ramsey and Andersen [129] by including a more elaborate description of the respiratory tract. Another related PBPK model was also developed to describe the disposition of nicotine in Sprague-Dawley rats [132]. The nicotine tissue kinetics predicted by the model were in agreement with those determined experimentally in the muscle, liver, skin, fat, and kidney of the rat. A more recent PBPK model was also developed for ingested and inhaled manganese, and provided consistent cross-species descriptions of manganese tracer kinetics across multiple exposure routes [133].

Regionally based PBPK models, such as the one developed by Schroeter et al. [26], rely on detailed descriptions of uptake mechanisms and information about the local aerosol deposition characteristics in various geometrical regions of the airways for model input. As mentioned in the previous sections, CFD models for aerosol transport, evolution, and deposition have the potential to provide detailed information of the amount and composition of deposited matter, for regional PBPK models.

4.2 Coupling of CFD and PBPK Models

CFD simulations of aerosol dynamics in the respiratory system give information about the amount, composition, and location of the deposited matter as output, and the PBPK model can use this information as input data. This natural coupling between deterministic CFD models and semiempirical PBPK models makes the coupled CFD–PBPK approach highly versatile for many applications, ranging from toxicological assessment to the potential linking of physical and chemical aerosol properties to certain biomarkers measured in the body. The hybrid CFD–PBPK models can be categorized into the two categories: *CFD-informed PBPK models* and *PK-informed CFD models.*

In *CFD-informed PBPK models*, the local mass transfer coefficients for the airways and flow rates are typically determined by CFD simulations and serve as inputs to traditional PBPK models based on well-mixed compartments. Such a hybrid CFD–PBPK model was used to link the inhaled chemicals with the biological response within the nasal cavity of rats [134]. In this model, CFD was used to calculate the fluxes of the inhaled chemicals, which were then used as input parameter in the PBPK model. A similar approach was also used by [135], where values of the local gas-phase mass transfer coefficients in the nasal cavities were determined by solving transport equations using CFD. The resulting mass transfer coefficients were thereafter used as input to a PBPK simulation of toxicant transport through a tissue stack, to determine the effect of spatially distributed vapor uptake within the nasal cavities on focal lesion formation. A more recent hybrid CFD-informed

PBPK model, describing nasal dosimetry of diacetyl and butyric acid, was developed by Morris and Hubbs [136]. The uptake of the vapors predicted by this model was validated by measurements in Sprague-Dawley rats and the model was thereafter used to extrapolate the vapor uptake in humans. This study inspired the development of a similar PBPK model for the dosimetry of diacetyl applied to a rat lung geometry [137]. In this work, the respiratory tract was modeled as a series of airways: nose, trachea, main bronchi, large bronchi, small bronchi, bronchioles, and alveoli with tissue dimensions obtained from the literature. Airborne vapor was allowed to absorb in tissues based on mass transfer coefficients. The transfer of vapor within the tissues was based on molecular diffusivity, and the direct reaction with tissue substrates and/or metabolism was allowed in each tissue compartment. To determine the unknown parameters for the diacetyl metabolism kinetics and the direct reaction rates, in vitro studies were performed.

PK-informed CFD models attempt to use the strengths of pharmacokinetic modeling together with the site-specific predictions of CFD models. Instead of using mass transfer coefficients computed by CFD, PK information is used to improve the deposition prediction of the CFD simulations. One of the first PK-informed CFD models, was developed for inhaled acrolein in nasal geometries of rats and humans [138]. This model structure was then extended to include anatomically accurate CFD models of animals to simulate the uptake of inhaled formaldehyde in rats, monkeys, and humans [26]. In this model, inhalation of incompressible air was simulated together with a concentration transport equation for the formaldehyde vapor. A mass-transfer coefficient was calculated based on the physicochemical and pharmacokinetic properties of formaldehyde in air and in nasal tissues, and was used with mass flux boundary conditions for the concentration equation to allow for vapor deposition. In addition, first-order differential equations were numerically solved on the surface mesh to determine the formaldehyde concentration in the submucosal layer. Moreover, the model structure developed by [138] for acrolein was adapted by [102] to include anatomically correct airway models of rat, monkey and human respiratory tracts, down to 17 ± 7, 19 ± 9 and 9 ± 2 airway generations, respectively.

Integration of CFD and PBPK models requires ongoing efforts to understand the range of applicability and the validity of certain assumptions and results generated by the hybrid CFD–PBPK model. From the viewpoint of CFD, emphasis should be placed on the accuracy and quality of the delivered deterministic computational results of the deposition, whereas PBPK modeling approaches require further cross-validation studies to determine the applicability for a wide range of applications.

5 Conclusions

The different ways to model aerosol transport, evolution, and deposition in the respiratory system using CFD have been outlined and put into the context of aerosol dosimetry modeling. A brief overview of experimental and whole-lung deposition modeling approaches has also been given to highlight the importance of CFD modeling for the prediction of local particle deposition in the respiratory system. In contrast to existing reviews on the topic of aerosol deposition modeling in the airways, where all modeling approaches are reviewed and specific results are shown, this review focused instead on explaining the methodology and requirements behind CFD modeling of aerosol dynamics and points out the necessary elements required to perform high-quality predictions of aerosol deposition.

To understand the complexity of the problem at hand, the physical and chemical mechanisms governing the transport, evolution, and deposition of aerosols have been outlined and explained. CFD approaches with different representation of the aerosol, from the *Eulerian–Lagrangian* method to various *Eulerian–Eulerian* methods, such as the *Eulerian chemical species method* and the *Eulerian drift-flux method*, have been discussed, including their governing equations, advantages, limitations, and current usage in the literature. In addition to the treatment of the aerosol system and the governing conservation and transport equations for multicomponent aerosols, the topics of turbulence and aerosol modeling have also been discussed to complete the overview of aerosol dynamics modeling.

The mathematical representation of aerosol dynamics is only one element of a CFD model. The other elements, such as the generation of a geometrical representation of the respiratory system and the related computational mesh, have been discussed together with the simulation conditions, numerical algorithms, and the important topics of verification and validation. For example, different geometrical representations of the respiratory tract can be realized, ranging from traditional geometrical models based on morphometric measurements to realistic geometries generated from modern imaging techniques like CT and MRI, or from a combination of both. Computational meshes based on the geometrical model can then be generated using different types of control volume shape, structures, and resolutions. The necessary attributes related to the simulation conditions of the respiratory system, such as the *aerosol characteristics*, *environmental conditions*, and *inhalation pattern*, have also been discussed. Moreover, a brief overview of algorithms suitable for aerosol transport, evolution, and deposition in the respiratory system is given together with some notes on the important topics of verification and validation to generate high quality and predictive CFD results.

Because accurate prediction of the deposition directly relates to biological effects because of the exposure, the coupling between CFD models for aerosol deposition with PBPK models for biokinetic effects is also briefly discussed. This type of hybrid CFD–PBPK model has great potential in using state-of-the-art computational methods to determine the toxicity of an inhaled aerosol.

Despite the fact that extensive research on the topic of aerosol dosimetry modeling using CFD has been carried out over the past decades, in the author's opinion, the presented conceptual way of treating the aerosol physics in the context of deposition in the airways is closer to the beginning of its exploration rather than the end. Current CFD modeling of aerosol dynamics applied to the respiratory tract is generally performed for research purposes to investigate certain phenomena in specific parts of the airways. With growing computational capabilities, more detailed and accurate simulations of multicomponent aerosols in complex geometries and for realistic breathing patterns will be performed in the future. Validation of the CFD models remains a challenge, but is nevertheless a crucial element for CFD to become a reliable and predictive tool for aerosol dosimetry. Moreover, bridging CFD with multidisciplinary biological models like PBPK will continue to develop and find its place in the toxicological assessment of inhaled aerosols. Finally, the success of CFD as a predictive tool in toxicological assessment relies not only on the technical developments related to simulation capability, but also on critical reviews of the current modeling assumptions and their consequences on the predicted quantities.

Acknowledgments

This work is dedicated to the memory of our late colleague Dr. Matthias Schorp, who initiated and inspired the writing of this book chapter. The authors would also like to thank Dr. Jeffry D. Schroeter for critical review and insightful comments on the book chapter.

References

1. Hinds HC (1999) Aerosol Technology: Properties, Behavior, and Measurement of Airborne Particles. John Wiley & Sons Inc., New York, 2nd edition
2. Kreyling WG, Hirn S, Schleh C (2010) Nanoparticles in the lung. Nat Biotechnol 28(12):1275–1276
3. Rostami AA (2009) Computational modeling of aerosol deposition in respiratory tract: a review. Inhal Toxicol 21:262–290
4. Darquenne C (2012) Aerosol deposition in health and disease. J Aerosol Med Pulm Drug Deliv 25(3):140–147
5. Jaques PA, Kim CS (2000) Measurement of total lung deposition of inhaled ultrafine particles in healthy men and women. Inhal Toxicol 12:715–731
6. Kim CS, Hu SC (2006) Total respiratory tract deposition of fine micrometer-sized particles in healthy adults: empirical equations

for sex and breathing pattern. J Appl Physiol 101:401–412
7. Kim CS (2009) Deposition of aerosol particles in human lungs: in vivo measurement and modelling. Biomarkers 14(S1):54–58
8. Kim CS, Hu SC (1998) Regional deposition of inhaled particles in human lungs: comparison between men and women. J Appl Physiol 84:1834–1844
9. Kim CS, Jaques PA (2000) Respiratory dose of inhaled ultrafine particles in healthy humans. Phil Trans R Soc A 358: 2693–2701
10. Stahlhofen W, Rudolf G, James A (1989) Intercomparison of experimental regional aerosol deposition data. J Aerosol Med Pulm Drug Deliv 2:285–308
11. Mendez LB, Gookin G, Phalen RF (2010) Inhaled aerosol particle dosimetry in mice: a review. Inhal Toxicol 22(12):1032–1037
12. Kuehl PJ et al (2012) Regional particle size dependent deposition of inhaled aerosols in rats and mice. Inhal Toxicol 24(1):27–35
13. Geiser M, Kreyling WG (2010) Deposition and biokinetics of inhaled nanoparticles. Part Fibre Toxicol 7(2)
14. Cheng Y-S, Zhou Y, Chen BT (1999) Particle deposition in a cast of human oral airways. Aerosol Sci Tech 31:286–300
15. Phalen RF et al (1997) Deposition of fluorescent particles in replica casts of distal human airways: comparison of theory with experiment. Ann Occup Hyg 41(1):565–570
16. Sosnowski TR, Moskal A, Gradon L (2006) Dynamics of oropharyngeal aerosol transport and deposition with the realistic flow pattern. Inhal Toxicol 18:773–780
17. Sosnowski TR, Moskal A, Gradon L (2007) Mechanisms of aerosol particle deposition in oro-pharynx under non-steady airflow. Ann Occup Hyg 51(1):19–25
18. Robinson RJ et al (2006) Experimental and numerical smoke deposition in a multi-generation human replica tracheobronchial model. Ann Biomed Eng 34(3):373–383
19. Tippe A, Heinzmann U, Roth C (2002) Deposition of fine and ultrafine aerosol particles during exposure at the air/cell interface. J Aerosol Sci 33(2):207–218
20. Comouth A et al (2013) Modelling and measurement of particle deposition for cell exposure at the air-liquid interface. J Aerosol Sci 63:103–114
21. Paur H-R et al (2011) In vitro cell exposure studies for the assessment of nanoparticle toxicity in the lung – a dialog between aerosol science and biology. J Aerosol Sci 42(10): 668–692
22. Hofmann W (2009) Modelling particle deposition in human lungs: modelling concepts and comparison with experimental data. Biomarkers 14(S1):59–62
23. Hofmann W (2011) Modelling inhaled particle deposition in the human lung – a review. J Aerosol Sci 42:693–724
24. Longest PW, Holbrook LT (2012) In silico models of aerosol delivery to the respiratory tract – development and applications. Adv Drug Deliv Rev 64:296–311
25. Phalen RF, Mendez LB, Oldham MJ (2010) New developments in aerosol dosimetry. Inhal Toxicol 22(S2):6–14
26. Schroeter JD et al (2014) Effects of endogenous formaldehyde in nasal tissues on inhaled formaldehyde dosimetry predictions in the rat, monkey, and human nasal passages. Toxicol Sci 138(2):412–424
27. Kleinstreuer C, Zhang Z (2010) Airflow and particle transport in the human respiratory system. Annu Rev Fluid Mech 42:301–334
28. Li A, Ahmadi G (1995) Computer simulation of particle deposition in the upper tracheobronchial tree. Aerosol Sci Tech 23:201–223
29. Li Z, Kleinstreuer C, Zhang Z (2007) Particle deposition in human tracheobronchial airways due to transient inspiratory flow patterns. J Aerosol Sci 38:626–644
30. Longest PW, Xi J (2008) Condensational growth may contribute to the enhanced deposition of cigarette smoke particles in the upper respiratory tract. Aerosol Sci Tech 42(8):579–602
31. Svenningsson B et al (2006) Hygroscopic growth and critical supersaturations for mixed aerosol particles of inorfanic and organic compounds of atmospheric relevance. Atmos Chem Phys 6:1937–1952
32. Millage KK et al (2010) A review of inhalability fraction models: discussion and recommendations. Inhal Toxicol 22(2):151–159
33. Asgharian B et al (2012) A lung dosimetry model of vapor uptake and tissue disposition. Inhal Toxicol 24(3):182–193
34. Hofmann W et al (2009) Deposition of combustion aerosols in the human respiratory tract: comparison of theoretical predictions with experimental data considering nonspherical shape. Inhal Toxicol 21(14):1154–1164
35. Asgharian B et al (2011) Derivation of mass transfer coefficients for transient uptake and tissue disposition of soluble and reactive vapors in lung airways. Ann Biomed Eng 39(6):1788–1804
36. Bird RB, Stewart WE, Lightfoot EN (2007) Transport Phenomena. John Wiley & Sons Inc., New York, revised second edition

37. Zhao B et al (2008) Particle dispersion and deposition in ventilated rooms: testing and evaluation of different Eulerian and Lagrangian models. Build Environ 43:388–397
38. Elghobashi S (1991) Particle-laden turbulent flows: direct simulation and closure models. Appl Sci Res 48(3-4):301–314
39. Kleinstreuer C, Feng Y (2013) Lung deposition analyses of inhaled toxic aerosols in conventional and less harmful cigarette smoke: a review. Int J Env Res Public Health 10:4454–4485
40. Zhao B et al (2004) Comparison of indoor aerosol particle concentration and deposition in different ventilated rooms by numerical method. Build Environ 36:1–8
41. Beghein C, Jiang Y, Chen Q (2005) Using large eddy simulation to study particle motions in a room. Indoor Air 15:281–290
42. Zhang Z, Chen Q (2006) Experimental measurements and numerical simulation of fine particle transport and distribution in ventilated rooms. Atmos Environ 40:3396–3408
43. Hofmann W, Balashazy I (1991) Particle deposition within airway bifurcation – solution of 3D Navier-Stokes equation. Radiat Prot Dosim 38:57–63
44. Balashazy I, Hofmann W (1995) Deposition of aerosols in asymmetric airway bifurcations. J Aerosol Sci 26:273–292
45. Weibel ER (1963) Morphometry of the human lung. Springer Verlag and Academic Press, Heidelberg
46. van Ertbruggen C, Hirsch C, Paiva M (2005) Anatomically based three dimensional model airways to simulate flow and particle transport using computational fluid dynamics. J Appl Physiol 98:970–980
47. Longest PW, Xi J (2007) Effectiveness of direct Lagrangian tracking models for simulating nanoparticle deposition in the upper airways. Aerosol Sci Tech 41(4):380–397
48. Longest PW, Xi J (2007) Computational investigation of particle inertia effects on submicron aerosol deposition in the respiratory tract. Aerosol Sci Tech 38(1):111–130
49. Frederix EMA et al (2015) Extension of the compressible PISO algorithm to single-species aerosol formation and transport. Int J Multiphase Flow (In Press). Available online: doi: 10.1016/j.ijmultiphaseflow.2015.04.015
50. Broday DM (2004) Deposition of ultrafine particles at carinal ridges of the upper bronchial airways. Aerosol Sci Tech 38:991–1000
51. Shi H et al (2004) Nanoparticle transport and deposition in bifurcating tubes with different inlet conditions. Phys Fluids 16(7):2199–2213
52. Zhang Z, Kleinstreuer C (2004) Airflow structures and nano-particle deposition in a human upper airway model. J Comput Phys 198(1):178–210
53. Longest PW, Oldham MJ (2008) Numerical and experimental deposition of fine respirators aerosols: development of a two-phase drift flux model with near-wall velocity corrections. J Aerosol Sci 39:48–70
54. Manninen M, Taivassalo V, Kallio S (1996) On the mixture model for multiphase flow. VTT Publications 288, Technical Research Center of Finland
55. Murakami S et al (1992) Diffusion characteristics of airborne particles with gravitational settling in a convection dominant indoor flow field. ASHRAE Trans 98:82–97
56. Shimada M et al (1996) Numerical simulation and experiment on the transport of fine particles in a ventilated room. Aerosol Sci Tech 25:242–255
57. Holmberg S, Li Y (1998) Modelling of the indoor environment-particle dispersion and deposition. Indoor Air 8:113–122
58. Holmberg S, Chen Q (2003) Air flow and particle control with different ventilation systems in a classroom. Indoor Air 13:200–204
59. Zhao B, Wu J (2005) Numerical study of particle deposition in two different ventilated rooms. Indoor Built Environ 14:469–479
60. Chen F, Yu SCM, Lai ACK (2006) Modeling particle distribution and deposition in indoor environments with a new drift-flux model. Atmos Environ 40:357–367
61. Chen F, Lai ACK (2004) An Eulerian model for particle deposition under electrostatic and turbulent conditions. J Aerosol Sci 35:47–62
62. Wang JB, Lai ACK (2006) A new drift-flux model for particle transport and deposition in human airways. J Biomech Eng 128:97–105
63. Xi J, Longest PW (2008) Numerical predictions of submicrometer aerosol deposition in the nasal cavity using a novel drift flux approach. Int J Heat Mass Transfer 51(23–24):5562–5577
64. Sanyal J et al (1999) Numerical simulation of gas-liquid dynamics in cylindrical bubble column reactors. Chem Eng J 54:5071–5083
65. Moser RD, Moin P (1987) The effects of curvature in wall-bounded turbulent flows. J Fluid Mech 175:479–510
66. Longest PW, Vinchurkar S (2007) Effects of mesh style and grid convergence on particle deposition in bifurcating airway models with comparisons to experimental data. Med Eng Phys 29:350–366

67. Pope SB (2000) Turbulent flows. Cambridge University Press, Cambridge
68. Tsinober A (2009) An informal conceptual introduction to turbulence: second edition of an informal introduction to turbulence. Springer, Dordrecht
69. Deissler RG (1998) Turbulent fluid motion. Taylor & Francis, Philadelphia, PA
70. Geurts BJ (2003) Elements of direct and large-eddy simulation. R.T. Edwards, Inc., Philadelphia, PA
71. Germano M et al (1991) A dynamic subgrid-scale eddy viscosity model. Phys Fluids A3:1760–1765
72. Peng S-H, Haase W (eds) (2008) Advances in hybrid RANS-LES modelling. Papers contributed to the 2007 symposium of hybrid RANS-LES methods, Corfu, Greece, 17–18 June 2007. Springer, Berlin
73. Spalart PR et al (1998) Comments on the feasibility of LES for wings, and on a hybrid RANS/LES approach. Advances in DNS/LES. Greyden Press, Dayton, OH
74. Katz IM, Davis BM, Martonen TB (1999) A numerical study of particle motion within the human larynx and trachea. J Aerosol Sci 30(2):173–183
75. Stapleton KW et al (2000) On the suitability of k-epsilon turbulence modeling for aerosol deposition in the mouth and throat: a comparison with experiment. J Aerosol Sci 31:739–749
76. Zhang Z, Kleinstreuer C (2003) Low-Reynolds-number turbulent flows in locally constricted conduits: a comparison study. AIAA J 41(5):831–840
77. Matida EA et al (2004) Improved numerical simulation of aerosol deposition in an idealized mouth-throat. J Aerosol Sci 35:1–19
78. Mylavarapu G et al (2009) Validation of computational fluid dynamics methodology used for human upper airway flow simulations. J Biomech 42(10):1553–1559
79. Luo XY et al (2004) LES modelling of flow in a simple airway model. Med Eng Phys 26:403–413
80. Matida EA et al (2006) Improving prediction of aerosol deposition in an idealized mouth using large-eddy simulation. J Aerosol Med Pulm Drug Deliv 19(3):290–300
81. Jayaraju ST, Verbanck S, Lacor C (2010) LES and DES study of fluid-particle dynamics in a human mouth-throat geometry, turbulence and interactions. Notes Number Fluid Mech Multidiscip Des 110:183–189
82. Talukdar SS, Swihart MT (2004) Aerosol dynamics modeling of silicon nanoparticle formation during silane pyrolysis: a comparisin of three solution methods. J Aerosol Sci 35: 889–908
83. Gelbard F, Tambour Y, Seinfeld JH (1980) Sectional representations for simulating aerosol dynamics. J Colloid Interface Sci 76(2): 541–556
84. Gelbard F, Seinfeld JH (1980) Simulation of multicomponent aerosol dynamics. J Colloid Interface Sci 78(2):485–501
85. Gelbard F (1990) Modeling multicomponent aerosol-particle growth by vapor condensation. Aerosol Sci Tech 12(2):399–412
86. Jacobson MZ (1997) Development and application of a new air pollution modeling system 3. Aerosol-phase simulations. Atmos Environ 31(4):587–608
87. Jacobson MZ (1997) Development and application of a new air pollution modeling system 2. Aerosol module structure and design. Atmos Environ 31(2):131–144
88. Zhang Y et al (2000) A comparative review of inorganic aerosol thermodynamic equilibrium modules: similarities, differences, and their likely causes. Atmos Environ 34(1):117–137
89. Sun Z, Axelbaum RL, Huertas JI (2004) Monte Carlo simulation of multicomponent aerosols undergoing simultaneous coagulation and condensation. Aerosol Sci Tech 38(10):963–971
90. Whitby ER, McMurry PH (1997) Modal aerosol dynamics modeling. Aerosol Sci Tech 27(6):673–688
91. Seigneur C et al (1986) Simulation of aerosol dynamics – a comparative review of mathematical-models. Aerosol Sci Tech 5(2):205–222
92. Binkowski FS, Shankar U (1995) The regional particulate matter model 1. Model description and preliminary results. J Geohys Res D 100(D12):26191–26209
93. Horsfield K, Cumming G (1968) Morphology of bronchial tree in man. J Appl Physiol 24(3):373–383
94. Koblinger L, Hafmann W (1985) Analysis of human lung morphometric data for stochastic aerosol deposition calculations. Phys Med Biol 30:541–556
95. Kitaoka H, Takaki R, Suki B (1999) A three dimensional model for human airway tree. J Appl Physiol 76(6):2207–2217
96. Clinkenbeard RE et al (2002) Replication of human tracheobronchial hollow airway models using a selective laser sintering rapid prototyping technique. AIHA J 63:141–150
97. Sauret V et al (2002) Study of three dimensional geometry of central conducting airways

in man using computed tomographic (CT) images. J Anat 200:123–134
98. McRobbie DW, Pritchard S, Quest RA (2003) Studies of the human oropharyngeal airspaces using magnetic resonance imaging. J Aerosol Med Pulm Drug Deliv 16(4):401–415
99. Jayaraju ST et al (2007) Fluid flow and particle deposition analysis in a realistic extrathoracic airway model using unstructured grids. J Aerosol Sci 38:494–508
100. Ahmed A et al (2012) Development of an in vitro model to assess deposition of aerosol particles in a representative replica of the rat's respiratory tract. J Aerosol Med Pulm Drug Deliv 25(3):169–178
101. Lizal F et al (2012) Development of a realistic human airway model. Proc Inst Mech Eng H J Eng Med 35:84–92
102. Corley RA et al (2012) Comparative computational modeling of airflows and vapor dosimetry in the respiratory tracts of rat, monkey, and human. Toxicol Sci 128(2): 500–516
103. Zhang Z, Kleinstreuer C, Kim CS (2002) Aerosol transport and deposition in a triple bifurcation bronchial airway model. J Aerosol Sci 33:257–281
104. Tsuda A, Henry FS, Butler JP (1995) Chaotic mixing of alveolated duct flow in rhythmically expanding pulmonary acinus. J Appl Physiol 79(3):1055–1063
105. Ferziger JH, Peric M (2001) Computational Methods for Fluid Dynamics. Springer-Verlag, Berlin, 2nd edition
106. European Research Community on Flow, Turbulence and Combustion. Special Interest Group on Quality and Trust in Industrial CFD, Casey M, Wintergerste T (2000) Best practice guidelines. ERCOFTAC
107. Pauling BE, Prausnitz JM, O'Connell JP (2001) The properties of gases and liquids. McGraw-Hill, New York, NY
108. Jedelsky J, Lizal F, Jicha M (2012) Characteristics of turbulent particle transport in human airways under steady and cyclic flows. Int J Heat Fluid Flow 35:84–92
109. Patankar SV, Spalding DB (1972) A calculation procedure for heat, mass and momentum transfer in three-dimensional parabolic flows. Int J Heat Mass Transfer 15(10):1787–1806
110. Issa R (1986) Solution of the implicitly discretised fluid flow equations by operator-splitting. J Comput Phys 62:40–65
111. Patankar SV (1980) Numerical heat transfer and fluid flow. Hemisphere Publishing Corporation, Taylor & Francis Group, New York, NY
112. Van Doormal JP, Raithby GD (1980) Enhancements of the SIMPLE method for predicting incompressible fluid flows. Numerical Heat Transfer 7(2):147–163
113. Issa R et al (1991) Solution of the implicitly discretised reacting flow equations by operator-splitting. J Comput Phys 93:388–410
114. Wang DM, Watkins AP, Cant RS (1993) Three-dimensional diesel engine combustion simulation with a modified EPISO procedure. Numerical Heat Transfer A Appl 24:249–272
115. Issa R, Gosman A, Watkins A (1986) The computation of compressible and incompressible recirculating flows by a non-iterative implicit scheme. J Comput Phys 62:66–82
116. Bressloff N (2001) A parallel pressure implicit splitting of operators algorithm applied to flows at all speeds. Int J Numer Meth Fluids 36:497–518
117. Barton IE (1998) Comparison of SIMPLE- and PISO-type algorithms for transient flows. Int J Numer Meth Fluids 26:459–483
118. Wanik A, Schnell U (1989) Some remarks on the PISO and SIMPLE algorithms for steady turbulent flow problems. Comput Fluids 17:555–570
119. Oberkampf WL, Roy CJ (2010) Verification and validation in scientific computing. Cambridge University Press, New York, NY
120. Roache PJ (1997) Quantification of uncertainty in computational fluid dynamics. Annu Rev Fluid Mech 29:123–160
121. Pilou M et al (2011) Inertial particle deposition in a 90° laminar flow bend: An Eulerian fluid particle approach. Aerosol Sci Tech 45(11):1376–1387
122. Zhang Z, Kleinstreuer C, Kim CS (2002) Micro-particle transport and deposition in human oral airway model. J Aerosol Sci 33(2):1635–1652
123. Reddy M et al (2005) Physiologically based pharmacokinetic modeling: science and applications. Wiley, New York, NY
124. Jongeneelen FJ, Ten Berge WF (2011) A generic, cross-chemical predictive PBTK model with multiple entry routes running as application in MS Excel; design of the model and comparison of predictions with experimental results. Ann Occup Hyg 55(8):841–864
125. Jones HM, Rowland-Yeo K (2013) Basic concepts in physiologically based pharmacokinetic modeling in drug discovery and development. CPT Pharmacometr Syst Pharmacol 2(8), e63
126. De Buck SS et al (2007) Prediction of human pharmacokinetics using physiologically based modeling: a retrospective analysis

of 26 clinically tested drugs. Drug Metab Dispos 35:1766–1780
127. Chen Y et al (2012) Application of IVIVE and PBPK modeling in prospective prediction of clinical pharmacokinetics: strategy and approach during the drug discovery phase with four case studies. Biopharm Drug Dispos 33:85–98
128. Jones RM (2012) A new methodology for predicting human pharmacokinetics for inhaled drugs from oratracheal pharmacokinetic data in rats. Xenobiotica 42(1):75–85
129. Ramsey JC, Andersen ME (1984) A physiologically based description of the inhalation pharmacokinetics of styrene in rats and humans. Toxicol Appl Pharmacol 73:159–175
130. Andersen ME et al (1987) Physiologically based pharmacokinetics and the risk assessment process for methylene chloride. Toxicol Appl Pharmacol 87:185–205
131. Sarangapani R, Teeguarden JG (2002) Physiologically based pharmacokinetic modeling of styrene and styrene oxide respiratory-tract dosimetry in rodents and humans. Inhal Toxicol 14:789–834
132. Plowchalk DR, Andersen ME, DeBethizy JD (1992) A physiologically based pharmacokinetic model for nicotine disposition in the Sprague-Dawley rat. Toxicol Appl Pharmacol 116:177–188
133. Schroeter JD et al (2011) Analysis of manganese tracer kinetics and target tissue dosimetry in monkeys and humans with multi-route physiologically based pharmacokinetic models. Toxicol Sci 120(2):481–498
134. Andersen ME et al (2000) Application of a hybrid CFD-PBPK nasal dosimetry model in an inhalation risk assessment: an example with acrylic acid. Toxicol Sci 57(2):312–325
135. Bush ML et al (1998) A CFD-PBPK hybrid model for simulating gas and vapor uptake in the rat nose. Toxicol Appl Pharmacol 150(1): 133–145
136. Morris JB, Hubbs AF (2009) Inhalation dosimetry of diacetyl and butyric acid, two components of butter flavoring vapors. Toxicol Sci 108(1):173–183
137. Gloede E et al (2011) A validated hybrid computational fluid dynamics-physiologically based pharmacokinetic model for respiratory tract vapor absorption in the human and rat and its application to inhalation dosimetry of diacetyl. Toxicol Sci 123(1):231–246
138. Schroeter JD et al (2008) Application of physiological computational fluid dynamics model to predict interspecies nasal dosimetry of inhaled acrolein. Inhal Toxicol 20(3): 227–243

INDEX

Julia Hoeng and Manuel C. Peitsch (eds.), *Computational Systems Toxicology*, Methods in Pharmacology and Toxicology, DOI 10.1007/978-1-4939-2778-4,

T

X

MIX
Papier aus verantwortungsvollen Quellen
Paper from responsible sources
FSC® C105338

If you have any concerns about our products,
you can contact us on
ProductSafety@springernature.com

In case Publisher is established outside the EU,
the EU authorized representative is:
Springer Nature Customer Service Center GmbH
Europaplatz 3, 69115 Heidelberg, Germany

Printed by Libri Plureos GmbH
in Hamburg, Germany